Problems and Solutions in

Mathematical Olympiad

High School 1

Other Related Titles from World Scientific

Problems and Solutions in Mathematical Olympiad
Secondary 3
by Jun Ge
translated by Huan-Xin Xie
ISBN: 978-981-122-982-4
ISBN: 978-981-123-141-4 (pbk)

Problems and Solutions in Mathematical Olympiad
High School 2
by Shi-Xiong Liu
translated by Jiu Ding
ISBN: 978-981-122-988-6
ISBN: 978-981-123-143-8 (pbk)

Problems and Solutions in Mathematical Olympiad
High School 3
by Hong-Bing Yu
translated by Fang-Fang Lang and Yi-Chao Ye
ISBN: 978-981-122-991-6
ISBN: 978-981-123-144-5 (pbk)

Problems and Solutions in

Mathematical Olympiad

High School 1

Editors-in-Chief

Zun Shan *Nanjing Normal University, China*

Bin Xiong *East China Normal University, China*

Original Authors

Bin Xiong *East China Normal University, China*

Zhi-Gang Feng *Shanghai High School, China*

English Translator

Tian-You Zhou *Shanghai High School, China*

Copy Editors

Ming Ni *East China Normal University Press, China*

Ling-Zhi Kong *East China Normal University Press, China*

Lei Rui *East China Normal University Press, China*

East China Normal University Press

World Scientific

Published by

East China Normal University Press
3663 North Zhongshan Road
Shanghai 200062
China

and

World Scientific Publishing Co. Pte. Ltd.
5 Toh Tuck Link, Singapore 596224
USA office: 27 Warren Street, Suite 401-402, Hackensack, NJ 07601
UK office: 57 Shelton Street, Covent Garden, London WC2H 9HE

British Library Cataloguing-in-Publication Data
A catalogue record for this book is available from the British Library.

PROBLEMS AND SOLUTIONS IN MATHEMATICAL OLYMPIAD
High School 1

ISBN 978-981-122-985-5 (hardcover)
ISBN 978-981-123-142-1 (paperback)
ISBN 978-981-122-986-2 (ebook for institutions)
ISBN 978-981-122-987-9 (ebook for individuals)

For any available supplementary material, please visit
https://www.worldscientific.com/worldscibooks/10.1142/12087#t=suppl

Desk Editor: Tan Rok Ting

Typeset by Stallion Press
Email: enquiries@stallionpress.com

Printed in Singapore

Editorial Board

Board Members

Preface

It is said that in many countries, especially the United States, children are afraid of mathematics and regard mathematics as an "unpopular subject." But in China, the situation is very different. Many children love mathematics, and their math scores are also very good. Indeed, mathematics is a subject that the Chinese are good at. If you see a few Chinese students in elementary and middle schools in the United States, then the top few in the class of mathematics are none other than them.

At the early stage of counting numbers, Chinese children already show their advantages.

Chinese people can express integers from 1 to 10 with one hand, whereas those in other countries would have to use two.

The Chinese have long had the concept of digits, and they use the most convenient decimal system (many countries still have the remnants of base 12 and base 60 systems).

Chinese characters are all single syllables, which are easy to recite. For example, the multiplication table can be quickly mastered by students, and even the "stupid" people know the concept of "three times seven equals twenty one." But for foreigners, as soon as they study multiplication, their heads get bigger. Believe it or not, you could try and memorise the multiplication table in English and then recite it, it is actually much harder to do so in English.

It takes the Chinese one or two minutes to memorize $\pi = 3.14159\cdots$ to the fifth decimal place. However, in order to recite these digits, the Russians wrote a poem. The first sentence contains three words and the second sentence contains one $\cdots$ To recite π, recite poetry first. In our opinion, this just simply asks for trouble, but they treat it as a magical way of memorization.

Application problems for the four arithmetic operations and their arithmetic solutions are also a major feature of Chinese mathematics. Since ancient times, the Chinese have compiled a lot of application questions, which has contact or close relations with reality and daily life. Their solutions are simple and elegant as well as smart and diverse, which helps increase students' interest in learning and enlighten students'. For example:

"There are one hundred monks and one hundred buns. One big monk eats three buns and three little monks eat one bun. How many big monks and how many little monks are there?"

Most foreigners can only solve equations, but Chinese have a variety of arithmetic solutions. As an example, one can turn each big monk into 9 little monks, and 100 buns indicate that there are 300 little monks, which contain 200 added little monks. As each big monk becomes a little monk 8 more little monks are created, so $200/8 = 25$ is the number of big monks, and naturally there are 75 little monks. Another way to solve the problem is to group a big monk and three little monks together, and so each person eats a bun on average, which is exactly equal to the overall average. Thus the big monks and the little monks are not more and less after being organized this way, that is, the number of the big monks is $100/(3 + 1) = 25$.

The Chinese are good at calculating, especially good at mental arithmetic. In ancient times, some people used their fingers to calculate (the so-called "counting by pinching fingers"). At the same time, China has long had computing devices such as counting chips and abaci. The latter can be said to be the prototype of computers.

In the introductory stage of mathematics – the study of arithmetic, our country has obvious advantages, so mathematics is often the subject that our smart children love.

Geometric reasoning was not well-developed in ancient China (but there were many books on the calculation of geometric figures in our country), and it was slightly inferior to the Greeks. However, the Chinese are good at learning from others. At present, the geometric level of middle school students in our country is far ahead of the rest of the world. Once a foreign education delegation came to a junior high school class in our country. They thought that the geometric content taught was too in-depth for students to comprehend, but after attending the class, they had to admit that the content was not only understood by Chinese students, but also well mastered.

The achievements of mathematics education in our country are remarkable. In international mathematics competitions, Chinese contestants have won numerous medals, which is the most powerful proof. Ever since our country officially sent a team to participate in the International Mathematical Olympiad in 1986, the Chinese team has won 14 team championships, which can be described as very impressive. Professor Shiing-Shen Chern, a famous contemporary mathematician, once admired this in particular. He said, "One thing to celebrate this year is that China won the first place in the international math competition ⋯ Last year it was also the first place." (Shiing-Shen Chern's speech, *How to Build China into a Mathematical Power*, at Cheng Kung University in Taiwan in October 1990)

Professor Chern also predicted: "China will become a mathematical power in the 21st century."

It is certainly not an easy task to become a mathematical power. It cannot be achieved overnight. It requires unremitting efforts. The purpose of this series of books is: (1) To further popularize the knowledge of mathematics, to make mathematics be loved by more young people, and to help them achieve good results; (2) To enable students who love mathematics to get better development and learn more knowledge and methods through the series of books.

"The important things in the world must be done in detail." We hope and believe that the publication of this series of books will play a role in making our country a mathematical power. This series was first published in 2000. According to the requirements of the curriculum reform, each volume is revised to different degrees.

Well-known mathematician, academician of the Chinese Academy of Sciences, and former chairman of the Chinese Mathematical Olympiad Professor Yuan Wang, served as a consultant to this series of books and wrote inscriptions for young math enthusiasts. We express our heartfelt thanks. We would also like to thank East China Normal University Press, and in particular Mr. Ming Ni and Mr. Lingzhi Kong. Without them, this series of books would not have been possible.

Zun Shan and Bin Xiong
May 2018

Contents

Chapter 1

Concepts and Operations of Sets

1.1 Key Points of Knowledge and Basic Methods

Set is one of the most important concepts s in mathematics, and the theory of sets is the foundation of modern mathematics. Set is a primitive concept that is not defined. In general, some objects are put together to form a set, and each object in the set is called an element of the set. For a given set, its elements are definite, distinct, and order-irrelevant.

Expressing a set correctly is a basis for learning mathematics. Description is an important method to represent a set. It is based on the following generalization principle.

Generalization Principle: Given any property p, there exists a unique set S that consists of all the objects satisfying the property p. In other words,

$$S = \{x|p(x)\},$$

where $p(x)$ is the abbreviation for "the property p that the object x satisfies."

Relations between two sets can be reflected with subsets, intersections, and unions. When dealing with problems , one usually begins the discussion from the perspective of elements. Here it reflects the mathematical thought of "from part to whole."

Apart from intersections and unions, operations of sets also include complements and differences. The set consisting of all elements of A that are not in B is called the difference of A and B, denoted $A\backslash B$ (or sometimes

$A - B$), namely $A \backslash B = \{x | x \in A \text{ and } x \notin B\}$. Operations of sets satisfy the following rules:

Distributive Laws:

$$A \cap (B \cup C) = (A \cap B) \cup (A \cap C),$$

$$A \cup (B \cap C) = (A \cup B) \cap (A \cup C).$$

De Morgan's Laws:

$$\complement_U(\bar{A} \cup B) = (\complement_U \bar{A}) \cap (\complement_U B),$$

$$\complement_U(\bar{A} \cap B) = (\complement_U \bar{A}) \cup (\complement_U B).$$

1.2 Illustrative Examples

Example 1. Let sets $A = \{2, 0, 1, 3\}$ and $B = \{x | -x \in A, 2 - x^2 \notin A\}$. Compute the sum of all elements in B.

Solution. By the assumption, $B \subseteq \{2, 0, 1, 3\}$. If $x = -2$ or -3, then $2 - x^2 = -2$ or -7, so $2 - x^2 \notin A$. If $x = 0$ or -1, then $2 - x^2 = 2$ or 1, so $2 - x^2 \in A$. Therefore, from the definition of B, we know that $B = \{-2, -3\}$, and the sum of all elements in B is -5.

Example 2. A proper subset S of $\mathbf{R}$ (the set of all real numbers) is said to be closed under addition and subtraction if for any x, $y \in S$, we have $x + y \in S$ and $x - y \in S$.

(1) Find an example of S that is closed under addition and subtraction.
(2) Show that if S_1 and S_2 are two proper subsets of $\mathbf{R}$ that are closed under addition and subtraction then there exists $c \in \mathbf{R}$ such that $c \notin S_1 \cup S_2$.

Solution. (1) Examples include $S = \{2k | k \in \mathbf{Z}\}$.

(2) Since S_1 is a proper subset of $\mathbf{R}$, there exists $a \in \mathbf{R}$ such that $a \notin S_1$. If $a \notin S_2$, then the statement holds; otherwise $a \in S_2$. Similarly, there exists $b \in \mathbf{R}$ such that $b \notin S_2$, and the statement holds unless $b \in S_1$. Let $c = a + b$, and we will show that $c \notin (S_1 \cup S_2)$. Assume for contradiction that $c \in (S_1 \cup S_2)$. Then, without loss of generality, we may assume that $c \in S_1$, that is, $a + b \in S_1$. Since $b \in S_1$, we also have $a + b - b = a \in S_1$, which leads to a contradiction.

Remark. The choice of c is not unique. Letting $c = a - b$ or $c = b - a$ also works.

Example 3 (2011 China West Mathematical Invitational). Let $M \subseteq \{1, 2, \ldots, 2011\}$ be such that for any three elements of M, we can find two of them a and b such that $a|b$ or $b|a$. Find the maximum value of $|M|$ (here $|M|$ denotes the number of elements in the set M).

Solution. Let $M = \{1, 2, 2^2, 2^3, \ldots, 2^{10}, 3, 3 \times 2, 3 \times 2^2, \ldots, 3 \times 2^9\}$, which satisfies the requirements, and $|M| = 21$.

For any M that satisfies the reauirements of the problem, if $|M| \geq 22$, then suppose the elements of M are $a_1 < a_2 \ldots < a_k (k \geq 22)$.

First we observe that $a_{n+2} \geq 2a_n$ for each n, because otherwise $a_n < a_{n+1} < a_{n+2} < 2a_n$, and the triplet (a_n, a_{n+1}, a_{n+2}) contradicts the rule. With this property, we have

$$a_4 \geq 2a_2 \geq 4,$$
$$a_6 \geq 2a_4 \geq 8,$$
$$\cdots$$
$$a_{22} \geq 2a_{20} \geq 2^{11} > 2011,$$

which gives a contradiction. Therefore, the maximum value of $|M|$ is 21.

Example 4. Let a function $f(x) = x^2 + ax + b (a, b \in \mathbf{R})$ and sets

$$A = \{x | x = f(x), x \in \mathbf{R}\}, \; B = \{x | x = f(f(x)), \; x \in \mathbf{R}\}.$$

(1) Show that $A \subseteq B$.
(2) If $A = \{-1, 3\}$, find B.

Solution. (1) For any $x_0 \in A$, we have $x_0 = f(x_0)$, so

$$x_0 = f(x_0) = f(f(x_0)),$$

and thus $x_0 \in B$. Hence $A \subseteq B$.

(2) If $A = \{-1, 3\}$, then $-1 = f(-1)$ and $3 = f(3)$, namely

$$\begin{cases} (-1)^2 + a(-1) + b = -1, \\ 3^2 + a \cdot 3 + b = 3. \end{cases}$$

Solving the equations gives that $a = -1$ and $b = -3$. Therefore, $f(x) = x^2 - x - 3$. If $x = f(f(x))$, then

$$(x^2 - x - 3)^2 - (x^2 - x - 3) - x - 3 = 0.$$

By (1), we know that -1 and 3 belong to B, so they are roots of the above equation. This allows us to factorize the left side of the equation, resulting in

$$(x^2 - 2x - 3)(x^2 - 3) = 0,$$

so its roots are $x = -1, 3$, and $\pm\sqrt{3}$. Therefore, $B = \{-1, 3, -\sqrt{3}, \sqrt{3}\}$.

Example 5. Sets A, B, and C (not necessarily distinct) have the union $A \cup B \cup C = \{1, 2, \ldots, 10\}$. Find the number of the ordered triplets (A, B, C) with such a property.

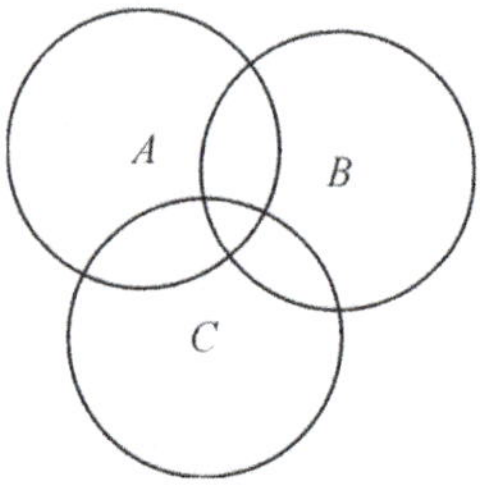

Solution. In the Venn diagram shown on the right, the sets A, B, and C generate seven disjoint regions, whose union is $A \cup B \cup C$. Each element in $\{1, 2, \ldots, 10\}$ belongs to one of the seven regions, and the sets A, B, and C are uniquely determined by the locations of these elements. By the multiplication principle, there are 7^{10} such triplets.

Remark. When determining the number of triplets, we made use of the assumption that A, B, and C are not necessarily distinct and they are ordered triplets. This is an interesting application of Venn diagrams.

Example 6. Let $n \in \mathbf{N}^*$ (the set of all positive integers), $n \geq 15$, and A and B be proper subsets of $I = \{1, 2, \ldots, n\}$, with $A \cap B = \emptyset$ and $A \cup B = I$. Show that either A or B contains two distinct numbers, whose sum is a perfect square.

Solution. Since $A \cap B = \emptyset$ and $A \cup B = I$, we may assume without loss of generality that $1 \in A$, and prove the statement by contradiction.

Suppose neither A nor B has two elements whose sum is a perfect square. Then $3 \in B$, so $6 \in A$ and $10 \in B$. Next, if $15 \in B$, then $10 + 15 = 25$ is a perfect square, and if $15 \in A$, then $1 + 15 = 16$ is a perfect square. Thus, 15 cannot belong to A or B, which is a contradiction. Therefore, the proposition holds.

Remark. "Assuming without loss of generality that $1 \in A$" is a technique that should be mastered. Artificially making reasonable assumtions can simplify the problem when dealing with objects in problems with symmetry.

Example 7 (2015 China Mathematical Competition). Let a_1, a_2, a_3, and a_4 be four rational numbers such that

$$\{a_ia_j|1 \le i < j \le 4\} = \left\{-24, -2, -\frac{3}{2}, -\frac{1}{8}, 1, 3\right\}.$$

Find the value of $a_1 + a_2 + a_3 + a_4$.

Solution. Observe that $a_ia_j (1 \le i < j \le 4)$ are six distinct numbers and no two of them sum to 0, so no two of a_1, a_2, a_3, and a_4 sum to 0. Without loss of generality, we assume that $|a_1| < |a_2| < |a_3| < |a_4|$. Then among $|a_i||a_j| (1 \le i < j \le 4)$, the smallest and second smallest numbers are $|a_1||a_2|$ and $|a_1||a_3|$, respectively, and the largest and second largest numbers are $|a_3||a_4|$ and $|a_2||a_4|$, respectively. So it is necessary that

$$\begin{cases} a_1a_2 = -\dfrac{1}{8}, \\ a_1a_3 = 1, \\ a_2a_4 = 3, \\ a_3a_4 = -24, \end{cases}$$

and hence $a_2 = -\frac{1}{8a_1}$, $a_3 = \frac{1}{a_1}$, and $a_4 = \frac{3}{a_2} = -24a_1$. Therefore,

$$\{a_2a_3, a_1a_4\} = \left\{-\frac{1}{8a_1^2}, -24a_1^2\right\} = \left\{-2, -\frac{3}{2}\right\}.$$

Together with the fact that $a_1 \in \mathbf{Q}$ (the set of rational numbers), we have $a_1 = \pm\frac{1}{4}$.

Consequently, $(a_1, a_2, a_3, a_4) = \left(\frac{1}{4}, -\frac{1}{2}, 4, -6\right)$ or $\left(-\frac{1}{4}, \frac{1}{2}, -4, 6\right)$. It is simple to verify that both solutions satisfy the requirements. Therefore, $a_1 + a_2 + a_3 + a_4 = \pm\frac{9}{4}$.

Example 8 (2010 China Mathematical Olympiad). Let m and n be given integers greater than 1 and $a_1 < a_2 < \cdots < a_m$ be integers. Show that there exists a subset T of the set of integers such that

$$|T| \le 1 + \frac{a_m - a_1}{2n + 1},$$

and for each $i \in \{1, 2, \ldots, m\}$, there exist $t \in T$ and $s \in [-n, n]$ such that $a_i = t + s$.

Solution. Let $a_1 = a$ and $a_m = b$. Using the division with remainder, we can write $b - a = (2n+1)q + r$, where q, $r \in Z$ and $0 \leq r \leq 2n$.

Choose $T = \{a + n + (2n+1)k | k = 0, 1, \ldots, q\}$. Then $|T| = q + 1 \leq 1 + \frac{b-a}{2n+1}$, and the set

$$B = \{t + s | t \in T,\ s = -n, -n+1, \ldots, n\}$$
$$= \{a, a+1, \ldots, a + (2n+1)q + 2n\}.$$

Note that $a + (2n+1)q + 2n \geq a + (2n+1)q + r = b$, so each a_i is in B, and the assertion is valid.

Example 9. Let A be a set of positive integers with the following properties:

(1) A has at least three elements.
(2) If $a \in A$, then all the positive divisors of a belong to A.
(3) If $a \in A$, $b \in A$, and $1 < a < b$, then $1 + ab \in A$.

Solve the following problems:

(1) Show that 1, 2, 3, 4, and 5 are all elements of A.
(2) Is 2005 an element of A?

Solution. (1) First, apparently $1 \in A$. Suppose $a \in A$, $b \in A$, and $1 < a < b$. If at least one of a and b is even, then $2 \in A$, and if both a and b are odd, then $1 + ab$ is even. Thus, $2 \in A$.

Suppose 1, 2, $a \in A (a > 2)$. Then

$$1 + 2a \in A,\ 1 + 2(1 + 2a) = 3 + 4a \in A,$$
$$1 + (1 + 2a)(3 + 4a) = 4 + 10a + 8a^2 \in A.$$

If a is even, then $4 | (4 + 10a + 8a^2)$, so $4 \in A$. If a is odd, then $4 + 10a + 8a^2$ is even, and replacing a with $4 + 10a + 8a^2$ in the discussion above still gives $4 \in A$.

Also, $1+2\times 4 = 9 \in A$, so $3 \in A$, $1+2\times 3 = 7 \in A$, and $1+2\times 7 = 15 \in A$, which gives $5 \in A$. Therefore, 1,2,3,4, and 5 are all elements of A.

(2) Since $1 + 3 \times 5 = 16$, we have $8 \in A$, and hence

$$1 + 4 \times 8 = 33,\ 1 + 3 \times 33 = 100,$$
$$1 + 5 \times 100 = 501,\ 1 + 4 \times 501 = 2005,$$

which all belong to A. Therefore, 2005 is an element of A.

Remark. In fact, we can prove that $A = \mathbf{N}^*$.
We know from (1) that 1,2,3,4, and 5 all belong to A. Assume that $1, 2, \ldots, n \in A (n \geq 5)$. We show that $n + 1 \in A$.

If $n + 1 = 2k + 1$ is odd, then $3 \leq k < n$, so $n + 1 = 1 + 2k \in A$; if $n + 1 = 2k$ is even, then $3 \leq k < n$, thus $n = 2k - 1 \in A$ and $1 + 2k \in A$. Hence, $1+(2k-1)(2k+1) = 4k^2 \in A$, from which $2k \in A$, namely $n+1 \in A$. By the induction principle, we have proven that $A = \mathbf{N}^*$.

1.3 Exercises

Group A

1. (2012 China Mathematical Competition Test B) We call $b-a$ the length of the set $\{x|a \leq x \leq b\}$. Let sets
$$A = \{x|a \leq x \leq a + 1981\}, \quad B = \{x|b - 1014 \leq x \leq b\}$$
be subsets of $U = \{x|0 \leq x \leq 2012\}$. Then the minimal length of $A \cap B$ is __________.
2. (2011 China Mathematical Competition) Let $A = \{a_1, a_2, a_3, a_4\}$. If $B = \{-1, 3, 5, 8\}$ is the set of all the sums of three-element subsets of A, then $A =$ __________.
3. Suppose x, y, and z are nonzero real numbers. Find the set of all possible values of
$$\frac{x}{|x|} + \frac{y}{|y|} + \frac{z}{|z|} + \frac{xy}{|xy|} + \frac{xyz}{|xyz|}$$
by listing its elements.
4. Let
$$A = \{2, 4, a^3 - 2a^2 - a + 7\},$$
$$B = \{1, 5a - 5, -\frac{1}{2}a^2 + \frac{3}{2}a + 4, a^3 + a^2 + 3a + 7\}.$$
Prove or disprove that there exists $a \in \mathbf{R}$ such that $A \cap B = \{2, 5\}$.
5. Let X be the set of the real solutions to the equation $x^2 + px + q = 0$, and
$$A = \{1, 3, 5, 7, 9\}, B = \{1, 4, 7, 10\},$$
with $X \cap A = \emptyset$ and $X \cap B = X$. Find the values of p and q.
6. Suppose $A = \{(x, y)|y = ax + 2\}$ and $B = \{(x, y)|y = |x + 1|\}$, where a is a real number and $A \cap B$ is a one-element set. Find the value range of a.

7. Let

$$A = \{x | x = a^2 + 1,\ a \in \mathbf{N}^*\},$$
$$B = \{y | y = b^2 - 6b + 10,\ b \in \mathbf{N}^*\}.$$

What is the relationship between A and B?

8. Let $M = \{x, xy, \lg(xy)\}$ and $N = \{0, |x|, y\}$ be such that $M = N$. Find the value of

$$\left(x + \frac{1}{y}\right) + \left(x^2 + \frac{1}{y^2}\right) + \left(x^3 + \frac{1}{y^3}\right) + \cdots + \left(x^{2006} + \frac{1}{y^{2006}}\right).$$

9. Let M and n be the maximal and minimal elements of the set $\{\frac{3}{a} + b | 1 \leq a \leq b \leq 2\}$, respectively. Find $M - m$.

10. Let A be the set consisting of all sums of the squares of two integers, that is,

$$A = \{x | x = m^2 + n^2,\ m,\ n \in \mathbf{Z}\}$$

(here $\mathbf{Z}$ is the set of integers).

(1) Show that if s, $t \in A$, then $st \in A$.

(2) Show that if s, $t \in A$ with $t \neq 0$, then $\frac{s}{t} = p^2 + q^2$, where p and q are rational numbers.

11. Suppose S is a subset of $\{1, 2, \ldots, 50\}$ and the sum of the squares of any two elements of S is not divisible by 7. Find the maximum value of $|S|$.

12. Let M be a subset of the set of positive integers such that $1 \in M$, $2006 \in M$, and $2007 \notin M$. Suppose M has the property that if a, $b \in M$, then $\left[\sqrt{\frac{a^2+b^2}{2}}\right] \in M$. How many nonempty subsets does M have? (The notation $[x]$ denotes the greatest integer not exceeding x.)

13. Let S_1, S_2, and S_3 be three nonempty sets of integers. Suppose for i, j, k being any permutation of 1, 2, 3, if $x \in S_i$ and $y \in S_j$, then $x - y \in S_k$. Show that two of S_1, S_2, and S_3 are identical.

14. Suppose $A = \{(x, y) | ax + y = 1\}$, $B = \{(x, y) | x + ay = 1\}$, and $C = \{(x, y) | x^2 + y^2 = 1\}$, where a is a real number.

(1) For what values of a is $(A \cup B) \cap C$ a two-element set?

(2) For what values of a is $(A \cup B) \cap C$ a three-element set?

15. Suppose $A = \{a_1, a_2, a_3, a_4, a_5\}$ and $B = \{a_1^2, a_2^2, a_3^2, a_4^2, a_5^2\}$, where $a_i (1 \leq i \leq 5)$ are positive integers, with $a_1 < a_2 < a_3 < a_4 < a_5$ and $a_1 + a_4 = 10$. If $A \cap B = \{a_1, a_4\}$ and the sum of elements in $A \cup B$ is 224, find A.

Group B

16. Suppose A is a subset of $\{1, 2, \ldots, 2000\}$ with the following property: no element of A is equal to five times another element of A. What is the maximum value of $|A|$?
17. Suppose a collection of sets A_1, A_2,..., A_n has the following properties:

    (1) Each A_i has 30 elements.
    (2) For each pair i and j with $1 \leq i < j \leq n$, the set $A_i \cap A_j$ has exactly one element.
    (3) $A_1 \cap A_2 \cap \ldots \cap A_n = \emptyset$.

    Find the maximum value of n for such a collection of sets to exist.
18. Suppose $A \subseteq \{1, 2, \ldots, 2000\}$ and the difference of any two elements in A is not equal to 4 or 7. Find the maximum value of $|A|$.
19. Let $A = \{a_1, a_2, a_3\}$ and T be a collection of subsets of A, containing $\emptyset$ and A. Suppose the intersection and union of any two elements of T also belong to T. Find the number of all such T.
20. (2011 China West Mathematical Invitational) Let an integer $n \geq 2$ be given.

    (1) Show that there exists a permutation of all the subsets of $\{1, 2, \ldots, n\}$, namely A_1, $A_2, \ldots, A_{2^n}$, such that the difference of the numbers of elements in A_i and A_{i+1} is exactly 1 for each $i = 1, 2, \ldots, 2^n$, where $A_{2^n+1} = A_1$.
    (2) Suppose A_1, $A_2, \ldots, A_{2^n}$ satisfy the condition in (1). Find all possible values of $\sum_{i=1}^{2^n} (-1)^i S(A_i)$, where $S(A_i) = \sum_{x \in A_i} x$ and $S(\emptyset) = 0$.

Chapter 2

Number of Elements in a Finite Set

2.1 Key Points of Knowledge and Basic Methods

A set can be classified as a finite or infinite set, depending on whether it has a finite number of elements. If a set A is a finite set, we use $|A|$ to denote the number of elements in it.

A. Maps

For any two sets A and B, if there is a rule f such that for every element x in A, there is a unique element $f(x)$ in B that corresponds to x, then we call $f : A \to B$ a map (from A to B). Here $f(x)$ is called the image of x, and x is called a preimage of $f(x)$.

If $f : A \to B$ is a map and for any $x, y \in A$ with $x \neq y$, we have $f(x) \neq f(y)$, then f is called an injection from A to B. It is clear that if A and B are finite sets and there exists an injection from A to B, then $|A| \leq |B|$.

If $f : A \to B$ is a map and for any $y \in B$, there exists $x \in A$ such that $y = f(x)$, then f is called a surjection from A onto B. Also, if A and B are finite sets and there exists a surjection from A onto B, then $|A| \geq |B|$.

If $f : A \to B$ is a map that is both an injection and a surjection, then f is called a bijection (or a one-to-one correspondence) from A to B. If A and B are finite sets and there exists a bijection from A to B, then $|A| = |B|$.

B. The inclusion-exclusion principle

Theorem 1. *If A and B are finite sets, then*

$$|A \cup B| = |A| + |B| - |A \cap B|.$$

Theorem 2. *If A, B, and C are finite sets, then*

$$|A \cup B \cup C| = |A| + |B| + |C| - |A \cap B| - |B \cap C| - |C \cap A| + |A \cap B \cap C|.$$

2.2 Illustrative Examples

Example 1. How many integers among $1, 2, \ldots, 1000$ are neither divisible by 2 nor divisible by 5?

Solution. Let $S = \{1, 2, \ldots, 1000\}$, $A_2 = \{a | a \in S,\ 2|a\}$, and $A_5 = \{a | a \in S,\ 5|a\}$.

Then $|A_2| = \frac{1000}{2} = 500$, $|A_5| = \frac{1000}{5} = 200$, and $|A_2 \cap A_5| = \frac{1000}{10} = 100$. Hence,

$$\begin{aligned}|\complement_S \bar{A}_2 \cap \complement_S \bar{A}_5| &= |S| - (|A_2| + |A_5|) + |A_2 \cap A_5| \\ &= 1000 - (500 + 200) + 100 = 400.\end{aligned}$$

Example 2. Let A be a set consisting of positive integers, whose smallest element is 1 and largest element is 100. Apart from 1, every element of A is the sum of two other elements (not necessarily distinct) of A. Find the minimum value of $|A|$.

Solution. Let $A = \{1, a_1, a_2, \ldots, a_n, 100\}$, where $1 < a_1 < a_2 < \cdots < a_n < 100$. Then $a_1 = 1 + 1 = 2$.

If $n = 6$, then $a_2 \leq 2 + 2 = 4$, $a_3 \leq 4 + 4 = 8$, $a_4 \leq 8 + 8 = 16$, $a_5 \leq 16 + 16 = 32$, and $a_6 \leq 32 + 32 = 64$. Since $a_5 + a_6 \leq 32 + 64 = 96 < 100$, we have $100 = 2a_6$, so $a_6 = 50$. Also, $a_4 + a_5 \leq 16 + 32 = 48 < 50$, thus $50 = 2a_5$, that is $a_5 = 25$. Further, $a_3 + a_4 \leq 8 + 16 = 24 < 25$, from which $25 = 2a_4$, a contradiction.

If $n \leq 5$, then similarly $a_n \leq 32$, so $2a_n \leq 64 < 100$, which is impossible.

Therefore, $|A| \geq 9$. Observing that the set $A = \{1, 2, 3, 5, 10, 20, 25, 50, 100\}$ satisfies the requirements and $|A| = 9$, we conclude that the minimum value of $|A|$ is 9.

Example 3. Let A and B be two subsets of $\{1, 2, 3, \ldots, 100\}$ such that they have the same number of elements and $A \cap B$ is the empty set. Suppose for every $n \in A$, we have $2n + 2 \in B$. Find the maximum value of $|A \cup B|$.

Solution. First we prove that $|A \cup B| \leq 66$, or equivalently $|A| \leq 33$. It suffices to show that if A is a 34-element subset of $\{1, 2, 3, \ldots, 49\}$, then there exists some $n \in A$ such that $2n + 2 \in A$.

We partition the set $\{1, 2, 3, \dots, 49\}$ into 33 subsets:

$$\{1,4\}, \{3,8\}, \{5,12\}, \dots, \{23,48\}, \{2,6\}, \{10,22\}, \dots,$$
$$\{18,38\}, \{25\}, \{27\}, \dots, \{49\}, \{26\}, \{34\}, \{42\}, \{46\}.$$

Since A is a 34-element subset, by the pigeonhole principle it must contain two elements in one of the above 33 subsets. This shows that there is some $n \in A$ such that $2n+2 \in A$, a contradiction, and hence $|A| \le 33$.

On the other hand, if we take $A = \{1, 3, 5, \dots, 23, 2, 10, 14, 18, 25, 27, \dots, 49, 26, 34, 42, 46\}$ and $B = \{2n+2 | n \in A\}$, then these two sets satisfy the conditions, and $|A| = |B| = 33$ with $|A \cup B| = 66$. Therefore, the maximum value of $|A \cup B|$ is 66.

Example 4. Let $S = \{1, 2, 3, 4\}$. An n-term sequence of numbers $a_1, a_2, \dots, a_n$ has the following property: for every nonempty subset B of S (its number of elements is $|B|$), there exists $|B|$ consecutive terms in the sequence that constitute B. Find the minimum value of n.

Solution. First, we show that each element of S appears at least twice in $a_1, a_2, \dots, a_n$. In fact, if some element x appears only once, then since there are 3 two-element subsets containing x, while there are at most two ways to choose two consecutive terms containing x in the sequence, some two-element subset of S does not appear as two consecutive terms in the sequence, which leads to a contradiction.

As each element of S appears at least twice in the sequence, we have $n \ge 8$.

Next, we give an eight-term sequence that satisfies the conditions:

$$3,\ 1,\ 2,\ 3,\ 4,\ 1,\ 2,\ 4.$$

Therefore, the minimum value of n is 8.

Example 5. Let $A = \{1, 2, 3, \dots, 2n, 2n+1\}$ and B be a subset of A such that for any three distinct elements x, y, and z of B, we have $x + y \ne z$. Find the maximum value of $|B|$.

Solution. Let $O = \{1, 3, \dots, 2n+1\}$ and $E = \{2, 4, \dots, 2n\}$. Then $A = O \cup E$. Suppose $B = \{b_1, \dots, b_s,\ c_1, \dots, c_t\}$, where $b_1, \dots, b_s \in O$, $c_1, \dots, c_t \in E$, and $b_1 < b_2 < \cdots < b_s$. By the assumption, $b_s - b_i \ne c_j$ for any $i = 1, 2, \dots, s-1$ and $j = 1, 2, \dots, t$ (otherwise $b_i + c_j = b_s$, and b_i, c_j, and b_s all belong to B), and $2 \le b_s - b_i \le 2n$, in other words,

$$b_s - b_i \in E, \quad i = 1, 2, \ldots, s-1.$$

Therefore, $b_s - b_1, b_s - b_2, \ldots, b_s - b_{s-1}, c_1, \ldots, c_t$ are distinct elements of E, so

$$(s-1) + t \le |E| = n, \quad |B| = s + t \le n + 1.$$

If we let $B = \{n+1, n+2, \ldots, 2n+1\}$, then for every two elements x and y in B, we have $x + y \ge 2n+3$, so B satisfies the conditions. Therefore, the maximum value of $|B|$ is $n+1$.

Remark. In Examples 25, we first found an upper bound or lower bound for the problem (namely $|A| \ge 9$ in Example 2, $|A| \le 33$ in Example 3, $n \ge 8$ in Example 4, and $|B| \le n+1$ in Example 5), and then constructed a specific instance to show that the upper bound or lower bound can be reached, which is a common method for extreme value problems. When dealing with such problems, we usually guess a bound based on specific instances and then try to prove it.

Example 6. The students of a class have taken midterm exams in three subjects, namely Literature, Mathematics, and English. The results are as follows: 18 students got full marks in at least one subject, 9 students got full marks in Literature, 11 students got full marks in Mathematics, 8 students got full marks in English, 5 students got full marks in both Literature and Mathematics, 3 students got full marks in both Mathematics and English, and 4 students got full marks in both Literature and English. Answer the following questions:

(1) How many students got full marks in at least one of Literature and Mathematics?
(2) How many students got full marks in all three subjects?

Solution. Let A, B, and C be the set of students who got full marks in Literature, Mathematics, and English, respectively. Then according to the problem,

$$|A| = 9, \quad |B| = 11, \quad |C| = 8.$$

$$|A \cap B| = 5, \quad |B \cap C| = 3, \quad |C \cap A| = 4, \quad |A \cup B \cup C| = 18.$$

(1) The number of students who got full marks in at least one of Literature and Mathematics is

$$|A \cup B| = |A| + |B| - |A \cap B| = 9 + 11 - 5 = 15.$$

(2) The number of students who got full marks in all three subjects is

$$
\begin{aligned}
|A \cap B \cap C| &= |A \cup B \cup C| - |A| - |B| - |C| + |A \cap B| \\
&\quad + |B \cap C| + |C \cap A|. \\
&= 18 - 9 - 11 - 8 + 5 + 3 + 4 = 2.
\end{aligned}
$$

Example 7 (2016 China Mathematical Competition Test B). Let A be an 11-element set of real numbers, and let $B = \{uv | u,\ v \in A,\ u \neq v\}$. Find the minimum value of $|B|$.

Solution. First we show that $|B| \geq 17$. Since switching the sign of each element in A does not change B, we can assume without loss of generality that there are at least as many positive numbers as negative numbers in A. Now, we consider the following cases:

(1) There are no negative numbers in A.
Let $a_1 < a_2 < \cdots < a_{11}$ be all the elements in A, where $a_1 \geq 0$ and $a_2 > 0$. Then

$$a_1a_2 < a_2a_3 < a_2a_4 < \cdots < a_2a_{11} < a_3a_{11} < \cdots < a_{10}a_{11},$$

all of which belong to B, so B has at least $1 + 9 + 8 = 18$ elements.

(2) There is at least one negative number in A.
Let $b_1,\ b_2, \ldots, b_k$ be all the nonnegative elements and $c_1,\ c_2, \ldots, c_l$ be all the negative elements in A, such that

$$c_l < \cdots < c_1 < 0 \leq b_1 < \cdots < b_k,$$

where k and l are positive integers with $k + l = 11$ and $k \geq l$ so that $k \geq 6$. Then

$$c_1b_1 > c_1b_2 > \cdots > c_1b_k > c_2b_k > \cdots > c_lb_k,$$

which are $k + l - 1 = 10$ distinct nonpositive elements in B, and also

$$b_2b_3 < b_2b_4 < b_2b_5 < b_2b_6 < b_3b_6 < b_4b_6 < b_5b_6,$$

which are seven distinct positive elements of B. This means that $|B| \geq 10 + 7 = 17$.

Therefore, we always have $|B| \geq 17$.

If we take $A = \{0, \pm 1,\ \pm 2,\ \pm 2^2,\ \pm 2^3,\ \pm 2^4\}$, then $B = \{0, -1,\ \pm 2,\ \pm 2^2,\ \pm 2^3, \ldots, \pm 2^6,\ \pm 2^7,\ -2^8\}$, and $|B| = 17$. Overall, the minimum value of $|B|$ is 17.

Example 8. Let $A_1, A_2, \ldots, A_k$ be a partition of the set $S = \{1, 2, \ldots, 36\}$ (so no two of them intersect and their union is S). Suppose for each A_i $(i = 1, 2, \ldots, k)$, the sum of any two elements of A_i is not a perfect square. Find the minimum value of k.

Solution. First we consider the numbers 6, 19, and 30. Since $6+19 = 5^2$, $6+30 = 6^2$, and $19+30 = 7^2$, we see that no two of the three numbers belong to the same A_i. This implies that $k \geq 3$.

On the other hand, we can partition S into 3 subsets A_1, A_2, and A_3 in the following way:

$$A_1 = \{4k+3 | 0 \leq k \leq 8\} \cup \{4, 8, 16, 24, 36\},$$

$$A_2 = \{4k+1 | 0 \leq k \leq 8\} \cup \{6, 14, 18, 26, 34\},$$

$$A_3 = \{2, 10, 12, 20, 22, 28, 30, 32\}.$$

Since the remainder of a perfect square modulo 4 is either 0 or 1, we can easily verify the validity of the partition. Therefore, the minimum value of k is 3.

Example 9 (2017 China Mathematical Competition Test B). Let $a_1, a_2, \ldots, a_{20} \in \{1, 2, \ldots, 5\}$, $b_1, b_2, \ldots, b_{20} \in \{1, 2, \ldots, 10\}$, and $X = \{(i, j) | 1 \leq i < j \leq 20,\ (a_i - a_j)(b_i - b_j) < 0\}$. Find the maximum value of $|X|$.

Solution. Consider a list of integers $(a_1, a_2, \ldots, a_{20}, b_1, b_2, \ldots, b_{20})$ that satisfies the conditions. For $k = 1, 2, \ldots, 5$, let t_k be the number of the terms in $a_1, \ldots, a_{20}$ that are equal to k. By the definition of X, if $a_i = a_j$ then $(i, j) \notin X$, so at least $\sum_{k=1}^{5} C_{t_k}^2$ pairs of (i, j) are not in X. Observing that $\sum_{k=1}^{5} t_k = 20$, by Cauchy's inequality we have

$$\sum_{k=1}^{5} C_{t_k}^2 = \frac{1}{2} \cdot \left(\sum_{k=1}^{5} t_k^2 - \sum_{k=1}^{5} t_k\right) \geq \frac{1}{2} \cdot \left(\frac{1}{5}\left(\sum_{k=1}^{5} t_k\right)^2 - \sum_{k=1}^{5} t_k\right)$$

$$= \frac{1}{2} \cdot 20 \cdot \left(\frac{20}{5} - 1\right) = 30.$$

Consequently, the number of elements in X cannot exceed $\mathrm{C}_{20}^2 - 30 = 190 - 30 = 160$.

On the other hand, if we choose $a_{4k-3} = a_{4k-2} = a_{4k-1} = a_{4k} = k (k = 1, 2, \ldots, 5)$ and $b_i = 6 - a_i$ $(i = 1, 2, \ldots, 20)$, then for each i and

j $(1 \leq i < j \leq 20)$,

$$(a_i - a_j)(b_i - b_j) = (a_i - a_j)((6 - a_i) - (6 - a_j)) = -(a_i - a_j)^2 \leq 0,$$

where the equality is satisfied if and only if $a_i = a_j$, which occurs exactly $5\mathrm{C}_4^2 = 30$ times. In this case the number of elements in X is exactly $\mathrm{C}_{20}^2 - 30 = 160$.

Therefore, the maximum value of $|X|$ is 160.

2.3 Exercises

Group A

1. Let $A = \{a_1, a_2, a_3\}$ and $B = \{b_1, b_2, b_3, b_4\}$.
   (1) Write down a map $f : A \to B$, which is an injection, and then find the number of injections from A to B.
   (2) Write down a map $f : A \to B$, which is not an injection, and then find the number of non-injective maps from A to B.
   (3) Is it possible for a map from A to B to be a surjection?
2. Let A and B be sets such that $A \cup B = \{a_1, a_2, a_3\}$. If $(A,\ B)$ and $(B,\ A)$ are considered different pairs when $A \neq B$, find the number of such pairs of $(A,\ B)$.
3. If three distinct nonzero real numbers a, b, and c satisfy the equality $\frac{1}{a} + \frac{1}{b} = \frac{2}{c}$, then we call them harmonic. If $a + c = 2b$, then we call them arithmetic. Let

$$M = \{x \| x| \leq 2013,\ x \in \mathbf{Z}\}$$

   and P be a three-element subset of M. We call P a "good subset" if its three elements are both harmonic and arithmetic (in some order). The number of distinct "good subsets" is ____________.
4. Let

$$A = \{(x, y) \| x| + |y| = a,\ a > 0\},$$
$$B = \{(x, y) \| xy| + 1 = |x| + |y|\}.$$

   If $A \cap B$ is the set of vertices of a regular octagon in the plane, find the value of a.
5. Let $M = \{1, 2, 3, \ldots, 1995\}$ and $A \subseteq M$ be such that for each $x \in A$, we have $19x \notin A$. Find the maximum value of $|A|$.
6. How many integers from 1 to 100 can we choose at most, so that none of the chosen numbers are equal to 3 times another chosen number?

7. For a nonempty subset of $\{1, 2, \ldots, n\}$, we define its alternating sum as follows: we arrange the numbers in the set in descending order, and alternately add or subtract its numbers (starting from the largest number). For example, the alternating sum of the set $\{1, 2, 4, 6, 9\}$ is $9 - 6 + 4 - 2 + 1 = 6$, while the alternating sum of $\{5\}$ is 5. If $n = 7$, find the sum of all these alternating sums.
8. Let S be a subset of $\{1, 2, \ldots, 9\}$ such that the sums of each pair of different numbers in S are pairwise distinct. How many elements can S have at most?
9. Let $A = \{a | 1 \leq a \leq 2000,\ a = 4k + 1,\ k \in \mathbf{Z}\}$ and $B = \{b | 1 \leq b \leq 3000,\ b = 3k - 1,\ k \in \mathbf{Z}\}$. Find $|A \cap B|$.
10. Consider the sums of elements in all the subsets of $M = \{1, 2, 3, \ldots, 100\}$. Compute the sum of all these sums.
11. Let $M = \{1, 2, 3, \ldots, 1995\}$ and A be its subset such that for each $x \in A$, we have $15x \notin A$. Find the maximum value of $|A|$.
12. A middle school has 120 teachers. Among them, 40 can teach Literature, 50 can teach Mathematics, 45 can teach English, 15 can teach both Mathematics and English, 10 can teach both Mathematics and Literature, 8 can teach both English and Literature, and 4 can teach all three subjects. How many of the teachers can teach none of the three subjects?

Group B

13. Let $E = \{1, 2, 3, \ldots, 200\}$ and $G = \{a_1, a_2, \ldots, a_{100}\}$ be a proper subset of E. Suppose G has the following two properties:

    (1) For all $1 \leq i \leq j \leq 100$, we have $a_i + a_j \neq 201$.
    (2) $a_1 + a_2 + \cdots + a_{100} = 10080$.

    Prove that the number of odd integers in G is divisible by 4, and that the sum of the squares of numbers in G is a fixed number (i.e., independent of the choice of G).
14. Let A be a finite set with a map $f : \mathbf{N}^* \to A$ such that for $i, j \in \mathbf{N}^*$, if $|i - j|$ is a prime number, then $f(i) \neq f(j)$. Find the minimum value of $|A|$.
15. Let $A = \{0, 1, 2, \ldots, 9\}$ and $B_1, B_2, \ldots, B_k$ be its nonempty subsets. Suppose for each pair $i \neq j$, the set $B_i \cap B_j$ has at most two elements. Find the maximum value of k.

16. Let $S = \{1, 2, \ldots, 100\}$. Find the smallest positive integer n such that every n-element subset of S contains four elements that are pairwise coprime.
17. Suppose n four-element sets $A_1, A_2, \ldots, A_n$ satisfy the following properties:

    (1) $|A_i \cap A_j| = 1$ with $1 \leq i < j \leq n$.
    (2) $|A_1 \cup A_2 \cup \cdots \cup A_n| = n$.

    Find the maximum value of n.

Chapter 3

Quadratic Functions

3.1 Key Points of Knowledge and Basic Methods

A. The function $f(x) = ax^2 + bx + c$ $(a \neq 0)$ is called a quadratic function. It is also written as $f(x) = a(x-k)^2 + m(a \neq 0)$ (the vertex form) or $f(x) = a(x - x_1) \cdot (x - x_2)(a \neq 0)$ (the factored form).

B. Properties of $f(x) = ax^2 + bx + c$ $(a \neq 0)$

(1) Symmetry: for every real number x,

$$f\left(-\frac{b}{2a} + x\right) = f\left(-\frac{b}{2a} - x\right).$$

(2) $f(0) = c$.

(3) If $\Delta = b^2 - 4ac \geq 0$, then $f\left(\frac{-b \pm \sqrt{\Delta}}{2a}\right) = 0$.

(4) If $a > 0$, then $f(x)$ is decreasing on the interval $\left(-\infty, -\frac{b}{2a}\right]$, and increasing on the interval $\left[-\frac{b}{2a} + \infty\right)$; if $a < 0$, then $f(x)$ is increasing on $\left(-\infty, -\frac{b}{2a}\right]$, and decreasing on $\left[-\frac{b}{2a}, +\infty\right)$.

(5) If $a > 0$, then $f(x)$ has the minimum value:

$$f_{\min}(x) = f\left(-\frac{b}{2a}\right) = \frac{4ac - b^2}{4a}.$$

If $a < 0$, then $f(x)$ has the maximum value:

$$f_{\max}(x) = f\left(-\frac{b}{2a}\right) = \frac{4ac - b^2}{4a}.$$

C. The graph of $f(x) = ax^2 + bx + c(a \neq 0)$

(1) Axis of symmetry: The graph of $f(x) = ax^2 + bx + c(a \neq 0)$ is a parabola that is symmetric with respect to the line $x = -\frac{b}{2a}$.

(2) Vertex: The intersection point of the graph and its axis of symmetry, or $\left(-\frac{b}{2a}, \frac{4ac-b^2}{4a}\right)$ in the coordinate form, is its vertex. The vertex is the lowest point of the parabola when $a > 0$ and the highest point when $a < 0$.
(3) Opening: The parabola opens upwards when $a > 0$ and downwards when $a < 0$, and the magnitude of opening depends on $|a|$.
(4) The relative position of a parabola and the x-axis: When $\Delta > 0$, the parabola and the $x-$axis intersect at two different points; when $\Delta = 0$, the parabola and the x-axis are tangent to each other and have one common point; when $\Delta < 0$, the parabola and the $x-$axis have no common point.

3.2 Illustrative Examples

Example 1. Suppose $f(x) = x^2 + ax + b$ and

$$f(0) > 1, \quad f(4b - a) = f(3 - a^2).$$

Find the minimum value of $f(x)$.

Solution. If $4b-a = 3-a^2$, then $b = \frac{1}{4}(-a^2+a+3) = \frac{13}{16} - \frac{1}{4}\left(a - \frac{1}{2}\right)^2 \leq \frac{13}{16}$, contradictory to $f(0) = b > 1$. Therefore, $4b - a \neq 3 - a^2$.

Since $f(4b - a) = f(3 - a^2)$ and $4b - a \neq 3 - a^2$, the two points $(4b - a, f(4b - a))$ and $(3 - a^2, f(3 - a^2))$ are symmetric with respect to the axis of symmetry $x = -\frac{a}{2}$. Consequently,

$$(4b - a) + (3 - a^2) = 2 \times \left(-\frac{a}{2}\right),$$

and hence $4b - a^2 = -3$. Therefore, the minimum value of $f(x)$ is $b - \frac{a^2}{4} = -\frac{3}{4}$.

Example 2. Suppose a, b, $c \in R$, $f(x) = ax^2 + bx + c$, and $g(x) = ax + b$. Given that $|f(x)| \leq 1$ when $x \in [-1, 1]$, answer the following questions:

(1) Show that $|c| \leq 1$.
(2) Show that $|g(x)| \leq 2$ for $x \in [-1, 1]$.
(3) Further suppose $a > 0$ and the maximum value of $g(x)$ for $x \in [-1, 1]$ is 2. Find $f(x)$.

Solution. (1) $|c| = |f(0)| \leq 1$.

(2) Since $g(x)$ is a linear function, and hence a monotonic function, it suffices to show that $|g(1)| \le 2$ and $|g(-1)| \le 2$. We have the following:

$$\left.\begin{aligned} f(0) &= c \\ f(1) &= a+b+c \end{aligned}\right\} \Rightarrow |g(1)| = |a+b| = |f(1)-f(0)|$$

$$\le |f(1)| + |f(0)| \le 2,$$

$$\left.\begin{aligned} f(-1) &= a-b+c \\ f(0) &= c \end{aligned}\right\} \Rightarrow |g(-1)| = |a-b| = |f(-1)-f(0)|$$

$$\le |f(-1)| + |f(0)| \le 2.$$

(3) Since $a > 0$, we have $g(x)_{\max} = g(1) = a + b = 2$.

Note that $2 = g(1) = a + b = f(1) - c$, while $|f(1)| \le 1$ and $|c| \le 1$, so necessarily $c = f(0) = -1$.

Also, $|f(x)| \le 1$ for $x \in [-1, 1]$, thus $f(x) \ge f(0) = c = -1$ when $x \in [-1, 1]$, which means that $f(0)$ is the minimum value of $f(x)$, and $x = 0$ is the axis of symmetry of the parabola $y = f(x)$. It follows that $b = 0$ and $a = 2$. Therefore, $f(x) = 2x^2 - 1$.

Example 3. Suppose the parabola graph of $y = x^2 - (k-1)x - k - 1$ intersects the x-axis at two points A and B, and its vertex is C. Find the minimal area of the triangle ABC.

Solution. First, since

$$\Delta = (k-1)^2 + 4(k+1) = k^2 + 2k + 5 = (k+1)^2 + 4 > 0,$$

the parabola always intersects the x-axis at two points, no matter what k we choose.

Let x_1 and x_2 be the x-coordinates of A and B, respectively. Then

$$|AB| = |x_2 - x_1| = \sqrt{(x_2 - x_1)^2} = \sqrt{(x_1 + x_2)^2 - 4x_1x_2} = \sqrt{k^2 + 2k + 5}.$$

Since the vertex is $C\left(\frac{k-1}{2}, -\frac{k^2+2k+5}{4}\right)$,

$$S_{\Delta ABC} = \frac{1}{2}\sqrt{k^2+2k+5} \cdot \left|-\frac{k^2+2k+5}{4}\right| = \frac{1}{8}\sqrt{(k^2+2k+5)^3}$$

$$= \frac{1}{8}\sqrt{[(k+1)^2+4]^3} \ge \frac{1}{8}\sqrt{4}^3 = 1,$$

where the equality is valid if and only if $k = -1$.

Therefore, the minimal area of the triangle ABC is 1.

Example 4. Suppose the quadratic function $f(x) = ax^2 + (2b+1)x - a - 2$ ($a, b \in R$, and $a \neq 0$) has at least one zero in the interval [3,4]. Find the minimum value of $a^2 + b^2$.

Solution 1. Let s be a zero of the function in [3,4]. Then

$$as^2 + (2b+1)s - a - 2 = 0,$$

$$(2-s)^2 = [a(s^2-1) + 2bs]^2 \leq (a^2+b^2)[(s^2-1)^2 + 4s^2]$$
$$= (a^2+b^2)(1+s^2)^2,$$

and hence $a^2 + b^2 \geq \left(\frac{s-2}{1+s^2}\right)^2 = \frac{1}{\left(s-2+\frac{5}{s-2}+4\right)^2}$. (The inequality above follows from Cauchy's inequality.)

Since $t - 2 + \frac{5}{t-2}$ is a positive and decreasing function when $t \in [3,4]$, the expression $\frac{1}{\left(t-2+\frac{5}{t-2}+4\right)^2}$ has its minimum value $\frac{1}{100}$ when $t = 3$, and the equality holds when $a = -\frac{2}{25}$ and $b = -\frac{3}{50}$. Therefore, the minimum value of $a^2 + b^2$ is $\frac{1}{100}$.

Solution 2. We view the equation as a linear equation in a and b:

$$(x^2-1)a + 2xb + x - 2 = 0,$$

and use the fact that the minimal distance from a point on the line to the origin is the distance from the origin to the line. In other words, $\sqrt{a^2+b^2} \geq \frac{|x-2|}{\sqrt{(x^2-1)^2+(2x)^2}}$, and then argue similarly as in Solution 1.

Example 5. Suppose the parabola graph of $y = x^2 + ax + 2$ has two (different) common points with the line segment connecting $M(0,1)$ and $N(2,3)$ (including the endpoints). Find the value range of a.

Solution. The line passing through M and N is $y = x + 1$, and to say that the parabola graph of $y = x^2 + ax + 2$ intersects the line segment MN at two different points is to say that the equation $x^2 + ax + 2 = x + 1$ has two distinct zeros in the interval $[0, 2]$.

Let $f(x) = x^2 + (a-1)x + 1$, so that its axis of symmetry is $x = -\frac{a-1}{2}$. Then

$$\begin{cases} 0 < -\dfrac{a-1}{2} < 2, \\ \Delta = (a-1)^2 - 4 > 0, \\ f(0) = 1 \geq 0, \\ f(2) = 2a + 3 \geq 0. \end{cases}$$

Solving the system of inequalities gives $-\frac{3}{2} \le a < -1$, which is exactly the value range of a.

Remark. The graph of a quadratic function is a convenient tool to determine the distribution of the roots of its corresponding quadratic equation.

Example 6. Let $f(x) = ax^2 + bx + c$ $(a > 0)$, and let x_1 and x_2 be the roots of the equation $f(x) = x$, with properties $x_1 > 0$ and $x_2 - x_1 > \frac{1}{a}$. Suppose $0 < t < x_1$. Determine which of $f(t)$ and x_1 is greater.

Solution. Since x_1 and x_2 are the roots of the function $ax^2 + bx + c = x$,

$$x_1 + x_2 = -\frac{b-1}{a}, \quad x_1x_2 = \frac{c}{a}, ax_1^2 + bx_1 + c = x_1,$$

and hence $f(t) - x_1 = (at^2 + bt + c) - (ax_1^2 + bx_1 + c) = a(t + x_1)(t - x_1) + b(t - x_1) = a(t - x_1)\left(t + x_1 + \frac{b}{a}\right)$.

Further from $t + x_1 + \frac{b}{a} = t + \left(\frac{1}{a} - x_2\right) = \left(t + \frac{1}{a}\right) - x_2 < \left(x_1 + \frac{1}{a}\right) - x_2 < 0$, and $a > 0$ and $t - x_1 < 0$, we get that $f(t) - x_1 > 0$. Therefore, if $0 < t < x_1$, then $f(t) > x_1$.

Example 7 (2017 China Mathematical Competition). Suppose k and m are real numbers such that the inequality $|x^2 - kx - m| \le 1$ holds for all $x \in [a, b]$. Show that $b - a \le 2\sqrt{2}$.

Solution. Let $f(x) = x^2 - kx - m$ be such that $f(x) \in [-1, 1]$ when $x \in [a, b]$. Then

$$f(a) = a^2 - ka - m \le 1, \text{①}$$

$$f(b) = b^2 - kb - m \le 1, \text{②}$$

$$f\left(\frac{a+b}{2}\right) = \left(\frac{a+b}{2}\right)^2 - k \cdot \frac{a+b}{2} - m \ge -1. \text{③}$$

By ①+②− 2 × ③ we have

$$\frac{(a-b)^2}{2} = f(a) + f(b) - 2f\left(\frac{a+b}{2}\right) \le 4.$$

Therefore, $b - a \le 2\sqrt{2}$.

Example 8. Let $f(x) = 2px^2 + qx - p + 1$ be a quadratic function such that for $|x| \le 1$, we always have $f(x) \ge 0$. Find the maximum value of $p + q$.

Solution. Since

$$f\left(-\frac{1}{2}\right) = 2p\left(-\frac{1}{2}\right)^2 + q\left(-\frac{1}{2}\right) - p + 1 \geq 0,$$

we see that $p + q \leq 2$.

On the other hand, when $p = \frac{2}{3}$ and $q = \frac{4}{3}$,

$$f(x) = \frac{4}{3}x^2 + \frac{4}{3}x + \frac{1}{3} = \frac{1}{3}(2x+1)^2 \geq 0.$$

Therefore, the maximum value of $p + q$ is 2.

Example 9 (2017 Russian Mathematical Olympiad). Suppose a, b, and c are three different positive integers. Determine whether there exists a quadratic function $f(x)$ with integer coefficients such that its leading coefficient is positive and there exist integers x_1, x_2, and x_3 satisfying $f(x_1) = a^3$, $f(x_2) = b^3$, and $f(x_3) = c^3$.

Solution. The answer is affirmative.

Let $f(x) = x^3 - (x-a)(x-b)(x-c) = (a+b+c)x^2 - (ab+bc+ca)x + abc$.

Apparently the leading coefficient of $f(x)$ is $a+b+c \in N^*$, and $f(a) = a^3$, $f(b) = b^3$, and $f(c) = c^3$.

This shows the existence of such a function.

3.3 Exercises

Group A

1. Fill in the blanks:

   (1) Translate the graph of the function $y = 2x^2$ in such a way that its vertex lies on the graph of $y = -4x$ and the distance between its intersection points with the x-axis is 2. Then the formula of the translated function is ____________.

   (2) Suppose a can be any real number between 0 and 5 (including the endpoints). Then the number of integer values of b such that $3b = a(a-8)$ for some a is ____________.

   (3) Suppose $f(x) = ax^2 + bx$, and $1 \leq f(-1) \leq 2$ and $2 \leq f(1) \leq 4$. Then the value range of $f(-2)$ is ____________.

   (4) Suppose the quadratic function $f(x) = ax^2 + bx + c$ attains the maximum value 10 at $x = 3$ and the distance between its intersection points with the x-axis is 4. Then $f(1) =$ ____________.

(5) Suppose $P_1(x_1, 1994)$ and $P_2(x_2, 1994)$ both lie on the graph of

$$f(x) = ax^2 + bx + 7 \ (a \neq 0).$$

Then $f(x_1 + x_2) =$ ____________.

2. Let $f(x) = ax^2 + bx + c$ be such that $f(0) = 2$ and $f(1) = -1$, and the distance between its intersection points with the x-axis is $2\sqrt{2}$. Find the formula of $f(x)$.
3. Suppose a and b are unequal real numbers, and the function $f(x) = x^2 + ax + b$ satisfies $f(a) = f(b)$. Find the value of $f(2)$.
4. Suppose $y = f(x) = ax^2 + bx$ $(a > 0)$, where a and b are integers, and $y < 0$ when $x = 15$, while $y > 0$ when $x = 16$.

   (1) Show that $a \nmid b$ and b is a negative integer.
   (2) Among all integer values of x, which one gives the minimal $f(x)$?
5. If the parabola $y = 2x^2 - px + 4p + 1$ passes through a fixed point for every real number p, find the coordinates of the fixed point.
6. Suppose the function $y = x^2 - 2x + 3$ is defined on the interval $[0, a]$. For what values of a, the minimum and maximum values of y are 2 and 3, respectively?
7. Suppose a and b are positive real numbers with $a < b$, and when $x \in [a, b]$, the function $y = x^2 - 4x + 6$ has the minimum value a and the maximum value b. Find a and b.
8. If the minimum value and maximum value of $f(x) = -\frac{1}{2}x^2 + \frac{13}{2}$ on the interval $[a, b]$ are $2a$ and $2b$, respectively, find $[a, b]$.
9. Let b and c be real numbers such that $b < 2 < c$, and suppose when $b \leq x \leq c$, the function $y = x^2 - 4|x| + 4$ has the maximum value $4c$ and minimum value b. Find the value of $b + c$.
10. Suppose α and β are the roots of the equation $7x^2 - (p + 13)x + p^2 - p - 2 = 0$ (x is the unknown), and $0 < \alpha < 1 < \beta < 2$. Find the value range of p.

Group B

11. Let $f(x)$ be a quadratic function such that

    (1) $f(-1) = 0$;
    (2) the inequality $x \leq f(x) \leq \frac{1+x^2}{2}$ holds for all real values of x.

    Find the formula of $f(x)$.
12. Let A and B be two points on the parabola C: $y^2 = 4x$, such that A lies in the first quadrant and B lies in the fourth quadrant. Let l_1 and

l_2 be the tangent lines to the parabola at A and B, respectively, which intersect at P.

(1) If AB passes through the focus F of the parabola, show that P lies on a fixed line, and find the equation of this line.

(2) Let C and D be the intersection points of l_1 and l_2 with the line $x = 4$, respectively. Find the minimal area of the triangle PCD.

13. Let $f(x) = ax^2 + bx + c$ $(a > 0)$ and let x_1 and x_2 be the roots of the equation $f(x) = x$. Suppose $0 < x_1 < x_2 < \frac{1}{a}$.

(1) Show that $x < f(x) < x_1$ when $x \in (0, x_1)$.

(2) If the graph of $f(x)$ is symmetric with respect to the line $x = x_0$, show that $x_0 < \frac{x_1}{2}$.

14. Suppose a and b are real numbers such that one root of the equation $x^2 - ax + b = 0$ lies in the interval $[-1, 1]$, while the other lies in $[1, 2]$. Find the value range of $a - 2b$.

15. Suppose $ax^2 + bx + c = 0$ has two different real roots. Show that the equation

$$ax^2 + bx + c + k\left(x + \frac{b}{2a}\right) = 0 \quad (k \neq 0)$$

has at least one root that lies between the two roots of the first equation.

16. Suppose the graph of $y = ax^2 + bx + c$ passes through the points $(0, -1)$, $(1, 2)$, and $(-3, 2)$.

(1) Find the points on the parabola that are equidistant from the two coordinate axes.

(2) Find the area of the polygon whose vertices are exactly the points in (1).

17. Suppose the graph of $y = -x^2 + 2ax + b$ has its vertex on the line $mx - y - 2m + 1 = 0$, and has at least one common point with the graph of $y = x^2$. Find the value range of m.

18. (2013 China Mathematical Competition) Find all pairs of positive real numbers (a, b) such that the function $f(x) = ax^2 + b$ has the following property: the inequality

$$f(xy) + f(x + y) \geq f(x)f(y)$$

holds for all real numbers x and y.

Chapter 4

Graphs and Properties of Functions

4.1 Key Points of Knowledge and Basic Methods

A. Maps and functions

If $f : X \to Y$ is a map, where both X and Y are nonempty sets of numbers, then we call f a function from X to Y, denoted as

$$f : X \to Y, \quad x \mapsto y,$$

where X is the domain of f. For each x in X, the number in Y corresponding to x under f is called the value of f at the point x, denoted as $f(x)$. The set of all values of f,

$$f(X) = \{y | y = f(x),\ x \in X\} \subseteq Y,$$

is called the range of f.

B. Graphs of functions

The set of all points of the form $(x,\ f(x))$, or $\{(x, y) | y = f(x),\ x \in D\}$, is called the graph of the function $y = f(x)$, where D is its domain.

(1) The graph of $y = f(x + k)$ $(k \neq 0)$ can be obtained by translating the graph of $y = f(x)$ to the left (when $k > 0$) or to the right (when $k < 0$) by $|k|$ units.

(2) The graph of $y = f(x) + h$ $(h \neq 0)$ can be obtained by translating the graph of $y = f(x)$ upwards (when $k > 0$) or downwards (when $k < 0$) by $|k|$ units.
(3) The graph of $y = f(x+k)+h$ can be obtained by translating the graph of $y = f(x)$ twice, first horizontally by $|k|$ units and then vertically by $|h|$ units, or in the reverse order.
(4) The graphs of the functions $y = -f(x)$ and $y = f(x)$ are symmetric with respect to the x-axis.
(5) The graphs of the functions $y = f(-x)$ and $y = f(x)$ are symmetric with respect to the y-axis.
(6) The graphs of the functions $y = -f(-x)$ and $y = f(x)$ are centrally symmetric with respect to the origin.
(7) The graphs of the functions $y = f^{-1}(x)$ and $y = f(x)$ are symmetric with respect to the line $y = x$.
(8) The graph of $y = |f(x)|$ can be obtained by retaining the part of $y = f(x)$ above the $x-$axis, and folding the part below the x-axis symmetrically along the x-axis.

C. Properties of functions

(1) Odd and even functions:

Let $f(x)$ be a function with domain D, which is symmetric with respect to the origin. If $f(-x) = -f(x)$ for all $x \in D$, then f is called an odd function; if $f(-x) = f(x)$ for all $x \in D$, then f is called an even function.

(2) Monotonicity:

If a function $f(x)$ satisfies the following property on an interval I: for each $x_1, x_2 \in I$ with $x_1 < x_2$, we have $f(x_1) \leq f(x_2)$ (resp. $f(x_1) \geq f(x_2)$), then we call $f(x)$ an increasing (resp. decreasing) function on I, and I is called a monotonically increasing (resp. decreasing) interval of $f(x)$. If the inequality in the definition is strict, then we call $f(x)$ an strictly increasing (resp. decreasing) function on I.

(3) Periodicity:

For a function $f(x)$, if there exists a positive real number T such that $f(x + T) = f(x)$ for every x in the domain (in particular, $f(x + T)$ is defined), then we call $f(x)$ a periodic function, and T is called a period of $f(x)$. If there is a minimum value in all the periods of $f(x)$, which is T_0, then T_0 is called the least positive period of $f(x)$.

4.2 Illustrative Examples

A. Graphs of functions

Example 1. Graph the following functions:

$$(1)\ y = |x^2 + x - 2|; \quad (2)\ y = |x|^2 + |x| - 2.$$

Solution. (1) First draw the graph of $y = x^2 + x - 2$, and then fold the part below the x-axis symmetrically along the x-axis, as shown in Figure 4.1 (here the solid part is the graph of the function).

(2) Since $y = |x|^2 + |x| - 2$ is an even function, its graph is symmetric with respect to the y-axis, so we can first draw the graph for $x \geq 0$, and then draw its reflection across the y-axis, as shown in Figure 4.2.

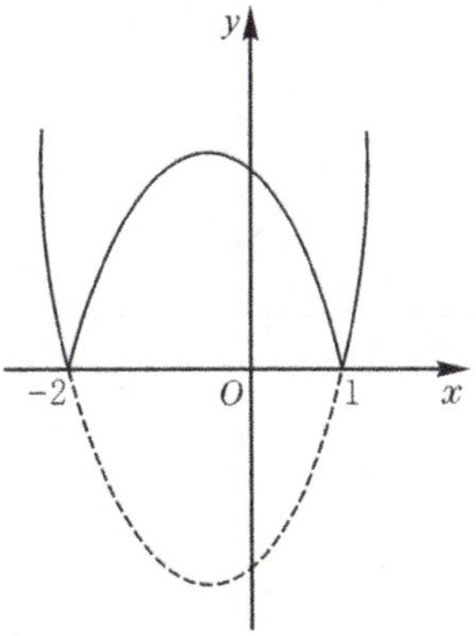

Figure 4.1

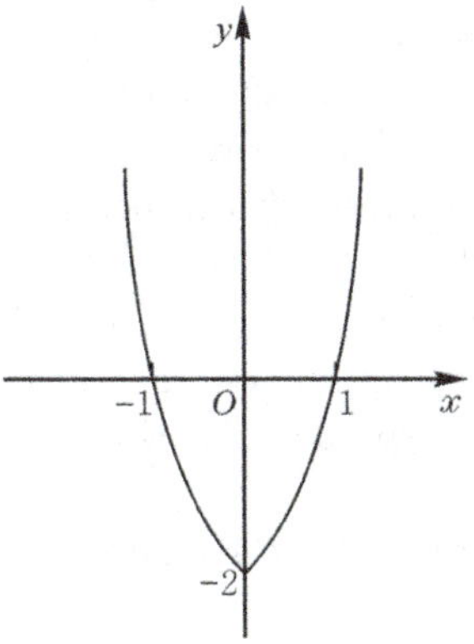

Figure 4.2

Example 2. Let $f_0(x) = |x|$, $f_1(x) = |f_0(x) - 1|$, and $f_2(x) = |f_1(x) - 2|$. Find the area of the closed region between the graph of $y = f_2(x)$ and the x-axis.

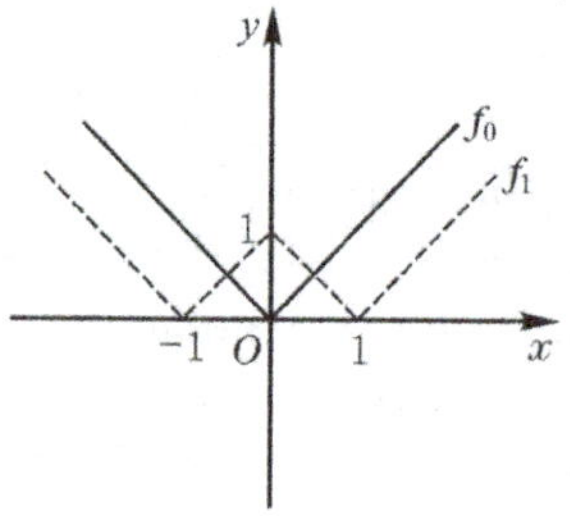

Figure 4.3

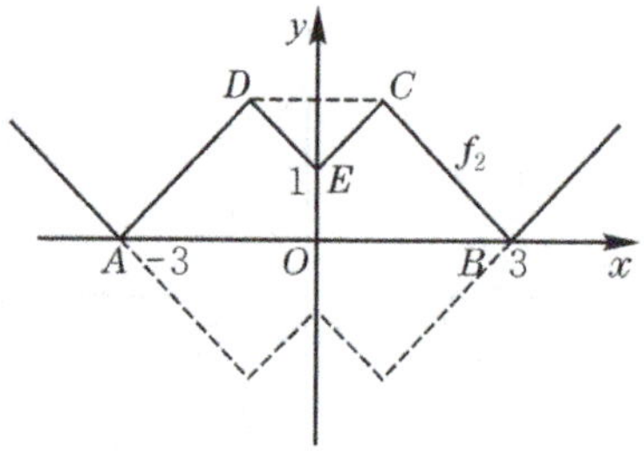

Figure 4.4

Solution. We first draw the graph of $f_2(x)$.

As shown in Figure 4.3, the graph of $f_0(x) = |x|$ consists of two rays (the solid part in the figure), while the graph of $f_1(x) = |f_0(x) - 1|$ is obtained by translating the first graph downwards by 1 unit, and then reflecting the part below the $x-$axis upwards with respect to the x-axis, shown as the dotted part in Figure 4.3.

Further, the graph of $f_2(x) = |f_1(x) - 2|$ is obtained by translating the graph of $f_1(x)$ downwards by 2 units and then reflecting the part below the x-axis upwards with respect to the x-axis, shown as the solid part in Figure 4.4.

Therefore, the area of the closed region is

$$S_{ABCD} - S_{\Delta CDE} = \frac{1}{2}(2+6) \times 2 - \frac{1}{2} \times 2 \times 1 = 7.$$

Example 3. Let $f(x)$ and $g(x)$ be functions defined on $\mathbf{R}$, such that the graph of $f(x)$ is symmetric with respect to the line $x = 1$ and the graph of $g(x)$ is centrally symmetric with respect to the point $(1, -2)$. Suppose

$$f(x) + g(x) = 9^x + x^3 + 1.$$

Find the value of $f(2)g(2)$.

Solution. It is immediate that

$$f(0) + g(0) = 2, ①$$

$$f(2) + g(2) = 81 + 8 + 1 = 90. ②$$

By the symmetry of the graphs of $f(x)$ and $g(x)$, we get that $f(0) = f(2)$ and $g(0) + g(2) = -4$. Combining them with ①, we have

$$f(2) - g(2) - 4 = f(0) + g(0) = 2. ③$$

Solving the equations ② and ③, we obtain that $f(2) = 48$ and $g(2) = 42$, so $f(2)g(2) = 48 \times 42 = 2016$.

B. Domains and ranges

Example 4. Let $f(x)$ be a function with domain $[-1, 1]$. Find the domain of the function $g(x) = f(ax) + f\left(\frac{x}{a}\right)$, where $a > 0$.

Solution. The domain of $g(x)$ is the intersection of the following two sets:

$$D_1 = \{x| - 1 \le ax \le 1\} = \left[-\frac{1}{a}, \frac{1}{a}\right],$$

$$D_2 = \left\{x \,\middle|\, -1 \le \frac{x}{a} \le 1\right\} = [-a, a].$$

If $a \ge 1$, then $a \ge \frac{1}{a}$ and $-a \le -\frac{1}{a}$, so $D_1 \cap D_2 = D_1$.

If $0 < a < 1$, then $\frac{1}{a} > a$ and $-\frac{1}{a} < -a$, so $D_1 \cap D_2 = D_2$.

Therefore, the domain of $g(x)$ is $[-a, a]$ (if $0 < a < 1$) or $\left[-\frac{1}{a}, \frac{1}{a}\right]$ (if $a \ge 1$).

Example 5. Suppose $f(x) = \sqrt{x + 2} + k$, and there exists a and b $(a < b)$ such that the range of $f(x)$ for $x \in [a, b]$ is exactly $[a, b]$. Find the value range of k.

Solution. First we see that $f(x)$ is defined when $x \ge -2$. By monotonicity, it is equivalent to say that the equation $f(x) = \sqrt{x + 2} + k = x$ has two different solutions in the domain, that is, the equation

$$x^2 - (2k+1)x + k^2 - 2 = 0 \ (x \geq -2)$$

has two different roots. This implies that $\Delta = (2k+1)^2 - 4(k^2-2) > 0$, hence $k > -\frac{9}{4}$.

Also, $f(x) = \sqrt{x+2} + k = x$ implies that $x \geq k$. Consequently

$$\frac{2k+1-\sqrt{(2k+1)^2 - 4(k^2-2)}}{2} \geq k,$$

which gives that $k \leq -2$. On the other hand, if $k \in \left(-\frac{9}{4}, -2\right]$, then the inequality

$$\frac{2k+1-\sqrt{(2k+1)^2 - 4(k^2-2)}}{2} \geq -2$$

always holds, so the conditions are satisfied, and hence the value range of k is $\left(-\frac{9}{4}, -2\right]$.

Example 6. Suppose $f(x) = x^2 + ax + b\cos x$ and

$$\{x | f(x) = 0, \ x \in \mathbf{R}\} = \{x | f(f(x)) = 0, \ x \in \mathbf{R}\} \neq \emptyset.$$

Find all possible values of real numbers a and b.

Solution. Suppose $x_0 \in \{x | f(x) = 0, \ x \in \mathbf{R}\}$. Then

$$b = f(0) = f(f(x_0)) = 0,$$

so $f(x) = x(x+a)$ and

$$f(f(x)) = f(x)(f(x)+a) = x(x+a)(x^2+ax+a).$$

Apparently $a = 0$ is a solution. Suppose $a \neq 0$. Then since the roots of $x^2 + ax + a = 0$ cannot be 0 or $-a$, the only possible case is that $x^2 + ax + a = 0$ has no real roots. Equivalently, $\Delta = a^2 - 4a < 0$, and hence $0 < a < 4$.

Therefore, all possible values of a and b are $0 \leq a < 4$ and $b = 0$.

Example 7. Find the range of the function $y = x^2 + x\sqrt{x^2-1}$.

Solution. First we see that the domain of the function is $\{x | x \geq 1$ or $x \leq -1\}$.

(1) Since $y = x^2 + x\sqrt{x^2-1}$ is increasing when $x \in [1, +\infty)$ and y tends to infinity as x tends to infinity, the range of the function when $x \geq 1$ is $y \geq 1$.

(2) When $x \leq -1$, we have

$$y = x(x + \sqrt{x^2-1}) = \frac{x(x+\sqrt{x^2-1})(x-\sqrt{x^2-1})}{x-\sqrt{x^2-1}}$$

$$= \frac{x}{x-\sqrt{x^2-1}} = \frac{1}{1-\frac{\sqrt{x^2-1}}{x}} = \frac{1}{1+\sqrt{1-\frac{1}{x^2}}},$$

and using $x \leq -1$ we get

$$0 \leq 1 - \frac{1}{x^2} < 1, \quad 1 \leq 1 + \sqrt{1-\frac{1}{x^2}} < 2,$$

so $\frac{1}{2} < \frac{1}{1+\sqrt{1-\frac{1}{x^2}}} \leq 1$, and hence $\frac{1}{2} < y \leq 1$.

Combining the above, we conclude that the range of the function $y = x^2 + x\sqrt{x^2-1}$ is $\left(\frac{1}{2}, +\infty\right)$.

C. Properties of functions

Example 8. Show that every function whose domain is symmetric with respect to the origin can be written as the sum of an odd function and an even function.

Solution. Suppose the domain of $f(x)$ is D, which is symmetric with respect to the origin. Then $f(x)$ is defined if and only if $f(-x)$ is defined. The following identity follows easily:

$$f(x) = \frac{f(x)+f(-x)}{2} + \frac{f(x)-f(-x)}{2}.$$

Let $f_1(x) = \frac{f(x)+f(-x)}{2}$ and $f_2(x) = \frac{f(x)-f(-x)}{2}$. We claim that $f_1(x)$ is an even function and $f_2(x)$ is an odd function. In fact,

$$f_1(-x) = \frac{f(-x)+f(x)}{2} = f_1(x),$$

$$f_2(-x) = \frac{f(-x)-f(x)}{2} = -\frac{f(x)-f(-x)}{2} = -f_2(x).$$

Therefore, the proposition follows.

Remark. How did we find the functions $f_1(x)$ and $f_2(x)$ in this problem? In fact, they can be solved by equations. Let $f(x) = f_1(x) + f_2(x)$, where $f_1(x)$ is even and $f_2(x)$ is odd. Then

$$f(x) = f_1(x) + f_2(x), \quad f(-x) = f_1(x) - f_2(x).$$

It follows that $f_1(x) = \frac{f(x)+f(-x)}{2}$ and $f_2(x) = \frac{f(x)-f(-x)}{2}$.

Example 9. Let $f(x)$ be a function whose domain is $\mathbf{R}$, and suppose $f(x)+x^2$ is an odd function while $f(x) + 2^x$ is an even function. Find the value of $f(1)$.

Solution. By the assumption,

$$f(1) + 1 = -(f(-1) + (-1)^2) = -f(-1) - 1,$$
$$f(1) + 2 = f(-1) + \frac{1}{2}.$$

Adding the two equations to eliminate $f(-1)$, we get $2f(1) + 3 = -\frac{1}{2}$, so $f(1) = -\frac{7}{4}$.

Example 10. Suppose $f(x) = x^5 + ax^3 + bx + c\sqrt[3]{x} + 8$ (here a, b, and c are real constants) and $f(-2) = 10$. Find the value of $f(2)$.

Analysis. The condition $f(-2) = 10$ is not sufficient to determine the values of a, b, and c, so in order to find $f(2)$, we have to consider the relationship between $f(-2)$ and $f(2)$. It would be convenient if $f(x)$ is odd or even, but it is not. However, $f(x) - 8$ is an odd function, which shows a way towards the solution.

Solution. Let $g(x) = f(x) - 8$. Then $g(x)$ is an odd function and

$$g(-2) = f(-2) - 8 = 10 - 8 = 2,$$

so $g(2) = -g(-2) = -2$, from which $f(2) = g(2) + 8 = 6$.

Remark. This problem can also be solved in the following way:

Since $f(x) = x^5+ax^3+bx+c\sqrt[3]{x}+8$ and $f(-x) = -x^5-ax^3-bx-c\sqrt[3]{x}+8$, adding the two expressions gives that $f(x) + f(-x) = 16$ for every real number x. Therefore, $f(2) + f(-2) = 16$ and $f(2) = 6$.

Example 11. Let x and y be real numbers such that

$$\begin{cases} (x-1)^3 + 2018(x-1) = -1, \\ (y-1)^3 + 2018(y-1) = 1. \end{cases}$$

Find the value of $x + y$.

Solution. The system of equations above can be written in the following way:

$$\begin{cases} (x-1)^3 + 2018(x-1) = -1, \\ (1-y)^3 + 2018(1-y) = -1. \end{cases}$$

Let $f(t) = t^3 + 2018t$. Then $f(t)$ is strictly increasing on $(-\infty, +\infty)$. The system of equations tells us that

$$f(x-1) = f(1-y),$$

so necessarily $x - 1 = 1 - y$, or equivalently $x + y = 2$.

Example 12. Suppose $f(x)$ is a function such that for every real number x,

$$f(2-x) = f(2+x), \quad f(7-x) = f(7+x),$$

and $f(0) = 0$. Show that the equation $f(x) = 0$ has at least 13 solutions in the interval $[-30, 30]$ and $f(x)$ is a periodic function with period 10.

Solution. It follows from the conditions that the graph of $f(x)$ is symmetric with respect to the lines $x = 2$ and $x = 7$. Therefore,

$$f(4) = f(2+2) = f(2-2) = f(0) = 0,$$
$$f(10) = f(7+3) = f(7-3) = f(4) = 0,$$

so the equation $f(x) = 0$ has at least two solutions in $(0, 10]$.

On the other hand,

$$f(x+10) = f(7+3+x) = f(7-3-x) = f(4-x) = f(2+2-x)$$
$$= f(2-2+x) = f(x),$$

which implies that $f(x)$ is periodic with period 10. Therefore, the equation $f(x) = 0$ has at least $6 \times 2 + 1 = 13$ solutions in the interval $[-30, 30]$ (note that the endpoint -30 is not counted in the intervals).

Remark. For every x such that $a - x \in D$, we have

$$a + x \in D \quad \text{and} \quad f(a + x) = f(a - x),$$

where a is a constant, then the graph of $f(x)$ is symmetric with respect to the line $x = a$.

Example 13. Given the function $f(x) = \left|1 - \frac{1}{x}\right|$.

(1) Determine whether there exist real numbers a and $b (a < b)$ such that the range of $f(x)$ for $x \in [a, b]$ is exactly $[a, b]$.
(2) Suppose a and b are real numbers such that the range of $f(x)$ for $x \in [a, b]$ is $[ma, mb]$ $(m \neq 0)$. Find the value range of m.

Solution. (1) The answer is negative.

Suppose such a and b $(a < b)$ exist. Then by $f(x) \geq 0$ we have $a \geq 0$, and since 0 is not in the domain of $f(x)$, we get $a > 0$. Thus,

$$f(x) = \begin{cases} 1 - \dfrac{1}{x}, & x \geq 1, \\ \dfrac{1}{x} - 1, & 0 < x < 1. \end{cases}$$

If a, $b \in (0, 1)$, then since $f(x) = \frac{1}{x} - 1$ is decreasing on $(0, 1)$,

$$\begin{cases} f(a) = b \\ f(b) = a, \end{cases} \quad \text{or} \quad \begin{cases} \dfrac{1}{a} - 1 = b \\ \dfrac{1}{b} - 1 = a. \end{cases}$$

Solving the equations gives that $a = b$, a contradiction.

If a, $b \in [1, +\infty)$, then since $f(x) = 1 - \frac{1}{x}$ is increasing on $[1, +\infty)$,

$$\begin{cases} f(a) = a \\ f(b) = b, \end{cases} \quad \text{or} \quad \begin{cases} 1 - \dfrac{1}{a} = a \\ 1 - \dfrac{1}{b} = b. \end{cases}$$

Then a and b are solutions to the equation $x^2 - x + 1 = 0$, which has no real roots as $\Delta = 1 - 4 < 0$, a contradiction.

If $a \in (0,1)$ and $b \in [1,+\infty)$, then $1 \in [a,b]$, and so $f(1) = 0$. Thus $0 \in [a,b]$, a contradiction.

Therefore, such a and b do not exist.

(2) Suppose a and b ($a < b$) are real numbers and the range of $f(x)$ for $x \in [a,b]$ is $[ma, mb]$. Then by $f(x) \geq 0$ and $x \neq 0$ we have $ma > 0$. Also, by $mb > ma$ and $b > a$ we have $m > 0$, so $a > 0$. Thus,

$$f(x) = \begin{cases} 1 - \dfrac{1}{x}, & x \geq 1, \\ \dfrac{1}{x} - 1, & 0 < x < 1. \end{cases}$$

If a, $b \in (0,1)$, then as in (1) we get

$$\begin{cases} \dfrac{1}{a} - 1 = mb, \\ \dfrac{1}{b} - 1 = ma, \end{cases}$$

which shows that $a = b$, a contradiction.

If a, $b \in [1,+\infty)$, then

$$\begin{cases} 1 - \dfrac{1}{a} = ma, \\ 1 - \dfrac{1}{b} = mb, \end{cases}$$

so a and b are the two roots of the equation $mx^2 - x + 1 = 0$, which are both greater than 1. (If $a = 1$, then $m = 0$, which is impossible.) This implies that

$$\begin{cases} \Delta = 1 - 4m > 0, \\ \dfrac{1 + \sqrt{1-4m}}{2m} > 1, \\ \dfrac{1 - \sqrt{1-4m}}{2m} > 1, \end{cases}$$

and hence $0 < m < \frac{1}{4}$.

If $a \in (0,1)$ and $b \in [1,+\infty)$, then $1 \in [a,b]$ and $f(1) = 0 \in [ma, mb]$, a contradiction.

In summary, the value range of m is $0 < m < \frac{1}{4}$.

4.3 Exercises

Group A

1. Fill in the blanks

   (1) A function $y = f(x)$ has domain $\mathbf{R}$ and satisfies $f(x-1) = f(2-x)$ for all x. Then an axis of symmetry for the graph of this function is __________.

   (2) The function $y = \frac{1}{x-1}$ $(x \neq \pm 1)$ can be written as the sum of an even function $f(x)$ and an odd function $g(x)$. Then $f(x) =$ __________.

   (3) Suppose there are three functions. The first one is $y = \varphi(x)$, and its inverse function is the second function. The graph of the third function and the graph of the second function are symmetric with respect to the line $x + y = 0$. Then the third function is __________.

   (4) Given $f(x) = \sqrt{1-x^2}$, the domain of the function $y = f(1-x^2)$ is __________.

2. If the function $y = f(x) + x^3$ is even and $f(10) = 15$, find the value of $f(-10)$.
3. Let $f(x) = \frac{9^x}{9^x+3}$. Compute the value of the following expression:

$$f\left(\frac{1}{2000}\right) + f\left(\frac{2}{2000}\right) + f\left(\frac{3}{2000}\right) + \cdots + f\left(\frac{1999}{2000}\right).$$

4. Suppose $a > 0$, and $f(x) = |x|^a - a^x$ is an even function defined on $\mathbf{R}$. Find the solutions to the equation $f(x) = a$.
5. Suppose $f(x)$ is an odd function whose domain is $(-1, 1)$, and decreasing in its domain. For what values of m does the inequality $f(1-m) + f(1-m^2) < 0$ hold?
6. For $x \in [-1, 1]$, find the range of the following function:

$$f(x) = \frac{x^4 + 4x^3 + 17x^2 + 26x + 106}{x^2 + 2x + 7}.$$

7. Suppose $f(x)$ is a function such that $f(x) - 2f\left(\frac{1}{x}\right) = x$ for all $x \neq 0$. Find the range of $f(x)$.
8. Let $f(x)$ be a decreasing function defined on $(0, +\infty)$. If the inequality

$$f(2a^2 + a + 1) < f(3a^2 - 4a + 1)$$

   holds, find the value range of a.

9. Suppose a function $y = f(x)$ satisfies $f(3+x) = f(3-x)$ for every real number x, and the equation $f(x) = 0$ has exactly six different solutions. Find the sum of all these solutions.
10. Let $f(x)$ and $g(x)$ be functions defined for all nonzero real numbers. Suppose $f(x)$ is an even function while $g(x)$ is an odd function, and $f(x) + g(x) = \frac{1}{x^2-x+1}$ for all x in the domain. Find the value range of $\frac{f(x)}{g(x)}$.
11. Let $f(x)$ be an increasing function defined on $(-\infty, +\infty)$. Suppose the inequality

$$f(1 - kx - x^2) < f(2 - k)$$

holds for all $x \in [0, 1]$. Find the value range of k.
12. Let $f(x) = a \sin x - \frac{1}{2} \cos 2x + a - \frac{3}{a} + \frac{1}{2}$, where $a \in \mathbf{R}$ and $a \neq 0$.
    (1) If $f(x) \leq 0$ for all $x \in \mathbf{R}$, find the value range of a.
    (2) If $a \geq 2$ and there exists some $x \in \mathbf{R}$ such that $f(x) \leq 0$, find the value range of a.

Group B

13 Let $f(x)$ be defined on $\mathbf{R}$ such that

$$f(x+2)(1 - f(x)) = 1 + f(x).$$

(1) Show that $f(x)$ is a periodic function.
(2) If $f(1) = 2 + \sqrt{3}$, find the values of $f(1997)$ and $f(2001)$.

14. Suppose $f(x)$ is defined on $\mathbf{R}$ such that $f(0) = 1008$, and the following inequalities hold for all $x \in \mathbf{R}$:

$$f(x+4) - f(x) \leq 2(x+1),$$

$$f(x+12) - f(x) \geq 6(x+5).$$

Find the value of $\frac{f(2016)}{2016}$.
15. Show that the function $f(x) = 3x^2$ is the difference of two increasing polynomial functions.
16. Find the range of the function $y = 2x - 3 + \sqrt{x^2 - 12}$.
17. Given the function

$$f(x) = \begin{cases} x + \dfrac{1}{2}, & 0 \leq x \leq \dfrac{1}{2} \\ 2(1-x), & \dfrac{1}{2} \leq x \leq 1 \end{cases},$$

and define $f_n(x) = \underbrace{f(f \ldots f}_{n \text{ iteration}}(x)\ldots))$ for $n \in \mathbf{N}^*$.

(1) Find the value of $f_{2004}\left(\frac{2}{15}\right)$.

(2) Let $B = \{x | f_{15}(x) = x,\ x \in [0, 1]\}$. Show that B has at least nine elements.

18. Let $f(x) = a\sin^2 x + b\sin x + c$, where $a, b,$and c are nonzero real numbers. Ann and Bob are playing the following game: they take turns to determine the coefficients a, b, and c, and they are free to choose which of the remaining coefficients to determine, with Ann taking the first turn (for example, first Ann lets $b = 1$; hen Bob lets $a = -2$; finally Ann lets $c = 3$). When all the coefficients are determined, if $f(x) \neq 0$ for every real number x, then Ann wins the game. Conversely, if $f(x) = 0$ for some real number x, then Bob wins the game. Determine whether Ann has a winning strategy and explain the reasons.

If a, b, and c can be any real numbers, will the result be different? Explain.

Chapter 5

Power Functions, Exponential Functions, and Logarithmic Functions

5.1 Key Points of Knowledge and Basic Methods

A. Power functions

Functions of the form $y = x^a$ $(a \in \mathbf{R})$ are called power functions. In middle school and high school we usually deal with the case when $a \in \mathbf{Q}$.

B. Exponential functions

Functions of the form $y = a^x$ ($a > 0$ and $a \neq 1$) are called exponential functions. Such a function has domain $\mathbf{R}$ and range $(0, +\infty)$. If $0 < a < 1$, then $y = a^x$ is a strictly decreasing function; if $a > 1$, then $y = a^x$ is a strictly increasing function.

C. Logarithmic functions

Functions of the form $y = \log_a x$ $(a > 0, a \neq 1)$ are called logarithmic functions. Such a function has domain $(0, +\infty)$ and range $\mathbf{R}$. If $0 < a < 1$, then $y = \log_a x$ is a strictly decreasing function; if $a > 1$, then $y = \log_a x$ is an strictly increasing function.

The functions $y = \log_a x$ and $y = a^x$ are inverse to each other.

5.2 Illustrative Examples

A. Simplification and evaluation

Example 1. (1) Let $a = \frac{\log_7 4(\log_7 5 - \log_7 2)}{\log_7 25(\log_7 8 - \log_7 4)}$. Find the value of 5^a.

(2) Let $b = \frac{\log_{77} 4(\log_{77} 5 - \log_{77} 2)}{\log_{77} 25(\log_{77} 8 - \log_{77} 4)}$ Find the value of 5^b.

Solution. (1) Since

$$a = \frac{\log_7 2^2 \cdot \log_7 \frac{5}{2}}{\log_7 5^2 \cdot \log_7 2} = \frac{2\log_7 2 \cdot \log_7 \frac{5}{2}}{2\log_7 5 \cdot \log_7 2} = \frac{\log_7 \frac{5}{2}}{\log_7 5} = \log_5 \frac{5}{2},$$

we have $5^a = 5^{\log_5 \frac{5}{2}} = \frac{5}{2}$.

(2) The computation in (1) does not depend on the base of logarithms, so $5^b = \frac{5}{2}$.

Example 2. Let u, v, and w be positive real numbers that are not equal to 1. Suppose

$$\log_u vw + \log_v w = 5, \quad \log_v u + \log_w v = 3.$$

Find the value of $\log_w u$.

Solution. Let $\log_u v = a$ and $\log_v w = b$. Then

$$\log_v u = \frac{1}{a}, \quad \log_w v = \frac{1}{b}, \quad \log_u vw = \log_u v + \log_u v \cdot \log_v w = a + ab.$$

Therefore, by the conditions, $a + ab + b = 5$, $\frac{1}{a} + \frac{1}{b} = 3$, and $ab = \frac{5}{4}$. Hence

$$\log_w u = \log_w v \cdot \log_v u = \frac{1}{ab} = \frac{4}{5}.$$

Example 3. Let $f(x) = \frac{a^x}{a^x + \sqrt{a}}$, where a is a positive real number. Find the value of

$$f\left(\frac{1}{101}\right) + f\left(\frac{2}{101}\right) + \cdots + f\left(\frac{100}{101}\right).$$

Solution. Since

$$f(x)+f(1-x)=\frac{a^x}{a^x+\sqrt{a}}+\frac{a^{1-x}}{a^{1-x}+\sqrt{a}}=\frac{a^x}{a^x+\sqrt{a}}+\frac{a}{a+\sqrt{a}\cdot a^x}$$
$$=\frac{a^x}{a^x+\sqrt{a}}+\frac{\sqrt{a}}{\sqrt{a}+a^x}=1,$$

pairing the terms in the expression, we get that

$$f\left(\frac{1}{101}\right)+f\left(\frac{2}{101}\right)+\cdots+f\left(\frac{100}{101}\right)$$
$$=\left[f\left(\frac{1}{101}\right)+f\left(\frac{100}{101}\right)\right]+\left[f\left(\frac{2}{101}\right)+f\left(\frac{99}{101}\right)\right]$$
$$+\cdots+\left[f\left(\frac{50}{101}\right)+f\left(\frac{51}{101}\right)\right]=50.$$

Example 4. Let $f(x)$ be defined on $\mathbf{R}$ such that for every real number x,

$$f(x+3)\cdot f(x-4)=-1.$$

Also, suppose for $0\le x<7$, we have $f(x)=\log_2(9-x)$. Find the value of $f(-100)$.

Solution. Since $f(x+3)\cdot f(x-4)=-1$, we see that $f(x+14)=-\frac{1}{f(x+7)}=f(x)$, so

$$f(-100)=f(-100+14\times 7)=f(-2)$$
$$=-\frac{1}{f(5)}=-\frac{1}{\log_2 4}=-\frac{1}{2}.$$

B. Graphs and properties

Example 5. Let a and b be the solutions to the equations $\log_2 x+x-3=0$ and $2^x+x-3=0$, respectively. Find the values of $a+b$ and $\log_2 a+2^b$.

Solution. We draw the graphs of the functions $y=2^x$ and $y=\log_2 x$ in the coordinate plane, and then draw the lines $y=x$ and $y=-x+3$, as shown in Figure 5.1.

Since the functions $y=2^x$ and $y=\log_2 x$ are inverse to each other, their graphs are symmetric with respect to the line $y=x$. The solution to the equation $\log_2 x+x-3=0$ is the x-coordinate of A (the intersection point of the graphs of $y=\log_2 x$ and $y=-x+3$), while the solution to $2^x+x-3=0$ is the x-coordinate of B (the intersection point of the graphs of $y=2^x$ and $y=-x+3$).

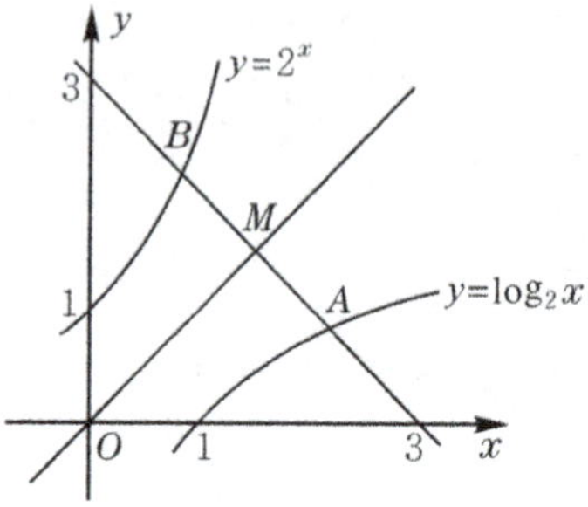

Figure 5.1

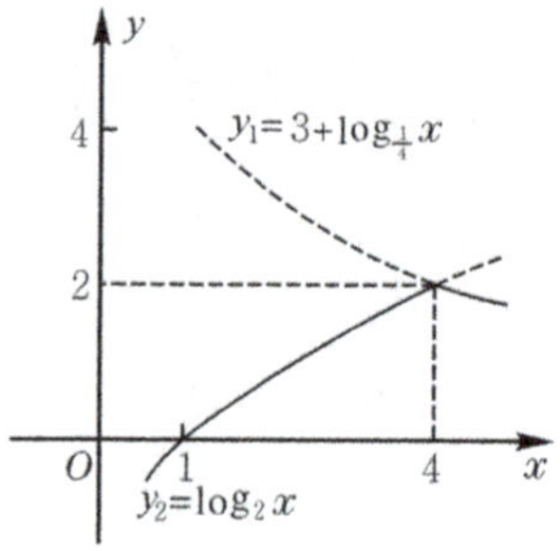

Figure 5.2

Let M be the intersection point of $y = -x + 3$ and $y = x$. Then M has the coordinates $\left(\frac{3}{2}, \frac{3}{2}\right)$. Therefore,

$$a + b = 2x_M = 3,$$

$$\log_2 a + 2^b = 2y_M = 3.$$

Example 6. Let $f(x) = \min\{3 + \log_{\frac{1}{4}} x,\ \log_2 x\}$, where $\min\{p, q\}$ is the minimum of p and q. Find the maximum value of $f(x)$.

Solution. Apparently the domain of $f(x)$ is $(0, +\infty)$.

The function $y_1 = 3+\log_{1/4} x$ is decreasing on $(0, +\infty)$, while $y_2 = \log_2 x$ is increasing on $(0, +\infty)$, and $y_1 = y_2$ (or $3 + \log_{\frac{1}{4}} x = \log_2 x$) if and only if $x = 4$. Hence, the graphs of y_1 and y_2 (shown in Figure 5.2) imply that

$$f(x) = \begin{cases} 3 + \log_{\frac{1}{4}} x, & x \geq 4 \\ \log_2 x, & 0 < x < 4 \end{cases}.$$

Therefore, $f(x)$ attains its maximum value 2 when $x = 4$.

Remark. In the solution above we first found the formula for $f(x)$, which is a piecewise function, and then used its graph to determine the maximum value. Next, we show a different approach. Since

$$f(x) \leq 3 + \log_{\frac{1}{4}} x = 3 - \frac{1}{2}\log_2 x, ①$$

$$f(x) \leq \log_2 x, ②$$

eliminating $\log_2 x$ in ①$\times 2 +$②, we get that $3f(x) \leq 6$, namely $f(x) \leq 2$. As $f(4) = 2$, we conclude that the maximum value of $f(x)$ is 2.

Here we first found an upper bound for $f(x)$, and then found an instance to show that the bound is reachable. This is a common method for "composite" extreme value problems like this one.

Example 7. Let $f(x)$ be a function defined on R such that $f(0) = 2008$. Suppose for each $x \in \mathbf{R}$,

$$f(x+2) - f(x) \leq 3 \cdot 2^x,$$

$$f(x+6) - f(x) \geq 63 \cdot 2^x.$$

Find the value of $f(2008)$.

Solution 1. It follows from the conditions that

$$\begin{aligned} f(x+2) - f(x) &= -(f(x+4) - f(x+2)) - (f(x+6) \\ &\quad - f(x+4)) + (f(x+6) - f(x)) \\ &\geq -3 \cdot 2^{x+2} - 3 \cdot 2^{x+4} + 63 \cdot 2^x = 3 \cdot 2^x. \end{aligned}$$

Thus, $f(x+2) - f(x) = 3 \cdot 2^x$.

Therefore,

$$\begin{aligned} f(2008) &= f(2008) - f(2006) + f(2006) - f(2004) \\ &\quad + \cdots + f(2) - f(0) + f(0) \\ &= 3 \cdot (2^{2006} + 2^{2004} + \cdots + 2^2 + 1) + f(0) \\ &= 3 \cdot \frac{4^{1004} - 1}{4 - 1} + f(0) = 2^{2008} + 2007. \end{aligned}$$

Solution 2. Let $g(x) = f(x) - 2^x$. Then

$$g(x+2) - g(x) = f(x+2) - f(x) - 2^{x+2} + 2^x \leq 3 \cdot 2^x - 3 \cdot 2^x = 0,$$

$$g(x+6) - g(x) = f(x+6) - f(x) - 2^{x+6} + 2^x \geq 63 \cdot 2^x - 63 \cdot 2^x = 0,$$

so $g(x+2) \leq g(x)$ and $g(x+6) \geq g(x)$. Hence,

$$g(x) \leq g(x+6) \leq g(x+4) \leq g(x+2) \leq g(x),$$

which shows that $g(x)$ is periodic with period 2. Therefore,

$$f(2008) = g(2008) + 2^{2008} = g(0) + 2^{2008} = f(0) - 1 + 2^{2008} = 2^{2008} + 2007.$$

C. Equations and inequalities

Example 8. Solve the equation:

$$x + \log_2(2^x - 31) = 5.$$

Solution. The given equation is equivalent to $\log_2 2^x + \log_2(2^x - 31) = 5$, that is, $\log_2[2^x(2^x - 31)] = 5$. Taking the exponential with base 2, we have $(2^x)^2 - 31 \cdot 2^x = 2^5$, and factorizing it gives $(2^x + 1)(2^x - 32) = 0$.

Since $2^x + 1 > 0$, the above equation is reduced to $2^x - 32 = 0$, so $x = 5$.

Example 9. Suppose $f(x)$ is an odd function, and for $x \in [0, +\infty)$ we have $f(x) = \lg(1+x)$. Solve the inequality $f(x) + f(2x) > f(4x)$.

Solution. If $x \geq 0$, then from the assumption we have $\lg(1+x) + \lg(1+2x) > \lg(1+4x)$, so

$$(1+x)(1+2x) > 1 + 4x.$$

Solving this inequality, we get $x > \frac{1}{2}$.

If $x < 0$, then

$$-\lg(-x) - \lg(1-2x) > -\lg(1-4x),$$

From which $(1-x)(1-2x) < 1 - 4x$. Hence $-\frac{1}{2} < x < 0$.

Therefore, the solution set to the inequality is $\left(-\frac{1}{2}, 0\right) \cup \left(\frac{1}{2}, +\infty\right)$.

Example 10. Let $a > 0$ and $a \neq 1$. Show that the equation $a^x + a^{-x} = 2a$ has no solution in the interval $[-1, 1]$.

Solution. Let $t = a^x$. Then the equation reduces to

$$t^2 - 2at + 1 = 0. \text{①}$$

If it has real roots, then $\Delta = 4a^2 - 4 \geq 0$, so $|a| \geq 1$, namely $a \geq 1$ since a is positive.

When $a \geq 1$, the two real roots of ① are

$$t = a \pm \sqrt{a^2 - 1},$$

so $a^x = a \pm \sqrt{a^2 - 1}$, from which $x = \log_a(a \pm \sqrt{a^2 - 1})$.

By $a > 1$ and $a + \sqrt{a^2 - 1} > a$,

$$\begin{aligned} x_1 &= \log_a\left(a + \sqrt{a^2 - 1}\right) > \log_a a = 1, \\ x_2 &= \log_a\left(a - \sqrt{a^2 - 1}\right) = \log_a \frac{1}{a + \sqrt{a^2 - 1}} \\ &= -\log_a(a + \sqrt{a^2 - 1}) < -\log_a a = -1. \end{aligned}$$

Therefore, the equation has real roots when $a \geq 1$, but its roots never lie in $[-1, 1]$.

Example 11. Solve the inequality

$$\log_2(x^{12} + 3x^{10} + 5x^8 + 3x^6 + 1) < 1 + \log_2(x^4 + 1).$$

Solution 1. Since $1 + \log_2(x^4 + 1) = \log_2(2x^4 + 2)$ and $\log_2 y$ is increasing for $y \in (0, +\infty)$, the given inequality is equivalent to

$$x^{12} + 3x^{10} + 5x^8 + 3x^6 + 1 < 2x^4 + 2,$$

or equivalently $x^{12} + 3x^{10} + 5x^8 + 3x^6 - 2x^4 - 1 < 0$.

Rewriting the above by grouping, we get that

$$\begin{aligned} &x^{12} + x^{10} - x^8 \\ &\quad + 2x^{10} + 2x^8 - 2x^6 \\ &\quad\quad + 4x^8 + 4x^6 - 4x^4 \\ &\quad\quad\quad + x^6 + x^4 - x^2 \\ &\quad\quad\quad\quad + x^4 + x^2 - 1 < 0, \end{aligned}$$

and factorizing the left-hand side gives $(x^8 + 2x^6 + 4x^4 + x^2 + 1)(x^4 + x^2 - 1) < 0$. Hence, $x^4 + x^2 - 1 < 0$, in other words,

$$\left(x^2 - \frac{-1 - \sqrt{5}}{2}\right)\left(x^2 - \frac{-1 + \sqrt{5}}{2}\right) < 0.$$

Therefore, $x^2 < \frac{-1+\sqrt{5}}{2}$, and the solution set to the inequality is

$$\left(-\sqrt{\frac{-1+\sqrt{5}}{2}},\ \sqrt{\frac{\sqrt{5}-1}{2}}\right).$$

Solution 2. Since $1+\log_2(x^4+1) = \log_2(2x^4+2)$ and $\log_2 y$ is increasing for $y \in (0,+\infty)$, the given inequality is equivalent to

$$x^{12}+3x^{10}+5x^8+3x^6+1 < 2x^4+2.$$

If $x=0$, then the inequality is satisfied obviously. Suppose $x \neq 0$. Then

$$\frac{2}{x^2}+\frac{1}{x^6} > x^6+3x^4+3x^2+1+2x^2+2 = (x^2+1)^3+2(x^2+1),$$

or equivalently $\left(\frac{1}{x^2}\right)^3 + 2\left(\frac{1}{x^2}\right) > (x^2+1)^3+2(x^2+1)$.

Let $g(t) = t^3+2t$. Then the inequality reduces to

$$g\left(\frac{1}{x^2}\right) > g(x^2+1).$$

Apparently $g(t)$ is increasing on $\mathbf{R}$, so the inequality is equivalent to $\frac{1}{x^2} > x^2+1$, namely

$$(x^2)^2+x^2-1<0.$$

Solving this equation (together with the solution $x=0$) gives $-\sqrt{\frac{-1+\sqrt{5}}{2}} < x < \sqrt{\frac{-1+\sqrt{5}}{2}}$.

5.3 Exercises

Group A

1. Fill in the blanks:
   (1) $5^{lg20} \cdot \left(\frac{1}{2}\right)^{\lg 0.5} =$ ____________.
   (2) Let $f(x) = \log_a x$ ($a>0$ and $a \neq 1$). If $f(x_1) - f(x_2) = 1$, then $f(x_1^2) - f(x_2^2) =$ ____________.
   (3) The solution(s) to the equation $\log_{4x}\sqrt{4x^2-5x+2} = \frac{1}{2}$ is(are) ____________.
   (4) The number of proper subsets of $\left\{x \mid -1 \leq \log_{\frac{1}{x}} 10 < -\frac{1}{2},\ x \in \mathbf{N}^*\right\}$ is ____________.
   (5) If $f(x) = 3^x + \log_3 x + 2$, then $f^{-1}(30) =$ ____________.

(6) If the graph of $f(x) = 25^{-|x+1|} - 4 \times 5^{-|x+1|} - m$ intersects the x-axis, then the value range of the real number m is __________.

2. If the range of the function $y = \lg(ax^2 + 2x + 1)$ is the set of real numbers, find the value range of a.
3. If the solution set to the inequality $\log_a(x^2 + 2x + 5) < \log_a 3$ is $\mathbf{R}$, find the value range of a.
4. Let $\log_7(2\sqrt{2} - 1) + \log_2(\sqrt{2} + 1) = a$. Find the value of $\log_7(2\sqrt{2} + 1) + \log_2(\sqrt{2} - 1)$.
5. Let $f(x) = \begin{cases} a - x, & x \in [0, 3], \\ a \log_2 x, & x \in (3, +\infty), \end{cases}$ where a is a real number. If $f(2) < f(4)$, find the value range of a.
6. Suppose $\frac{1}{\log_x 3} + \frac{1}{\log_y 3} \geq 4$. Find the minimum value of $u = 2^x + 2^y$.
7. Let $f(x) = 2^x$ and $g(x) = x^2$. Solve the inequality $f(g(x)) < 3g(f(x))$.
8. If the equation (for x) $9^{-x^2+x-1} - 2 \times 3^{-x^2+x+1} - m = 0$ has a positive real root, find the value range of m.
9. If the domain of $f(x) = \left(\log_2 \frac{x}{2}\right)\left(\log_2 \frac{x}{4}\right)$ is exactly the solution set to the inequality

$$2(\log_{\frac{1}{2}} x)^2 + 7\log_{\frac{1}{2}} x + 3 \leq 0,$$

find the maximum and minimum values of $f(x)$.

10. Suppose the function$f(x)$ satisfies $f(2^x) = x^2 - 2ax + a^2 - 1$ for all x (a is a real constant), and the range of $f(x)$ for $x \in [2^{a-1}, 2^{a^2-2a+2}]$ is $[-1, 0]$. Find the value range of a.

Group B

11. Given $a > 0$ and $a \neq 1$, find the value range of the real numbers k such that the equation (for x) $\log_a(x - ak) = \log_{a^2}(x^2 - a^2)$ has at least one solution.
12. Let n be a positive integer and a be a real number such that $a > 1$. Solve the inequality for x:

$$\log_a x - 4\log_{a^2} x + 12\log_{a^3} x + \cdots + n(-2)^{n-1}\log_a nx$$
$$> \frac{1}{3}[1 - (-2)^n]\log_a(x^2 - a).$$

13. For what values of a, the inequality (for x)

$$\log_{\frac{1}{a}}(\sqrt{x^2 + ax + 5} + 1) \cdot \log_5(x^2 + ax + 6) + \log_a 3 \geq 0$$

has exactly one solution?

14. Suppose $a = \lg z + \lg[x(yz)^{-1} + 1]$, $b = \lg x^{-1} + \lg(xyz + 1)$, and $c = \lg y + \lg[(xyz)^{-1} + 1]$, where x, y, and z are positive real numbers. Let M be the maximum in a, b, and c. Find the minimum value of M.
15. Suppose $0 < a < 1$ and $x < 0$. Show that

$$\ln(\sqrt{x^2+1}+x) < \frac{x(a^x-1)}{(a^x+1)\log_a(\sqrt{x^2+1}-x)}.$$

Chapter 6

Functions with Absolute Values

6.1 Key Points of Knowledge and Basic Methods

Retain the part of the graph of $y = f(x)$ above the x-axis and flip the part below the x-axis to above the x-axis. We thus get the graph of $y = |f(x)|$.

Retain the part of the graph of $y = f(x)$ right to the y-axis, and replace the part left to the y-axis by the reflection of the right part with respect to the y-axis. We thus get the graph of $y = f(|x|)$.

6.2 Illustrative Examples

Example 1. Draw the graph of $y = |\lg|x||$.

Solution.

$$\text{Since } y = \begin{cases} \lg(-x), & (-\infty < x \leq -1), \\ -\lg(-x), & (-1 < x < 0), \\ -\lg x, & (0 < x < 1), \\ \lg x, & (1 \leq x < +\infty), \end{cases}$$

we can draw the graph of the function as in Figure 6.1.

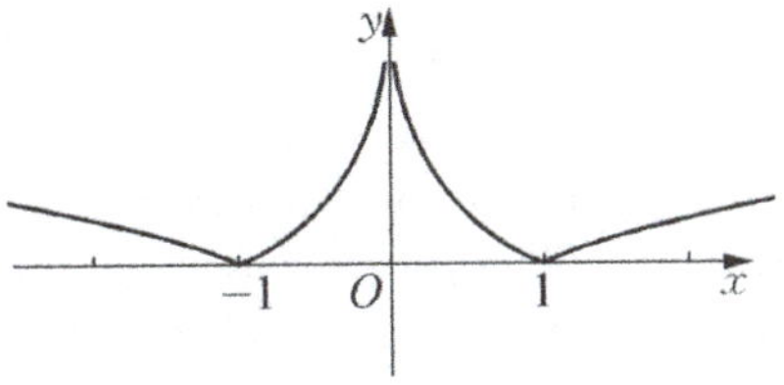

Figure 6.1

Remark. The formula of the function involves absolute values, so we tried to remove the absolute value signs. This can be done by restricting the sign of the expression inside the absolute value symbol.

Example 2. Suppose the function $f(x) = x^2 + a|x-1|$ is monotonically increasing on $[0, +\infty)$. Find the value range of a.

Solution. Since $f(x) = x^2 + ax - a$ for $x \in [1, +\infty)$, the monotonicity condition is equivalent to $-\frac{a}{2} \leq 1$, namely $a \geq -2$. For $x \in [0, 1]$, the function $f(x) = x^2 - ax + a$ is monotonically increasing, so $\frac{a}{2} \leq 0$, which is equivalent to $a \leq 0$.

Therefore, the value range of a is $[-2, 0]$.

Example 3. Given $0 < k < 1$, try to determine the number of solutions to the equation $|1 - x^2| = kx + k$.

Solution. We examine the graphs of $y = |1 - x^2|$ and $y = kx + k$ $(0 < k < 1)$, and determine the number of their common points.

First draw the graph of the function $y = |1 - x^2|$, which is the same as $y = |x^2 - 1|$. Then draw the graph of $y = kx + k$ $(0 < k < 1)$ as in Figure 6.2. Note that the line passes through a fixed point $(-1, 0)$, and its y-intercept k satisfies $0 < k < 1$.

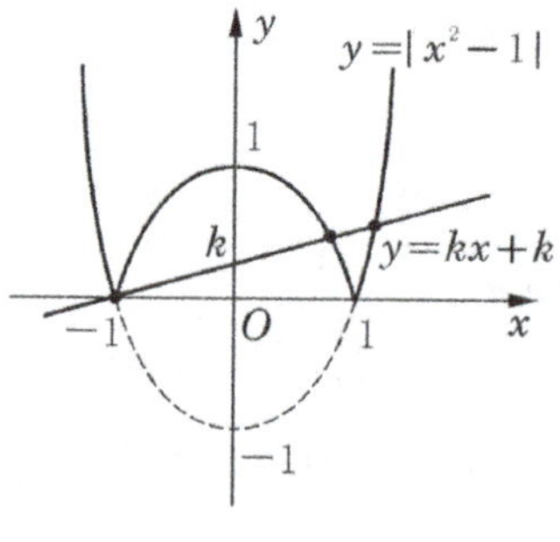

Figure 6.2

It follows from the above figure that the two graphs have three common points, so the equation $|1 - x^2| = kx + k$ has 3 solutions.

Example 4. Let a_1, a_2, a_3, and a_4 be real constants such that $a_1 \leq a_2 \leq a_3 \leq a_4$. Find the minimum values of the following functions.

(1) $f(x) = |x - a_1| + |x - a_2|$;
(2) $f(x) = |x - a_1| + |x - a_2| + |x - a_3|$;
(3) $f(x) = |x - a_1| + |x - a_2| + |x - a_3| + |x - a_4|$.

Analysis. In order to find the minimum value of the function (1), we can remove the absolute value signs to get a piecewise linear function, and then solve for its minimum value. When dealing with (2) and (3), we can apply the result obtained in (1).

Solution. (1) By the definition of absolute values of real numbers,

$$f(x) = |x - a_1| + |x - a_2| = \begin{cases} a_1 + a_2 - 2x, & x < a_1, \\ a_2 - a_1, & a_1 \le x \le a_2, \\ 2x - a_1 - a_2, & x > a_2. \end{cases}$$

Let $g(x) = a_1 + a_2 - 2x$ $(x \le a_1)$. Then the minimum value of $g(x)$ is $g(a_1) = a_2 - a_1$. Let $h(x) = 2x - a_1 - a_2$ $(x \ge a_2)$. Then the minimum value of $h(x)$ is $h(a_2) = a_2 - a_1$. Therefore, the minimum value of $f(x)$ is $a_2 - a_1$, which is attained when $a_1 \le x \le a_2$.

(2) By the result in (1), the function $|x-a_1|+|x-a_3|$ attains the minimum value a_3-a_1 when $a_1 \le x \le a_3$. Since apparently $|x-a_2|$ takes its minimum value 0 when $x = a_2$, which lies inside the interval $[a_1, a_3]$, the function

$$f(x) = |x - a_1| + |x - a_2| + |x - a_3|$$

also takes the minimum value $a_3 - a_1$ when $x = a_2$.

(3) By the result in (1), the function $|x - a_1| + |x - a_4|$ takes its minimum value $a_4 - a_1$ when $a_1 \le x \le a_4$, and the function $|x-a_2|+|x-a_3|$ takes its minimum value $a_3 - a_2$ when $a_2 \le x \le a_3$. Note that $a_1 \le a_2 \le a_3 \le a_4$, and so the minimum value of the function

$$f(x) = |x - a_1| + |x - a_2| + |x - a_3| + |x - a_4|$$

is $a_4 + a_3 - a_2 - a_1$, attained when $a_2 \le x \le a_3$.

Remark. In (1) we divided $f(x)$ into pieces to apply the method for finding the maximum (or minimum) value of a linear function. If we view $|x - a_1|$ as the distance between the points x and a_1 on the real line, then we do not really need to remove the absolute value signs. In fact, it is easy to see that the sum of the distances from x to a_1 and a_2 is at least $a_2 - a_1$, which is attained when x lies between a_1 and a_2 (including the endpoints).

As in (2) and (3), we can generalize the problem in the following way:

Let $a_1 \le a_2 \le \cdots \le a_n$. Find the minimum value of $f(x) = |x - a_1| + |x - a_2| + \cdots + |x - a_n|$.

In addition, since the numbers $a_1, a_2, \ldots, a_n$ are allowed to be equal, we can also find the minimum value of the following function in the same way:

$$f(x) = k_1|x - a_1| + k_2|x - a_2| + \cdots + k_n|x - a_n|,$$

where $k_1, k_2, \ldots, k_n$ are positive integers.

Example 5 (2017 China Mathematical Competition). If the inequality $|2^x - a| < |5 - 2^x|$ holds for all $x \in [1, 2]$, find the value range of a.

Solution. Let $t = 2^x$. Then $t \in [2, 4]$. Hence, $|t - a| < |5 - t|$ for all $t \in [2, 4]$. Note that

$$\begin{aligned} |t - a| < |5 - t| &\Leftrightarrow (t - a)^2 < (5 - t)^2 \\ &\Leftrightarrow (2t - a - 5)(5 - a) < 0. \end{aligned}$$

For the given constant a, the function $f(t) = (2t - a - 5)(5 - a)$ is linear, so $f(t) < 0$ for all $t \in [2, 4]$ if and only if

$$\begin{cases} f(2) = (-1 - a)(5 - a) < 0, \\ f(4) = (3 - a)(5 - a) < 0, \end{cases}$$

which gives that $3 < a < 5$. Therefore, the value range of a is $(3, 5)$.

Example 6. Let $x, y, z \in [0, 1]$. Find the maximum value of

$$M = \sqrt{|x - y|} + \sqrt{|y - z|} + \sqrt{|z - x|}.$$

Solution. Without loss of generality we assume that $0 \le x \le y \le z \le 1$. Then

$$M = \sqrt{y - x} + \sqrt{z - y} + \sqrt{z - x}.$$

Since

$$\sqrt{y - x} + \sqrt{z - y} \le \sqrt{2[(y - x) + (z - y)]} = \sqrt{2(z - x)},$$

we have

$$M \le \sqrt{2(z-x)} + \sqrt{z-x} = (\sqrt{2}+1)\sqrt{z-x} \le \sqrt{2}+1.$$

The equality is satisfied if and only if $y-x=z-y$, $x=0$, and $z=1$, or equivalently, $x=0$, $y=\frac{1}{2}$, and $z=1$.

Therefore, $M_{\max} = \sqrt{2}+1$.

Example 7. Find the minimum value of the function

$$f(x) = |x-1| + |x-3| + |x-5| + |x-7|.$$

Solution. As shown in Figure 6.3, let A, B, C, and D denote the points 1, 3, 5, and 7 on the real line, respectively. Let P denote the number x. Then

$$f(x) = |PA| + |PB| + |PC| + |PD|.$$

By the triangle inequality $|PA| + |PD| \ge |AD| = 6$, where the equality holds if and only if P lies on the line segment AD (including the endpoints). Similarly, $|PB| + |PC| \ge |BC| = 2$, where the equality holds if and only if P lies on the line segment BC (including the endpoints).

Therefore, $f(x) \ge 8$, where the equality holds if and only if P lies on the line segment BC (including the endpoints). Equivalently, the function $f(x)$ attains the minimum value 8 when $3 \le x \le 5$.

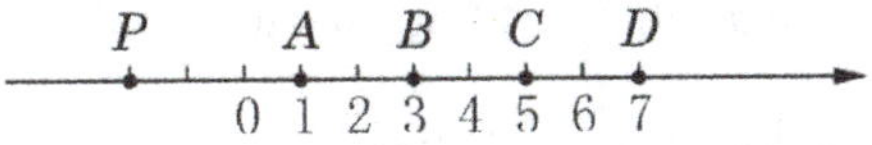

Figure 6.3

Example 8. Let a be a real number and $f(x) = x^2 + |x-a| + 1$ for $x \in \mathbf{R}$.

(1) Determine whether $f(x)$ is odd or even.
(2) Find the minimum value of $f(x)$.

Solution. (1) If $a=0$, then $f(-x) = (-x)^2 + |-x| + 1 = f(x)$, so $f(x)$ is an even function.

If $a \ne 0$, then $f(a) = a^2+1$, $f(-a) = a^2 + 2|a| + 1$, $f(-a) \ne f(a)$, and $f(-a) \ne -f(a)$. In this case $f(x)$ is neither odd nor even.

(2) (i) For $x \le a$,

$$f(x) = x^2 - x + a + 1 = \left(x - \frac{1}{2}\right)^2 + a + \frac{3}{4}.$$

If $a \leq \frac{1}{2}$, then $f(x)$ is monotonically decreasing on $(-\infty, a]$, so the minimum value of $f(x)$ on $(-\infty, a]$ is $f(a) = a^2 + 1$.

If $a > \frac{1}{2}$, then the minimum value of $f(x)$ on $(-\infty, a]$ is $f\left(\frac{1}{2}\right) = \frac{3}{4} + a$, and $f\left(\frac{1}{2}\right) \leq f(a)$.

(ii) For $x \geq a$,

$$f(x) = x^2 + x - a + 1 = \left(x + \frac{1}{2}\right)^2 - a + \frac{3}{4}.$$

If $a \leq -\frac{1}{2}$, then the minimum value of $f(x)$ on $[a, +\infty)$ is $f\left(-\frac{1}{2}\right) = \frac{3}{4} - a$, and $f\left(-\frac{1}{2}\right) \leq f(a)$.

If $a > -\frac{1}{2}$, then $f(x)$ is monotonically increasing on $[a, +\infty)$, so the minimum value of $f(x)$ on $[a, +\infty)$ is $f(a) = a^2 + 1$.

In summary, if $a \leq -\frac{1}{2}$, then the minimum value of $f(x)$ is $\frac{3}{4} - a$.

If $-\frac{1}{2} < a \leq \frac{1}{2}$, then the minimum value of $f(x)$ is $a^2 + 1$.

If $a > \frac{1}{2}$, then the minimum value of $f(x)$ is $a + \frac{3}{4}$.

Example 9. Let $\max\{a, b\}$ denote the maximum of a and b. For example,

$$\max\{0.1, -2\} = 0.1, \max\{2, 2\} = 2.$$

Find the minimum value of the function $f(x) = \max\{|x+1|, |x^2 - 5|\}$, and determine the value of x when $f(x)$ attains its minimum value.

Analysis. The function $f(x)$ is defined as the maximum of $|x + 1|$ and $|x^2 - 5|$. Since the graphs of $f_1(x) = |x + 1|$ and $f_2(x) = |x^2 - 5|$ can be drawn easily, we can also determine the graph of $y = f(x)$, which is the upper part of the two graphs. The minimum value of $f(x)$ can be found with the help of its graph.

Solution. We draw the graphs of $f_1(x) = |x+1|$ and $f_2(x) = |x^2-5|$ in the same coordinate plane (Figure 6.4). The two graphs have four intersection points A, B, C, and D, whose x-coordinates are determined by the equation $|x+1| = |x^2-5|$. Removing the absolute value signs, we have $x^2-5 = x+1$ or $x^2 - 5 = -(x+1)$, and solving the equations we get $x_1 = 3$, $x_2 = -2$, $x_3 = \frac{-1+\sqrt{17}}{2}$, and $x_4 = \frac{-1-\sqrt{17}}{2}$. Hence, the x-coordinates of A, B, C, and D are $-\frac{1+\sqrt{17}}{2}$, -2, $\frac{-1+\sqrt{17}}{2}$, and 3, respectively.

By the definition of $f(x)$, its graph is the solid part in Figure 6.4. Since the y-coordinate of the point B is the minimum value of the function $f(x)$, we conclude that the minimum value is $f(-2) = 1$.

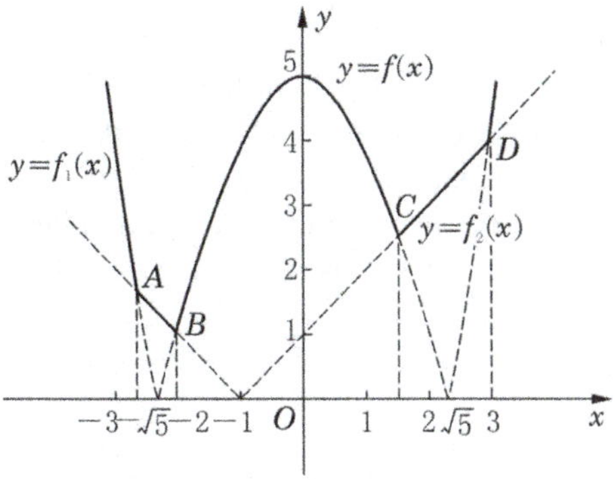

Figure 6.4

6.3 Exercises

Group A

1. Draw the graph of $y = |x+1| + |x-2|$.
2. Suppose $f(x)$ is a periodic function defined on $\mathbf{R}$ with period 2, and is also an even function. If $f(x) = x$ for $x \in [2,3]$, then the formula of $f(x)$ when $x \in [-2,0]$ is ( ).

   (A) $f(x) = x+4$ (B) $f(x) = 2-x$

   (C) $f(x) = 3-|x+1|$ (D) $f(x) = 2+|x+1|$

3. The number of solutions to the equation $|x^2-1| = (4-2\sqrt{3})(x+2)$ is ( ).

   (A) 1 (B) 2 (C) 3 (D) 4

4. Draw the graph of $y = x^2 - 2|x| - 1$.
5. Find the number of solutions to the equation $|x - |2x+1|| = 3$.
6. Let $f(x) = 1 - |1-2x|$ and $g(x) = (x-1)^2$ be defined on $[0,1]$, and let $F(x) = \min\{f(x), g(x)\}$.

   (1) Find the expression for $F(x)$.
   (2) Find the maximum value of $F(x)$.
   (3) Solve the equation $F(x) = \frac{1}{3}$.

7. (1) Sketch the graph of $y = \frac{|x|+1}{|x+1|}$.

   (2) Find the number of solutions to the equation $|x^2 - 4|x| + 3| = a$.

8. Let x and y be positive real numbers. Find the minimum value of the expression $x + y + \frac{|x-1|}{y} + \frac{|y-1|}{x}$.

Group B

9. Find the minimum value of the function $f(x) = \frac{|x-a|}{x^2-ax+1}$ $(|a| < 2)$.
10. If there are exactly six real numbers x such that $||x^2-6x-16|-10| = a$, find the value of the real number a.
11. Let $f(x) = x^2+px+q$, where p and q are real numbers. If the maximum value of $|f(x)|$ for $-1 \le x \le 1$ is M, find the minimum value of M.

Chapter 7

Maximum and Minimum Values of Functions

7.1 Key Points of Knowledge and Basic Methods

A. Basic concepts

Let D be the domain of a function $f(x)$. If there exists $x_0 \in D$ such that $f(x) \leq f(x_0)$ for all $x \in D$, then we call $f(x_0)$ the maximum value of $f(x)$ on D, which can be denoted as $f_{\max}$. If there exists $y_0 \in D$ such that $f(x) \geq f(y_0)$ for all $x \in D$, then we call $f(y_0)$ the maximum value of $f(x)$ on D, which can be denoted as $f_{\min}$.

B. Extrema of monotonic functions on closed intervals

If a function $f(x)$ is increasing on an interval $[a, b]$, then its minimum and maximum values on this interval are $f(a)$ and $f(b)$, respectively. If $f(x)$ is decreasing on an interval $[a, b]$, then its minimum and maximum values on this interval are $f(b)$ and $f(a)$, respectively.

C. Extrema of linear functions

Let $f(x) = ax + b$ be defined on an interval $[\alpha, \beta]$. If $a > 0$, then

$$f_{\min}(x) = f(\alpha), \quad f_{\max}(x) = f(\beta);$$

if $a < 0$, then

$$f_{\min}(x) = f(\beta), \quad f_{\max}(x) = f(\alpha).$$

D. Extrema of quadratic functions

Let $f(x) = ax^2 + bx + c$ $(a \neq 0)$. If $a > 0$, then $f(x)$ has its minimum value, and $f_{\min}(x) = f\left(-\frac{b}{2a}\right) = \frac{4ac-b^2}{4a}$;

if $a < 0$, then $f(x)$ has its maximum value, and $f_{\max}(x) = f\left(-\frac{b}{2a}\right) = \frac{4ac-b^2}{4a}$.

E. Extrema of quadratic functions on closed intervals

(1) If $a > 0$, then the maximum value of $f(x)$ is attained at an endpoint of the interval:

$$f_{\max}(x) = \max\{f(\alpha),\ f(\beta)\}.$$

The minimum value of $f(x)$ has the following two cases:
① If $-\frac{b}{2a} \in [\alpha, \beta]$, then

$$f_{\min}(x) = f\left(-\frac{b}{2a}\right) = \frac{4ac-b^2}{4a}.$$

② If $-\frac{b}{2a} \notin [\alpha, \beta]$, then

$$f_{\min}(x) = \min\{f(\alpha),\ f(\beta)\}.$$

(2) If $a < 0$, then the minimum value of $f(x)$ is attained at an endpoint of the interval:

$$f_{\min}(x) = \min\{f(\alpha),\ f(\beta)\}.$$

The maximum value of $f(x)$ has the following two cases:
① If $-\frac{b}{2a} \in [\alpha, \beta]$, then

$$f_{\max}(x) = f\left(-\frac{b}{2a}\right) = \frac{4ac-b^2}{4a}.$$

② If $-\frac{b}{2a} \notin [\alpha, \beta]$, then

$$f_{\max}(x) = \max\{f(\alpha),\ f(\beta)\}.$$

F. Common methods for finding maximum and minimum values of functions

(1) Completing the square: Write the function in terms of several nonnegative expressions and a constant, in order to estimate a lower bound of $f(x)$ and find its minimum value.

(2) Discriminant: Express the coefficients of a quadratic equation in terms of the given function, and use the discriminant to find lower or upper bounds of the function, thus finding the extrema of the function.
(3) Monotonicity: Determine the extrema of a function by inspecting its monotonicity.
(4) Inequalities: Use basic inequalities (the AM-GM inequality, etc.) to find the minimum or maximum value of a function.
(5) Changing variables: Use auxiliary variables to transform the function so that its extrema are easier to find.

7.2 Illustrative Examples

Example 1. Suppose $f(x) = x + g(x)$, where $g(x)$ is a function defined on $\mathbf{R}$ that is periodic with the least positive period 2. If the maximum value of $f(x)$ on the interval $[2, 4)$ is 1, find its maximum value on $[10, 12)$.

Solution. The conditions show that

$$f(x+2) = (x+2) + g(x+2) = x + g(x) + 2 = f(x) + 2.$$

Since the maximum value of $f(x)$ on $[2, 4)$ is 1, its maximum value is 3 on $[4, 6)$, 5 on $[6, 8)$, 7 on $[8, 10)$, and 9 on $[10, 12)$.

Example 2. Let x be a positive real number. Find the minimum value of $y = x^2 + x + \frac{3}{x}$.

Solution. We first estimate a lower bound of y:

$$\begin{aligned} y &= (x^2 - 2x + 1) + 3\left(x + \frac{1}{x} - 2\right) + 5 \\ &= (x-1)^2 + 3\left(\sqrt{x} - \frac{1}{\sqrt{x}}\right)^2 + 5 \geq 5. \end{aligned}$$

When $x = 1$, we have $y = 5$. Therefore, the minimum value of y is 5.

Remark. In this problem we used the method of "completing the square" to find a lower bound of y, and then found an instance to show that the lower bound is in fact the minimum value. The instance is indispensable, since otherwise the minimum value may not be correct. For example, we can give an alternative estimation in this problem:

$$\begin{aligned} y &= (x^2 - 2x + 1) + 3\left(x + \frac{1}{x} + 2\right) - 7 \\ &= (x-1)^2 + 3\left(\sqrt{x} + \frac{1}{\sqrt{x}}\right)^2 - 7 \geq -7. \end{aligned}$$

However, there is no such x that makes $y = -7$. This means that -7 is not the minimum value of y.

Example 3. Let $f(x)$ be defined as follows: for each $x \in \mathbf{R}$, the value $f(x)$ is the minimum of the three numbers $x + 2$, $4x + 1$, and $-2x + 4$. Find the maximum value of $f(x)$.

Solution. We draw three lines in the coordinate plane, namely $y = x + 2$, $y = 4x + 1$, and $y = -2x + 4$ (as in Figure 7.1). From the graph of $f(x)$ we see that the point P is where the function attains its minimum value.

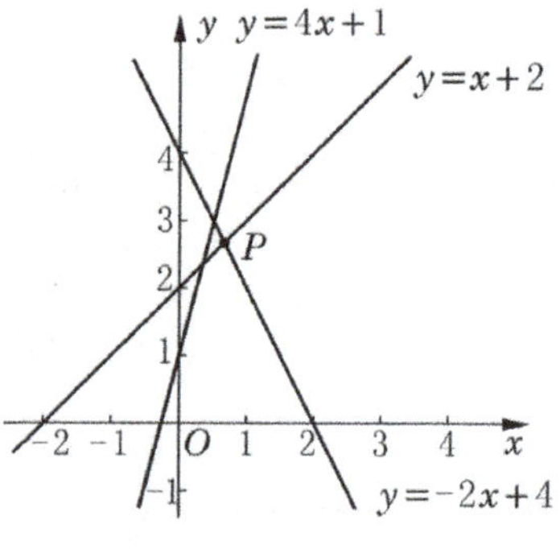

Figure 7.1

Solving the system of equations $\begin{cases} y = x + 2, \\ y = -2x + 4, \end{cases}$ we find that the coordinates of P are $\left(\frac{2}{3}, \frac{8}{3}\right)$, so the minimum value of $f(x)$ is $\frac{8}{3}$.

Example 4. Given the function $f(x) = \log_2(x + 1)$, suppose when a point (x, y) moves along the graph of $y = f(x)$, the point $\left(\frac{x}{3}, \frac{y}{2}\right)$ moves along the graph of $y = g(x)$. Find the maximum value of $p(x) = g(x) - f(x)$.

Solution. Since (x, y) lies on the graph of $y = f(x)$, we have $y = \log_2(x + 1)$. Since $\left(\frac{x}{3}, \frac{y}{2}\right)$ lies on the graph of $y = g(x)$, we see that $\frac{y}{2} = g\left(\frac{x}{3}\right)$.

Hence,

$$g\left(\frac{x}{3}\right) = \frac{1}{2}\log_2(x+1), \quad g(x) = \frac{1}{2}\log_2(3x+1);$$
$$p(x) = g(x) - f(x) = \frac{1}{2}\log_2(3x+1) - \log_2(x+1) = \frac{1}{2}\log_2\frac{3x+1}{(x+1)^2}.$$

Let $u = \frac{3x+1}{(x+1)^2}$. Then

$$u = \frac{3(x+1)-2}{(x+1)^2} = -\frac{2}{(x+1)^2} + \frac{3}{x+1}$$
$$= -2\left(\frac{1}{x+1} - \frac{3}{4}\right)^2 + \frac{9}{8} \le \frac{9}{8}.$$

When $\frac{1}{x+1} = \frac{3}{4}$, or equivalently $x = \frac{1}{3}$, we have $u = \frac{9}{8}$, so $u_{\max} = \frac{9}{8}$.
Therefore, $p_{\max}(x) = \frac{1}{2}\log_2\frac{9}{8}$.

Example 5. Let p, $q \in \mathbf{R}$, and suppose the graph of

$$f(x) = x^3 + px^2 + (p+q)x + q + 1$$

is centrally symmetric with respect to $(1, 0)$. Find the maximum value of the function $g(x) = f(x) - x^3 - px - q$.

Solution. Apparently $f(x)$ is a cubic function. Since the graph of $f(x)$ is centrally symmetric with respect to $(1, 0)$, we have $f(1) = 0$. Note that

$$f(-1) = -1 + p - (p+q) + q + 1 = 0,$$

so by the symmetry we get $f(3) = 0$.

Since $f(x)$ has the leading coefficient 1,

$$f(x) = (x-1)(x+1)(x-3) = x^3 - 3x^2 - x + 3.$$

Comparing the coefficients we get $p = -3$ and $q = 2$. Hence,

$$g(x) = f(x) - x^3 + 3x - 2 = -3x^2 + 2x + 1 = -3\left(x - \frac{1}{3}\right)^2 + \frac{4}{3}.$$

Therefore, the maximum value of $g(x)$ is $\frac{4}{3}$, attained at $x = \frac{1}{3}$.

Example 6. Find the maximum and minimum values of the function $y = \frac{x^2+x+2}{2x^2-x+1}$.

Solution. Removing the denominator, we get

$$(2y-1)x^2-(y+1)x+y-2=0.$$

If $y \neq \frac{1}{2}$, then the equation above can be viewed as a quadratic equation for x. Since x and y are real numbers,

$$\Delta=(y+1)^2-4(2y-1)(y-2) \geq 0,$$

$$7y^2-22y+7 \leq 0,$$

so $\frac{11-6\sqrt{2}}{7} \leq y \leq \frac{11+6\sqrt{2}}{7}$.

On the other hand, when $x=-1-\sqrt{2}$, we have $y=\frac{11-6\sqrt{2}}{7}$, and when $x=-1+\sqrt{2}$, we have $y=\frac{11+6\sqrt{2}}{7}$. Apparently $\frac{1}{2}$ lies between these two values, hence $y=\frac{1}{2}$ cannot be an extremum value.

Therefore, $y_{\min}=\frac{11-6\sqrt{2}}{7}$ and $y_{\max}=\frac{11+6\sqrt{2}}{7}$.

Remark. In this solution we have applied the method of discriminant.

Example 7. Suppose the minimum and maximum values of the function $y=\frac{ax^2+bx+6}{x^2+2}$ are 2 and 6, respectively. Find the values of a and b.

Solution. Removing the denominator, we have

$$(a-y)x^2+bx+(6-2y)=0.$$

If $y \equiv a$, then y is a constant, so its maximum and minimum values have to be equal, which is impossible. Therefore, $y \not\equiv a$, and hence

$$\Delta=b^2-4(a-y)(6-2y) \geq 0,$$

or equivalently $y^2-(a+3)y+3a-\frac{b^2}{8} \leq 0$. ①

Since the minimum and maximum values of y are 2 and 6, respectively, the value range of y is determined by

$$(y-2)(y-6) \leq 0,$$

namely $y^2-8y+12 \leq 0$. ②

Comparing ① and ② we have

$$\begin{cases} a+3=8, \\ 3a-\dfrac{b^2}{8}=12. \end{cases}$$

Solving for a and b, we get that $a=5$ and $b=\pm 2\sqrt{6}$.

Example 8. Find the minimum and maximum values of the function

$$f(x) = \sqrt{8x - x^2} - \sqrt{14x - x^2 - 48}.$$

Solution. We first determine the domain of $f(x)$. Since

$$\begin{cases} 8x - x^2 \geq 0, \\ 14x - x^2 - 48 \geq 0, \end{cases}$$

we see that $6 \leq x \leq 8$. The function can be written as

$$f(x) = \sqrt{8-x}\left(\sqrt{x} - \sqrt{x-6}\right) = \frac{6\sqrt{8-x}}{\sqrt{x} + \sqrt{x-6}}, \quad x \in [6, 8].$$

As x increases in the interval [6, 8], the value of $\sqrt{x} + \sqrt{x-6}$ increases and the value of $\sqrt{8-x}$ decreases. This implies that $f(x)$ is a decreasing function on [6, 8]. Therefore,

$$f_{\min}(x) = f(8) = 0, \quad f_{\max}(x) = f(6) = 2\sqrt{3}.$$

Example 9. Find the minimum and maximum values of the function $y = \frac{x-1}{x^2-2x+5}$ for $\frac{3}{2} \leq x \leq 2$.

Solution. Since $x \neq 1$,

$$y = \frac{x-1}{(x-1)^2+4} = \frac{1}{x - 1 + \frac{4}{x-1}}, \quad x \in \left[\frac{3}{2}, 2\right].$$

Let $f(t) = t + \frac{4}{t}$ with $t \in \left[\frac{1}{2}, 1\right]$. For $\frac{1}{2} \leq t_1 < t_2 \leq 1$,

$$\begin{aligned} f(t_2) - f(t_1) &= (t_2 - t_1) + \left(\frac{4}{t_2} - \frac{4}{t_1}\right) \\ &= (t_2 - t_1)\left(1 - \frac{4}{t_1 t_2}\right) < 0, \end{aligned}$$

so the function $f(t) = t + \frac{4}{t}$ is decreasing on $\left[\frac{1}{2}, 1\right]$. Thus,

$$f_{\min}(t) = f(1) = 5, \quad f_{\max}(t) = f\left(\frac{1}{2}\right) = \frac{17}{2}.$$

Therefore, $y_{\min} = \frac{2}{17}$ and $y_{\max} = \frac{1}{5}$.

Remark. The solutions of Examples 8 and 9 both used the monotonicity of a function to find its extrema. Rationalizing the numerator as in Example 8 is a useful tool in such problems. The method of discriminant does not apply immediately to Example 9, and the reason is explained here: Consider the equation

$$yx^2 - (2y+1)x + 5y + 1 = 0.$$

If $y = 0$, then $x = 1$, which is not in the domain, so $y \neq 0$.
For $y \neq 0$,

$$\Delta = (2y+1)^2 - 4y(5y+1) \geq 0,$$

so $-\frac{1}{4} \leq y \leq \frac{1}{4}$.

However, $y = -\frac{1}{4}$ when $x = -1$ and $y = \frac{1}{4}$ when $x = 3$, while both -1 and 3 lie outside the domain, hence the extrema of y cannot be found through this approach.

Example 10. Let x and y be nonnegative real numbers such that

$$\sqrt{1 - \frac{x^2}{4}} + \sqrt{1 - \frac{y^2}{16}} = \frac{3}{2}.$$

Find the maximum value of xy.

Solution 1. Let $u = \sqrt{1 - \frac{x^2}{4}}$ and $v = \sqrt{1 - \frac{y^2}{16}}$. Then $0 \leq u, v \leq 1$ and $u + v = \frac{3}{2}$. Thus,

$$x^2y^2 = 4(1-u^2) \cdot 16(1-v^2) = 64[1 - (u^2+v^2) + u^2v^2].$$

Let $uv = t$. Then $u^2 + v^2 = (u+v)^2 - 2uv = \frac{9}{4} - 2t$, and

$$x^2y^2 = 64\left[1 - \left(\frac{9}{4} - 2t\right) + t^2\right] = 64\left[(t+1)^2 - \frac{9}{4}\right].$$

From $2\sqrt{uv} \leq u + v = \frac{3}{2}$ we get that $0 \leq t = uv \leq \left(\frac{3}{4}\right)^2 = \frac{9}{16}$, thus

$$x^2y^2 = 64\left[(t+1)^2 - \frac{9}{4}\right] \leq 64\left[\left(\frac{9}{16} + 1\right)^2 - \frac{9}{4}\right] = \frac{49}{4},$$

and hence $xy \leq \frac{7}{2}$.

When $x = \frac{\sqrt{7}}{2}$ and $y = \sqrt{7}$, the equality holds. Therefore, the maximum value of xy is $\frac{7}{2}$.

Solution 2. By $(a+b)^2 \le 2(a^2+b^2)$, we see that

$$\frac{9}{4} = \left(\sqrt{1-\frac{x^2}{4}} + \sqrt{1-\frac{y^2}{16}}\right)^2 \le 2\left(1-\frac{x^2}{4}+1-\frac{y^2}{16}\right),$$

so $\frac{x^2}{4}+\frac{y^2}{16} \le \frac{7}{8}$, that is $4x^2+y^2 \le 14$. ①

Also, $4xy \le 4x^2+y^2$. ②

Comparing ① and ②, we obtain that $xy \le \frac{7}{2}$. ③

The equality in ① holds if and only if $\sqrt{1-\frac{x^2}{4}} = \sqrt{1-\frac{y^2}{16}} = \frac{3}{4}$, namely $x = \frac{\sqrt{7}}{2}$ and $y = \sqrt{7}$, and the equality in ② holds if and only if $2x = y$.

Therefore, the equality in ③ holds if and only if $x = \frac{\sqrt{7}}{2}$ and $y = \sqrt{7}$, and the maximum value of xy is $\frac{7}{2}$.

Example 11. Let the function

$$f(x) = \begin{cases} x, & \text{if } x \text{ is irrational}, \\ \dfrac{q+1}{p}, & \text{if } x = \dfrac{q}{p}, \text{ where } p, q \in \mathbf{N}^*, \ \gcd(p,q) = 1, \ p > q. \end{cases}$$

Find the maximum value of $f(x)$ on the interval $\left(\frac{7}{8}, \frac{8}{9}\right)$.

Solution. If x is rational and $x \in \left(\frac{7}{8}, \frac{8}{9}\right)$, let $x = \frac{a}{a+\lambda} \in \left(\frac{7}{8}, \frac{8}{9}\right)$ $(a, \lambda \in \mathbf{N}^*)$. By $\frac{7}{8} < \frac{a}{a+\lambda} < \frac{8}{9}$,

$$\begin{cases} 9a < 8a + 8\lambda, \\ 7a + 7\lambda < 8a, \end{cases} \quad \Rightarrow \quad 7\lambda < a < 8\lambda.$$

If $\lambda = 1$, then a does not exist.

If $\lambda = 2$, then $a = 15$ is unique, and $x = \frac{15}{17}$ and $f(x) = \frac{16}{17}$.

If $\lambda \ge 3$, then $a = 7\lambda + m$, where $1 \le m \le \lambda - 1$ and $m \in \mathbf{N}^*$. In this case

$$f(x) = \frac{7\lambda + m + 1}{8\lambda + m}.$$

Since

$$\frac{16}{17} - \frac{7\lambda+m+1}{8\lambda+m} = \frac{9\lambda - m - 17}{17(8\lambda+m)} = \frac{(\lambda-m)+(8\lambda-17)}{17(8\lambda+m)} > 0,$$

we see that the maximum value of $f(x)$ for rational numbers x is $\frac{16}{17}$, attained at $x = \frac{15}{17}$.

If x is irrational, then $f(x) = x < \frac{8}{9} < \frac{16}{17}$. Therefore, the maximum value of $f(x)$ for $x \in \left(\frac{7}{8}, \frac{8}{9}\right)$ is $\frac{16}{17}$.

7.3 Exercises

Group A

1. Fill in the blanks:

   (1) Set the function $y = |x-a| + |x+19| + |x-a-96|$, where a is a constant such that $19 < a < 96$. For $a \le x \le 96$, the maximum value of y is __________.

   (2) The sum of the maximum and minimum values of $y = x^2 - 4x + 7$ for $x \in [0, 3]$ is __________.

   (3) The maximum value of the function $y = \frac{x+1}{x^2+3}$ is __________.

   (4) When $|x+1| \le 6$, the maximum value of the function $y = x|x| - 2x + 1$ is __________.

   (5) The maximum value of the function $y = -2x + 5\sqrt{x+1}$ for $x \in [0, 1]$ is __________.

   (6) If $3x^2 + 2y^2 = 2x$, then the maximum value of $x^2 + y^2$ is __________.

2. Let x take any positive real number. Find the minimum value of the function $y = x^2 - x + \frac{1}{x}$.
3. Let M and m be the maximum and minimum values of the function $f(x) = \frac{(x+\sqrt{2013})^2 + \sin 2013x}{x^2+2013}$, respectively. Find the value of $M + m$.
4. Let x and y be positive real numbers. Find the minimum value of
$$f(x, y) = \frac{x^4}{y^4} + \frac{y^4}{x^4} - \frac{x^2}{y^2} - \frac{y^2}{x^2} + \frac{x}{y} + \frac{y}{x}.$$
5. For real numbers x and y, find the minimum value of $f(x, y) = x^2 + xy + y^2 - x - y$.
6. Find the maximum and minimum values of the function $y = (x+1)(x+2)(x+3)(x+4) + 5$ on the interval $[-6, 6]$.
7. Find the maximum and minimum values of the function $y = \frac{x^4+x^2+5}{(x^2+1)^2}$.
8. Let x and y be real numbers such that $x^2 - 2xy + y^2 - \sqrt{2}x - \sqrt{2}y + 6 = 0$. Find the minimum value of $u = x + y$.
9. Let x_1 and x_2 be two real roots of the equation $x^2 - (k-2)x + (k^2 + 3k + 5) = 0$ (k is a constant). Find the maximum and minimum values of $x_1^2 + x_2^2$.

10. Let a be a real number and let m be the minimum value of the function $y = x^2 - 4ax + 5a^2 - 3a$. For all the values of a such that

$$0 \le a^2 - 4a - 2 \le 10,$$

find the maximum value of m.
11. For $x \neq 0$, find the maximum value of the function $f(x) = \frac{\sqrt{x^4+x^2+1}-\sqrt{x^4+1}}{x}$.
12. Let x and y be real numbers. Find the minimum value of $u = x^2 + xy + y^2 - x - 2y + 3$.
13. Suppose the maximum and minimum values of the function $y = \frac{ax+b}{x^2+1}$ are 4 and -1, respectively. Find the values of a and b.
14. Let $f(x) = ax^2 + bx + c(0 < 2a < b)$ be such that $f(x) \ge 0$ for all $x \in \mathbf{R}$. Find the minimum value of $\frac{f(1)}{f(0)-f(-1)}$.

Group B

15. Find the minimum value of $f(x) = \sqrt{2x^2 - 3x + 4} + \sqrt{x^2 - 2x}$.
16. Suppose the function $f(x) = -9x^2 - 6ax - a^2 + 2a$ for $x \in \left[-\frac{1}{3}, \frac{1}{3}\right]$ (a is a real constant) has the maximum value -3. Find the value of a.
17. Find the maximum value of $f(x) = \left|\frac{1}{x} - \left[\frac{1}{x} + \frac{1}{2}\right]\right|$ where $[a]$ denotes the greatest integer that does not exceed a.
18. Find the maximum value of $f(x) = \sqrt{x^4 - 3x^2 - 6x + 13} - \sqrt{x^4 - x^2 + 1}$.
19. Let $f(x) = ax^2 + 8x + 3$ $(a < 0)$. For a given negative number a, let $l(a)$ be the greatest positive number such that the inequality $|f(x)| \le 5$ holds for all $x \in [0, l(a)]$. Find the maximum value of $l(a)$ and the value of a when $l(a)$ reaches the maximum value.
20. Suppose $\omega > 0$ is a real number such that the least positive period of the function $f(x) = \sin^2 \omega x + \sqrt{3} \sin \omega x \cdot \sin\left(\omega x + \frac{\pi}{2}\right)$ is $\frac{\pi}{2}$. Find the maximum and minimum values of $f(x)$ on the interval $\left[\frac{\pi}{8}, \frac{\pi}{4}\right]$.
21. Let α and $\beta(\alpha < \beta)$ be the two real roots of the equation (for x) $2x^2 - tx - 2 = 0$.

   (1) If x_1 and x_2 are two distinct numbers in the interval $[\alpha, \beta]$, show that

   $$4x_1x_2 - t(x_1 + x_2) - 4 < 0.$$

   (2) Let $f(x) = \frac{4x-t}{x^2+1}$ and let its maximum and minimum values on the interval $[\alpha, \beta]$ be $f_{\max}$ and $f_{\min}$, respectively. If $g(t) = f_{\max} - f_{\min}$, find the minimum value of $g(t)$.

Chapter 8

Properties of Inequalities

8.1 Key Points of Knowledge and Basic Methods

A. Properties of inequalities

$$a > b \Leftrightarrow a - b > 0,$$

$$a = b \Leftrightarrow a - b = 0,$$

$$a < b \Leftrightarrow a - b < 0,$$

$$a > b \Leftrightarrow b < a,$$

$$a > b \quad \text{and} \quad b > c \Rightarrow a > c,$$

$$a > b \Leftrightarrow a + c > b + c,$$

$$a + b > c \Leftrightarrow a > c - b,$$

$$a > b \quad \text{and} \quad c > d \Rightarrow a + c > b + d,$$

$$a > b \quad \text{and} \quad c > 0 \Rightarrow ac > bc,$$

$$a > b \quad \text{and} \quad c < 0 \Rightarrow ac < bc,$$

$$a > b \quad \text{and} \quad ab > 0 \Rightarrow \tfrac{1}{a} < \tfrac{1}{b},$$

$$a > b > 0 \quad \text{and} \quad c > d > 0 \ \Rightarrow \ ac > bd \quad \text{and} \quad \tfrac{d}{a} < \tfrac{c}{b},$$

$$a > b > 0 \ \Rightarrow \ a^n > b^n \quad \text{and} \quad \sqrt[n]{a} > \sqrt[n]{b}\ (n \in \mathbf{N} \text{ and } n > 1).$$

B. Comparing real numbers

(1) We can apply properties of inequalities, basic inequalities, and monotonicity of functions to compare two real numbers.

(2) We can also use the following methods to compare real numbers (they are derived from properties of inequalities):

Comparison by difference: $a - b > 0 \Rightarrow a > b$;
Comparison by ratio: $\frac{a}{b} > 1$ and $b > 0 \Rightarrow a > b$.

8.2 Illustrative Examples

Example 1. Let a and b be nonzero real numbers, and let $A = \frac{b}{a^2} + \frac{a}{b^2}$ and $B = \frac{1}{a} + \frac{1}{b}$. Compare the numbers A and B.

Solution. We use comparison by difference:

$$A - B = \frac{b}{a^2} + \frac{a}{b^2} - \left(\frac{1}{a} + \frac{1}{b}\right) = \frac{b-a}{a^2} + \frac{a-b}{b^2}$$
$$= \left(\frac{1}{a^2} - \frac{1}{b^2}\right)(b-a) = (a+b)\left(\frac{a-b}{ab}\right)^2.$$

Therefore, if $a = \pm b\ (\neq 0)$, then $A = B$; if $a > -b$ and $a \neq b$, then $A > B$; if $a < -b$ and $a \neq b$, then $A < B$.

Remark. In general, we usually compare two polynomials or fractions of polynomials by computing their difference, and then factorize or complete the square to determine its sign. When comparing two numbers A and B, we need to find the condition that guarantees $A > B$, $A = B$, or $A < B$.

Example 2. Let $a > 0$ with $a \neq 1$. For $x \in (0, 1)$, compare the numbers $|\log_a(1-x)|$ and $|\log_a(1+x)|$.

Solution 1. We use comparison by ratio:

Since $0 < x < 1$, we have $1 < 1 + x < 2$, $0 < 1 - x < 1$, and $|\log_a(1+x)| \neq 0$. Hence,

$$\frac{|\log_a(1-x)|}{|\log_a(1+x)|} = |\log_{(1+x)}(1-x)|$$
$$= -\log_{(1+x)}(1-x) = \log_{(1+x)}\frac{1}{1-x}.$$

Since $1 + x - \frac{1}{1-x} = \frac{-x^2}{1-x} < 0$, we get $\log_{(1+x)}\frac{1}{1-x} > 1$, or equivalently

$$\frac{|\log_a(1-x)|}{|\log_a(1+x)|} > 1.$$

Therefore, $|\log_a(1-x)| > |\log_a(1+x)|$.

Remark. Some exponential and logarithmic expressions can be compared by ratios. In the solution above, we also applied comparison by difference to the base expression of the logarithm and and the expression after the logarithm symbol: $1 + x - \frac{1}{1-x} < 0$.

In fact, we can also use comparison by difference directly in this problem, but we need to discuss different cases based on the value of a. However, since $|\log_a(1-x)|$ and $|\log_a(1+x)|$ are both nonnegative, by properties of inequalities we can compare their squares, and thus avoiding the discussion of various cases on the base a.

Solution 2. We calculate the difference of the squares of the two numbers:

$$\begin{aligned}
&|\log_a(1-x)|^2 - |\log_a(1+x)|^2 \\
&= \log_a(1-x^2)\log_a\frac{1-x}{1+x} \\
&= \frac{1}{(\lg a)^2}\lg(1-x^2)\lg\frac{1-x}{1+x}.
\end{aligned}$$

Since $0 < x < 1$, we have $0 < 1-x^2 < 1$ and $0 < \frac{1-x}{1+x} < 1$, so $\lg(1-x^2) < 0$ and $\lg\frac{1-x}{1+x} < 0$. Thus,

$$|\log_a(1-x)|^2 - |\log_a(1+x)|^2 > 0.$$

Therefore, $|\log_a(1-x)| > |\log_a(1+x)|$.

Example 3. Let $a, b, c > 0$. Compare the numbers $a^{2a}b^{2b}c^{2c}$ and $a^{b+c}b^{c+a}c^{a+b}$.

Solution. We compare them by their ratio:

$$\begin{aligned}
\frac{a^{2a}b^{2b}c^{2c}}{a^{b+c}b^{c+a}c^{a+b}} &= a^{2a-b-c}\cdot b^{2b-c-a}\cdot c^{2c-a-b} \\
&= \left(\frac{a}{b}\right)^{a-b}\cdot\left(\frac{b}{c}\right)^{b-c}\cdot\left(\frac{c}{a}\right)^{c-a}.
\end{aligned}$$

By symmetry, we can assume that $a \geq b \geq c$ (> 0). Then

$$\frac{a}{b} \geq 1, \quad \frac{b}{c} \geq 1, \quad 0 < \frac{c}{a} \leq 1, \quad a-b \geq 0, \quad b-c \geq 0, \quad c-a \leq 0.$$

Hence, $\left(\frac{a}{b}\right)^{a-b} \geq 1$, $\left(\frac{b}{c}\right)^{b-c} \geq 1$, and $\left(\frac{c}{a}\right)^{c-a} \geq 1$, from which

$$\frac{a^{2a}b^{2b}c^{2c}}{a^{b+c}b^{c+a}c^{a+b}} \geq 1,$$

where the equality is true if and only if $a = b = c$.

Therefore, if $a = b = c$, then $a^{2a}b^{2b}c^{2c} = a^{b+c}b^{c+a}c^{a+b}$, and otherwise $a^{2a}b^{2b}c^{2c} > a^{b+c}b^{c+a}c^{a+b}$.

Remark. In the solution above we have used the monotonicity of exponential functions. Sometimes we can use the methods for comparing two real numbers to prove some inequalities. Such is called the method of comparison.

Example 4 (2016 Western Mathematics Invitational). Let a, b, c, and d be real numbers such that $abcd > 0$. Show that there exists a permutation of a, b, c, d labeled as x, y, z, w, such that $2(xz + yw)^2 > (x^2 + y^2)(z^2 + w^2)$.

Solution 1. Suppose the inequality

$$2(xz + yw)^2 \leq (x^2 + y^2)(z^2 + w^2)$$

holds for every permutation x, y, z, w. Then for the permutation a, c, b, d,

$$2(ab + cd)^2 \leq (a^2 + c^2)(b^2 + d^2);$$

for the permutation a, d, c, b,

$$2(ac + db)^2 \leq (a^2 + d^2)(c^2 + b^2);$$

for the permutation a, b, d, c,

$$2(ad + bc)^2 \leq (a^2 + b^2)(d^2 + c^2).$$

Adding the three inequalities, we get that

$$\begin{aligned} &2(a^2b^2 + c^2d^2 + a^2c^2 + b^2d^2 + a^2d^2 + b^2c^2) + 12abcd \\ &\leq 2(a^2b^2 + c^2d^2 + a^2c^2 + b^2d^2 + a^2d^2 + b^2c^2), \end{aligned}$$

namely $abcd \leq 0$, contradictory to the condition that $abcd > 0$.

Solution 2. We let x and z be the greatest and second greatest numbers among a, b, c, and d, and let y and w be the other two. Next, we show that such a permutation satisfies the requirement.

In fact, since

$$(x^2+y^2)(z^2+w^2)-(xz+yw)^2=(xw-yz)^2,$$

it suffices to prove that

$$(xz+yw)^2>(xw-yz)^2,$$

or equivalently

$$|xz+yw|>|xw-yz|. \quad (*)$$

Since $xyzw>0$, we see that x and z have the same sign, and y and w have the same sign. Note that if we change the signs of x and z (or y and w) simultaneously, both sides of the inequality $(*)$ stay the same, so we can assume that x, y, z, and w are all positive. Then

$$|xz+yw|=xz+yw>xz>\max\{xw,yz\}>|xw-yz|,$$

which means that $(*)$ holds. Therefore, the proposition is proven.

Example 5. Let a,b and c be positive real numbers. Show that

$$a^3+b^3+c^3+3abc\geq ab(a+b)+bc(b+c)+ca(c+a).$$

Solution. Without loss of generality we assume that $c=\min\{a,b,c\}>0$. Then

$$\begin{aligned}\text{LHS}-\text{RHS}&=[a^3+b^3-ab(a+b)+2abc-c(a^2+b^2)]\\&\quad+[c^3+abc-c^2(a+b)]\\&=[(a+b)(a-b)^2-c(a-b)^2]+c(c^2+ab-ca-cb)\\&=(a+b-c)(a-b)^2+c(a-c)(b-c)\geq 0.\end{aligned}$$

Therefore, $a^3+b^3+c^3+3abc\geq ab(a+b)+bc(b+c)+ca(c+a)$.

Remark. In the solution above we have used the method of comparison by difference. However, it is impossible to factorize the difference or write it as the sum of squares. In the solution here we wrote the difference as the sum of two expressions whose signs are easy to determine (both positive or both negative). In order to determine the signs, we also made use of the symmetry of the inequality to assume that $c=\min\{a,b,c\}$.

Example 6. Suppose a and b are real numbers with $|a|\leq 1$ and $|a+b|\leq 1$. Show that

$$-2\leq(a+1)(b+1)\leq\frac{9}{4}.$$

Solution. Let $t = a + b$. Then $|t| \leq 1$. Note that $(a+1)(b+1) = (a+1)(t-a+1)$ can be viewed as a linear function in t, whose leading coefficient is $a + 1 \geq 0$. Thus, the maximum and minimum values of this expression are attained when $t = 1$ and $t = -1$, respectively.

If $t = 1$, then $(a+1)(b+1) = (a+1)(2-a)$, whose maximum value is $\frac{9}{4}$, attained when $a = \frac{1}{2}$.

If $t = -1$, then $(a+1)(b+1) = (a+1)(-a)$, whose minimum value is -2, attained when $a = 1$.

Therefore,

$$-2 \leq (a+1)(b+1) \leq \frac{9}{4}.$$

Example 7. Suppose $a > b > 0$ and $\frac{x}{a} < \frac{y}{b}$. Show that

$$\frac{1}{2}\left(\frac{x}{a} + \frac{y}{b}\right) > \frac{x+y}{a+b}.$$

Solution. Since $a > b > 0$ and $\frac{x}{a} < \frac{y}{b}$,

$$a - b > 0, \quad \frac{y}{b} - \frac{x}{a} > 0.$$

Thus, $(a-b)\left(\frac{y}{b} - \frac{x}{a}\right) > 0$, that is $b \cdot \frac{x}{a} + a \cdot \frac{y}{b} > x + y$, so

$$b \cdot \frac{x}{a} + a \cdot \frac{y}{b} + x + y > 2(x+y),$$

$$\left(\frac{x}{a} + \frac{y}{b}\right)(a+b) > 2(x+y).$$

Therefore,

$$\frac{1}{2}\left(\frac{x}{a} + \frac{y}{b}\right) > \frac{x+y}{a+b}.$$

Remark. It is perhaps more natural to use comparison by difference in this problem. The interested readers may give a try.

Example 8. Let $f(x) = |\sin x|$ for $x \in \mathbf{R}$. Show that

$$\sin 1 \leq f(x) + f(x+1) \leq 2\cos\frac{1}{2}.$$

Solution. Let $g(x) = f(x) + f(x+1) = |\sin x| + |\sin(x+1)$. Then $g(x)$ is a periodic function with period $T = \pi$. Now, it suffices to consider the case $x \in [0, \pi]$.

If $x \in [0, \pi - 1]$, then

$$g(x) = \sin x + \sin(x+1) = 2\sin\left(x + \frac{1}{2}\right)\cos\frac{1}{2}.$$

Since $x + \frac{1}{2} \in \left[\frac{1}{2}, \pi - \frac{1}{2}\right]$, we have $\sin\left(x + \frac{1}{2}\right) \in \left[\sin\frac{1}{2},\ 1\right]$ and

$$g(x) \in \left[2\sin\frac{1}{2}\cos\frac{1}{2}, 2\cos\frac{1}{2}\right] = \left[\sin 1, 2\cos\frac{1}{2}\right].$$

If $x \in [\pi - 1, \pi]$, then

$$g(x) = \sin x - \sin(x+1) = -2\sin\frac{1}{2}\cos\left(x + \frac{1}{2}\right).$$

Since $x + \frac{1}{2} \in \left[\pi - \frac{1}{2}, \pi + \frac{1}{2}\right]$, we see that $\cos\left(x + \frac{1}{2}\right) \in \left[-1, -\cos\frac{1}{2}\right]$ and

$$g(x) \in \left[\sin 1, 2\sin\frac{1}{2}\right] \subset \left[\sin 1, 2\cos\frac{1}{2}\right].$$

Therefore, $\sin 1 \leq f(x) + f(x+1) \leq 2\cos\frac{1}{2}$.

8.3 Exercises

Group A

1. Compare the numbers $\sin 1$ and $\log_3\sqrt{7}$ without calculating their exact values.
2. Suppose $m > 1$, $a = \sqrt{m+1} - \sqrt{m}$, and $b = \sqrt{m} - \sqrt{m-1}$. Compare a and b.
3. For $0 < x < 1$, let $f(x) = \frac{x}{lgx}$. Arrange the numbers $f(x)$, $f(x^2)$, and $f^2(x)$ in increasing order as ____________.
4. For an integer n ($n \geq 2$) and positive real numbers a and b such that $a^n = a + 1$ and $b^{2n} = b + 3a$, compare the numbers a, b, and 1.
5. Let a, b, p, $q > 0$, $A = \frac{a^{p+q}+b^{p+q}}{2}$, and $B = \left(\frac{a^p+b^p}{2}\right)\left(\frac{a^q+b^q}{2}\right)$. Compare A and B.
6. Let a, b, and c be positive real numbers. Compare the following pairs of numbers:

   (1) $2a^4b^4 + a^2 + b^2 + c^2 + 1$ and $4ab - 2bc + 2ca$.
   (2) $a^ab^bc^c$ and $(abc)^{\frac{a+b+c}{3}}$.

7. Let $a > b > c$. Show that $a^2(b-c) + c^2(a-b) > b^2(a-c)$.

8. Suppose $a_1 > 1$, $a_2 > 1$, and $a_3 > 1$ such that $a_1 + a_2 + a_3 = S$. If $\frac{a_i^2}{a_i - 1} > S$ for $i = 1, 2$, and 3, show that

$$\frac{1}{a_1 + a_2} + \frac{1}{a_2 + a_3} + \frac{1}{a_3 + a_1} > 1.$$

9. Let x, y, and z be positive numbers such that the difference between any two of them is not greater than 2. Show that

$$\sqrt{xy+1} + \sqrt{yz+1} + \sqrt{zx+1} > x + y + z.$$

Group B

10. Let a, b, $c \in \mathbf{R}_+$. Show that for all real numbers x, y, and z,

$$x^2 + y^2 + z^2 \geq 2\sqrt{\frac{abc}{(a+b)(b+c)(c+a)}} \times \left(\sqrt{\frac{a+b}{c}}xy + \sqrt{\frac{b+c}{a}}yz + \sqrt{\frac{c+a}{b}}zx\right).$$

11. Let a, b, and c be the side lengths of a triangle. Show that $a^4 + b^4 + c^4 < 2(a^2b^2 + b^2c^2 + c^2a^2)$.
12. Let a, b, and c be positive numbers. Show that

$$\frac{a^2}{(a+b)(a+c)} + \frac{b^2}{(b+c)(b+a)} + \frac{c^2}{(c+a)(c+b)} \geq \frac{3}{4}.$$

13. Let $a_1, a_2, \ldots, a_n$ be positive real numbers whose sum is 1. Show that

$$\frac{a_1^2}{a_1 + a_2} + \frac{a_2^2}{a_2 + a_3} + \cdots + \frac{a_{n-1}^2}{a_{n-1} + a_n} + \frac{a_n^2}{a_n + a_1} \geq \frac{1}{2}.$$

14. Suppose a, b, c, d, e, and f are positive integers such that

$$\frac{a}{b} > \frac{c}{d} > \frac{e}{f}, \quad af - be = 1.$$

Show that $d \geq b + f$.
15. Suppose x, y, and z are real numbers with $xy + yz + zx = -1$. Show that $x^2 + 5y^2 + 8z^2 \geq 4$, and find a sufficient and necessary condition for the inequality to be an equality.

Chapter 9

Basic Inequalities

9.1 Key Points of Knowledge and Basic Methods

A. Basic inequalities

(1) If a and $b \in \mathbf{R}$, then

$$a^2 + b^2 \geq 2ab,$$

where the equality holds if and only if $a = b$.

(2) If a, b, $c \in \mathbf{R}$, then

$$a^2 + b^2 + c^2 \geq ab + bc + ca,$$

where the equality holds if and only if $a = b = c$.

B. The AM-GM inequality

(1) **Theorem:** For two positive real numbers a and b, their arithmetic mean $\frac{a+b}{2}$ is greater than or equal to their geometric mean $\sqrt{ab}$. In other words, if a, $b \in \mathbf{R}_+$, then

$$\frac{a+b}{2} \geq \sqrt{ab},$$

where the equality holds if and only if $a = b$.

Similar results are valid for three or four numbers:

(2) If a, b, $c \in \mathbf{R}_+$, then

$$\frac{a+b+c}{3} \geq \sqrt[3]{abc},$$

where the equality holds if and only if $a = b = c$.

(3) If a, b, c, $d \in \mathbf{R}_+$, then

$$\frac{a+b+c+d}{4} \geq \sqrt[4]{abcd},$$

where the equality holds if and only if $a = b = c = d$.

C. Cauchy's inequality in 2 and 3 dimensions

(1) If a_1, a_2, b_1, $b_2 \in \mathbf{R}$, then

$$(a_1^2 + a_2^2)(b_1^2 + b_2^2) \geq (a_1b_1 + a_2b_2)^2,$$

where the equality holds if and only if $a_1b_2 = a_2b_1$.

(2) If a_1, a_2, a_3, b_1, b_2, $b_3 \in \mathbf{R}$, then

$$(a_1^2 + a_2^2 + a_3^2)(b_1^2 + b_2^2 + b_3^2) \geq (a_1b_1 + a_2b_2 + a_3b_3)^2,$$

where the equality holds if and only if $b_i = 0$ ($i = 1, 2$, and 3) or there exists some real number k, such that $a_i = kb_i$ ($i = 1, 2$, and 3).

D. The triangle inequality

(1) The triangle inequality for absolute values

If a, $b \in \mathbf{R}$, then

$$|a + b| \leq |a| + |b|,$$

where the equality holds if and only if $ab \geq 0$.

Corollary 1. *If a, $b \in \mathbf{R}$, then*

$$|a| - |b| \leq |a \pm b| \leq |a| + |b|.$$

Corollary 2. *If a, b, $c \in \mathbf{R}$, then*

$$|a - c| \leq |a - b| + |b - c|,$$

where the equality holds if and only if $(a - b)(b - c) \geq 0$.

(2) The triangle inequality in 2 dimensions

If x_1, x_2, y_1, $y_2 \in \mathbf{R}$, then

$$\sqrt{x_1^2 + y_1^2} + \sqrt{x_2^2 + y_2^2} \geq \sqrt{(x_1 - x_2)^2 + (y_1 - y_2)^2}.$$

E. Two remarks

(1) The AM-GM inequality, Cauchy's inequality, and the triangle inequality can be generalized to the case of more variables.

(2) When solving problems we usually apply variants of the above basic inequalities. For example, we may use the following variant of the AM-GM inequality for 2 variables: $\left(\frac{a+b}{2}\right)^2 \geq ab$.

9.2 Illustrative Examples

We start from the proof of the AM-GM inequality for 3 and 4 variables.

Example 1. Let a, b, c, and d be positive numbers. Show that

(1) $\frac{a+b+c}{3} \geq \sqrt[3]{abc}$;
(2) $\frac{a+b+c+d}{4} \geq \sqrt[4]{abcd}$.

And point out the condition for each inequality to be an equality.

Analysis. We may use the inequality $\frac{a+b}{2} \geq \sqrt{ab}$ for positive numbers a and b to complete the proof. Observe that the inequality (2) can be written as

$$\frac{\frac{a+b}{2} + \frac{c+d}{2}}{2} \geq \sqrt{\sqrt{ab} \cdot \sqrt{cd}},$$

so we can prove (2) first.

Solution. (2) Since a, b, c, and d are positive,

$$\frac{a+b}{2} \geq \sqrt{ab}, \quad \frac{c+d}{2} \geq \sqrt{cd},$$

so

$$\frac{a+b+c+d}{4} \geq \frac{\sqrt{ab}+\sqrt{cd}}{2} \geq \sqrt{\sqrt{ab} \cdot \sqrt{cd}} = \sqrt[4]{abcd}.$$

If each inequality above holds as an equality, then

$$\begin{cases} a = b(> 0), \\ c = d(> 0), \\ \sqrt{ab} = \sqrt{cd}. \end{cases}$$

Solving the equations we get $a = b = c = d$, which is the sufficient and necessary condition for the equality to hold.

(1) We can apply the inequality (2) to prove (1).
Since a, b, $c > 0$,

$$\frac{\mathrm{a}+\mathrm{b}+\mathrm{c}+\sqrt[3]{abc}}{4} \geq \sqrt[4]{abc\sqrt[3]{abc}} = \sqrt[3]{abc},$$

and after simplification we get $\frac{a+b+c}{3} \geq \sqrt[3]{abc}$.

Here the equality holds if and only if $a = b = c = \sqrt[3]{abc}(>0)$, which is the same as $a = b = c$.

Remark. The proof in (2) applied the additivity and transitivity of inequalities. We should bear in mind that the proof of an inequality should always be based on properties of inequalities, theorems, and known or proven facts. We can also prove (2) based on (1). The interested readers may have a try.

Example 2. Let $x_1, x_2, \ldots, x_n$ be positive numbers. Show that

$$\frac{x_1^2}{x_2} + \frac{x_2^2}{x_3} + \cdots + \frac{x_{n-1}^2}{x_n} + \frac{x_n^2}{x_1} \geq x_1 + x_2 + \cdots + x_n.$$

Analysis. Don't be intimidated by the fractions on the left side. Associating the AM-GM inequality for two variables, we can do the following trick: We add $x_1 + x_2 + \cdots + x_n$ on both sides, so the inequality is equivalent to

$$\left(\frac{x_1^2}{x_2} + x_2\right) + \left(\frac{x_2^2}{x_3} + x_3\right) + \cdots + \left(\frac{x_n^2}{x_1} + x_1\right) \geq 2x_1 + 2x_2 + \cdots + 2x_n.$$

The proof of this inequality can be divided into n inequalities of the form $\frac{a^2}{b} + b \geq 2a$ (a and b are positive). This is one example of mathematical proof where a little trick goes a long way.

Solution. Since $x_1, x_2, \ldots, x_n > 0$, by the AM-GM inequality,

$$\frac{x_1^2}{x_2} + x_2 \geq 2\sqrt{\frac{x_1^2}{x_2} \cdot x_2} = 2x_1,$$

$$\frac{x_2^2}{x_3} + x_3 \geq 2x_2,$$

$$\cdots$$

$$\frac{x_{n-1}^2}{x_n} + x_n \geq 2x_{n-1},$$

$$\frac{x_n^2}{x_1} + x_1 \geq 2x_n.$$

We add both sides of all the inequalities above, and simplify to get

$$\frac{x_1^2}{x_2} + \frac{x_2^2}{x_3} + \cdots + \frac{x_{n-1}^2}{x_n} + \frac{x_n^2}{x_1} \geq x_1 + x_2 + \cdots + x_n.$$

Remark. This problem has many proofs. The interested readers may search for different proofs after reading subsequent chapters.

Example 3. Let a, b, and c be nonnegative real numbers such that $a+b+c=1$. Show that

$$(1+a)(1+b)(1+c) \geq 8(1-a)(1-b)(1-c).$$

Analysis. From the condition, $1+a=(1-b)+(1-c)$. Since

$$(1-b)+(1-c) \geq 2\sqrt{(1-b)(1-c)},$$

the subsequent steps follow naturally.

Solution. Since a, b, and c are nonnegative and $a+b+c=1$, we have $0 \leq a, b, c \leq 1$. Hence,

$$\begin{aligned}
1+a &= 1-b+1-c \geq 2\sqrt{(1-b)(1-c)}, \\
1+b &= 1-a+1-c \geq 2\sqrt{(1-a)(1-c)}, \\
1+c &= 1-a+1-b \geq 2\sqrt{(1-a)(1-b)}.
\end{aligned}$$

Therefore, multiplying both sides of the inequalities above gives that $(1+a)(1+b)(1+c) \geq 8(1-a)(1-b)(1-c)$.

Remark. The proof above used the property of "inequalities of the same direction with both sides positive can be multiplied."

Example 4. Let a, b, and c be positive such that $a > b^2$ and $b > c^2$. Find the maximum value of the expression $(a-b^2)(b-c^2)(c-a^2)$.

Solution. If $c \leq a^2$, then $(a-b^2)(b-c^2)(c-a^2) \leq 0$.

If $c > a^2$, then by the AM-GM inequality,

$$\begin{aligned}
(a-b^2)(b-c^2)(c-a^2) &\leq \left(\frac{(a-b^2)+(b-c^2)+(c-a^2)}{3}\right)^3 \\
&= \left(\frac{(a-a^2)+(b-b^2)+(c-c^2)}{3}\right)^3.
\end{aligned}$$

For any real number x, we have $x-x^2=-\left(x-\frac{1}{2}\right)^2+\frac{1}{4} \leq \frac{1}{4}$, so

$$(a-b^2)(b-c^2)(c-a^2) \leq \frac{1}{64},$$

where the equality holds when $a=b=c=\frac{1}{2}$.

Therefore, the maximum value of $(a-b^2)(b-c^2)(c-a^2)$ is $\frac{1}{64}$.

Example 5. Show that for three nonnegative real numbers a, b, and c that are not all equal,

$$\frac{(a-bc)^2+(b-ca)^2+(c-ab)^2}{(a-b)^2+(b-c)^2+(c-a)^2} \geq \frac{1}{2},$$

and find the necessary and sufficient condition for the equality to hold.

Solution. For such a, b, and c, the inequality in the problem is equivalent to

$$2(a-bc)^2+2(b-ca)^2+2(c-ab)^2 \geq (a-b)^2+(b-c)^2+(c-a)^2.$$

It can also be written as

$$2a^2b^2+2b^2c^2+2c^2a^2-12abc \geq -2ab-2bc-2ca,$$

or equivalently $a^2b^2+b^2c^2+c^2a^2+ab+bc+ca \geq 6abc$. ①

Since a^2b^2, b^2c^2, c^2a^2, ab, bc, $ca \geq 0$, by the AM-GM inequality,

$$a^2b^2+b^2c^2+c^2a^2+ab+bc+ca \geq 6\sqrt[6]{a^2b^2\cdot b^2c^2\cdot c^2a^2\cdot ab\cdot bc\cdot ca} = 6abc, ②$$

and the result is proven.

Note that the equality in ② holds if and only if $a^2b^2 = b^2c^2 = c^2a^2 = ab = bc = ca$.

If $abc \neq 0$, then $ab = bc = ca$, and hence $a = b = c$, a contradiction.

If $abc = 0$, then necessarily $ab = bc = ca = 0$. Since a, b, and c are not all 0, we may assume that $a \neq 0$. Then $b = c = 0$, in which case the equality in ② holds obviously. The other two cases are similar. Therefore, the equality holds if and only if one of a, b, and c is positive and the other two are 0.

Example 6. Suppose real numbers a, b, and c satisfy the inequalities

$$|a+b+c| \leq 1, \quad |a-b+c| \leq 1, \quad |c| \leq 1.$$

Show that $|4a+2b+c| \leq 7$.

Solution. By the triangle inequality for absolute values,

$$\begin{aligned} |4a+2b+c| &= |3(a+b+c)+(a-b+c)-3c| \\ &\leq |3(a+b+c)|+|a-b+c|+|-3c| \\ &= 3|a+b+c|+|a-b+c|+3|c| \\ &\leq 3+1+3=7. \end{aligned}$$

Example 7. Let $a \in (0,1)$ and $b \in (0,1)$. Show that

$$\sqrt{a^2+b^2}+\sqrt{(1-a)^2+b^2}+\sqrt{a^2+(1-b)^2}+\sqrt{(1-a)^2+(1-b)^2} \geq 2\sqrt{2}.$$

Analysis. The structure of this inequality can be associated with the triangle inequality in 2 dimensions.

Solution. By the triangle inequality in 2 dimensions,

$$\begin{aligned}
&\sqrt{a^2+b^2}+\sqrt{(1-a)^2+b^2}+\sqrt{a^2+(1-b)^2}+\sqrt{(1-a)^2+(1-b)^2} \\
&\quad = \left(\sqrt{(a-1)^2+(b-1)^2}+\sqrt{a^2+b^2}\right) \\
&\qquad + \left(\sqrt{(a-1)^2+b^2}+\sqrt{a^2+(b-1)^2}\right) \\
&\quad \geq \sqrt{(a-1-a)^2+(b-1-b)^2}+\sqrt{(a-1-a)^2+(b-b+1)^2} \\
&\quad = \sqrt{2}+\sqrt{2} = 2\sqrt{2}.
\end{aligned}$$

Example 8. Let $x, y, z \in \left[\frac{1}{2}, 2\right]$, and let a, b, c be a permutation of x, y, z. Show that

$$\frac{60a^2-1}{4xy+5z}+\frac{60b^2-1}{4yz+5x}+\frac{60c^2-1}{4zx+5y} \geq 12.$$

Solution. Since $(x-2)\left(y-\frac{1}{2}\right) \leq 0$,

$$xy-\frac{1}{2}x-2y+1 \leq 0. \text{①}$$

Since $\left(x-\frac{1}{2}\right)(y-2) \leq 0$,

$$xy-2x-\frac{1}{2}y+1 \leq 0. \text{②}$$

By ① and ②,

$$4xy+5z \leq 5(x+y+z)-4,$$

and similarly

$$4yz+5x \leq 5(x+y+z)-4,$$
$$4zx+5y \leq 5(x+y+z)-4.$$

Thus, it suffices to show that $\frac{60(a^2+b^2+c^2)-3}{5(x+y+z)-4} \geq 12$, namely $60(x^2+y^2+z^2)-3 \geq 60(x+y+z)-48$, which is equivalent to $(2x-1)^2+(2y-1)^2+(2z-1)^2 \geq 0$.

Therefore, the result holds.

Example 9. For all real numbers a and b, show that there exist real numbers x and y in the interval $[0, 1]$ such that

$$|xy - ax - by| \geq \frac{1}{3}. \text{①}$$

Also, if the number $\frac{1}{3}$ is replaced by $\frac{1}{2}$ or 0.33334, does the result still hold?

Analysis. It seems hard to find such x and y that make the inequality ① hold for all a and b. If we consider using the contradiction method, then we only need to show that there do not exist real numbers a and b such that $|xy - ax - by| < \frac{1}{3}$ for all $x, y \in [0, 1]$. Since we can take arbitrary values of x and y, this allows us much freedom when dealing with the problem.

Solution. We use the contradiction method. Suppose there exist real numbers a and b such that $|xy - ax - by| < \frac{1}{3}$ for all $x, y \in [0, 1]$. Then choose $(x, y) = (1, 0)$, $(0, 1)$, and $(1, 1)$ in succession, and we get

$$|a| < \frac{1}{3}, \quad |b| < \frac{1}{3}, \quad |1 - a - b| < \frac{1}{3}.$$

However, by $|a| < \frac{1}{3}$ and $|b| < \frac{1}{3}$,

$$|1 - a - b| \geq 1 - |a + b| \geq 1 - |a| - |b| > \frac{1}{3},$$

which leads to a contradiction. Therefore, the result holds.

Next, we show that if $\frac{1}{3}$ is replaced by any number c greater than $\frac{1}{3}$, then the proposition in the problem does not hold. Equivalently, we show that there exist real numbers a and b such that for all x, $y \in [0, 1]$,

$$|xy - ax - by| < c.$$

In fact, we can choose $a = b = \frac{1}{3}$. Since

$$xy - ax - by = \frac{1}{3}[xy - x(1 - y) - y(1 - x)],$$

for all x, $y \in [0, 1]$,

$$|xy - ax - by| \leq \frac{1}{3}\max[xy, x(1 - y) + y(1 - x)].$$

Also, by $0 \leq xy \leq 1$ and $0 \leq x(1-y) + y(1-x) \leq x + (1-x) = 1$, we get

$$|xy - ax - by| \leq \frac{1}{3} < c.$$

Therefore, the number $\frac{1}{3}$ is optimal in this problem.

9.3 Exercises

Group A

1. Suppose x and y lie in the interval $(-2, 2)$ such that $xy = -1$ and $\frac{4}{4-x^2} + \frac{9}{9-y^2} = \frac{12}{5}$. Then the values of x and y are ____________.
2. Let a, b, c, d, m, and n be positive numbers, and

$$P = \sqrt{ab} + \sqrt{cd}, \quad Q = \sqrt{ma + nc} \cdot \sqrt{\frac{b}{m} + \frac{d}{n}}.$$

   Then the necessary and sufficient condition for $Q > P$ is ____________.
3. The maximum value of the function $f(x) = 2\sqrt{x-3} + \sqrt{5-x}$ is ____________.
4. Let a, b, c, x, y, and z be real numbers such that $a^2 + b^2 + c^2 = 25$, $x^2 + y^2 + z^2 = 36$, and $ax + by + cz = 30$. Then the value of $\frac{a+b+c}{x+y+z}$ is ____________.
5. Let A, B, and C denote the three angles of a triangle ΔABC (in radian). Show that

$$\frac{1}{A} + \frac{1}{B} + \frac{1}{C} \geq \frac{9}{\pi}, \quad A^2 + B^2 + C^2 \geq \frac{\pi^2}{3}.$$

6. Find the real solutions to the following system of equations:

$$\begin{cases} (x-2)^2 + \left(y + \frac{3}{2}\right)^2 + (z-6)^2 = 64, \\ (x+2)^2 + \left(y - \frac{3}{2}\right)^2 + (z+6)^2 = 25. \end{cases}$$

7. Suppose a, b, and c are positive real numbers such that $(1+a)(1+b)(1+c) = 8$. Show that $abc \leq 1$.
8. Let $x, y, z \in \mathbf{R}$. Show that

$$\sqrt{x^2 + xy + y^2} + \sqrt{x^2 + xz + z^2} \geq \sqrt{y^2 + yz + z^2}.$$

Group B

9. Let $a, b, c \in \mathbf{R}_+$ and $abc = 1$. Show that

$$\frac{1}{a^3(b+c)} + \frac{1}{b^3(a+c)} + \frac{1}{c^3(a+b)} \geq \frac{3}{2}.$$

10. Suppose $x + y + z = 0$. Show that

$$6(x^3 + y^3 + z^3)^2 \leq (x^2 + y^2 + z^2)^3.$$

11. Let a, b, $c \in \mathbf{R}$ with the property that $|ax^2 + bx + c| \leq 1$ for all $-1 \leq x \leq 1$. Show that

$$|cx^2 + bx + a| \leq 2 \quad \text{for all } -1 \leq x \leq 1.$$

12. Let $a_1, a_2, \ldots, a_n$ be positive real numbers such that $a_1 a_2 \ldots a_k \geq 1$ for each $1 \leq k \leq n$. Show that

$$\frac{1}{1+a_1} + \frac{2}{(1+a_1)(1+a_2)} + \cdots + \frac{n}{(1+a_1)\cdots(1+a_n)} < 2.$$

Chapter 10

Solutions of Inequalities

10.1 Key Points of Knowledge and Basic Methods

A. Solutions to linear inequalities of one variable

Any linear inequality for x can be simplified to the standard form $ax > b$ or $ax < b$. For the inequality $ax > b$:

If $a > 0$, then its solution set is $\left\{x \,\middle|\, x > \frac{b}{a}\right\}$.
If $a < 0$, then its solution set is $\left\{x \,\middle|\, x < \frac{b}{a}\right\}$.
If $a = 0$ and $b < 0$, then its solution set is $\mathbf{R}$; if $a = 0$ and $b \geq 0$, then its solution set is the empty set $\emptyset$.

The solution set to $ax < b$ (or $ax \geq b$, or $ax \leq b$) involves a similar discussion.

B. Solutions to quadratic inequalities of one variable

When solving quadratic inequalities, we first change the inequality into the standard form $ax^2 + bx + c > 0$ (or < 0) or $ax^2 + bx + c \geq 0$ (or ≤ 0), where $a \neq 0$.

Let Δ be the discriminant of the quadratic equation $ax^2 + bx + c = 0$. If $\Delta \geq 0$, let x_1 and x_2 ($x_1 \leq x_2$) be its real roots.

Assume that $a > 0$. Then the solution set to the inequality is shown in the following table:

	$ax^2 + bx + c > 0$	$ax^2 + bx + c \geq 0$	$ax^2 + bx + c < 0$	$ax^2 + bx + c \leq 0$
$\Delta > 0$	$\{x \| x < x_1 \text{ or } x > x_2\}$	$\{x \| x \leq x_1 \text{ or } x \geq x_2\}$	$\{x \| x_1 < x < x_2\}$	$\{x \| x_1 \leq x \leq x_2\}$
$\Delta = 0$	$\{x \| x \neq -\frac{b}{2a},\ x \in \mathbf{R}\}$	$\mathbf{R}$	$\emptyset$	$\left\{x \,\middle\|\, x = -\frac{b}{2a}\right\}$
$\Delta < 0$	$\mathbf{R}$	$\mathbf{R}$	$\emptyset$	$\emptyset$

There is an intuitive explanation for the solution set of a quadratic inequality using the graph of the quadratic function.

C. Solutions to univariate inequalities of higher degrees

When solving univariate inequalities of higher degrees, we usually factorize the left side of the inequality

$$a_n x^n + a_{n-1}x^{n-1} + \cdots + a_1 x + a_0 > 0 \text{ (or } < 0)①$$

(assuming that $a_n > 0$) into the product of several irreducible linear and quadratic factors. Removing the irreducible quadratic factors (they have constant signs) and the linear factors with even multiples, we reduce the inequality ① to

$$(x - x_1)(x - x_2)\cdots(x - x_m) > 0 \text{ (or } < 0),$$

where $x_1 < x_2 < \cdots < x_m$. Let

$$f(x) = (x - x_1)(x - x_2)\cdots(x - x_m).$$

Then the graph of $f(x)$ divides the x-axis into $m + 1$ intervals $(-\infty, x_1)$, $(x_1, x_2), \ldots, (x_m, +\infty)$. Since $a_n > 0$, the function is positive for $x \in (x_m, +\infty)$. Starting from the right, the function $f(x)$ changes its sign every time x goes to the next interval. Therefore, the solution set to ① is the union of the intervals where $f(x)$ takes positive (or negative) values. Note that we also need to examine whether the multiple roots eliminated in the first step belong to the solution set.

The discussion for $a_n x^n + a_{n-1}x^{n-1} + \cdots + a_1 x + a_0 \geq 0$ (or ≤ 0) is similar.

D. Solutions to inequalities with absolute values

$$|f(x)| < g(x) \Leftrightarrow -g(x) < f(x) < g(x);$$
$$|f(x)| > g(x) \Leftrightarrow f(x) < -g(x) \text{ or } f(x) > g(x);$$
$$|f(x)| > |g(x)| \Leftrightarrow f^2(x) > g^2(x).$$

For more complex inequalities with absolute values, we can solve them by using the definition of absolute values or its geometric interpretation.

E. Solutions to fractional inequalities

The fundamental idea for solving fractional inequalities is to transform them into univariate polynomial inequalities (possibly with higher degrees).

$$\frac{f(x)}{g(x)} > 0 \text{ (or } < 0) \Leftrightarrow f(x)g(x) > 0 \text{ (or } < 0);$$

$$\frac{f(x)}{g(x)} \geq 0 \text{ (or } \leq 0) \Leftrightarrow \begin{cases} f(x)g(x) \geq 0 \text{ (or } \leq 0), \\ g(x) \neq 0. \end{cases}$$

F. Solutions to irrational inequalities

The fundamental idea for solving irrational inequalities is to transform them into rational inequalities. In the case of univariate inequalities with square roots, there are three basic types:

$$\sqrt{f(x)} > \sqrt{g(x)} \Leftrightarrow f(x) > g(x) \geq 0;$$

$$\sqrt{f(x)} > g(x) \Leftrightarrow \begin{cases} f(x) \geq 0, \\ g(x) < 0, \end{cases} \text{or} \quad \begin{cases} g(x) \geq 0, \\ f(x) > [g(x)]^2; \end{cases}$$

$$\sqrt{f(x)} < g(x) \Leftrightarrow \begin{cases} f(x) \geq 0, \\ g(x) > 0, \\ f(x) < [g(x)]^2. \end{cases}$$

G. Solutions to exponential and logarithmic inequalities

When solving exponential or logarithmic inequalities, the fundamental idea is to use the monotonicity of exponential and logarithmic functions to reduce them to polynomial inequalities.

$$a^{f(x)} > a^{g(x)}(a > 1) \Leftrightarrow f(x) > g(x);$$

$$a^{f(x)} > a^{g(x)}(0 < a < 1) \Leftrightarrow f(x) < g(x);$$

$$\log_a f(x) > \log_a g(x)(a > 1) \Leftrightarrow f(x) > g(x) > 0;$$

$$\log_a f(x) > \log_a g(x)(0 < a < 1) \Leftrightarrow g(x) > f(x) > 0.$$

H. For inequalities that are not equivalent to the standard forms above, we may solve them with properties of inequalities, known theorems, graphs, changing variables, etc.

10.2 Illustrative Examples

Example 1. Solve the following inequalities of higher degrees:

(1) $x^5 - x^4 > x - 1$;
(2) $(x+1)^2(x-1)(x-2)(x-3) \leq 0$.

Solution. (1) $x^5 - x^4 > x - 1$

$$\Leftrightarrow x^5 - x^4 - x + 1 > 0$$

$$\Leftrightarrow (x^4 - 1)(x - 1) > 0$$

$$\Leftrightarrow (x+1)(x-1)^2(x^2+1) > 0$$

$$\Leftrightarrow (x+1)(x-1)^2 > 0.$$

Let $f(x) = (x+1)(x-1)^2$. If $f(x) = 0$, then $x = -1$ or 1. The signs of $f(x)$ can be sketched as in Figure 10.1. Therefore, the solution set to the inequality is $\{x | -1 < x < 1 \text{ or } x > 1\}$.

(2) Let $f(x) = (x+1)^2(x-1)(x-2)(x-3)$, whose roots are $x = -1, 1, 2$, and 3. The roots divide the real line into five intervals, as in Figure 10.2.

Therefore, the solution set to the inequality is $\{x | x \leq 1 \text{ or } 2 \leq x \leq 3\}$.

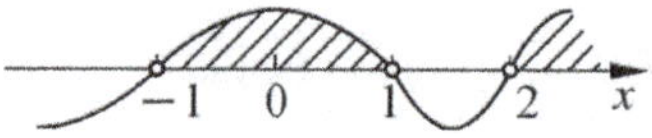

Figure 10.1

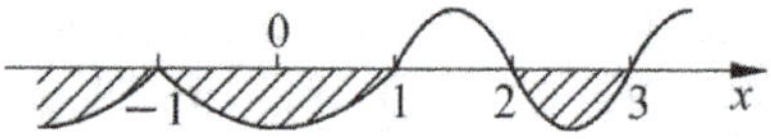

Figure 10.2

Remark. When solving polynomial inequalities of higher degrees, we should carefully examine whether each root of the polynomial belongs to the solution set. For example, in (2) $x = -1$ is in the solution set, while in (1) $x = 1$ is not.

Example 2. Solve the following fractional inequalities:

$$(1)\ \frac{x^3 + 2x^2 - 3x - 16}{x^2 + x - 12} > 1; \quad (2)\ \frac{(x-1)^2(x+1)}{(x-2)(x-3)} \leq 0.$$

Figure 10.3

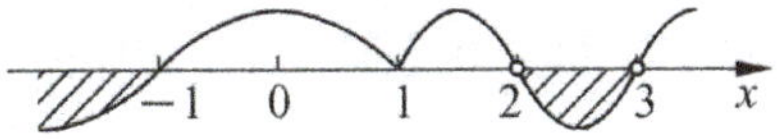

Figure 10.4

Solution. (1) $\dfrac{x^3+2x^2-3x-16}{x^2+x-12} > 1$

$$\Leftrightarrow \frac{x^3+x^2-4x-4}{x^2+x-12} > 0$$

$$\Leftrightarrow \frac{(x+1)(x+2)(x-2)}{(x+4)(x-3)} > 0$$

$$\Leftrightarrow (x+4)(x+2)(x+1)(x-2)(x-3) > 0.$$

Hence, the solution set to the inequality is $\{x | -4 < x < -2 \text{ or } -1 < x < 2 \text{ or } x > 3\}$ (Figure 10.3).

(2) $\dfrac{(x-1)^2(x+1)}{(x-2)(x-3)} \le 0$

$$\Leftrightarrow (x+1)(x-1)^2(x-2)(x-3) \le 0 \ (x \ne 2, 3).$$

Therefore, the solution set to the inequality is $\{x | x \le -1 \text{ or } x = 1 \text{ or } 2 < x < 3\}$ (Figure 10.4).

Remark. When solving fractional inequalities, we should always remember that the denominator cannot be 0.

Example 3. Solve the inequality $\frac{7x}{\sqrt{1+x^2}} + \frac{2-2x^2}{1+x^2} > 0$.

Analysis. Obviously $1 + x^2 > 0$, so the inequality above is equivalent to $7x\sqrt{1+x^2} > 2x^2 - 2$. If $x > 0$, then it reduces to the form $\sqrt{f(x)} > g(x)$; if $x < 0$, then it reduces to the form $\sqrt{f(x)} < g(x)$.

Solution 1. Since $1 + x^2 > 0$, the original inequality is equivalent to

$$7x\sqrt{1+x^2} > 2x^2 - 2.$$

(1) If $x \ge 0$, then since $7x\sqrt{1+x^2} \ge 7x \cdot x \ge 2x^2 > 2x^2 - 2$, the inequality holds constantly.

(2) If $x < 0$, then the inequality is equivalent to

$$\begin{cases} x < 0, \\ 2x^2 - 2 < 0, \\ (7x\sqrt{1+x^2})^2 < (2x^2 - 2)^2 \end{cases} \Leftrightarrow \begin{cases} -1 < x < 0, \\ 45x^4 + 57x^2 - 4 < 0. \end{cases}$$

Thus, $(3x^2 + 4)(15x^2 - 1) < 0$, or equivalently $15x^2 - 1 < 0$, and so

$$-\frac{\sqrt{15}}{15} < x < \frac{\sqrt{15}}{15}.$$

Therefore, combining (1) and (2), we obtain the solution set $\left\{x \,\middle|\, x > -\frac{\sqrt{15}}{15}\right\}$ to the original inequality.

Remark. When solving inequalities, we need to keep in mind the characteristics of the inequality after each transformation, and find the best way to solve it based on its characteristics. For example, in the solution above we noted that $7x\sqrt{1+x^2} \geq 2x^2$ when $x \geq 0$, so $x \geq 0$ is contained in the solution set. Similarly, if we notice the structure of the expressions $\sqrt{1+x^2}$ and $\frac{1-x^2}{1+x^2}$, we can also consider the change of variables $x = \tan t \left[t \in \left(-\frac{\pi}{2}, \frac{\pi}{2}\right)\right]$, and get the following alternative solution.

Solution 2. Let $x = \tan t \left[t \in \left(-\frac{\pi}{2}, \frac{\pi}{2}\right)\right]$. Then the inequality is equivalent to

$$\begin{aligned} &\frac{7\tan t}{\sqrt{1+\tan^2 t}} + \frac{2(1-\tan^2 t)}{1+\tan^2 t} > 0 \\ &\Leftrightarrow 7\sin t + 2\cos 2t > 0 \\ &\Leftrightarrow 4\sin^2 t - 7\sin t - 2 < 0 \\ &\Leftrightarrow (\sin t - 2)(4\sin t + 1) < 0. \end{aligned}$$

Since $\sin t - 2 < 0$, it is equivalent to

$$4\sin t + 1 > 0,$$

hence $-\frac{1}{4} < \sin t < 1$.

Since $t \in \left(-\frac{\pi}{2}, \frac{\pi}{2}\right)$, we get that $t \in \left(-\arcsin\frac{1}{4}, \frac{\pi}{2}\right)$, and so

$$\tan t \in \left(-\tan\left(\arcsin\frac{1}{4}\right), +\infty\right).$$

Therefore, the solution set to the inequality is $\left(-\frac{\sqrt{15}}{15}, +\infty\right)$.

Remark. We may further notice that $\frac{x^2}{1+x^2} = \left(\frac{x}{\sqrt{1+x^2}}\right)^2$, which leads to the third solution.

Solution 3. Let $t = \frac{x}{\sqrt{1+x^2}}(< 1)$. Then the inequality reduces to

$$7t + 2 - 4t^2 > 0,$$

$$\Leftrightarrow 4t^2 - 7t - 2 < 0,$$

$$\Leftrightarrow (4t+1)(t-2) < 0,$$

$$\Leftrightarrow -\tfrac{1}{4} < t < 2.$$

Thus, $-\frac{1}{4} < \frac{x}{\sqrt{1+x^2}} < 2$. Since $\frac{x}{\sqrt{1+x^2}} < 2$ always holds, we only need to solve $\frac{x}{\sqrt{1+x^2}} > -\frac{1}{4}$, namely $\sqrt{1+x^2} > -4x$. ①

Apparently ① holds when $x \geq 0$. If $x < 0$, then we can take the square of both sides of ① to get $x^2 < \frac{1}{15}$, which gives that $-\frac{\sqrt{15}}{15} < x < 0$.

Therefore, the solution set to the original inequality is $\left\{x \,\middle|\, x > -\frac{\sqrt{15}}{15}\right\}$.

Example 4. Solve the following inequalities:

(1) $4^x + 4^{-x} + 5 > 2^{x+2} + 2^{2-x}$;
(2) $x^{3-3\log_2 x-(\log_2 x)^2} - \frac{1}{8}x^2 > 0$.

Analysis. (1) This is an exponential inequality with different bases. However, we note that $4=2^2$, so our solution begins by unifying bases.

(2) This inequality involves both exponentials and logarithms. Considering the relationship between exponentials and logarithms, we can transform the inequality into $x^{3-3\log_2 x-(\log_2 x)^2} > \frac{1}{8}x^2$, and then take the logarithm with base 2 on both sides. Next, we only need to solve the logarithmic inequality.

Solution. (1) The inequality is equivalent to

$$(2^x + 2^{-x})^2 - 4(2^x + 2^{-x}) + 3 > 0,$$

that is, $(2^x + 2^{-x} - 1)(2^x + 2^{-x} - 3) > 0$.

Since $2^x + 2^{-x} \geq 2\sqrt{2^x \cdot 2^{-x}} = 2$, we have $2^x + 2^{-x} - 1 > 0$, so the original inequality reduces to

$$2^x + 2^{-x} - 3 > 0,$$

namely $(2^x)^2 - 3\cdot 2^x + 1 > 0$. Solving the inequality, we have $0 < 2^x < \frac{3-\sqrt{5}}{2}$ or $2^x > \frac{3+\sqrt{5}}{2}$, or equivalently $x < \log_2 \frac{3-\sqrt{5}}{2}$ or $x > \log_2 \frac{3+\sqrt{5}}{2}$.

Therefore, the solution set to the original inequality is

$$\left\{x \,\middle|\, x < \log_2 \frac{3-\sqrt{5}}{2} \text{ or } x > \log_2 \frac{3+\sqrt{5}}{2}\right\}.$$

(2) Apparently $x > 0$. Dividing x^2 on both sides, we see that the inequality is equivalent to

$$x^{1-3\log_2 x-(\log_2 x)^2} > \frac{1}{8}.$$

Then we take the logarithm with base 2 on both sides, and get

$$\log_2 x \cdot [1 - 3\log_2 x - (\log_2 x)^2] > -3,$$

or equivalently $(\log_2 x)^3 + 3(\log_2 x)^2 - \log_2 x - 3 < 0$, from which $(\log_2 x + 3)(\log_2 x + 1)(\log_2 x - 1) < 0$.

Hence, $\log_2 x < -3$ or $-1 < \log_2 x < 1$.

Therefore, the solution set to the original inequality is $\{x | 0 < x < \frac{1}{8}$ or $\frac{1}{2} < x < 2\}$.

Example 5. Solve the inequality $|x^2 - 4| < 3|x|$.

Solution. $|x^2 - 4| < 3|x|$

$$\Leftrightarrow -3|x| < x^2 - 4 < 3|x|$$

$$\Leftrightarrow \begin{cases} |x|^2 + 3|x| - 4 > 0, \\ |x|^2 - 3|x| - 4 < 0 \end{cases}$$

$$\Leftrightarrow \begin{cases} (|x| + 4)(|x| - 1) > 0, \\ (|x| + 1)(|x| - 4) < 0 \end{cases}$$

$$\Leftrightarrow 1 < |x| < 4$$

$$\Leftrightarrow -4 < x < -1 \text{ or } 1 < x < 4.$$

Therefore, the solution set to the inequality is $\{x | -4 < x < -1$ or $1 < x < 4\}$.

Remark. In this problem we can also remove the absolute value signs by squaring both sides: $(x^2 - 4)^2 < 9x^2 \Leftrightarrow (x + 4)(x + 1)(x - 1)(x - 4) < 0 \Leftrightarrow -4 < x < -1$ or $1 < x < 4$.

In addition, the inequality can also be solved with graphs. Consider the functions $y = |x^2 - 4|$ and $y = 3|x|$, and draw their graphs in the same coordinate plane (as in Figure 10.5). It is easy to see that $|x^2 - 4| = 3|x|$

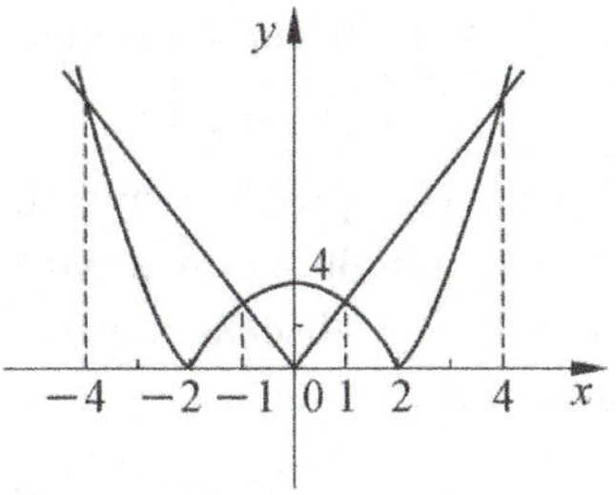

Figure 10.5

when $x = \pm 1$ and ± 4, where the two graphs intersect. The graph of $y = 3|x|$ lies above the graph of $y = |x^2 - 4|$ exactly when $-4 < x < -1$ or $1 < x < 4$. Therefore, the solution set is $\{x | -4 < x < -1 \text{ or } 1 < x < 4\}$.

The next example involves solving an inequality with a parameter.

Example 6. Let $a > 0$. Solve the inequality for x : $\sqrt{2ax - a^2} > 1 - x$.

Solution. The inequality is equivalent to

$$\text{(I)} \begin{cases} 1 - x \geq 0, \\ 2ax - a^2 > (1 - x)^2 \end{cases} \quad \text{or} \quad \text{(II)} \begin{cases} 2ax - a^2 \geq 0, \\ 1 - x < 0. \end{cases}$$

In (I) we have

$$\begin{cases} x \leq 1, \\ x^2 - 2(a + 1)x + 1 + a^2 < 0, \end{cases}$$

so $\begin{cases} x \leq 1, \\ a + 1 - \sqrt{2a} < x < a + 1 + \sqrt{2a}. \end{cases}$

In (II) we have

$$\begin{cases} x > 1, \\ x \geq \dfrac{a}{2}. \end{cases}$$

(1) If $0 < a \leq 2$, then $a + 1 + \sqrt{2a} > 1$, and $a + 1 - \sqrt{2a} \leq 1$ with $\frac{a}{2} \leq 1$.
If (I) holds, then $a + 1 - \sqrt{2a} < x \leq 1$; if (II) holds, then $x > 1$.
Hence, $x > a + 1 - \sqrt{2a}$.

(2) If $a > 2$, then $a + 1 + \sqrt{2a} > a + 1 - \sqrt{2a} > 1$ and $\frac{a}{2} > 1$.
If (I) holds, then $x \in \emptyset$; if (II) holds, then $x \geq \frac{a}{2}$.
Hence, $x \geq \frac{a}{2}$.

To summarize, if $0 < a \leq 2$, then the solution set is $\{x|x > a+1-\sqrt{2a}\}$; if $a > 2$, then the solution set is $\left\{x \,|x \geq \frac{a}{2}\right\}$.

Remark. The inequality in this problem has the form $\sqrt{f(x)} > g(x)$. It is not hard to get started, but it requires patience and insight to find a rigorous and convenient way to discuss, based on the value of a. This problem can also be solved with the help of graphs.

Example 7. Find all real numbers x and y such that the following system of inequalities holds:

$$\begin{cases} 4^{-x} + 27^{-y} = \dfrac{5}{6}, \\ \log_{27} y - \log_4 x \geq \dfrac{1}{6}, \\ 27^y - 4^x \leq 1. \end{cases}$$

Solution. Let $4^x = a$ and $27^y = b$ with $a, b > 0$. Then

$$\begin{cases} \dfrac{1}{a} + \dfrac{1}{b} = \dfrac{5}{6}, & ① \\ b - a \leq 1. & ② \end{cases}$$

By ① we have $b = \frac{6a}{5a-6}$ and $5a - 6 > 0$. Substituting into ② we get $\frac{6a}{5a-6} - a \leq 1$, or equivalently

$$\frac{5a^2 - 7a - 6}{5a - 6} \geq 0.$$

Solving the inequality gives $a \geq 2$, namely $4^x \geq 2$, so $x \geq \frac{1}{2}$.

Further by ①, $a = \frac{6b}{5b-6}$ and $5b - 6 > 0$. Substituting into ② we get $b - \frac{6b}{5b-6} \leq 1$, or equivalently

$$5b^2 - 17b + 6 \leq 0,$$

hence $\frac{2}{5} \leq b \leq 3$, so $\log_{27} \frac{2}{5} \leq y \leq \frac{1}{3}$.

Since $\log_{27} y - \log_4 x \geq \frac{1}{6}$, we have $\log_{27} y + \frac{1}{3} \geq \log_4 x + \frac{1}{2}$, thus

$$\log_{27} 3y \geq \log_4 2x.$$

As $x \geq \frac{1}{2}$ and $y \leq \frac{1}{3}$, we have $\log_4 2x \geq 0$ and $\log_{27} 3y \leq 0$. Therefore, $\log_{27} y - \log_4 x \geq \frac{1}{6}$ if and only if $x = \frac{1}{2}$ and $y = \frac{1}{3}$, and this is the only pair of real numbers that makes the system of inequalities hold.

Remark. Observing the inequalities, we discovered that changing variables in the first and third inequalities as $4^x = a$ and $27^y = b$ $(a, b > 0)$ gave us a simple equation and a simple inequality. By the equation we can write a and b in terms of each other, and by the inequality we can find bounds for a and b.

Example 8. Solve the inequality $|\log_2 x - 3| + |2^x - 8| \geq 9$.

Analysis. We can remove the absolute value signs by restricting the value of x. Since the zeros of $|\log_2 x - 3|$ and $|2^x - 8|$ are $x = 8$ and $x = 3$, respectively, we can discuss the three cases $x \in (0, 3]$, $(3, 8]$, and $(8, +\infty)$.

Solution. Obviously $x > 0$. Letting $|\log_2 x - 3| = 0$ or $|2^x - 8| = 0$ gives $x = 8$ or $x = 3$, respectively, so we discuss the three cases $x \in (0, 3]$, $(3, 8]$, and $(8, +\infty)$ separately.

(1) If $x \in (0, 3]$, then the inequality becomes

$$3 - \log_2 x + 8 - 2^x \geq 9,$$

or equivalently $2 \geq 2^x + \log_2 x$.

Let $f(x) = 2^x + \log_2 x$. Then $f(x)$ is an increasing function on $(0, 3]$, and $f(1) = 2$. Therefore, the solution set to the inequality is $(0, 1]$ in this case.

(2) If $x \in (3, 8]$, then the inequality becomes

$$3 - \log_2 x + 2^x - 8 \geq 9,$$

or equivalently $2^x \geq 14 + \log_2 x$.

We draw the graphs of the functions $y = 2^x$ and $y = 14 + \log_2 x$ in the same coordinate plane (Figure 10.6). The two graphs have two common points, one of which is $(4, 16)$. When $x \geq 4$, the graph of $y = 2^x$ lies above

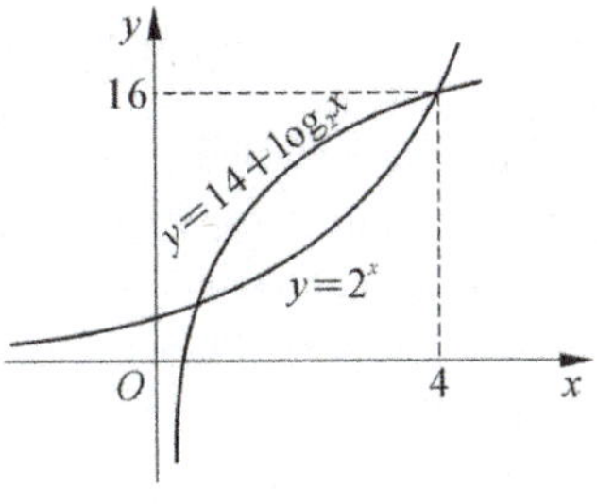

Figure 10.6

the graph of $y = 14 + \log_2 x$, so the solution set to the inequality is $[4, 8]$ in this case.

(3) If $x \in (8, +\infty)$, then the inequality becomes

$$\log_2 x - 3 + 2^x - 8 \geq 9,$$

or equivalently $2^x + \log_2 x \geq 20$.

Let $h(x) = 2^x + \log_2 x$. Then $h(x)$ is an increasing function on $[8, +\infty)$, and $h(8) = 259 > 20$. Therefore, the solution set is $(8, +\infty)$ in this case.

In summary, the solution set to the original inequality is $\{x | 0 < x \leq 1$ or $x \geq 4\}$.

Remark. The second case above can be solved with the help of derivatives. We first write the inequality in the following form:

$$2^x - \log_2 x \geq 14.$$

Let $g(x) = 2^x - \log_2 x$. Then $g'(x) = 2^x \cdot \ln 2 - \frac{1}{x \ln 2} = \ln 2 \cdot (2^x - \frac{1}{x \ln^2 2}) > 0$, so $g(x)$ is increasing on the interval $(3, 8]$ and $g(4) = 14$. Therefore, the solution set in this case is $[4, 8]$.

10.3 Exercises

Group A

1. The solution set to the system of inequalities $\begin{cases} 4x^2 - 4x - 15 < 0, \\ \sqrt{x-1} < x - 3 \end{cases}$ is ____________.

2. The solution set to the inequality $\sqrt{\log_2 x - 1} + \frac{1}{2} \log_{\frac{1}{2}} x^3 + 2 > 0$ is ____________.

3. Let the sets

$$A = \{x | x^2 - 4x + 3 < 0,\ x \in \mathbf{R}\},$$

$$B = \{x | 2^{1-x} + a \leq 0,\ x^2 - 2(a+7)x + 5 \leq 0,\ x \in \mathbf{R}\}.$$

If $A \subseteq B$, then the value range of real numbers a is ____________.

4. The solution set to the inequality $1 + 2^x < 3^x$ is ____________.

5. Suppose for a real variable x such that $x^2 + bx \leq -x (b < -1)$, the minimum value of $f(x) = x^2 + bx$ is $-\frac{1}{2}$. Then the value of b is ____________.

6. Solve the following inequalities:

(1) $x(3x+1)(2x+1)(x-2)(x-1)^2 \geq 0$.

(2) $\frac{x^2-4x+1}{3x^2-7x+2} \leq 1$.

7. Solve the following inequalities:

(1) $\sqrt{(x-1)(x+2)} > 2x^2 + 2x - 10$.

(2) $\frac{4x^2}{(1-\sqrt{1+2x})^2} < 2x + 9$.

(3) $\sqrt{3-x} - \sqrt{x+1} > \frac{1}{2}$.

8. Solve the following inequalities:

(1) $2\log_3(\log_3 x) + \log_{\frac{1}{3}}(\log_3(9\sqrt[3]{x})) \geq 1$.

(2) $\log_2(2^x - 1) \cdot \log_{\frac{1}{2}}(2^{x+1} - 2) > -2$.

9. Solve the following inequalities:

(1) $\left|\frac{1}{\log_{\frac{1}{2}} x} + 2\right| > \frac{3}{2}$;

(2) $||x+3| - |x-3|| > 3$.

10. Suppose a is a real number such that $a < 9a^3 - 11a < |a|$. Find the value range of a.

Group B

11. Solve the following inequalities for x:

(1) $a^{x^4-2x^2} > \left(\frac{1}{a}\right)^{a^2}$ ($a > 0$ and $a \neq 1$).

(2) $\sqrt{\frac{a(x+1)}{x+2}} \geq 1$.

12. Solve the following inequalities for x:

(1) $2\log_a(x-1) > \log_a[1 + a(x-2)](a > 1)$.

(2) $\log_{ax} x + \log_x(ax)^2 > 0$.

13. Given the following inequality for x

$$\log_{\frac{1}{a}}(\sqrt{x^2 - a|x| + 5} + 1) \cdot \log_5(x^2 - a|x| + 6) + \frac{1}{\log_3 a} \geq 0,$$

if its solution set contains exactly two elements, find the value of a.

14. If the inequality $x^3 - ax + 1 \geq 0$ holds for all $x \in [-1, 1]$, find the value range of a.

Chapter 11

Synthetical Problems of Inequalities

11.1 Key Points of Knowledge and Basic Methods

A. Find the value (or value range) of the parameter when the solution set is given to an inequality with a parameter

In such problems we usually construct a simple algebraic inequality based on the solution set and compare it with the original inequality to determine the value (or value range) of the parameter.

B. Inequalities related to a quadratic polynomial $f(x) = ax^2+bx+c\ (a \neq 0)$

(1) Properties of the function $f(x) = ax^2 + bx + c$

(i) The graph of $f(x)$ is symmetric with respect to $x = -\frac{b}{2a}$.

(ii) The maximum (minimum) value of $f(x)$:

if $a > 0$, then $f(x)_{\min} = f\left(-\frac{b}{2a}\right) = \frac{4ac-b^2}{4a}$, attained when $x = -\frac{b}{2a}$;

if $a < 0$, then $f(x)_{\max} = f\left(-\frac{b}{2a}\right) = \frac{4ac-b^2}{4a}$, attained when $x = -\frac{b}{2a}$.

(iii) The maximum (minimum) value of $f(x)$ on an interval $[\alpha, \beta]$:

$$\text{if } -\frac{b}{2a} \in [\alpha, \beta], \text{ then } f(x)_{\min} = \min\left\{f(\alpha),\ f(\beta), f\left(-\frac{b}{2a}\right)\right\},$$

$$f(x)_{\max} = \max\left\{f(\alpha),\ f(\beta),\ f\left(-\frac{b}{2a}\right)\right\};$$

$$\text{if } -\frac{b}{2a} \notin [\alpha, \beta], \text{ then } f(x)_{\min} = \min\{f(\alpha),\ f(\beta)\},$$
$$f(x)_{\max} = \max\{f(\alpha),\ f(\beta)\}.$$

(2) Real roots of the equation $f(x) = 0$

If $\Delta = b^2 - 4ac > 0$, then $f(x) = 0$ has two different real roots;
if $\Delta = 0$, then $f(x) = 0$ has two equal real roots;
if $\Delta < 0$, then $f(x) = 0$ has no real roots.

(3) Real roots of $f(x) = 0$ in the interval (α, β)

(i) $f(x) = 0$ has exactly one real root in (α, β) if and only if $f(\alpha) \cdot f(\beta) < 0$.

(ii) $f(x) = 0$ has two real roots in (α, β) if and only if

$$\begin{cases} \Delta = b^2 - 4ac \geq 0, \\ \alpha < -\dfrac{b}{2a} < \beta, \\ af(\alpha) > 0, \\ af(\beta) > 0. \end{cases}$$

(iii) $f(x) = 0$ has no real roots in (α, β) if and only if neither of the above two holds.

(iv) $f(x) = 0$ has one root greater than α and one root less than α if and only if $af(\alpha) < 0$.

For the case of closed intervals, we need to discuss the endpoints separately.

C. Inequalities with functions

(1) The domain and range of a function (omitted)
(2) Monotonicity, boundedness, extrema, and convexity of functions (omitted)
(3) Inequalities arising from functional equations

In such problems we usually examine the characteristics of the functional equation to find the specific properties that its solution needs to satisfy (monotonicity, symmetry, etc.), and then use these properties to solve the problem.

11.2 Illustrative Examples

Example 1. If the solution set to the inequality (for x) $\frac{x^2-8x+20}{mx^2-2mx+2m-4} < 0$ is $\mathbf{R}$, find the value range of the real number m.

Solution. Since $x^2 - 8x + 20 = (x-4)^2 + 4 > 0$, it is equivalent to say that the solution set to $mx^2 - 2mx + 2m - 4 < 0$ ① is $\mathbf{R}$.

If $m = 0$, then ① becomes $-4 < 0$, whose solution set is $\mathbf{R}$.
If $m \neq 0$, then ① is a quadratic inequality, and a necessary and sufficient condition for its solution set to be $\mathbf{R}$ is

$$\begin{cases} m < 0, \\ \Delta = 4m^2 - 4m(2m-4) < 0. \end{cases}$$

Solving the system of inequalities, we get $m < 0$.

Therefore, the value range of m is $\{m | m \leq 0\}$.

Remark. It is a common mistake to neglect the case $m = 0$, where ① is not a quadratic inequality.

Example 2. Let $f(x) = x^3 + x$. Suppose the inequality

$$f(m \cos \theta) + f(1-m) > 0$$

holds for all $0 \leq \theta \leq \frac{\pi}{2}$. Find the value range of m.

Solution. We observe that $f(x)$ is an odd increasing function on $\mathbf{R}$.

Since $f(m\cos\theta) + f(1-m) > 0$,

$$f(m\cos\theta) > -f(1-m) = f(m-1),$$

so $m\cos\theta > m - 1$ for all $0 \leq \theta \leq \frac{\pi}{2}$.

Let $t = \cos\theta$. Then $0 \leq t \leq 1$. Now, $mt - m + 1 > 0$ for all $0 \leq t \leq 1$, so

$$\begin{cases} -m + 1 > 0, \\ m - m + 1 > 0, \end{cases}$$

which gives that $m < 1$.

Therefore, the value range of m is $(-\infty, 1)$.

Example 3. Suppose the solution set to the inequality $\sqrt{x} > ax + \frac{3}{2}$ is $4 < x < c$. Find a and c.

Solution. Let $\sqrt{x} = t (\geq 0)$. Then the inequality becomes

$$at^2 - t + \frac{3}{2} < 0. \text{ ①}$$

Since the solution set to the original inequality is $(4, c)$, it follows that $a > 0$ and the solution set to ① is $(2, \sqrt{c})$. Equivalently, $(t-2)(t-\sqrt{c}) < 0$, in other words

$$t^2 - (2+\sqrt{c})t + 2\sqrt{c} < 0. \text{ ②}$$

Compare the coefficients in ① and ②. Since they have the same solution set, we get that

$$\frac{a}{1} = \frac{-1}{-(2+\sqrt{c})} = \frac{\frac{3}{2}}{2\sqrt{c}},$$

hence $a = \frac{1}{8}$ and $c = 36$.

Remark. Apart from the solution above, apparently $t = 2$ and $t = \sqrt{c}$ are roots of the equation $at^2 - t + \frac{3}{2} = 0$, which easily shows that $a = \frac{1}{8}$ and $c = 36$.

Example 4. Let $f(x) = \min\{x^2 - 1,\ x+1,\ -x+1\}$, where $\min\{x, y, z\}$ is the smallest number in x, y, and z. If $f(a+2) > f(a)$, find the value range of a.

Solution. If $a+2 \le -1$, then $a < a+2 \le -1$, and $f(a) < f(a+2)$ in this case.

If $-1 < a+2 < 0$, then $-3 < a < -2$, and

$$f(a) \le f(-2) = -1 < f(a+2).$$

If $0 \le a+2 \le 1$, then $-2 \le a < -1$, and $f(a) \ge f(a+2)$.
If $1 < a+2 < 2$, then $-1 < a < 0$, and $f(a) < f(a+2)$.
If $a+2 \ge 2$, then $a \ge 0$, and $f(a) \ge f(a+2)$.

In summary, the value range of a is $(-\infty, -2) \cup (-1, 0)$.

Example 5. Let $f(x) = \frac{a}{x} - x$ such that for each $x \in (0,1)$,

$$f(x) \cdot f(1-x) \ge 1.$$

Find the value range of a.

Solution. Let $y = 1 - x \in (0,1)$. Then

$$\begin{aligned} f(x) \cdot f(1-x) = f(x)f(y) &= \left(\frac{a}{x} - x\right)\left(\frac{a}{y} - y\right) \\ &= \frac{x^2y^2 - a(x^2+y^2) + a^2}{xy} \\ &= \frac{(xy)^2 - a[(x+y)^2 - 2xy] + a^2}{xy}. \end{aligned}$$

Let $xy = t$. Then $t \in \left(0, \frac{1}{4}\right]$. Now,

$$g(t) = f(x) \cdot f(1-x) = \frac{t^2 - a(1-2t) + a^2}{t} \geq 1,$$

or equivalently $t^2 + (2a-1)t + a^2 - a \geq 0$.

(1) If $\frac{1}{2} - a \leq 0$, then $\frac{1}{2} \leq a$, and it reduces to $g(0) \geq 0$, namely $a^2 - a \geq 0$, so $a \geq 1$ in this case.
(2) If $0 < \frac{1}{2} - a \leq \frac{1}{4}$ so that $\frac{1}{4} \leq a < \frac{1}{2}$, then it reduces to $g\left(\frac{1}{2} - a\right) \geq 0$, which has no real solutions.
(3) If $\frac{1}{2} - a > \frac{1}{4}$ so that $a < \frac{1}{4}$, then it reduces to $g\left(\frac{1}{4}\right) \geq 0$, or equivalently

$$\left(a - \frac{1}{4}\right)^2 \geq \frac{1}{4},$$

which gives that $a \leq -\frac{1}{4}$.

In summary, the value range of a is $\{a | a \geq 1 \text{ or } a \leq -\frac{1}{4}\}$.

Example 6. Let $f(x)$ be an odd function defined on $[-1, 1]$ with $f(1) = 1$. Suppose for

$$a,\ b \in [-1, 1], \quad a + b \neq 0,$$

we have $\frac{f(a)+f(b)}{a+b} > 0$.

(1) Solve the inequality $f\left(x + \frac{1}{2}\right) < f\left(\frac{1}{x-1}\right)$.
(2) If $f(x) \leq m^2 - 2am + 1$ for all $x \in [-1, 1]$ and $a \in [-1, 1]$, find the value range of m.

Analysis. It is natural to examine the monotonicity of $f(x)$ due to the nature of both problems.

Solution. Let $-1 \leq x_1 < x_2 \leq 1$. Then $x_1 - x_2 \neq 0$ and $\frac{f(x_1)+f(-x_2)}{x_1+(-x_2)} > 0$. Since $f(x)$ is an odd function on $[-1, 1]$,

$$\begin{aligned} f(x_1) - f(x_2) &= f(x_1) + f(-x_2) \\ &= \frac{f(x_1) + f(-x_2)}{x_1 + (-x_2)}(x_1 - x_2) < 0. \end{aligned}$$

Hence, $f(x_1) < f(x_2)$, so $f(x)$ is increasing on $[-1, 1]$.

(1) Since $f\left(x+\frac{1}{2}\right)<f\left(\frac{1}{x-1}\right)$,

$$-1 \leq x+\frac{1}{2}<\frac{1}{x-1} \leq 1.$$

Solving the inequality gives that $-\frac{3}{2} \leq x<-1$.

Therefore, the solution set to the original inequality is $\left\{x \mid -\frac{3}{2} \leq x<-1\right\}$.

(2) Since $f(x)$ is increasing on $[-1,1]$, we have $f(x)_{\max}=f(1)=1$. It follows that for all $a \in[-1,1]$, the inequality

$$m^2-2am+1 \geq f(1)=1,$$

or equivalently $2am-m^2 \leq 0$ holds.

Let $g(a)=2m \cdot a-m^2$, whose graph is a line segment over $[-1,1]$. Then

$$\begin{cases} m=0, \\ -m^2 \leq 0 \end{cases} \quad \text{or} \quad \begin{cases} m>0, \\ g(1)=2m-m^2 \leq 0 \end{cases}$$

$$\text{or} \quad \begin{cases} m<0, \\ g(-1)=-2m-m^2 \leq 0. \end{cases}$$

Solve the inequalities, and we get that $m=0$ or $m \geq 2$ or $m \leq -2$.

Therefore, the value range of m is $\{m \mid m \leq -2 \text{ or } m=0 \text{ or } m \geq 2\}$.

Example 7. Let $f(x)=ax^2+bx+c(a, b, c \in \mathbf{R}$ and $a \neq 0)$ with the following properties:

(1) For all $x \in \mathbf{R}$, we have $f(x-4)=f(2-x)$ and $f(x) \geq x$.
(2) For $x \in(0,2)$, it is true that $f(x) \leq\left(\frac{x+1}{2}\right)^2$.
(3) The minimum value of $f(x)$ on $\mathbf{R}$ is 0.

Find the maximum value of the real numbers m such that for some real number t, the inequality $f(x+t) \leq x$ holds for all $x \in[1, m]$

Solution. By $f(x-4)=f(2-x)$ for $x \in \mathbf{R}$, the quadratic function $f(x)$ has the axis of symmetry $x=-1$. Also, by (3) we see that $f(x)$ opens upwards, that is $a>0$. Since the minimum value of $f(x)$ is 0,

$$f(x)=a(x+1)^2(a>0).$$

By (1), the value $f(1) \geq 1$, and $f(1) \leq \left(\frac{1+1}{2}\right)^2 = 1$ from (2). Hence, $f(1) = 1$, namely $1 = a(1+1)^2$, so $a = \frac{1}{4}$. Consequently, $f(x) = \frac{1}{4}(x+1)^2$.

Now, the graph of $f(x) = \frac{1}{4}(x+1)^2$ opens upwards, and the graph of $y = f(x+t)$ is obtained by translating the graph of $y = f(x)$ by $|t|$ units. If the graph of $y = f(x+t)$ lies below the graph of $y = x$ on the interval $[1, m]$, and m is the greatest number that has this property, then 1 and m need to be the roots of the equation (for x)

$$\frac{1}{4}(x+t+1)^2 = x. \quad ①$$

Let $x = 1$ in ①, and we get $t = 0$ or $t = -4$. If $t = 0$, then the solutions to ① are $x_1 = x_2 = 1$, contradictory to $m > 1$. If $t = -4$, then the solutions to ① are $x_1 = 1$ and $x_2 = 9$, so $m = 9$.

On the other hand, if we let $t = -4$, then for all $x \in [1, 9]$,

$$(x-1)(x-9) \leq 0,$$
$$\tfrac{1}{4}(x-4+1)^2 \leq x.$$

Thus, $f(x-4) \leq x$. Therefore, the maximum value of m is 9.

Remark. In general, if $f(x-a) = f(b-x)$ for $x \in \mathbf{R}$, then

$$f\left(x + \frac{b-a}{2}\right) = f\left(x + \frac{b+a}{2} - a\right) = f\left(b - x - \frac{b+a}{2}\right)$$
$$= f\left(\frac{b-a}{2} - x\right),$$

so the graph of $f(x)$ is symmetric with respect to the line $x = \frac{b-a}{2}$.

Example 8. Let $f(x) = ax^2 + bx + c (a, b, c \in \mathbf{R})$ such that $|f(-1)| \leq 1$, $|f(0)| \leq 1$, and $|f(1)| \leq 1$. Show that $|f(x)| \leq \frac{5}{4}$ for all $x \in [-1, 1]$.

Solution. Since $\begin{cases} f(0) = c, \\ f(-1) = a - b + c, \\ f(1) = a + b + c, \end{cases}$ it follows that

$$\begin{cases} a = \dfrac{1}{2}(f(1) + f(-1)) - f(0), \\ b = \dfrac{1}{2}(f(1) - f(-1)), \\ c = f(0). \end{cases}$$

Hence, for $x \in [-1, 1]$,

$$\begin{aligned}
|f(x)| &= \left|\left(\frac{1}{2}(f(1)+f(-1)) - f(0)\right)x^2 + \frac{1}{2}(f(1) - f(-1)x + f(0))\right| \\
&= \left|\frac{1}{2}(x^2+x)f(1) + (1-x^2)f(0) + \frac{1}{2}(x^2-x)f(-1)\right| \\
&\leq \frac{1}{2}|x| \cdot |x+1| \cdot |f(1)| + |1-x^2| \cdot |f(0)| + \frac{1}{2}|x| \cdot |x-1| \cdot |f(-1)| \\
&\leq \frac{1}{2}|x|(1+x) + 1 - x^2 + \frac{1}{2}|x|(1-x) \\
&= -x^2 + |x| + 1 = -\left(|x| - \frac{1}{2}\right)^2 + \frac{5}{4} \leq \frac{5}{4}.
\end{aligned}$$

Example 9 (2014 China Mathematical Competition). Let a, b, and c be real numbers such that $a + b + c = 1$ and $abc > 0$. Show that

$$ab + bc + ca < \frac{\sqrt{abc}}{2} + \frac{1}{4}.$$

Solution 1. If $ab + bc + ca \leq \frac{1}{4}$, then the inequality holds trivially.

If $ab + bc + ca > \frac{1}{4}$, then we assume that $a = \max\{a, b, c\}$ without loss of generality. By $a + b + c = 1$ we have $a \geq \frac{1}{3}$. It follows that

$$ab + bc + ca - \frac{1}{4} \leq \frac{(a+b+c)^2}{3} - \frac{1}{4} = \frac{1}{12} \leq \frac{a}{4}, \text{①}$$

$$ab + bc + ca - \frac{1}{4} = a(b+c) - \frac{1}{4} + bc$$

$$= a(1-a) - \frac{1}{4} + bc \leq \frac{1}{4} - \frac{1}{4} + bc = bc, \text{②}$$

where the equality in ① holds when $a = \frac{1}{3}$, and the equality in ② holds when $a = \frac{1}{2}$. Note that the equalities in ① and ② cannot hold simultaneously.

Since $ab + bc + ca - \frac{1}{4} > 0$, multiplying ① and ② gives

$$\left(ab + bc + ca - \frac{1}{4}\right)^2 < \frac{abc}{4},$$

that is $ab + bc + ca - \frac{1}{4} < \frac{\sqrt{abc}}{2}$, so $ab + bc + ca < \frac{\sqrt{abc}}{2} + \frac{1}{4}$.

Solution 2. Since $abc > 0$, either the numbers a, b, and c are all positive or one of them is positive and the other two are negative.

For the second case, we assume that $a > 0$ and b, $c < 0$. Then

$$ab + bc + ca = b(a + c) + ca < b(a + c) = b(1 - b) < 0,$$

so the proposition holds obviously.

For the first case, we assume that $a \geq b \geq c$. Then $a \geq \frac{1}{3}$ and $c \leq \frac{1}{3}$. We have

$$\begin{aligned} ab + bc + ca - \frac{\sqrt{abc}}{2} &= c(a+b) + \sqrt{ab}\left(\sqrt{ab} - \frac{\sqrt{c}}{2}\right) \\ &= c(1-c) + \sqrt{ab}\left(\sqrt{ab} - \frac{\sqrt{c}}{2}\right). \end{aligned}$$

Since $\sqrt{ab} \geq \sqrt{\frac{b}{3}} \geq \sqrt{\frac{c}{3}} > \frac{\sqrt{c}}{2}$ and $\sqrt{ab} \leq \frac{a+b}{2} = \frac{1-c}{2}$,

$$\begin{aligned} c(1-c) + \sqrt{ab}\left(\sqrt{ab} - \frac{\sqrt{c}}{2}\right) &\leq c(1-c) + \frac{1-c}{2}\left(\frac{1-c}{2} - \frac{\sqrt{c}}{2}\right) \\ &= \frac{1}{4} - \frac{3c^2}{4} + \frac{c\sqrt{c}}{4} + \frac{c}{2} - \frac{\sqrt{c}}{4}. \end{aligned}$$

Thus, it suffices to show that $\frac{3c^2}{4} - \frac{c\sqrt{c}}{4} - \frac{c}{2} + \frac{\sqrt{c}}{4} > 0$, or equivalently

$$3c\sqrt{c} - c - 2\sqrt{c} + 1 > 0. \text{ ①}$$

Since $0 < c \leq \frac{1}{3}$, we have $\frac{1}{3} - c \geq 0$. ②

By the AM-GM inequality,

$$3c\sqrt{c} + \frac{1}{3} + \frac{1}{3} \geq 3\left(3c\sqrt{c} \cdot \frac{1}{3} \cdot \frac{1}{3}\right)^{\frac{1}{3}} = \sqrt[3]{9}\sqrt{c} > 2\sqrt{c}. \text{ ③}$$

Adding ② and ③, we get ① and hence completing the proof.

11.3 Exercises

Group A

1. The maximal real number k such that the inequality (for x) $\sqrt{x-3} + \sqrt{6-x} \geq k$ has a real solution is ____________.
2. If the inequality $1 + 2^x + (a - a^2) \cdot 4^x > 0$ holds for all $x \in (-\infty, 1]$, then the value range of a is ____________.
3. For $x \in \left(0, \frac{\pi}{2}\right)$, the integer part of $M = 3^{\cos^2 x} + 3^{\sin^2 x}$ is ____________. (The integer part of a real number is the greatest integer not exceeding that number.)
4. Let S be the area of a triangle inscribed in a circle with radius 1. Then the minimum value of $4S + \frac{9}{S}$ is ____________.

5. Let $f(x) = 4\sin x \sin^2\left(\frac{\pi}{4}+\frac{x}{2}\right) + \cos 2x$. If $\frac{\pi}{6} \le x \le \frac{2\pi}{3}$ is a sufficient condition for $|f(x)-m| < 2$, then the value range of m is ____________.
6. If the solution set to the inequality (for x) $|ax+b| < 2 (a \neq 0)$ is $\{x|2 < x < 6\}$, find the values of a and b.
7. Among all real numbers x and y such that $\log_{(x^2+y^2)}(x+y) \ge 1$, find the maximum value of y.
8. Find 100 real numbers $x_1, x_2, x_3, \ldots, x_{100}$ such that

$$x_1 = 1, \quad 0 \le x_k \le 2x_{k-1}, \quad k = 2, 3, \ldots, 100,$$

and $S = x_1 - x_2 + x_3 - x_4 + \cdots + x_{99} - x_{100}$ is maximal.

Group B

9. Prove the nonexistence of positive integersx_1, $x_2, \ldots, x_m$ with $m \ge 2$ and $x_1 < x_2 < \cdots < x_m$ such that

$$\frac{1}{x_1^3} + \frac{1}{x_2^3} + \cdots + \frac{1}{x_m^3} = 1.$$

10. Let $f(x) = ax^2 + bx + c$, where a, b, and c are real numbers with $a > 100$. How many integer solutions does the inequality $|f(x)| \le 50$ have at most?
11. Let $a > 0$ with $a \neq 1$ and let $f(x)$ be a function such that $f(\log_a x) = \frac{a(x^2-1)}{x(a^2-1)}$. Show that $f(n) > n$ for all positive integers n greater than 1.
12. Let $f(x) = x^2 + px + q$. If the inequality $|f(x)| > 2$ has no solution in $[1, 5]$, find all such real numbers p and q.
13. Let a, b, $c \in \mathbf{R}$. Suppose the equation $ax^2 + bx + c = 0$ has two real roots and

$$|a(b-c)| > |b^2 - ac| + |c^2 - ab|.$$

Show that the equation above has at least one root in the interval $(0, 2)$.
14. Let $y = f(x)$ be defined on $\mathbf{R}$ such that $f(x) > 1$ for all $x > 0$ and $f(x+y) = f(x)f(y)$ for all x, $y \in \mathbf{R}$. Solve the inequality $f(x) \le \frac{1}{f(x+1)}$.

Chapter 12

Concepts and Properties of Trigonometric Functions

12.1 Key Points of Knowledge and Basic Methods

Consider a point $P(x, y)$ lying on the terminal side of an angle α. When P moves along the terminal side, its coordinates x and y, and the number $r(=\sqrt{x^2+y^2})$ keep a certain proportion. This phenomenon leads to the notion of trigonometric functions:

$$\sin\alpha = \frac{y}{r}, \quad \cos\alpha = \frac{x}{r} \quad \tan\alpha = \frac{y}{x}, \quad \cot\alpha = \frac{x}{y}; \quad \sec\alpha = \frac{r}{x}, \quad \csc\alpha = \frac{r}{y}.$$

The definitions above immediately determine the domains and some simple rules of trigonometric functions. In addition, we have the following important properties:

A. Boundedness of sine and cosine functions

For any angle α, we have $|\sin\alpha| \le 1$ and $|\cos\alpha| \le 1$. This means that the sine function and the cosine function are both bounded functions. Some variants of this property are $1 \pm \sin\alpha \ge 0$, $|A\sin\alpha + B\cos\alpha| \le \sqrt{A^2+B^2}$, etc.

B. Odd/even functions and symmetry of graphs

By definitions of odd and even functions, the concept of trigonometric functions, and the induction formulas, the sine, tangent, and cotangent functions are odd functions, and their graphs are centrally symmetric with respect to the origin. The cosine function is an even function, and its graph is symmetric with respect to the y-axis. In addition, the graph of $f(x) = \sin x$ is also symmetric with respect to the lines $x = k\pi + \frac{\pi}{2}$ $(k \in \mathbf{Z})$, and the graph of $f(x) = \cos x$ is symmetric with respect to the lines $x = k\pi$ $(k \in \mathbf{Z})$.

C. Monotonicity

By the definition and properties of trigonometric functions, we can derive the monotonic intervals of each trigonometric function. For example, the function $f(x) = \sin x$ is increasing on the intervals $\left[2k\pi - \frac{\pi}{2}, 2k\pi + \frac{\pi}{2}\right]$, and decreasing on the intervals $\left[2k\pi + \frac{\pi}{2}, 2k\pi - \frac{\pi}{2}\right]$, where k is any integer.

D. Periodicity

Periodicity is a property of trigonometric functions that quadratic functions, power functions, exponential functions, or logarithmic functions do not possess.

In general, for a function $y = f(x)$, if there exists a positive number T such that for any x in its domain, $x + T$ also lies in the domain and $f(x + T) = f(x)$, then we call $f(x)$ a periodic function, and T is called a period of $f(x)$.

Apparently, if $y = f(x)$ is a periodic function with period T, then kT $(k \in \mathbf{N}^*)$ are also periods of $y = f(x)$. If there exists the smallest number among all periods of $f(x)$, then we call this number the least positive period of $f(x)$. Note that not every periodic function has the least positive period. For example, the constant function $f(x) = c$ for $x \in \mathbf{R}$ is periodic with arbitrary positive periods, and does not have the least positive period.

The trigonometric functions $y = \sin x$, $y = \cos x$, $y = \tan x$, and $y = \cot x$ are all periodic functions that have the least positive period.

12.2 Illustrative Examples

Example 1. Let $a (a > 0)$ and b be constants such that the function $y = a \sin x + b$ has the maximum value 4 and minimum value 2. Find the values of a and b.

Solution. The maximum and minimum values of the function $\sin x$ are 1 and -1, respectively. Together with $a > 0$ (so $f(t) = at + b$ is increasing) we can see that the maximum and minimum values of $y = a \sin x + b$ are $a + b$ and $-a + b$, respectively. Therefore,

$$\begin{cases} a + b = 4, \\ -a + b = 2, \end{cases}$$

hence $a = 14$ and $b = 3$.

Example 2. Find the range of the function $y = \frac{\sin 2x - 3}{\sin x + \cos x - 2}$.

Solution. Since

$$\begin{aligned} y &= \frac{\sin 2x - 3}{\sin x + \cos x - 2} \\ &= \frac{(1 + \sin 2x) - 4}{\sin x + \cos x - 2} \\ &= \frac{(\sin x + \cos x)^2 - 4}{\sin x + \cos x - 2} \\ &= \sin x + \cos x + 2 \end{aligned}$$

and $-\sqrt{2} \le \sin x + \cos x \le \sqrt{2}$, we have $2 - \sqrt{2} \le y \le 2 + \sqrt{2}$. Therefore, the range of y is $[2 - \sqrt{2}, 2 + \sqrt{2}]$.

Example 3. Let a, x, and y be real numbers with $x,\ y \in \left[-\frac{\pi}{4}, \frac{\pi}{4}\right]$ such that

$$x^3 + \sin x = 2a, \quad 4y^3 + \sin y \cos y = -a.$$

Find the value of $\cos(x + 2y)$.

Solution. It follows from the conditions that

$$\begin{aligned} x^3 + \sin x &= 2a, \\ (2y)^3 + \sin(2y) &= -2a. \end{aligned}$$

Let $f(t) = t^3 + \sin t$. Then $f(x) = 2a$ and $f(2y) = -2a$. Note that $f(t)$ is an odd function, so $f(x) = -f(2y) = f(-2y)$.

On the other hand, since $y = t^3$ and $y = \sin t$ are both increasing functions on $\left[-\frac{\pi}{2}, \frac{\pi}{2}\right]$, it follows that $f(t)$ is increasing on $\left[-\frac{\pi}{2}, \frac{\pi}{2}\right]$. By x, $y \in \left[-\frac{\pi}{4}, \frac{\pi}{4}\right]$ we have $x, -2y \in \left[-\frac{\pi}{2}, \frac{\pi}{2}\right]$. Thus, $f(x) = f(-2y)$ implies that $x = -2y$, namely $x + 2y = 0$. Therefore, $\cos(x + 2y) = 1$.

Example 4. Let $f(x)$ be defined on $\mathbf{R}$ such that

$$|f(x) + \cos^2 x| \le \frac{3}{4}, \quad |f(x) - \sin^2 x| \le \frac{1}{4}.$$

Find the formula of $f(x)$.

Solution. It follows from the conditions that

$$1 = \sin^2 x + \cos^2 x \le |f(x) + \cos^2 x| + |f(x) - \sin^2 x| \le \frac{3}{4} + \frac{1}{4} = 1.$$

Hence,

$$|f(x) + \cos^2 x| = \frac{3}{4}, \quad |f(x) - \sin^2 x| = \frac{1}{4},$$

so $f(x) = \sin^2 x - \frac{1}{4}$.

Example 5. Suppose the disk $x^2 + y^2 \leq k^2$ covers at least one maximum point and one minimum point on the graph of $f(x) = \sqrt{3}\sin\frac{\pi x}{k}$. Find the value range of k.

Solution. Since $f(x) = \sqrt{3}\sin\frac{\pi x}{k}$ is an odd function, its graph is centrally symmetric with respect to the origin. Since the disk is also centrally symmetric with respect to the origin, it is equivalent to say that the disk covers the extremum point on the graph of $f(x)$ that is closest to the origin.

Choose x such that $\frac{\pi x}{k} = \frac{\pi}{2}$. Then the maximum point on the graph of $f(x)$ closest to the origin is $P\left(\frac{k}{2}, \sqrt{3}\right)$.The disk$x^2 + y^2 \leq k^2$ covers P, so

$$k^2 \geq \left(\frac{k}{2}\right)^2 + (\sqrt{3})^2,$$

therefore $|k| \geq 2$, which is the value range of k.

Example 6. Suppose the least positive period of the function $f(x) = \sqrt{3}\sin wx \cos wx - \cos^2 wx (w > 0)$ is $\frac{\pi}{2}$. In a triangle ABC, let a, b, and c be the sides opposite to A, B, and C, respectively, and suppose $b^2 = ac$. Find the value range of $f(B)$.

Solution. The function $f(x)$ can be rewritten as

$$f(x) = \frac{\sqrt{3}}{2}\sin 2wx - \frac{\cos 2wx + 1}{2} = \sin\left(2wx - \frac{\pi}{6}\right) - \frac{1}{2}.$$

Since the least positive period of this function is $\frac{\pi}{2}$, we have $w = 2$, so $f(x) = \sin\left(4x - \frac{\pi}{6}\right) - \frac{1}{2}$.

Now, since

$$\cos B = \frac{a^2 + c^2 - b^2}{2ac} = \frac{a^2 + c^2 - ac}{2ac} \geq \frac{2ac - ac}{2ac} = \frac{1}{2},$$

we get $B \in \left(0, \frac{\pi}{3}\right]$. Hence, $4B - \frac{\pi}{6} \in \left(-\frac{\pi}{6}, \frac{7}{6}\pi\right]$, and $\sin\left(4B - \frac{\pi}{6}\right) \in \left[-\frac{1}{2}, 1\right]$. Therefore, $f(B) \in \left[-1, \frac{1}{2}\right]$.

Example 7. Show that the function $y = \sin(\cos x)$ is periodic. Also, determine whether it has the least positive period, and prove your result.

Solution. Note that for any real number x, we have $\sin(\cos(x + 2\pi)) = \sin(\cos x)$, so $y = \sin(\cos x)$ is periodic with period 2π.

Next, we show that 2π is the least positive period.

In fact, if there is $T \in (0, 2\pi)$ such that T is a period of $y = \sin(\cos x)$, then for all x,

$$\sin(\cos(x + T)) = \sin(\cos x).$$

Let $x = 0$ in the equation above, and we get $\sin(\cos T) = \sin 1$. Note that when $0 < T < 2\pi$,

$$\cos T \in [-1, 1) \subset \left[-\frac{\pi}{2}, \frac{\pi}{2}\right].$$

By monotonicity of the sine function on $\left[-\frac{\pi}{2}, \frac{\pi}{2}\right]$ we have $\sin(\cos T) < \sin 1$, a contradiction.

Therefore, 2π is the least positive period of $y = \sin(\cos x)$.

Remark. Periodicity is an important property of trigonometric functions. For a periodic function, we can study its behavior in one period to reveal its overall behavior.

Example 8. Let $f(x) = \sin^4 \frac{kx}{10} + \cos^4 \frac{kx}{10}$, where k is a positive integer. Suppose $\{f(x)|a < x < a+1\} = \{f(x)|x \in \mathbf{R}\}$ for all real numbers a. Find the minimum value of k.

Solution. Note that we can rewrite $f(x)$ as

$$\begin{aligned} f(x) &= \left(\sin^2 \frac{kx}{10} + \cos^2 \frac{kx}{10}\right)^2 - 2\sin^2 \frac{kx}{10} \cos^2 \frac{kx}{10} \\ &= 1 - \frac{1}{2}\sin^2 \frac{kx}{5} = \frac{1}{4}\cos \frac{2kx}{5} + \frac{3}{4}, \end{aligned}$$

so $f(x)$ attains its maximum value if and only if $x = \frac{5m\pi}{k} (m \in \mathbf{Z})$. It follows from the condition that any open interval $(a, a+1)$ of length 1 contains at least one maximum point of $f(x)$. This implies that $\frac{5\pi}{k} < 1$, namely $k > 5\pi$.

Conversely, if $k > 5\pi$, then any open interval $(a, a+1)$ of length 1 contains a full period of $f(x)$, so $\{f(x)|a < x < a+1\} = \{f(x)|x \in \mathbf{R}\}$.

Therefore, since k is a positive integer, its minimum value is $[5\pi] + 1 = 16$.

12.3 Exercises

Group A

1. Let α and β be acute angles such that $\alpha+\beta<\frac{\pi}{2}$ and $\sin\frac{\alpha}{2}=k\cos\beta$. Find the value range of k.
2. Suppose the temperature T (in degrees) of a room can be written in terms of time t (in hours) as $T=a\sin t+b\cos t$ for $t\in(0,+\infty)$, where a and b are positive. If the maximal difference of the temperatures in this room is 10 degrees, find the maximum value of $a+b$.
3. Let $f(x)=\frac{1}{x^2+1}$. Find the value of $f(\tan 5^\circ)+f(\tan 10^\circ)+\cdots+f(\tan 85^\circ)$.
4. Let $\alpha_1, \alpha_2, \ldots, \alpha_n$ be arbitrary real numbers. Show that

$$\cos\alpha_1\cos\alpha_2\ldots\cos\alpha_n+\sin\alpha_1\sin\alpha_2\ldots\sin\alpha_n\le\sqrt{2}.$$

5. Show that for all real numbers x and y,

$$\cos x^2+\cos y^2-\cos xy<3.$$

6. Suppose the domain of $f(x)=\sqrt{\sin^6 x+\cos^6 x+a\sin x\cos x}$ is $\mathbf{R}$. Find the value range of a.
7. (1) Show that both equations $\sin(\cos x)=x$ and $\cos(\sin x)=x$ for x have a unique solution in $\left(0,\frac{\pi}{2}\right)$.

   (2) Let a, b, $c\in\left(0,\frac{\pi}{2}\right)$ such that $\cos a=a$, $\sin(\cos b)=b$, and $\cos(\sin c)=c$. Arrange the numbers a, b, and c in increasing order.
8. Let a, b, c, $d\in(0,\pi)$ be such that $\frac{\sin a}{\sin b}=\frac{\sin c}{\sin d}=\frac{\sin(a-c)}{\sin(b-d)}$. Show that $a=b$ and $c=d$.
9. Suppose positive real numbers α, β, a, and b satisfy the following inequalities:

$$\alpha<\beta,\quad \alpha+\beta<\pi,\quad a+b<\pi,\quad \frac{\sin a}{\sin b}\le\frac{\sin\alpha}{\sin\beta}.$$

   Show that $a<b$.
10. Let w be a positive real number. If there exist a and $b(\pi\le a<b\le 2\pi)$ such that $\sin wa+\sin wb=2$, find the value range of w.
11. Suppose the inequality

$$x^2\cos\theta-x(1-x)+(1-x)^2\sin\theta>0$$

   holds for all $x\in[0,1]$. Find the value range of θ.

Group B

12. Let $A = \{t|0 < t < 2\pi,\ t \in R\}$ and let f be a map from A to the set of points in the plane, defined as $f\colon t \mapsto (\sin t, 2\sin t\cos t)$. Also, denote

$$B = \{f(t)|t \in A\}, \quad C(r) = \{(x,y)|x^2 + y^2 \leq r^2, \quad r > 0\}.$$

Find the minimal real number r such that $B \subseteq C(r)$.

13. For each real number b, let $f(b)$ be the maximum value of the function $\left|\sin x + \frac{2}{3} + \sin x + b\right|$. Find the minimum value of $f(b)$ and all b such that $f(b)$ reaches the minimum value.
14. Show that $f(x) = \sin(x^2)$ is not a periodic function.
15. Let a and b be real numbers such that $\cos(a\sin x) > \sin(b\cos x)$ for all real x. Show that $a^2 + b^2 < \frac{\pi^2}{4}$.

Chapter 13

Deformation via Trigonometric Identities

13.1 Key Points of Knowledge and Basic Methods

Deformation via trigonometric identities is an important kind of algebraic deformation. We need to master not only trigonometric formulas and their relationship, but also the ability to deal with complicated algebraic expressions. Furthermore, we need to have the consciousness of conversion and unified thoughts.

Deformation via trigonometric identities is involved in simplification and computation of trigonometric expressions, and also proofs of identities. It is especially important to use the right formula when dealing with trigonometric expressions. Common formulae include the relationship among trigonometric functions of the same angle, sum-angle formulae, double/half-angle formulae, sum-to-product/product-to-sum formulae, etc.

13.2 Illustrative Examples

Example 1. Simplify the following expressions:

(1) $\frac{1+\sin\theta-\cos\theta}{1-\sin\theta-\cos\theta}+\frac{1-\sin\theta-\cos\theta}{1+\sin\theta-\cos\theta}$;

(2) $\sin\alpha-\frac{1}{\sin\alpha}+\sqrt{\sin^2\alpha+\cot^2\alpha+3}$;

(3) $\cos\theta\cos 2\theta\cos 4\theta$.

Solution. (1) The expression equals

$$\frac{[1+(\sin\theta-\cos\theta)]^2+[1-(\sin\theta+\cos\theta)]^2}{(1-\cos\theta)^2-\sin^2\theta}$$

$$=\frac{2+2[(\sin\theta-\cos\theta)-(\sin\theta+\cos\theta)]+2(\sin^2\theta+\cos^2\theta)}{1-2\cos\theta+\cos^2\theta-\sin^2\theta}$$

$$=\frac{4-4\cos\theta}{2\cos^2\theta-2\cos\theta}=\frac{2}{\cos\theta}=-2\sec\theta$$

(2) By $1+\cot^2\alpha=\csc^2\alpha=\frac{1}{\sin^2\alpha}$, the expression equals

$$\sin\alpha-\frac{1}{\sin\alpha}+\sqrt{\sin^2\alpha+2+\frac{1}{\sin^2\alpha}}$$

$$=\sin\alpha-\frac{1}{\sin\alpha}+\sqrt{\left(\sin\alpha+\frac{1}{\sin\alpha}\right)^2}$$

$$=\sin\alpha-\frac{1}{\sin\alpha}+\left|\sin\alpha+\frac{1}{\sin\alpha}\right|$$

$$=\begin{cases}2\sin\alpha, & 2k\pi<\alpha<2k\pi+\pi,\ k\in\mathbf{Z};\\ -2\csc\alpha, & 2k\pi-\pi<\alpha<2k\pi,\ k\in\mathbf{Z}.\end{cases}$$

(3) Using the double/half angle formula, we see that the expression equals

$$\frac{8\sin\theta\cos\theta\cos2\theta\cos4\theta}{8\sin\theta}$$

$$=\frac{4\sin2\theta\cos2\theta\cos4\theta}{8\sin\theta}$$

$$=\frac{2\sin4\theta\cos4\theta}{8\sin\theta}=\frac{\sin8\theta}{8\sin\theta}.$$

Remark. Formulae we use when dealing with trigonometric expressions largely affect the amount of work involved, so we need to examine the structural characteristics of the expression such as symmetry, transitivity, and homogeneity, before choosing formulae to use.

The result in (3) can be used to simplify expressions such as $\cos\frac{\pi}{7}\cos\frac{2\pi}{7}\cos\frac{4\pi}{7}$, where we can let $\theta=\frac{\pi}{7}$ and get that $\cos\frac{\pi}{7}\cos\frac{2\pi}{7}\cos\frac{4\pi}{7}=\frac{\sin\frac{8\pi}{7}}{8\sin\frac{\pi}{7}}=-\frac{1}{8}$.

Example 2. Let $\alpha \in \mathbf{R}$ such that $\sin\alpha + \sin^2\alpha = 1$. Find the values of the following expressions:

(1) $\cos^2\alpha + \cos^4\alpha$;
(2) $\cos^2\alpha + \cos^6\alpha$;
(3) $\cos^2\alpha + \cos^6\alpha + \cos^8\alpha$.

Solution. Considering the given condition, we may try to lower the degree of the expression to make it closer to the condition.

(1) The condition implies that $\cos^2\alpha = 1 - \sin^2\alpha = \sin\alpha$, so

$$\cos^2\alpha + \cos^4\alpha = \sin\alpha + \sin^2\alpha = 1.$$

(2) By $\sin\alpha + \sin^2\alpha = 1$ we can get that $\cos^2\alpha = \sin\alpha$, thus $\sin\alpha = \frac{-1\pm\sqrt{5}}{2}$. Note that $|\sin\alpha| \leq 1$, so necessarily $\sin\alpha = \frac{-1+\sqrt{5}}{2}$, and consequently

$$\begin{aligned}\cos^2\alpha + \cos^6\alpha &= \sin\alpha + \sin^3\alpha = \sin\alpha(1 + \sin^2\alpha)\\ &= \sin\alpha(2 - \sin\alpha) = 2\sin\alpha - \sin^2\alpha = 3\sin\alpha - 1 = \frac{3\sqrt{5}-5}{2}.\end{aligned}$$

(3) Instead of using the result of (2), we can first try to lower the degree of the expression:

$$\begin{aligned}\cos^2\alpha + \cos^6\alpha + \cos^8\alpha &= \sin\alpha + \sin^3\alpha + \sin^4\alpha\\ &= \sin\alpha + \sin^2\alpha(\sin\alpha + \sin^2\alpha) = \sin\alpha + \sin^2\alpha = 1.\end{aligned}$$

Example 3. Given $\alpha = \frac{\pi}{12}$, $\beta = \frac{5\pi}{24}$, and $\gamma = \frac{7\pi}{24}$, find the value of $\sin(\alpha + \beta)\cos\gamma - \sin(\alpha+\gamma)\cos\beta$.

Solution. By calculation,

$$\begin{aligned}&\sin(\alpha+\beta)\cos\gamma - \sin(\alpha+\gamma)\cos\beta\\ &\quad= (\sin\alpha\cos\beta + \cos\alpha\sin\beta)\cos\gamma - (\sin\alpha\cos\gamma + \cos\alpha\sin\gamma)\cos\beta\\ &\quad= \cos\alpha(\sin\beta\cos\gamma - \sin\gamma\cos\beta) = \cos\alpha \cdot sin(\beta - \gamma)\\ &\quad= \cos\frac{\pi}{12}\cdot\sin\left(-\frac{\pi}{12}\right) = -\frac{1}{2}\sin\frac{\pi}{6} = -\frac{1}{4}.\end{aligned}$$

Example 4. Suppose $\sin\alpha + \cos\beta = \frac{3}{5}$ and $\cos\alpha + \sin\beta = \frac{4}{5}$. Find the value of $\cos\alpha\sin\beta$.

Solution. While this problem seems simple, it requires some techniques to do the deformation. Finding the exact values of $\sin\alpha$, $\cos\alpha$, $\sin\beta$, and $\cos\beta$ is not a good approach to this problem.

Since

$$\begin{cases} \sin\alpha + \cos\beta = \dfrac{3}{5}, & ① \\ \cos\alpha + \sin\beta = \dfrac{4}{5}, & ② \end{cases}$$

by $①^2 + ②^2$, we have $2 + 2(\sin\alpha\cos\beta + \cos\alpha\sin\beta) = 1$, or equivalently $\sin(\alpha+\beta) = -\frac{1}{2}$.

By $②^2 - ①^2$ we get $\cos 2\alpha - \cos 2\beta + 2(\cos\alpha\sin\beta - \sin\alpha\cos\beta) = \frac{7}{25}$. Using the sum-to-product formula on the left side, we have

$$-2\sin(\alpha+\beta)\sin(\alpha-\beta) - 2\sin(\alpha-\beta) = \frac{7}{25},$$

and combining it with $\sin(\alpha+\beta) = -\frac{1}{2}$, we get

$$\sin(\alpha-\beta) = -\frac{7}{25}.$$

Therefore,

$$\begin{aligned} \cos\alpha\sin\beta &= \frac{1}{2}(\sin(\alpha+\beta) - \sin(\alpha-\beta)) \\ &= \frac{1}{2}\left(-\frac{1}{2} + \frac{7}{25}\right) = -\frac{11}{100}. \end{aligned}$$

Example 5. Find the value of the following expression:

$$A = \cot 15^\circ \cot 25^\circ \cot 35^\circ \cot 85^\circ.$$

Solution. By the induction formula,

$$\begin{aligned} A &= \tan 75^\circ \tan 65^\circ \tan 55^\circ \tan 5^\circ \\ &= \tan 75^\circ \cdot \tan 5^\circ \cdot \frac{\sqrt{3} + \tan 5^\circ}{1 - \sqrt{3}\tan 5^\circ} \cdot \frac{\sqrt{3} - \tan 5^\circ}{1 + \sqrt{3}\tan 5^\circ} \\ &= \tan 75^\circ \cdot \tan 5^\circ \cdot \frac{3 - \tan^2 5^\circ}{1 - 3\tan^2 5^\circ} \\ &= \tan 75^\circ \cdot \tan 5^\circ \cdot \frac{3\cos^2 5^\circ - \sin^2 5^\circ}{\cos^2 5^\circ - 3\sin^2 5^\circ} \\ &= \tan 75^\circ \cdot \tan 5^\circ \cdot \frac{2\cos 10^\circ + 1}{2\cos 10^\circ - 1} \end{aligned}$$

$$= \tan 75^\circ \cdot \frac{2\sin 5^\circ \cos 10^\circ + \sin 5^\circ}{2\cos 5^\circ \cos 10^\circ - \cos 5^\circ}$$
$$= \tan 75^\circ \cdot \frac{\sin 15^\circ - \sin 5^\circ + \sin 5^\circ}{\cos 15^\circ + \cos 5^\circ - \cos 5^\circ}$$
$$= \tan 75^\circ \tan 15^\circ = 1.$$

Remark. The crucial part in this problem is to calculate $\tan 5^\circ \tan 55^\circ \tan 65^\circ$. In fact, with the method above we can get a triple-angle formula for the tangent function:

$$\tan 3\theta = \tan\theta \tan(60^\circ - \theta)\tan(60^\circ + \theta).$$

Example 6. Prove the following identity:

$$\cos\frac{2\pi}{2n+1} + \cos\frac{4\pi}{2n+1} + \cdots + \cos\frac{2n\pi}{2n+1} = -\frac{1}{2}, ①$$

where $n \in \mathbf{N}^*$.

Solution. Note that the arguments of the cosine function on the left side form an arithmetic sequence, so we have the following proof.

Let $\frac{\pi}{2n+1} = \alpha$. If we multiply the left side of ① by $2\sin\alpha$, then

$$2\sin\alpha\left(\cos\frac{2\pi}{2n+1} + \cos\frac{4\pi}{2n+1} + \cdots + \cos\frac{2n\pi}{2n+1}\right)$$
$$= 2\sin\alpha\cos 2\alpha + 2\sin\alpha\cos 4\alpha + \cdots + 2\sin\alpha\cos 2n\alpha$$
$$= (\sin 3\alpha - \sin\alpha) + (\sin 5\alpha - \sin 3\alpha)$$
$$+ \cdots + (\sin(2n+1)\alpha - \sin(2n-1)\alpha)$$
$$= \sin(2n+1)\alpha - \sin\alpha$$
$$= \sin\pi - \sin\alpha = -\sin\alpha.$$

Since $\sin\alpha \neq 0$, we can divide both sides by $2\sin\alpha$ and get the desired identity.

Remark. Here we have used a common method in summing sequences, named "summation by cancellation". We were led to this method by the fact that the arguments of the cosine function form an arithmetic sequence. In particular, when $n = 3$, the result says that

$$\cos\frac{2\pi}{7} + \cos\frac{4\pi}{7} + \cos\frac{6\pi}{7} = -\frac{1}{2},$$

or equivalently $\cos\frac{\pi}{7} + \cos\frac{3\pi}{7} + \cos\frac{5\pi}{7} = \frac{1}{2}$. This is usually seen as an evaluation problem.

Example 7 (2017 Russian Mathematical Olympiad). Suppose x is a real number such that $S = \sin 64x + \sin 65x$ and $C = \cos 64x + \cos 65x$ are both rational. Show that one of S and C has both rational summands.

Solution. Since

$$\begin{aligned} S^2 + C^2 &= (\sin^2 64x + \cos^2 64x) + (\sin^2 65x + \cos^2 65x) \\ &\quad + 2(\sin 64x \sin 65x + \cos 64x \cos 65x) \\ &= 2 + 2\cos(65x - 64x) = 2 + 2\cos x, \end{aligned}$$

which is rational, we see that $\cos x$ is rational. Note that $\cos 2a = 2\cos^2 a - 1$, so by induction, $\cos 2^k x$ is rational for every positive integer k. In particular, $\cos 64x$ is rational, and hence the two summands in C are both rational.

13.3 Exercises

Group A

1. Simplify the following expressions:

   (1) $\frac{(\tan\alpha+\cot\alpha)\sin\alpha\cos\alpha}{1/(1+\tan^2\alpha)+1/(1+\cot^2\alpha)}$;

   (2) $\frac{\sin\alpha+\sin\beta}{\cos\alpha+\cos\beta} + \frac{\cos\alpha-\cos\beta}{\sin\alpha-\sin\beta}$;

   (3) $\cos A\sin(B - C) + \cos B\sin(C - A) + \cos C\sin(A - B)$.

2. Calculate the values of the following expressions:

   (1) $\frac{1}{2}\csc 10^\circ - 2\sin 70^\circ$;

   (2) $\sin 20^\circ\cos^2 25^\circ - \sin 20^\circ\sin^2 25^\circ + \cos^2 50^\circ + \sin^2 20^\circ$;

   (3) $\frac{\sin 7^\circ+\sin 8^\circ\cos 15^\circ}{\cos 7^\circ-\sin 8^\circ\sin 15^\circ}$;

   (4) $(1 + \cot 46^\circ)(1 + \cot 47^\circ)\cdots(1 + \cot 89^\circ)$.

3. Suppose $\sin\theta - \cos\theta = \frac{1}{2}$. Find the value of $\sin^3\theta - \cos^3\theta$.
4. Suppose an angle α satisfies $\frac{\cos 3\alpha}{\cos\alpha} = \frac{1}{3}$. Find the value of $\frac{\sin 3\alpha}{\sin\alpha}$.
5. Let α and β be acute angles such that $\cos(\alpha+\beta) = \frac{4}{5}$ and $\cos(2\alpha+\beta) = \frac{3}{5}$. Find the value of $\cos\alpha$.
6. Let α, β, and γ be real numbers such that $\beta - \alpha = \gamma - \beta = \frac{\pi}{3}$. Find the value of $\tan\alpha\tan\beta + \tan\beta\tan\gamma + \tan\gamma\tan\alpha$.
7. Let α be an acute angle such that $\tan\alpha = \tan(\alpha + 10^\circ)\tan(\alpha + 20^\circ)\tan(\alpha + 30^\circ)$. Find all possible values of α.

8. In a triangle ABC,
   (1) if $\sin A$, $\sin B$, and $\sin C$ form an arithmetic sequence (in that order), show that $\cot\frac{A}{2}\cot\frac{C}{2}=3$;
   (2) if $\sin C=\frac{\sin A+\sin B}{\cos A+\cos B}$, show that the triangle ABC has a right angle;
   (3) if $\frac{\sin A+\sin B+\sin C}{\cos A+\cos B+\cos C}=\sqrt{3}$, show that one of the inner angles is $\frac{\pi}{3}$.
9. In a triangle ABC, prove the following identities:
   (1) $\sin A+\sin B+\sin C=4\cos\frac{A}{2}\cos\frac{B}{2}\cos\frac{C}{2}$;
   (2) $\cos A+\cos B+\cos C=1+4\sin\frac{A}{2}\sin\frac{B}{2}\sin\frac{C}{2}$;
   (3) $\tan A+\tan B+\tan C=\tan A\tan B\tan C$;
   (4) $\sin^2 A+\sin^2 B+\sin^2 C=2+2\cos A\cos B\cos C$;
   (5) $\cos^2 A+\cos^2 B+\cos^2 C=1-2\cos A\cos B\cos C$;
   (6) $\cot A\cot B+\cot B\cot C+\cot C\cot A=1$;
   (7) $\tan\frac{A}{2}\tan\frac{B}{2}+\tan\frac{B}{2}\tan\frac{C}{2}+\tan\frac{C}{2}\tan\frac{A}{2}=1$.

Group B

10. Let $\alpha\in\mathbf{R}$. Show that for each $n\in\mathbf{N}^*$,

$$\sin^3\frac{\alpha}{3}+3\sin^3\frac{\alpha}{3^2}+\cdots+3^{n-1}\sin^3\frac{\alpha}{3^n}=\frac{1}{4}\left(3^n\sin\frac{\alpha}{3^n}-\sin\alpha\right).$$

11. Let $\alpha\in\mathbf{R}$. Show that
   (1) $\tan 2\alpha+\frac{1}{2}\tan\alpha=\frac{1}{2}\cot\alpha-2\cot 4\alpha$. ①
   (2) For every positive even number n,

$$\tan\alpha+2\tan 2\alpha+\cdots+2^{n-1}\tan(2^{n-1}\alpha)=\cot\alpha-2^n\cot(2^n\alpha).\ ②$$

12. Let a,b, and c denote the three sides of a triangle ABC, and p be half its perimeter. Show that

$$(a+b)\cos C+(b+c)\cos A+(c+a)\cos B=2p.$$

13. Let A,B, and C be the three inner angles of a triangle. Suppose a real number x satisfies

$$\cos^3 x+\cos(x+A)\cos(x+B)\cos(x+C)=0.$$

Show that (1) $\tan x=\cot A+\cot B+\cot C$; (2) $\sec^2 x=\csc^2 A+\csc^2 B+\csc^2 C$.

14. Let $n\in\mathbf{N}^*$. Find the value of the product $\prod_{k=0}^{2n-1}\left(4\sin^2\frac{k\pi}{2^n}-3\right)$.

Chapter 14

Trigonometric Inequalities

14.1 Key Points of Knowledge and Basic Methods

When dealing with inequalities with trigonometric functions, we usually combine properties of trigonometric functions with general methods for proving inequalities.

When solving such problems, we need to make a good use of deformations of trigonometric expressions, as well as the boundedness and monotonicity of trigonometric functions. If the inequality involves the angles of a triangle, we also have the law of sines and law of cosines, together with some inequalities regarding the angles and sides of a triangle.

In addition, the following inequality is an important property of trigonometric functions, which is used in some problems:

If $x \in \left(0, \frac{\pi}{2}\right)$, then $\sin x < x < \tan x$. This inequality can be proved by using the unit circle model.

14.2 Illustrative Examples

Example 1. Let ABC be an acute triangle. Show that

$$\tan A + \tan B + \tan C > \cot A + \cot B + \cot C.$$

Solution. We apply the monotonicity of the tangent function on $\left(0, \frac{\pi}{2}\right)$.

Since ABC is an acute triangle, $0 < C = \pi - (A + B) < \frac{\pi}{2}$, so $A + B > \frac{\pi}{2}$, and

$$0 < \frac{\pi}{2} - B < A < \frac{\pi}{2}.$$

Hence, $\tan\left(\frac{\pi}{2} - B\right) < \tan A$, namely $\tan A > \cot B$.

Similarly, $\tan B > \cot C$ and $\tan C > \cot A$. Adding the three inequalities, we get the desired inequality.

Example 2. Let θ be an acute angle. Compare the two numbers $\frac{1-\sin\theta}{1-\cos\theta}$ and $\tan\theta$.

Solution. We use comparison by difference:

$$\frac{1-\sin\theta}{1-\cos\theta} - \tan\theta = \frac{\cos\theta(1-\sin\theta) - \sin\theta(1-\cos\theta)}{\cos\theta(1-\cos\theta)}$$
$$= \frac{\cos\theta - \sin\theta}{\cos\theta(1-\cos\theta)}.$$

Since θ is an acute angle, $\cos\theta(1-\cos\theta) > 0$.

If $0 < \theta < \frac{\pi}{4}$, then $\cos\theta = \sin\left(\frac{\pi}{2}-\theta\right) > \sin\theta$; if $\frac{\pi}{4} < \theta < \frac{\pi}{2}$, then $\cos\theta < \sin\theta$; if $\theta = \frac{\pi}{4}$, then $\cos\theta = \sin\theta$.

Therefore, the result is $\frac{1-\sin\theta}{1-\cos\theta} > \tan\theta$ when $0 < \theta < \frac{\pi}{4}$, $\frac{1-\sin\theta}{1-\cos\theta} = \tan\theta$ when $\frac{1-\sin\theta}{1-\cos\theta} = \tan\theta$, and $\frac{1-\sin\theta}{1-\cos\theta} < \tan\theta$ when $\frac{\pi}{4} < \theta < \frac{\pi}{2}$.

Remark. Note that the two numbers are both positive, so we can also use comparison by ratio.

Example 3. Suppose $\frac{2\pi}{3} < \theta < \frac{4\pi}{5}$. Show that $-\frac{4\sqrt{3}}{3} < \tan\theta + \cot\theta \leq -2$.

Solution. Let $\tan\theta = t$. Then $\tan\theta + \cot\theta = t + \frac{1}{t}$, which we denote as $f(t)$.

Note that $y = \tan x$ is increasing on $\left(\frac{\pi}{2}, \pi\right)$, so $t \in (-\sqrt{3}, -1]$ when $\theta \in \left(\frac{2\pi}{3}, \frac{3\pi}{4}\right]$, and consequently $f(t) \in (f(-\sqrt{3}), f(-1)]$; when $\theta \in \left(\frac{3\pi}{4}, \frac{4\pi}{5}\right)$, we have $t \in \left(-1, \tan\frac{4\pi}{5}\right)$, where $\tan\frac{4\pi}{5} < \tan\frac{5\pi}{6} = -\frac{\sqrt{3}}{3}$, and so

$$f(-1) > f(t) > f\left(-\frac{\sqrt{3}}{3}\right) = f(-\sqrt{3}).$$

Therefore, $f(t) \in (f(-\sqrt{3}), f(-1)] = \left(-\frac{4\sqrt{3}}{3}, -2\right]$, hence

$$-\frac{4\sqrt{3}}{3} < \tan\theta + \cot\theta \leq -2.$$

Example 4. Let x and y be acute angles and $\tan x = 3\tan y$. Show that $x - y \leq \frac{\pi}{6}$.

Solution. Since x and y are acute angles, $\tan x = 3\tan y > \tan y$, so $x - y > 0$. Further,

$$\tan(x-y) = \frac{\tan x - \tan y}{1 + \tan x \tan y} = \frac{2\tan y}{1 + 3\tan^2 y}$$
$$\leq \frac{2\tan y}{2\sqrt{3}\tan y} = \frac{\sqrt{3}}{3} = \tan\frac{\pi}{6}.$$

Therefore, by monotonicity $x - y \leq \frac{\pi}{6}$.

Remark. Here $1 + 3\tan^2 y \geq 2\sqrt{3}\tan y$ was obtained by the AM-GM inequality. Some famous inequalities are also useful in problems with trigonometric inequalities, such as the AM-GM inequality and Cauchy's inequality.

Example 5. Let A, B, and C denote the three inner angles of a triangle ABC. Show that

$$\sin\frac{A}{2}\sin\frac{B}{2}\sin\frac{C}{2} \leq \frac{1}{8}.$$

Solution. Note that $\sin\frac{A}{2} > 0$, $\sin\frac{B}{2} > 0$, and $\sin\frac{C}{2} > 0$, and so

$$\sin\frac{A}{2}\sin\frac{B}{2}\sin\frac{C}{2} = \frac{1}{2}\sin\frac{A}{2}\left(\cos\frac{B-C}{2} - \cos\frac{B+C}{2}\right)$$
$$\leq \frac{1}{2}\sin\frac{A}{2}\left(1 - \cos\frac{B+C}{2}\right) = \frac{1}{2}\sin\frac{A}{2}\left(1 - \sin\frac{A}{2}\right)$$
$$\leq \frac{1}{2}\left(\frac{\sin\frac{A}{2} + 1 - \sin\frac{A}{2}}{2}\right)^2 = \frac{1}{8}.$$

Therefore, the original inequality holds.

Remark. The proof above used two simple facts:

$$(1)\ \cos\frac{B-C}{2} \leq 1. \quad (2)\ \text{If } 0 \leq x \leq 1, \text{ then } x(1-x) \leq \frac{1}{4}.$$

The first one follows by boundedness of trigonometric functions, and the second one follows from the AM-GM inequality. The fact (1) is seen frequently in trigonometric inequalities.

Example 6. Let α and β be acute angles. Show that

$$\cos\alpha + \cos\beta + \sqrt{2}\sin\alpha\sin\beta \leq \frac{3\sqrt{2}}{2}.$$

Solution. This inequality becomes an equality when $\alpha = \beta = \frac{\pi}{4}$, and the steps below are based on this fact.

$$\begin{aligned}
&\cos\alpha + \cos\beta + \sqrt{2}\sin\alpha\sin\beta \\
&= 2\cos\frac{\alpha+\beta}{2}\cos\frac{\alpha-\beta}{2} + \frac{\sqrt{2}}{2}(\cos(\alpha-\beta) - \cos(\alpha+\beta)) \\
&\leq 2\cos\frac{\alpha+\beta}{2} + \frac{\sqrt{2}}{2}(1 - \cos(\alpha+\beta)) \\
&= 2\cos\frac{\alpha+\beta}{2} + \frac{\sqrt{2}}{2}\left(2 - 2\cos^2\left(\frac{\alpha+\beta}{2}\right)\right) \\
&= \sqrt{2} - \sqrt{2}\cos^2\frac{\alpha+\beta}{2} + 2\cos\frac{\alpha+\beta}{2} \\
&= \frac{3\sqrt{2}}{2} - \sqrt{2}\left(\cos\frac{\alpha+\beta}{2} - \frac{\sqrt{2}}{2}\right)^2 \leq \frac{3\sqrt{2}}{2}.
\end{aligned}$$

Therefore, the proposition is true.

Example 7. Suppose the inequality

$$\sin 2\theta - 2\sqrt{2}a\cos\left(\theta - \frac{\pi}{4}\right) - \frac{\sqrt{2}a}{\sin\left(\theta + \frac{\pi}{4}\right)} > -3 - a^2$$

is satisfied for all $\theta \in \left[0, \frac{\pi}{2}\right]$. Find the value range of a.

Solution. Let $x = \sin\theta + \cos\theta$ for $\theta \in \left[0, \frac{\pi}{2}\right]$. Then $x \in [1, \sqrt{2}]$ and

$$\sin 2\theta = x^2 - 1, \quad \sin\left(\theta + \frac{\pi}{4}\right) = \cos\left(\theta - \frac{\pi}{4}\right) = \frac{\sqrt{2}}{2}x.$$

The original inequality reduces to $x^2 - 2ax - \frac{2a}{x} + 2 + a^2 > 0$, or equivalently

$$(a - x)\left[a - \left(x + \frac{2}{x}\right)\right] > 0. ①$$

Let $f(x) = x + \frac{2}{x}$. Then $f(x)$ is decreasing on $[1, \sqrt{2}]$, so $f(x)_{\max} = f(1) = 3$.

If $1 \leq a \leq 3$, then ① does not hold for $x = 1$.
If $a < 1$ or $a > 3$, then $a - x$ and $a - \left(x + \frac{2}{x}\right)$ have the same sign for all $x \in [1, \sqrt{2}]$, and the inequality ① holds.

Therefore, the value range of a is $(-\infty, 1) \cup (3, +\infty)$.

Example 8. Let x, y, and z be real numbers such that $0 < x < y < z < \frac{\pi}{2}$. Show that

$$\frac{\pi}{2} + 2\sin x\cos y + 2\sin y\cos z > \sin 2x + \sin 2y + \sin 2z.$$

Solution. By the double-angle formula, the inequality is equivalent to

$$\frac{\pi}{4} + \sin x\cos y + \sin y\cos z > \sin x\cos x + \sin y\cos y + \sin z\cos z,$$

or equivalently

$$\frac{\pi}{4} > \sin x(\cos x - \cos y) + \sin y(\cos y - \cos z) + \sin z\cos z. \text{①}$$

Note that the right side of ① is the area of the shaded region in Figure 14.1, while $\frac{\pi}{4}$ is the area of the quarter circle in the first quadrant. Therefore, the inequality ① follows.

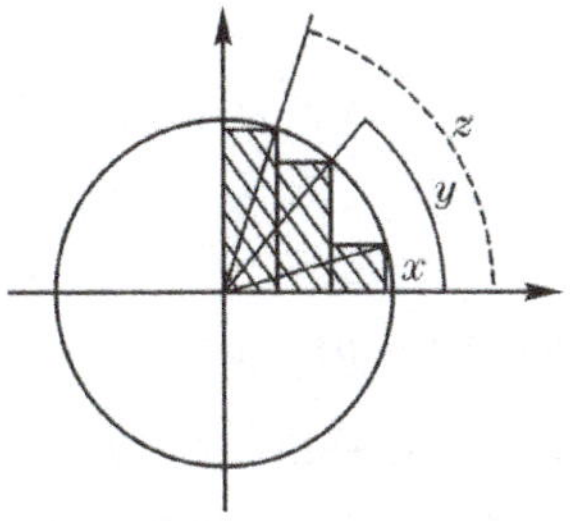

Figure 14.1

Remark. Trigonometric problems usually have geometric interpretations. Many plane geometry problems can be solved with trigonometry, and conversely we can also solve trigonometric problems by constructing geometric objects.

14.3 Exercises

Group A

1. Let $\theta \in \left(0, \frac{\pi}{2}\right)$ and $f(x) = \log_{\sin\theta} x$. Compare $f\left(\frac{\sin\theta+\cos\theta}{2}\right)$ and $f\left(\frac{\sin 2\theta}{\sin\theta+\cos\theta}\right)$.
2. Let α and β be two inner angles of a triangle ABC that are both acute and $f(x) = \left(\frac{\cos\alpha}{\sin\beta}\right)^x + \left(\frac{\cos\beta}{\sin\alpha}\right)^x$ such that $f(x) < 2$ for all $x > 0$. Show that the third inner angle of ABC is also acute.

3. Let α and β be acute angles such that $\sin^2\alpha = \cos(\alpha - \beta)$. Compare α and β.
4. Show that for all $x \in \mathbf{R}$,

$$2\sin^2\left(\frac{\pi}{4} - \frac{\sqrt{2}}{2}\right) \le \cos(\sin x) - \sin(\cos x) \le 2\sin^2\left(\frac{\pi}{4} + \frac{\sqrt{2}}{2}\right).$$

5. In a triangle ABC, show the following inequalities:
   (1) $\cos A \cos B \cos C \le \frac{1}{8}$;
   (2) $1 < \cos A + \cos B + \cos C \le \frac{3}{2}$;
   (3) $\sin A + \sin B + \sin C \le \frac{3\sqrt{3}}{2}$.
6. Let ABC be an acute triangle. Show that
   (1) $1 + \cos A + \cos B + \cos C < \sin A + \sin B + \sin C$.
   (2) For each $n \in \mathbf{N}^*$,

$$\tan^n A + \tan^n B + \tan^n C > 3 + \frac{3n}{2}.$$

7. Suppose

$$f(x) = 1 - a\cos x - b\sin x - A\cos 2x - B\sin 2x \ge 0$$

for all real numbers x. Show that $A^2 + B^2 \le 1$ and $a^2 + b^2 \le 2$.
8. Let A, B, and C be acute angles. Show that

$$\frac{\sin A\sin(A-B)\sin(A-C)}{\sin(B+C)} + \frac{\sin B\sin(B-A)\sin(B-C)}{\sin(C+A)} + \frac{\sin C\sin(C-A)\sin(C-B)}{\sin(A+B)} \ge 0.$$

9. Let A, B, and C be the inner angles of a triangle. Show that

$$\cos^2\frac{A-B}{2}\cos^2\frac{B-C}{2}\cos^2\frac{C-A}{2} \ge \left(8\sin\frac{A}{2}\sin\frac{B}{2}\sin\frac{C}{2}\right)^3.$$

Group B

10. Let a, b, and c be the lengths of the sides opposite to A, B, and C in a triangle ABC, respectively. Given $\angle C \ge 60°$, show that

$$(a+b)\left(\frac{1}{a} + \frac{1}{b} + \frac{1}{c}\right) \ge 4 + \frac{1}{\sin\frac{C}{2}}.$$

11. Let x, y, and z be real numbers and let A, B, and C be the inner angles of a triangle ABC. Show that

$$x^2 + y^2 + z^2 \geq 2yz\cos A + 2zx\cos B + 2xy\cos C.$$

12. Find all real numbers α such that every term of the sequence

$$\cos\alpha, \quad \cos 2\alpha, \quad \cos 4\alpha, \ldots, \cos 2^n\alpha, \ldots$$

is negative.

13. Let $x_1, x_2, \ldots$ and $y_1, y_2, \ldots$ be sequences of real numbers. Suppose $x_1 = y_1 = \sqrt{3}$ and

$$x_{n+1} = x_n + \sqrt{1 + x_n^2}, \quad y_{n+1} = \frac{y_n}{1 + \sqrt{1 + y_n^2}}, \qquad n = 1, 2, \ldots.$$

Show that $2 < x_n y_n < 3$ for all $n > 1$.

14. Let $0 < x, y < \frac{\pi}{2}$. Show that

$$\frac{x\csc x + y\csc y}{2} < \sec\frac{x+y}{2}.$$

15. Let $n \in \mathbf{N}^*$ for $\theta \in \mathbf{R}$. Show that

$$|\sin\theta\sin 2\theta \ldots \sin 2^n\theta| \leq \left(\frac{\sqrt{3}}{2}\right)^n.$$

Chapter 15

Extreme Value Problems of Trigonometric Functions

15.1 Key Points of Knowledge and Basic Methods

Compared to the general methods for extreme value problems of functions discussed in Chapter 7, the special characteristics of trigonometric functions, such as monotonicity, boundedness, and periodicity, make it quite different to solve extreme value problems of trigonometric functions.

On the other hand, finding extrema of functions usually involves two steps, namely proving an inequality and finding an instance to show that the bound is reachable. This pattern also applies to trigonometric expressions, and in addition, we need to consider the periodicity of trigonometric functions to determine the condition for the equality to be valid.

When dealing with such problems, a basic idea is to find a correct deformation for the trigonometric expression and reduce it to the simplest form, whose extrema can be found easily.

15.2 Illustrative Examples

Example 1. Let x and y be real numbers such that $x^2 + 2\cos y = 1$. Find the value range of $x - \cos y$.

Solution. Since $x^2 = 1 - 2\cos y \in [-1, 3]$, we have $x \in [-\sqrt{3}, \sqrt{3}]$. By $\cos y = \frac{1-x^2}{2}$,

$$x - \cos y = x - \frac{1 - x^2}{2} = \frac{1}{2}(x + 1)^2 - 1.$$

Hence, when $x = -1$, we know that $x - \cos y$ attains its minimum value, which is -1 (here y can take values such as $\frac{\pi}{2}$); when $x = \sqrt{3}$, the function $x - \cos y$ of y attains its maximum value, which is $\sqrt{3} + 1$ (here y can take values such as π). Since the value range of $\frac{1}{2}(x+1)^2 - 1$ is $[-1, \sqrt{3}+1]$, the value range of $x - \cos y$ is $[-1, \sqrt{3}+1]$.

Example 2. Let $x, y \in [0, 2\pi]$ be such that

$$2 \sin x \cos y + \sin x + \cos y = -\frac{1}{2}.$$

Find the maximum value of $x + y$.

Solution. Since $2 \sin x \cos y + \sin x + \cos y = -\frac{1}{2}$,

$$(2 \sin x + 1)(2 \cos y + 1) = 0.$$

This implies that $\sin x = -\frac{1}{2}$ or $\cos y = -\frac{1}{2}$. If $\sin x = -\frac{1}{2}$, then $x = \frac{7\pi}{6}$ or $\frac{11\pi}{6}$, and y can take any value in $[0, 2\pi]$; if $\cos y = -\frac{1}{2}$, then $y = \frac{2\pi}{3}$ or $\frac{4\pi}{3}$, and x can take any value in $[0, 2\pi]$.

Therefore, the maximum value of $x + y$ is $\frac{11\pi}{6} + 2\pi = \frac{23\pi}{6}$.

Example 3. Let A, B, and C be the inner angles of a triangle. Find the minimum value of $\cos A(\sin B + \sin C)$.

Solution. Note that if A is not an obtuse angle, then $\cos A(\sin B + \sin C) \geq 0$. Hence, the minimum value of this expression is attained when A is abtuse. In this case $\cos A < 0$ and

$$\begin{aligned}
&\cos A(\sin B + \sin C) \\
&= 2 \cos A \sin \frac{B+C}{2} \cos \frac{B-C}{2} \\
&\geq 2 \cos A \sin \frac{B+C}{2} = 2 \cos A \cos \frac{A}{2} \\
&= -2 \left(\sin^2 \frac{A}{2} - \cos^2 \frac{A}{2} \right) \cos \frac{A}{2} \\
&= -2 \sqrt{\left(\sin^2 \frac{A}{2} - \cos^2 \frac{A}{2} \right)^2 \cos^2 \frac{A}{2}} \\
&= -\sqrt{\left(\sin^2 \frac{A}{2} - \cos^2 \frac{A}{2} \right) \left(\sin^2 \frac{A}{2} - \cos^2 \frac{A}{2} \right) \left(4 \cos^2 \frac{A}{2} \right)}
\end{aligned}$$

$$\geq -\sqrt{\left(\frac{\left(\sin^2 \frac{A}{2} - \cos^2 \frac{A}{2}\right) \times 2 + 4\cos^2 \frac{A}{2}}{3}\right)^3}$$

$$= -\sqrt{\left(\frac{2}{3}\right)^3} = -\frac{2\sqrt{6}}{9}.$$

The equality holds when $B = C$, $\sin^2 \frac{A}{2} - \cos^2 \frac{A}{2} = 4\cos^2 \frac{A}{2}$, and A is an obtuse angle. Equivalently, the equality holds when $B = C$ and A is an obtuse angle that satisfies $\cos \frac{A}{2} = \frac{\sqrt{6}}{6}$. (Note that by monotonicity $\frac{A}{2} > \frac{\pi}{4}$, so A is obtuse.)

Therefore, the minimum value of $\cos A(\sin B + \sin C)$ is $-\frac{2\sqrt{6}}{9}$.

Remark. Here we have used the technique of scaling $\cos \frac{B-C}{2}$ to 1. In fact, we also have

$$\cos A(\sin B + \sin C) \leq 2$$

(since $\cos A \leq 1$ and $\sin B + \sin C \leq 2$), but the equality cannot hold, because if the equality holds then $A = 0$ and $B = C = \frac{\pi}{2}$. This means that $\cos A(\sin B + \sin C)$ has an upper bound, but no maximum value.

Example 4. Let $w > 0$. Suppose the function

$$f(x) = \sin^2 wx + \sqrt{3}\sin wx \cdot \sin\left(wx + \frac{\pi}{2}\right)$$

has the least positive period $\frac{\pi}{2}$. Find the maximum and minimum values of $f(x)$ on $\left[\frac{\pi}{8}, \frac{\pi}{4}\right]$

Solution. Since

$$\begin{aligned} f(x) &= \frac{1 - \cos 2wx}{2} + \frac{\sqrt{3}}{2}\sin 2wx \\ &= \frac{\sqrt{3}}{2}\sin 2wx - \frac{1}{2}\cos 2wx + \frac{1}{2} \\ &= \sin\left(2wx - \frac{\pi}{6}\right) + \frac{1}{2}, \end{aligned}$$

it follows that $T = \frac{2\pi}{2w} = \frac{\pi}{2}$, thus $w = 2$. Hence, $f(x) = \sin\left(4x - \frac{\pi}{6}\right) + \frac{1}{2}$.

If $\frac{\pi}{8} \leq x \leq \frac{\pi}{4}$, then $\frac{\pi}{3} \leq 4x - \frac{\pi}{6} \leq \frac{5\pi}{6}$, and so $\frac{1}{2} \leq \sin\left(4x - \frac{\pi}{6}\right) \leq 1$, or equivalently

$$1 \leq \sin\left(4x - \frac{\pi}{6}\right) + \frac{1}{2} \leq \frac{3}{2}.$$

Therefore, $f(x)$ has the maximum value $\frac{3}{2}$ when $x = \frac{\pi}{6}$ and the minimum value 1 when $x = \frac{\pi}{4}$.

Example 5. Let $\alpha,\ \beta \in \left[0, \frac{\pi}{4}\right]$. Find the maximum value of $\sin(\alpha-\beta) + 2\sin(\alpha+\beta)$.

Solution. Using the sum-angle formula, we have

$$\sin(\alpha-\beta)+2\sin(\alpha+\beta) = 3\sin\alpha\cos\beta+\cos\alpha\sin\beta.$$

Now, we view α as a constant, and by $A\cos\beta + B\sin\beta \le \sqrt{A^2+B^2}$,

$$3\sin\alpha\cos\beta+\cos\alpha\sin\beta \le \sqrt{(3\sin\alpha)^2+(\cos\alpha)^2} = \sqrt{8\sin^2\alpha+1}.$$

Since $\alpha \in \left[0, \frac{\pi}{4}\right]$, we have $\sqrt{8\sin^2\alpha+1} \le \sqrt{5}$, so

$$\sin(\alpha-\beta)+2\sin(\alpha+\beta) \le \sqrt{5}.$$

The equality holds when $\alpha = \frac{\pi}{4}$ and $\beta = \arctan\frac{1}{3}$. Therefore, the maximum value of the expression is $\sqrt{5}$.

Remark. If we view β as a constant instead, then

$$\begin{aligned} 3\sin\alpha\cos\beta+\cos\alpha\sin\beta &\le \sqrt{(3\cos\beta)^2+\sin^2\beta} = \sqrt{8\cos^2\beta+1} \\ &\le \sqrt{8+1} = 3. \end{aligned}$$

However, this bound is reached when $\beta = 0$ and $\sin\alpha = 1$, which is not possible for $\alpha,\ \beta \in \left[0, \frac{\pi}{4}\right]$. Therefore, whether a bound is reachable is very important, and sometimes we need certain foresight when estimating an expression.

Example 6. Let $g(\theta)$ denote the minimum value of the function

$$f(x) = (\sin x + 4\sin\theta + 4)^2 + (\cos x - 5\cos\theta)^2$$

Find the maximum value of $g(\theta)$ for all real θ.

Solution. Let $u = 5\cos\theta$ and $v = -4(1+\sin\theta)$. Then using $u\cos x + v\sin x \le \sqrt{u^2+v^2}$ we have

$$\begin{aligned} f(x) &= (\cos x - u)^2 + (\sin x - v)^2 \\ &= 1 - 2(u\cos x + v\sin x) + (u^2+v^2) \\ &\ge 1 - 2\sqrt{u^2+v^2} + (u^2+v^2) \\ &= (\sqrt{u^2+v^2} - 1)^2. \end{aligned}$$

The equality holds when $\tan x = \frac{v}{u}$. (Here if $u = 0$, then we take $x \in \left\{\frac{\pi}{2}, -\frac{\pi}{2}\right\}$ such that $v\sin x = |v|$.)

Therefore, the minimum value of$f(x)$ is $(\sqrt{u^2+v^2}-1)^2$, so

$$g(\theta)=(\sqrt{25\cos^2\theta+16(1+\sin\theta)^2}-1)^2.$$

In order to find the maximum value of $g(\theta)$, we examine the value range of $h(\theta)=25\cos^2\theta+16(1+\sin\theta)^2$. Since

$$h(\theta)=16+32\sin\theta+16\sin^2\theta+25\cos^2\theta=41+32\sin\theta-9\sin^2\theta,$$

combining it with $\sin\theta\in[-1,1]$ and the fact that $-9t^2+32t+41$ is monotonically increasing for $t<\frac{16}{9}$, we have

$$41+32\times(-1)-9\times(-1)^2\le h(\theta)\le 41+32\times 1-9\times 1^2,$$

that is $0\le h(\theta)\le 64$.

Therefore, the maximum value of $g(\theta)$ is $(8-1)^2=49$.

Remark. Since $f(x)$ is in fact the (squared) distance between two points, we can also solve this problem with analytic geometry.

Example 7. Let a and b be real numbers such that for each $x\in\mathbf{R}$,

$$a\cos 2x+b\cos x\ge -1. ①$$

Find the maximum value of $a+b$.

Solution. It is natural to view this problem as a quadratic inequality for $\cos x$, but the subsequent steps are still somewhat painful.

If we take another viewpoint and note that letting $x=\frac{2\pi}{3}$ in ① gives $a+b\le 2$, we might guess that 2 is exactly the maximum value of $a+b$.

In fact, let $a=\frac{2}{3}$ and $b=\frac{4}{3}$. Then $a+b=2$ and

$$\begin{aligned}3(a\cos 2x+b\cos x+1)&=2(2\cos^2 x-1)+4\cos x+3\\&=4\cos^2 x+4\cos x+1=(2\cos x+1)^2\ge 0.\end{aligned}$$

In this case, ① holds for all $x\in\mathbf{R}$. Therefore, the maximum value of $a+b$ is 2.

15.3 Exercises

Group A

1. Let α be an acute angle. Find the minimum value of $\tan\alpha+\cot\alpha+\sec\alpha+\csc\alpha$.
2. Let α and β be acute angles. Find the minimum value of $\frac{1}{\cos^2\alpha}+\frac{1}{\sin^2\alpha\sin^2\beta\cos^2\beta}$.

3. Let $\alpha \in \left[0, \frac{\pi}{2}\right]$ and $\beta \in \left(0, \frac{\pi}{2}\right]$. Find the minimum value of $\cos^2 \alpha \sin \beta + \frac{1}{\sin \beta}$.
4. Let x and y be real numbers such that $\sin x + \sin y = 1$. Find the maximum and minimum values of $\cos x + \cos y$.
5. Let a, $b \in \mathbf{R}$ with the property that $a \cos x + b \cos 3x \leq 1$ for all $x \in \mathbf{R}$. Find the maximum value of b.
6. Let m be a real number such that for any triangle ABC, its angles A, B, and C satisfy the following inequality:
$$\cot \frac{A}{2} + \cot \frac{B}{2} + \cot \frac{C}{2} - 2(\cot A + \cot B + \cot C) \geq m.$$
Find the maximum value of m.
7. Suppose the three angles of a triangle ABC satisfy
$$\sin A + \sin C = \sin B(\cos A + \cos C)$$
and the area of the triangle is 4. Find the minimum value of the perimeter of ABC.
8. Let x and y be real numbers such that $4x^2 - 5xy + 4y^2 = 19$. Find the maximum and minimum values of $x^2 + y^2$.
9. Find the maximum and minimum values of the function $f(x) = \sqrt{3x+6} + \sqrt{8-x}$.

Group B

10. Let n be a given positive integer and $A_1, A_2, \ldots, A_n$ be real numbers with
$$\tan A_1 \tan A_2 \ldots \tan A_n = 1.$$
Find the maximum value of $\sin A_1 \sin A_2 \ldots \sin A_n$.
11. Let A and B be real numbers and let $M(A, B)$ denote the maximum value of $F(x) = |\cos 2x + \sin 2x + Ax + B|$ for $x \in \left[0, \frac{3\pi}{2}\right]$. Find the minimum value of $M(A, B)$ as A and B vary and tell the values of A and B when the minimum value is reached.
12. In a triangle ABC, it is known that $\angle C = \frac{\pi}{2}$, $AB = 1$, $BC = a$, and $CA = b$. In another triangle DEF, given $\angle F = 90°$, $EF = (a + 1) \sin \frac{B}{2}$, and $FD = (b+1) \sin \frac{A}{2}$. Find the value range of the length of DE.
13. Let a, b, and c be positive real numbers with $a + b + c = abc$. Find the maximum value of
$$\frac{1}{\sqrt{1+a^2}} + \frac{1}{\sqrt{1+b^2}} + \frac{1}{\sqrt{1+c^2}}.$$

14. Let n be a given integer greater than or equal to 2. For $\theta \in \left(0, \frac{\pi}{2}\right)$, find the minimum value of $\left(\frac{1}{\sin^n \theta} - 1\right)\left(\frac{1}{\cos^n \theta} - 1\right)$.
15. Let α and β be acute angles. Find the maximum value of

$$A = \frac{\left(1 - \sqrt{\tan \frac{\alpha}{2} \tan \frac{\beta}{2}}\right)^2}{\cot \alpha + \cot \beta}.$$

Chapter 16

Inverse Trigonometric Functions and Trigonometric Equations

16.1 Key Points of Knowledge and Basic Methods

Since trigonometric functions are periodic, they do not have inverse functions on the whole domain. When we talk about inverse trigonometric functions, we only consider the inverse function of the trigonometric function defined on some monotonic interval (or in fact the inverse function of a trigonometric function restricted to some interval). Therefore, we must pay enough attention to the ranges of inverse trigonometric functions when solving problems.

Specifically, the following table lists the basic properties of inverse trigonometric functions:

Function	Domain	Range	Monotonicity
$y = \arcsin x$	$[-1, 1]$	$\left[-\frac{\pi}{2}, \frac{\pi}{2}\right]$	Increasing
$y = \arccos x$	$[-1, 1]$	$[0, \pi]$	Decreasing
$y = \arctan x$	$(-\infty, +\infty)$	$\left(-\frac{\pi}{2}, \frac{\pi}{2}\right)$	Increasing
$y = \operatorname{arccot} x$	$(-\infty, +\infty)$	$(0, \pi)$	Decreasing

In addition, the inverse sine function and the inverse tangent function are odd functions.

There are following identities among inverse trigonometric functions:

(1) $\arcsin x + \arccos x = \frac{\pi}{2}$ for $x \in [-1, 1]$;
(2) $\arctan x + \operatorname{arccot} x = \frac{\pi}{2}$ for $x \in \mathbf{R}$;
(3) $\arcsin(\sin x) = x$ for $x \in \left[-\frac{\pi}{2}, \frac{\pi}{2}\right]$;
(4) $\sin(\arcsin x) = x$ for $x \in [-1, 1]$.

Other inverse trigonometric functions also have identities similar to (3) and (4).

Note that the range of an inverse trigonometric function is a set of angles, and we also view the value of an inverse trigonometric function as an angle, which brings us some convenience when dealing with problems.

The basic idea of dealing with trigonometric equations is "reduction": try to convert the equations into the most basic trigonometric equations to solve.

The simplest form of trigonometric equations has the following cases:

(1) The equation $\sin x = a$ has no solution when $a \notin [-1, 1]$; if $a \in [-1, 1]$, then the solutions are $x = k\pi + (-1)^k \arcsin a$ with $k \in \mathbf{Z}$.
(2) The equation $\cos x = a$ has no solution when $a \notin [-1, 1]$; if $a \in [-1, 1]$, then the solutions are $x = 2k\pi \pm \arccos a$ with $k \in \mathbf{Z}$.
(3) The equations $\tan x = a$ and $\cot x = a$ have solutions for all $a \in \mathbf{R}$, and their solutions are $x = k\pi + \arctan a$ and $x = k\pi + \operatorname{arccot} a$ with $k \in \mathbf{Z}$, respectively.

16.2 Illustrative Examples

Example 1. Write the inverse function of $y = \cot x$ for $x \in (\pi, 2\pi)$ in terms of inverse trigonometric functions.

Solution. Let $t = x - \pi$. Then $t \in (0, \pi)$ and $\cot t = \cot x = y$. Hence, y is a function of t, whose inverse function is $t = \operatorname{arccot} y$, so $x - \pi = \operatorname{arccot} y$. Therefore, the inverse function of $y = \cot x$ for $x \in (\pi, 2\pi)$ is

$$y = \pi + \operatorname{arccot} x, \quad x \in \mathbf{R}.$$

Remark. Trigonometric functions have inverse functions when restricted to one monotonic interval, and inverse trigonometric functions are their inverse functions when restricted to specific intervals. We can write these two cases of inverse functions in terms of each other, using the induction formula.

Example 2. Find all real solutions to the equation

$$\sin^3 x + \cos^5 x = 1. \text{①}$$

Solution. Note that

$$\sin^3 x \le |\sin x|^3 \le |\sin x|^2 = \sin^2 x,$$
$$\cos^5 x \le |\cos x|^5 \le |\cos x|^2 = \cos^2 x,$$

so $\sin^3 x + \cos^5 x \le \sin^2 x + \cos^2 x = 1$. Combining it with ①, we have $\sin^3 x = \sin^2 x$ and $\cos^5 x = \cos^2 x$.

The first equality requires that $\sin x \in \{0, 1\}$ and the second equality requires that $\cos x \in \{0, 1\}$. Since $\sin^2 x + \cos^2 x = 1$, we get $(\sin x, \cos x) = (1, 0)$ or $(0, 1)$.

Therefore, $x = 2k\pi$ or $2k\pi + \frac{\pi}{2}$ with $k \in \mathbf{Z}$.

Remark. Inequalities can be used to solve equations. Such cases usually involve extreme value problems, especially the conditions for equality. In fact, for all m, $n \ge 2$ with m, $n \in \mathbf{N}^*$, the equation

$$\sin^m x + \cos^n x = 1$$

can be solved with the same method.

Example 3. Find all real solutions to the following equation:

$$\arccos\left|\frac{x^2-1}{x^2+1}\right| + \arcsin\left|\frac{2x}{x^2+1}\right| + \operatorname{arccot}\left|\frac{x^2-1}{2x}\right| = \pi.$$

Solution. The expression on the left side reminds us of the half angle formulas, so it is natural to change variables to get a trigonometric expression.

Let $x = \tan\alpha$ for $\alpha \in \left(-\frac{\pi}{2}, 0\right) \cup \left(0, \frac{\pi}{2}\right)$. Then the equation becomes

$$\arccos|\cos 2\alpha| + \arcsin|\sin 2\alpha| + \operatorname{arccot}|\cot 2\alpha| = \pi. \text{①}$$

If $|\tan\alpha| \ge 1$, then $-\frac{\pi}{2} < \alpha \le -\frac{\pi}{4}$ or $\frac{\pi}{4} \le \alpha < \frac{\pi}{2}$. If $-\frac{\pi}{2} < \alpha \le -\frac{\pi}{4}$, then ① reduces to

$$\pi + 2\alpha + \pi + 2\alpha + \pi + 2\alpha = \pi,$$

so $\alpha = -\frac{\pi}{3}$ and $x = -\sqrt{3}$. Similarly, if $\frac{\pi}{4} \le \alpha < \frac{\pi}{2}$, then $\alpha = \frac{\pi}{3}$, so $x = \sqrt{3}$.

If $|\tan\alpha| < 1$, then $-\frac{\pi}{4} < \alpha < 0$ or $0 < \alpha < \frac{\pi}{4}$. If $-\frac{\pi}{4} < \alpha < 0$, then ① reduces to

$$(-2\alpha) + (-2\alpha) + (-2\alpha) = \pi \Rightarrow \alpha = -\frac{\pi}{6}, \quad x = -\frac{\sqrt{3}}{3}.$$

Similarly, if $0 < \alpha < \frac{\pi}{4}$, then $2\alpha + 2\alpha + 2\alpha = \pi \Rightarrow \alpha = \frac{\pi}{6}$, thus $x = \frac{\sqrt{3}}{3}$.

Therefore, the solutions to the equation are $x = \pm\sqrt{3}$ and $\pm\frac{\sqrt{3}}{3}$.

Remark. In this problem, after changing variables we also used inequalities (the ranges of inverse trigonometric functions) to deal with the left side of ①. The mutual penetration between inequalities and equalities is a kind of harmony in mathematics!

Example 4. Find all real numbers α such that

$$\{\sin\alpha,\ \sin 2\alpha,\ \sin 3\alpha\} = \{\cos\alpha,\ \cos 2\alpha,\ \cos 3\alpha\}. \text{ ①}$$

Solution. It follows from ① that

$$\begin{aligned}
\sin\alpha + \sin 2\alpha + \sin 3\alpha &= \cos\alpha + \cos 2\alpha + \cos 3\alpha, \\
\sin 2\alpha + 2\sin 2\alpha\cos\alpha &= \cos 2\alpha + 2\cos 2\alpha\cos\alpha, \\
\sin 2\alpha(1 + 2\cos\alpha) &= \cos 2\alpha(1 + 2\cos\alpha), \\
(\sin 2\alpha - \cos 2\alpha)(1 + 2\cos\alpha) &= 0,
\end{aligned}$$

so $\sin 2\alpha - \cos 2\alpha = 0$ or $1 + 2\cos\alpha = 0$.

If $\sin 2\alpha - \cos 2\alpha = 0$, then we divide $\cos 2\alpha$ on both sides (note that $\cos 2\alpha \neq 0$ since otherwise $\sin 2\alpha = \pm 1$, which does not satisfy the equation), obtaining $\tan 2\alpha = 1$, so $\alpha = \frac{k\pi}{2} + \frac{\pi}{8}$ with $k \in \mathbf{Z}$.

If $1 + 2\cos\alpha = 0$, then $\cos\alpha = \cos 2\alpha = -\frac{1}{2}$, $\cos 3\alpha = 1$, and $\sin\alpha = \pm\frac{\sqrt{3}}{2}$, contradictory to ①.

Finally, we check ① for $\alpha = \frac{k\pi}{2} + \frac{\pi}{8}$ with $k \equiv 0, 1, 2, 3 \pmod 4$, and see that all such solutions are valid. Therefore, the solutions to the original equation are $\alpha = \frac{k\pi}{2} + \frac{\pi}{8}$ with $k \in \mathbf{Z}$.

Remark. In this problem we have obtained a trigonometric equation from a necessary condition, so we needed to check whether the conditions are satisfied for the obtained solutions. Here the idea that two identical sets have the same sum of elements plays an important role.

Example 5. Suppose $\arcsin x < \arccos x < \operatorname{arccot} x$. Find the value range of real numbers x.

Solution. By the definition of inverse trigonometric functions, $x \in [-1, 1]$. Since $y = \arcsin x$ is increasing while $y = \arccos x$ is decreasing on this domain, and they are equal for $x = \frac{\sqrt{2}}{2}$, necessarily $x \in \left[-1, \frac{\sqrt{2}}{2}\right)$.

The other inequality is $\arccos x < \operatorname{arccot} x$. Since both sides lie in $(0, \pi)$ and the cotangent function is decreasing on $(0, \pi)$, it is equivalent to say that

$$\cot(\arccos x) > x,$$

namely $\frac{\cos(\arccos x)}{\sin(\arccos x)} > x$, so $\frac{x}{\sqrt{1-x^2}} > x$, which means that $x \in (0, 1)$.

Therefore, the value range of x is $\left(0, \frac{\sqrt{2}}{2}\right)$.

Remark. We can also solve the inequality $\arcsin x < \arccos x$ by taking sine on both sides, but in this case we need to divide the domain into two intervals and discuss separately. If $x \in [-1, 0]$, then $\arcsin x \in \left[-\frac{\pi}{2}, 0\right]$ and $\arccos x \in \left[\frac{\pi}{2}, \pi\right]$, so in this case $\arcsin x < \arccos x$. If $x \in [0, 1]$, then $\arcsin x, \arccos x \in \left[0, \frac{\pi}{2}\right]$, and since the sine function is increasing on this interval, $\arcsin x < \arccos x$ if and only if $\sin(\arcsin x) < \sin(\arccos x)$, which gives that $x \in \left[0, \frac{\sqrt{2}}{2}\right)$.

When taking the values under the same trigonometric function on both sides of an inequality, we need to consider the monotonicity of this function to keep the inequality equivalent.

Example 6. Find all real constants c such that

$$f(x) = \arctan\frac{2-2x}{1+4x} + c$$

is an odd function defined on $\left(-\frac{1}{4}, \frac{1}{4}\right)$.

Solution. If $f(x)$ is an odd function on $\left(-\frac{1}{4}, \frac{1}{4}\right)$, then

$$f(0) = \arctan 2 + c = 0.$$

This implies that $c = -\arctan 2$.

On the other hand, we need to examine whether $f(x)$ is odd, since the above result is not sufficient.

For $x \in \left(-\frac{1}{4}, \frac{1}{4}\right)$, let $u(x) = \frac{2-2x}{1+4x} = \frac{1}{2}\left(\frac{5}{1+4x} - 1\right)$. Then $u(x) \in \left(\frac{3}{4}, +\infty\right)$ and

$$f(x) = \arctan u(x)$$
$$- \arctan 2 \in \left(\arctan\frac{3}{4} - \arctan 2, \frac{\pi}{2} - \arctan 2\right) \subset \left(-\frac{\pi}{2}, \frac{\pi}{2}\right).$$

Therefore, $f(x) \in \left(-\frac{\pi}{2}, \frac{\pi}{2}\right)$ for $x \in \left(-\frac{1}{4}, \frac{1}{4}\right)$, and also $-f(-x) \in \left(-\frac{\pi}{2}, \frac{\pi}{2}\right)$.

Now, in order to prove that $f(-x) = -f(x)$ for all $x \in \left(-\frac{1}{4}, \frac{1}{4}\right)$, it suffices to show that

$$\tan[-f(-x)] = \tan f(x). \text{ ①}$$

Note that

$$\begin{aligned}\tan f(x) &= \tan(\arctan u(x) - \arctan 2)\\ &= \frac{u(x) - 2}{1 + 2u(x)} = \frac{(2 - 2x) - 2(1 + 4x)}{1 + 4x + 2(2 - 2x)} = -2x,\\ \tan[-f(-x)] &= \tan(\arctan 2 - \arctan u(-x))\\ &= \frac{2 - u(-x)}{1 + 2u(-x)} = \frac{2(1 - 4x) - (2 + 2x)}{1 - 4x + 2(2 + 2x)} = -2x,\end{aligned}$$

so ① is valid.

Therefore, $c = -\arctan 2$ is the unique constant such that $f(x)$ is an odd function on $\left(-\frac{1}{4}, \frac{1}{4}\right)$.

Example 7. Let n be a given integer at least 3. Find all real solutions to the following equation for x:

$$\frac{1}{\cos x \cos 2x} + \frac{1}{\cos 2x \cos 3x} + \cdots + \frac{1}{\cos(n-1)x \cos nx} = 0. \text{ ①}$$

Solution. We multiply both sides of ① by $\sin x$. Note that for $2 \le k \le n$,

$$\begin{aligned}\frac{\sin x}{\cos(k-1)x \cos kx} &= \frac{\sin[kx - (k-1)x]}{\cos(k-1)x \cos kx}\\ &= \frac{\sin kx \cos(k-1)x - \sin(k-1)x \cos kx}{\cos(k-1)x \cos kx}\\ &= \tan kx - \tan(k-1)x.\end{aligned}$$

Hence, ① becomes

$$(\tan 2x - \tan x) + (\tan 3x - \tan 2x) + \cdots + (\tan nx - \tan(n-1)x) = 0,$$

so $\tan nx = \tan x$, and its solutions satisfy $nx = m\pi + x$, hence $x = \frac{m\pi}{n-1}$ with $m \in \mathbf{Z}$.

Note that we need to exclude the solutions that make at least one of $\cos x, \dots, \cos nx$ equal to zero, and also the false solutions from the case $\sin x = 0$.

If $\sin x = 0$, then either $\cos(k-1)x \cos kx = 1$ for all $2 \le k \le n$ or $\cos(k-1)x \cos kx = -1$ for all $2 \le k \le n$, and neither of these two satisfies ①. Therefore, the solutions of ① are

$$\left\{x \middle| x = \frac{m\pi}{n-1}, m \in \mathbf{Z}, \sin x \cos x \cos 2x \cdots \cos nx \neq 0\right\}.$$

Another interpretation of this set is

$$\left\{x \middle| x = \frac{m\pi}{n-1},\ m \in \mathbf{Z},\ \text{and } x \neq \frac{2l\pi \pm \frac{\pi}{2}}{k} \text{ or } l\pi \text{ for any } l \in \mathbf{Z},\ 1 \le k \le n\right\}.$$

Remark. In this problem we multiplied both sides by $\sin x$ and then sum by cancellation. This is a special technique that has been seen multiple times previously.

Example 8. Find the value of the following sum:

$$\arcsin\frac{\sqrt{3}}{2} + \arcsin\frac{\sqrt{8}-\sqrt{3}}{6} + \arcsin\frac{\sqrt{15}-\sqrt{8}}{12}$$
$$+ \cdots + \arcsin\frac{\sqrt{(n+1)^2-1}-\sqrt{n^2-1}}{n(n+1)},$$

where n is a given positive integer.

Solution. Let $S(n)$ denote the sum above. When calculating the sum, our idea is to sum by cancellation.

In fact, for $1 \le k \le n$ with k being a positive integer, we can prove that

$$\arcsin\frac{\sqrt{(k+1)^2-1}-\sqrt{k^2-1}}{k(k+1)} = \arcsin\frac{1}{k} - \arcsin\frac{1}{k+1}. \quad ①$$

It is easily seen that both sides of ① are acute angles, so it suffices to show that the sines of both sides are equal. Taking sine on both sides of ①, we have

$$\sin\left(\arcsin\frac{\sqrt{(k+1)^2-1}-\sqrt{k^2-1}}{k(k+1)}\right) = \frac{\sqrt{(k+1)^2-1}-\sqrt{k^2-1}}{k(k+1)};$$

$$\begin{aligned}
&\sin\left(\arcsin\frac{1}{k}-\arcsin\frac{1}{k+1}\right)\\
&=\sin\left(\arcsin\frac{1}{k}\right)\cos\left(\arcsin\frac{1}{k+1}\right)\\
&\quad-\cos\left(\arcsin\frac{1}{k}\right)\sin\left(\arcsin\frac{1}{k+1}\right)\\
&=\frac{1}{k}\cdot\sqrt{1-\left(\frac{1}{k+1}\right)^2}-\frac{1}{k+1}\cdot\sqrt{1-\left(\frac{1}{k}\right)^2}\\
&=\frac{\sqrt{(k+1)^2-1}}{k(k+1)}-\frac{\sqrt{k^2-1}}{(k+1)k}.
\end{aligned}$$

Hence, ① holds. By ①,

$$\begin{aligned}
S(n)&=\left(\arcsin\frac{1}{1}-\arcsin\frac{1}{2}\right)+\left(\arcsin\frac{1}{2}-\arcsin\frac{1}{3}\right)\\
&\quad+\left(\arcsin\frac{1}{3}-\arcsin\frac{1}{4}\right)+\cdots+\left(\arcsin\frac{1}{n}-\arcsin\frac{1}{n+1}\right)\\
&=\arcsin 1-\arcsin\frac{1}{n+1}=\frac{\pi}{2}-\arcsin\frac{1}{n+1}\\
&=\arccos\frac{1}{n+1}.
\end{aligned}$$

Therefore, the desired sum equals $\arccos\frac{1}{n+1}$.

Remark. While it is natural to think of summing by cancellation, it is not easy to find ① due to the inverse sine function. Discovering such an equality requires familiarity with trigonometric deformations as well as general algebraic deformations.

16.3 Exercises

Group A

1. Suppose the range of the function $f(x)=2\arcsin(x-2)$ is $\left[-\frac{\pi}{3},\pi\right]$. Find the domain of this function.
2. Find the real solutions to the following equation:

$$\log_{(1+\cos 2x)}(\sin x+\sin 3x)=1.$$

3. Find the range of the function $f(x)=\arcsin(1-x)+\arccos 2x$.

4. For a positive real number x, we calculate both $\sin x$ and $\sin x^\circ$. Find the minimum value of x such that the two results are equal.
5. Suppose $\arcsin(\sin\alpha+\sin\beta)+\arcsin(\sin\alpha-\sin\beta)=\frac{\pi}{2}$. Find the value of $\sin^2\alpha+\sin^2\beta$.
6. Suppose $f(x)=ax^2+bx+c$ satisfies that $f(2-x)=f(x)$ and $f\left(\arcsin\frac{2}{3}\right)>f\left(\arccos\frac{3}{4}\right)$ for all $x\in R$. Determine the signs of a and b.
7. Suppose a function $f(x)$ satisfies the property that for all $x\in\mathbf{R}$,
$$5f(\arctan x)+3f(-\arctan x)=\arctan x-\frac{\pi}{2}.$$
Find the formula of $f(x)$.
8. Let α be a real number such that $\cos\alpha-\cos 2\alpha-\cos 4\alpha=0$. Find all possible values of $\sin\alpha+\sin 2\alpha-\sin 4\alpha$.
9. Let $x\in(0,2\pi)$ such that $\cos 12x=5\sin 3x+9\tan^2 x+\cot^2 x$. Find the value of x.
10. Find all real numbers $x\in(0,\pi)$ such that $\left|\sin x-\frac{\sin x}{x}\right|+2\cos^2\left(\frac{\pi}{4}-\frac{x}{2}\right)+\frac{\sin x}{x}=3$.

Group B

11. Let $n\in\mathbf{N}^*$. Show that
$$\arctan\frac{1}{2}+\arctan\frac{1}{8}+\cdots+\arctan\frac{1}{2n^2}=\arctan\frac{n}{n+1}.$$
12. Let a be a positive integer. Suppose the equation for x
$$\cos^2\pi(a-x)-2\cos\pi(a-x)+\cos\frac{3\pi x}{2a}\cos\left(\frac{\pi x}{2a}+\frac{\pi}{3}\right)+2=0$$
has real solutions. Find the minimum value of a.
13. A sequence of real numbers $\{a_n\}$ is defined as follows:
$$a_1=t,\quad a_{n+1}=4a_n(1-a_n),\quad n=1,2,\ldots.$$
For a given integer $n(\geq 2)$, determine the number of different values of t such that $a_n=0$.
14. Find all real solutions to the equation
$$\cos^2 x+\cos^2 2x-2\cos x\cos 2x\cos 4x=\frac{3}{4}.$$

Chapter 17

The Law of Sines and the Law of Cosines

17.1 Key Points of Knowledge and Basic Methods

Trigonometry is an important tool in various branches of mathematics. In high school, it is usually involved in problems of plane geometry, analytic geometry, and complex numbers.

This chapter primarily concerns the applications of trigonometry in plane geometry, where the common method is to transform geometric problems concerning lengths, areas, etc. into trigonometric problems, and the law of sines and the law of cosines serve as a transition and bridge in this conversion.

17.2 Illustrative Examples

Example 1. In a triangle ABC, let a, b, and c denote the sides opposite to A, B, and C, respectively. If $\sin A = 2\sin C$ and b is the geometric mean of a and c, find the value of $\cos A$.

Solution. By the law of sines, $\frac{a}{c} = \frac{\sin A}{\sin C} = 2$, and since $b^2 = ac$, we have $a : b : c = 2 : \sqrt{2} : 1$. Hence, by the law of cosines,

$$\cos A = \frac{b^2 + c^2 - a^2}{2bc} = \frac{(\sqrt{2})^2 + 1^2 - 2^2}{2 \times \sqrt{2} \times 1} = -\frac{\sqrt{2}}{4}.$$

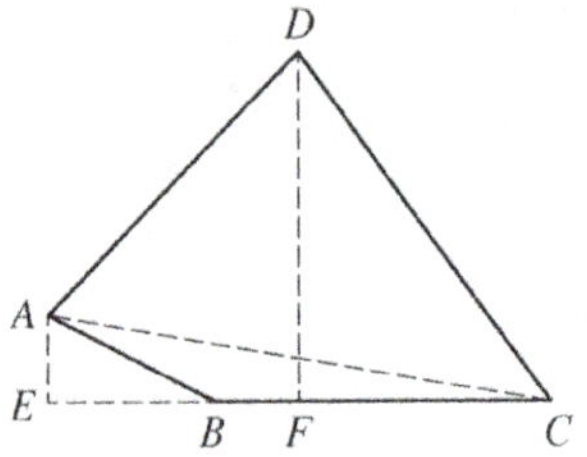

Figure 17.1

Example 2. In a convex quadrilateral $ABCD$, it is known that $AB = \sqrt{5}$, $BC = 4$, and $CD = 5$. Suppose

$$\tan \angle ABC = -\frac{1}{2}, \cos \angle BCD = \frac{3}{5}.$$

Find the value of $\sin \angle CDA$.

Solution. Let AE and DF be perpendicular to BC, and E and F lie on BC (Figure 17.1). Since $\tan \angle ABC = -\frac{1}{2}$, we have $\tan \angle ABE = \frac{1}{2}$, so by $AB = \sqrt{5}$ we get $AE = 1$ and $BE = 2$. Hence,

$$AC^2 = AE^2 + EC^2 = 1 + 6^2 = 37.$$

Also, since $FC = 3$ and $FD = 4$,

$$AD^2 = (DF - AE)^2 + EF^2 = 3^2 + 3^2 = 18,$$

so $\cos \angle CDA = \frac{CD^2+AD^2-AC^2}{2CD \cdot AD} = \frac{25+18-37}{2 \cdot 5 \cdot 3\sqrt{2}} = \frac{1}{5\sqrt{2}}$. Therefore, $\sin \angle CDA = \frac{7\sqrt{2}}{10}$.

Remark. We can also use $DF - AE = EF = 3$ to get that $\angle ADF = 45°$, hence

$$\sin \angle CDA = \sin(\angle FDC + 45°) = \frac{3}{5} \cdot \frac{\sqrt{2}}{2} + \frac{4}{5} \cdot \frac{\sqrt{2}}{2} = \frac{7\sqrt{2}}{10}.$$

Example 3. Let ABC be an acute triangle with $\angle A < 45°$ and let D be a point inside the triangle such that $\angle BDC = 4\angle A$ and $BD = CD$. Let E be the reflection of C with respect to AB and F be the reflection of B with respect to AC. Show that $AD \perp EF$.

Solution. By the inverse theorem of Pythagorean theorem, it suffices to show that

$$AE^2 - AF^2 = DE^2 - DF^2.$$

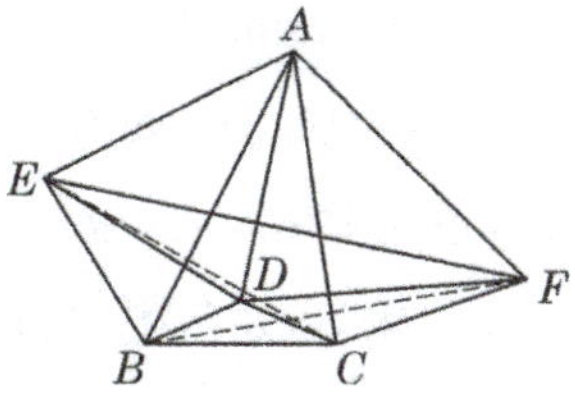

Figure 17.2

Since E and C are symmetric with respect to AB, we have $AE = AC$, and similarly $AF = AB$, so we only need to prove that

$$DE^2 - DF^2 = AC^2 - AB^2. \textcircled{1}$$

Next, we deal with the left side of ①. By symmetry, $BE = BC$ and $\angle EBC = 2B$. Since $BD = DC$ and $\angle BDC = 4A$, we see that $\angle DBC = 90° - 2A$, hence $\angle EBD = 2A + 2B - 90°$ (Figure 17.2). By the law of cosines,

$$\begin{aligned} DE^2 &= BD^2 + BE^2 - 2BD \cdot BE \cos \angle EBD \\ &= BD^2 + BC^2 - 2BD \cdot BC \cos(2A + 2B - 90°) \\ &= BD^2 + BC^2 - 2BD \cdot BC \sin(2A + 2B). \end{aligned}$$

Similarly,

$$DF^2 = CD^2 + BC^2 - 2CD \cdot BC \sin(2A + 2C).$$

Combining them with $BD = CD$, we get

$$\begin{aligned} DE^2 - DF^2 &= 2BD \cdot BC(\sin 2(A + C) - \sin 2(A + B)) \\ &= 2BD \cdot BC(\sin 2C - \sin 2B). \textcircled{2} \end{aligned}$$

Further, for convenience we assume that the radius of the circumcircle of ABC is 1, so by the law of sines $BC = 2\sin A$. Also, $BD = \frac{1}{2}BC \cdot \sec \angle DBC = \frac{\sin A}{\sin 2A} = \frac{1}{2\cos A}$, thus by ②,

$$\begin{aligned} DE^2 - DF^2 &= 2\tan A(\sin 2C - \sin 2B) \\ &= 4\tan A \cos(C + B)\sin(C - B) \\ &= -4\tan A \cos A \sin(C - B) \end{aligned}$$

$$
\begin{aligned}
&= -4\sin A\sin(C-B)\\
&= -4\sin(C+B)\sin(C-B)\\
&= -2(\cos 2B - \cos 2C)\\
&= -2[(1-2\sin^2 B)-(1-2\sin^2 C)]\\
&= -4(\sin^2 C - \sin^2 B)\\
&= (2\sin B)^2 - 2\sin C)^2\\
&= AC^2 - AB^2.
\end{aligned}
$$

Therefore, ① holds, hence proving the result.

Remark. Transforming a geometric problem into an algebraic one is an important method, in which the crucial part is to find appropriate algebraic tools to express the geometric structure. If we use trigonometric tools inappropriately in such problems, then the involved calculation may exceed the extent to which we can handle.

Example 4. Suppose there is a point P inside a triangle ABC such that $\angle PAB = \angle PBC = \angle PCA = 30°$. Show that ABC is an equilateral triangle.

Solution. Let the three sides of the triangle be $BC = a$, $CA = b$, and $AB = c$, and let the three angles be $\angle BAC = \alpha$, $\angle ABC = \beta$, and $\angle BCA = \gamma$. Then

$$
\begin{aligned}
\angle APB &= 180° - \angle ABP - \angle PAB\\
&= 180° - \angle ABP - \angle PBC = 180° - \angle ABC = \gamma + \alpha.
\end{aligned}
$$

Similarly, $\angle BPC = \alpha + \beta$ and $\angle CPA = \beta + \gamma$.

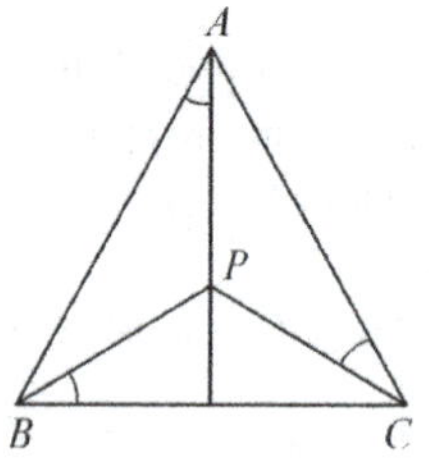

Figure 17.3

Apply the law of sines to the triangles ABP and BCP, and we have

$$BP = \frac{c\sin 30^\circ}{\sin(\gamma+\alpha)} = \frac{a\sin(\gamma-30^\circ)}{\sin(\alpha+\beta)}.$$

Hence,

$$\frac{1}{2}c\sin\gamma = a\sin(\gamma-30^\circ)\sin\beta.$$

Then apply the law of sines to the triangle ABC, and we get

$$\begin{aligned}\frac{1}{2}\sin^2\gamma = \sin\alpha\sin\beta\sin(\gamma-30^\circ) &\Leftrightarrow \sin^2\gamma \\ &= (\cos(\alpha-\beta)-\cos(\alpha+\beta))\cos(120^\circ-\gamma).\end{aligned}$$

Note that the conditions imply that α, β, $\gamma > 30^\circ$, so the equality above ensures that

$$\begin{aligned}&\sin^2\gamma \le (1-\cos(\alpha+\beta))\cos(120^\circ-\gamma)\\ &\Leftrightarrow -\cos^2\gamma \le (1+\cos\gamma)\cos(120^\circ-\gamma)\\ &\Leftrightarrow 1-\cos\gamma \le \cos(120^\circ-\gamma)\\ &\Leftrightarrow 1 \le \cos\gamma+\cos(120^\circ-\gamma) = 2\cos 60^\circ\cos(\gamma-60^\circ)\\ &\Leftrightarrow 1 \le \cos(\gamma-60^\circ).\end{aligned}$$

Therefore, $\gamma = 60^\circ$, and similarly $\alpha = \beta = 60^\circ$. This means that ABC is an equilateral triangle.

Example 5. Let P be a point that lies inside a triangle ABC or on its boundary, and the distances from P to the three sides of the triangle are PD, PE, and PF, respectively. Show that $PA+PB+PC \ge 2(PD+PE+PF)$ and the equality holds if and only if ABC is an equilateral trianglewith P as its center.

Solution. This is the famous Erdös-Mordell inequality in plane geometry.

As shown in Figure 17.4, for convenience we denote

$$PD = p, PE = q, PF = r.$$

Since $PD\perp BC$ and $PE\perp CA$, the points P, E, C, and D lie on the same circle, and PC is its diameter. By the law of sines,

$$DE = PC\cdot\sin C.$$

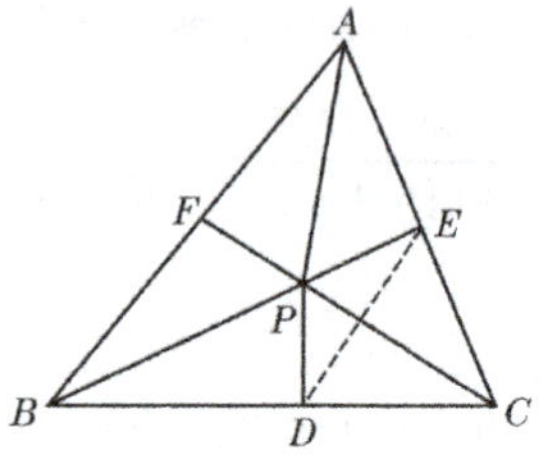

Figure 17.4

On the other hand, by the law of cosines, since $\angle DPE = 180° - \angle C = \angle A + \angle B$,

$$\begin{aligned} DE &= \sqrt{p^2 + q^2 - 2pq\cos(A+B)} \\ &= \sqrt{p^2 - 2pq(\cos A\cos B - \sin A\sin B) + q^2} \\ &= \sqrt{(p\sin B + q\sin A)^2 + (p\cos B - q\cos A)^2} \\ &\geq p\sin B + q\sin A. \end{aligned}$$

Hence,

$$PC = \frac{DE}{\sin C} \geq \frac{p\sin B + q\sin A}{\sin C}.$$

Similarly,

$$PA \geq \frac{q\sin C + r\sin B}{\sin A}, PB \geq \frac{r\sin A + p\sin C}{\sin B}.$$

Therefore,

$$\begin{aligned} PA + PB + PC \geq p\left(\frac{\sin B}{\sin C} + \frac{\sin C}{\sin B}\right) \\ + q\left(\frac{\sin C}{\sin A} + \frac{\sin A}{\sin C}\right) + r\left(\frac{\sin A}{\sin B} + \frac{\sin B}{\sin A}\right) \\ \geq 2(p+q+r) = 2(PD + PE + PF). \end{aligned}$$

If the equality holds, then $\sin A = \sin B = \sin C$ and

$$p : q : r = \cos A : \cos B : \cos C.$$

Therefore, the equality holds if and only if ABC is an equilateral triangle and P is its center.

Example 6. Let A, B, and C be the angles of a triangle ABC, and a, b, and c be the sides opposite to the angles, respectively. Also, let r and R denote the radii of its incircle and circumcircle, respectively, let p be its half perimeter, and let S be its area. Prove the following inequalities:

$$(1)\ \frac{1}{\sin\frac{A}{2}}+\frac{1}{\sin\frac{B}{2}}+\frac{1}{\sin\frac{C}{2}}\geq\frac{1}{r}\sqrt{\frac{a^2+b^2+c^2+4\sqrt{3}S}{2}};\ ①$$

$$(2)\ \sin\frac{A}{2}+\sin\frac{B}{2}+\sin\frac{C}{2}\geq\frac{1}{2}-\frac{r}{4R}+\frac{\sqrt{3}p}{4R}.\ ②$$

Solution. These are inequalities involving the sides and angles of a triangle, and we may try to prove them with geometric properties of triangles. While trigonometry can be used to solve geometric problems, many trigonometric problems also have geometric interpretations, and in this problem it is more convenient to restore the geometric background of the inequalities.

(1) As in Figure 17.5, let I be the incenter of the triangle ABC and let $AI = x$, $BI = y$, $CI = z$, $\angle BIC = \alpha$, $\angle CIA = \beta$, and $\angle AIB = \gamma$. Then (here $\sum$ denotes the cyclic sum)

$$① \Leftrightarrow \sum\frac{r}{\sin\frac{A}{2}}\geq\sqrt{\frac{\sum a^2+4\sqrt{3}S}{2}}.$$

Note that since I is the incenter,

$$\sin\frac{A}{2}=\frac{r}{x},\ \sin\frac{B}{2}=\frac{r}{y},\ \sin\frac{C}{2}=\frac{r}{z}.$$

Hence, $① \Leftrightarrow \sum x\geq\sqrt{\frac{\sum a^2+4\sqrt{3}S}{2}} \Leftrightarrow 2(\sum x)^2\geq\sum a^2+4\sqrt{3}S.$ ③

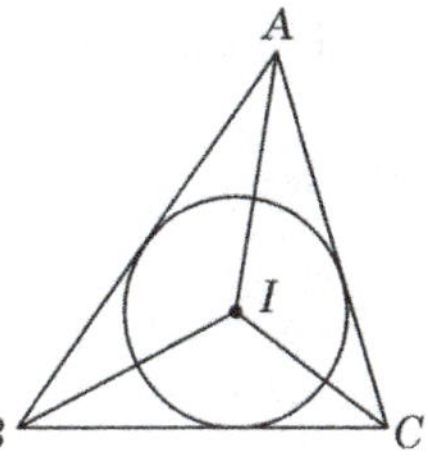

Figure 17.5

By the law of cosines and the area formula for triangles,

$$\sum a^2 = \sum (y^2 + z^2 - 2yz\cos\alpha),$$
$$S = \sum S_{\Delta BIC} = \frac{1}{2}\sum yz\sin\alpha.$$

Therefore,

$$\begin{aligned}
③ &\Leftrightarrow 2\left(\sum x\right)^2 \geq \sum (y^2 + z^2 - 2yz\cos\alpha) + 2\sqrt{3}\sum yz\sin\alpha \\
&\Leftrightarrow 4\sum yz \geq 2\sum yz(\sqrt{3}\sin\alpha - \cos\alpha) \\
&\Leftrightarrow \sum yz[2 - (\sqrt{3}\sin\alpha - \cos\alpha)] \geq 0 \\
&\Leftrightarrow 2\sum yz\left(1 - \sin\left(\alpha - \frac{\pi}{6}\right)\right) \geq 0.
\end{aligned}$$

Since $1 - \sin(\alpha - \frac{\pi}{6}) \geq 0$, the inequality above always holds, thus ① is true.

(2) By the law of sines, $\sum \sin A = \frac{p}{R}$. On the other hand, $S_{\Delta ABC} = \frac{1}{2}ab\sin C = rp$, so

$$\begin{aligned}
r = \frac{ab\sin C}{a+b+c} &= \frac{2R\sin A\sin B\sin C}{\sin A + \sin B + \sin C} \\
&= \frac{2R\sin A\sin B\sin C}{4\cos\frac{A}{2}\cos\frac{B}{2}\cos\frac{C}{2}} \\
&= 4R\sin\frac{A}{2}\sin\frac{B}{2}\sin\frac{C}{2} \\
&= R\left(\sum\cos A - 1\right)
\end{aligned}$$

(The trigonometric identities that we used here have appeared in Chapter 13.)

With this equality, we get that

$$\begin{aligned}
&\sum\sin\frac{A}{2} - \frac{1}{2} + \frac{r}{4R} - \frac{\sqrt{3}p}{4R} \\
&= \sum\sin\frac{A}{2} - \frac{1}{2} + \frac{1}{4}\left(\sum\cos A - 1\right) - \frac{\sqrt{3}}{4}\sum\sin A
\end{aligned}$$

$$= \sum \sin\frac{A}{2} - \frac{1}{4}\sum(1-\cos A) - \frac{\sqrt{3}}{2}\sum \sin\frac{A}{2}\cos\frac{A}{2}$$
$$= \sum\left(\sin\frac{A}{2} - \frac{1}{2}\sin^2\frac{A}{2} - \frac{\sqrt{3}}{2}\sin\frac{A}{2}\cos\frac{A}{2}\right)$$
$$= \sum \sin\frac{A}{2}\left[1-\left(\frac{1}{2}\sin\frac{A}{2}+\frac{\sqrt{3}}{2}\cos\frac{A}{2}\right)\right]$$
$$= \sum \sin\frac{A}{2}\left(1-\sin\left(\frac{A}{2}+\frac{\pi}{3}\right)\right),$$

where all the terms in the above expression are nonnegative, so ② is valid.

Example 7. Let ω be a circle centered at the incenter I of a triangle ABC. Let D be the intersection point of ω with the ray starting at I and perpendicular to BC, and define E and F similarly (Figure 17.6). Show that the lines AD, BE, and CF are concurrent.

Solution. As in Figure 17.6, let $\angle BAD = \alpha_1$, $\angle CAD = \alpha_2$, $\angle CBE = \beta_1$, $\angle ABE = \beta_2$, $\angle ACF = \gamma_1$, and $\angle BCF = \gamma_2$.

By the inverse theorem of Ceva's theorem (in the angular form), it suffices to show that $\frac{\sin\alpha_1 \sin\beta_1 \sin\gamma_1}{\sin\alpha_1 \sin\beta_2 \sin\gamma_2} = 1.$ ①

By the law of sines,

$$\frac{\sin\alpha_1}{\sin\alpha_2} = \frac{BD\sin\angle ABD}{AD}\cdot\frac{AD}{CD\sin\angle ACD} = \frac{BD}{CD}\cdot\frac{\sin\angle ABD}{\sin\angle ACD}, \quad ②$$

and similarly

$$\frac{\sin\beta_1}{\sin\beta_2} = \frac{CE}{AE}\cdot\frac{\sin\angle BCE}{\sin\angle BAE}, \frac{\sin\gamma_1}{\sin\gamma_2} = \frac{AF}{BF}\cdot\frac{\sin\angle CAF}{\sin\angle CBF}. \quad ③$$

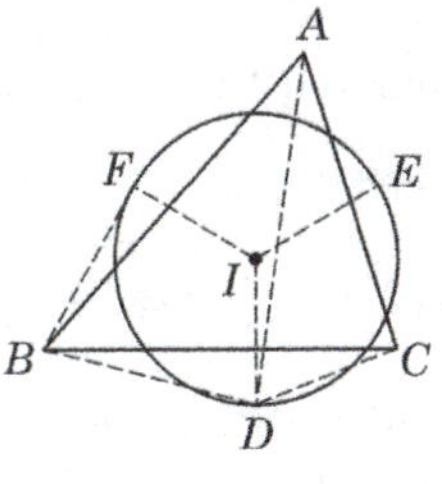

Figure 17.6

In the triangles BID and BIF,

$$BI = BI, ID = IF,$$

$$\angle BID = 90^\circ - \angle IBC = 90^\circ - \frac{B}{2} = 90^\circ - \angle ABI = \angle BIF,$$

so $\Delta BID \cong BIF$ and

$$BD = BF, \angle ABD = \angle IBD + \frac{B}{2} = \angle IBF + \frac{B}{2} = \angle CBF.$$

Similarly,

$$CD = CE, \angle ACD = \angle BCE;$$

$$AE = AF, \angle CAF = \angle BAE.$$

Combining the equalities above with ② and ③, we get ①. Therefore, AD, BE, and CF are concurrent.

Example 8. Suppose there are four points lying in a plane and all the distances between two of these points are integers. Show that at least one of these distances is divisible by 3.

Solution. We label the points A, B, C, and D properly such that $\angle BAD = \angle BAC + \angle CAD$, and let $\angle BAC = \alpha$, $\angle CAD = \beta$, and $\angle BAD = \gamma$. By the law of cosines,

$$\begin{cases} BC^2 = AB^2 + AC^2 - 2AB \cdot AC\cos\alpha, \\ CD^2 = AD^2 + AC^2 - 2AD \cdot AC\cos\beta, \\ BD^2 = AB^2 + AD^2 - 2AB \cdot AD\cos\gamma. \end{cases} \quad ①$$

If the proposition does not hold, then $AB^2 \equiv AC^2 \equiv AD^2 \equiv BC^2 \equiv BD^2 \equiv CD^2 \equiv 1 \pmod 3$, so $2AB \cdot AC\cos\alpha \equiv 2AD \cdot AC\cos\beta \equiv 2AB \cdot AD\cos\gamma \equiv 1 \pmod 3$, and

$$\begin{aligned} AC^2 \cdot AB \cdot AD\cos\alpha\cos\beta &\equiv 4AC^2 \cdot AB \cdot AD\cos\alpha\cos\beta \\ &= (2AB \cdot AC\cos\alpha)(2AD \cdot AC\cos\beta) \equiv 1 \pmod 3. \quad ② \end{aligned}$$

By ① we see that $\cos\alpha$, $\cos\beta$, and $\cos\gamma$ are all rational, and further, if

$$\cos\alpha = \frac{q}{p}, \cos\beta = \frac{s}{r},$$

where p and q as well as r and s are two pairs of coprime integers, then by ① none of p, q, r, and s are divisible by 3. This implies that $p^2 \equiv q^2 \equiv r^2 \equiv s^2 \equiv 1 \pmod 3$.

Note that $\cos\gamma = \cos(\alpha+\beta) = \cos\alpha\cos\beta - \sin\alpha\sin\beta$, so combining it with ① we have $2AC^2 \cdot AB \cdot AD\sin\alpha\sin\beta \in \mathbf{Z}$. Hence

$$2AC^2 \cdot AB \cdot AD\sin\alpha\sin\beta$$
$$= 2AC^2 \cdot AB \cdot AD\frac{\sqrt{p^2-q^2}\cdot\sqrt{r^2-s^2}}{|pr|} \equiv 0(\text{mod } 3).③$$

Here, we have used the fact that $p^2 - q^2 \equiv r^2 - s^2 \equiv 0$ (mod 3). On the other hand,

$$2AC^2 \cdot AB \cdot AD\cos\gamma \equiv 2AB \cdot AD\cos\gamma \equiv 1(\text{mod } 3).④$$

Adding ③ and ④ and combining the resulting equality with $\cos\gamma = \cos\alpha\cos\beta - \sin\alpha\sin\beta$ give that

$$2AC^2 \cdot AB \cdot AD\cos\alpha\cos\beta \equiv 1(\text{mod } 3),$$

which is contradictory to ②.

Therefore, the proposition is true.

17.3 Exercises

Group A

1. Suppose a painting is hung on the wall. The lower boundary of the painting is a meters above the level of the observer's eyes and its upper boundary is b meters above the level of the observer's eyes. Find the distance from the observer to the wall such that the angle of the painting is maximal.
2. Suppose in a quadrilateral with area 32, the sum of a pair of opposite sides and one diagonal is 16. Find all possible lengths of the other diagonal.
3. Let H and O be the orthocenter and circumcenter of an acute triangle ABC, respectively, and $AH = AO$. Find all possible values of $\angle A$.
4. Let AD, BE, and CF be the angle bisectors of a triangle ABC such that D, E, and F lie on the sides BC, CA, and AB, respectively. If $\angle EDF = 90°$, find the value of $\angle BAC$.
5. In a triangle ABC, suppose $\angle A + \angle C = 2\angle B$ and $c - a$ equals the altitude h above AC. Find the value of $\sin\frac{C-A}{2}$.
6. In a triangle ABC, suppose $c - a$ equals the altitude h above AC. Find the value of $\sin\frac{C-A}{2} + \cos\frac{C+A}{2}$.

7. In a triangle ABC, let K, L, and M be the midpoints of BC, CA, and AB, respectively. Let $\overparen{AB}$, $\overparen{BC}$, and $\overparen{CA}$ be arcs on the circumcircle of ABC such that each arc does not contain the third vertex, and let X, Y, and Z be the midpoints of the arcs $\overparen{BC}$, $\overparen{CA}$, and $\overparen{AB}$, respectively. In addition, let R and r denote the radii of the circumcircle and incircle of the triangle, respectively. Show that

$$r + KX + LY + MZ = 2R.$$

8. Let K and N be points on the sides AB and AD of a square $ABCD$ respectively such that $AK \cdot AN = 2BK \cdot DN$. Also, let CK and CN intersect BD at L and M, respectively. Show that the five points K, L, M, N, and A lie on the same circle.
9. Suppose a, b, c, and d are the side lengths of a quadrilateral that has an incircle, and the area of the quadrilateral satisfies $S = \sqrt{abcd}$. Show that this quadrilateral also has a circumcircle.
10. In a quadrilateral $ABCD$, suppose $\angle ABC + \angle BCD < 180°$, and let the rays BA and CD intersect at E. Show that $\angle ABC = \angle ADC$ if and only if $AC^2 = CD \cdot CE\text{-}AB \cdot AE$.

Group B

11. Let a, b, and c be the side lengths of a triangle ABC, and let r and R be the radii of its incircle and circumcircle, respectively. Show that

$$\frac{r}{2R} \leq \frac{abc}{\sqrt{2(a^2+b^2)(b^2+c^2)(c^2+a^2)}}.$$

12. Let Γ be the incircle of a triangle ABC and D be the tangent point on BC. Let DF be a diameter of Γ, and AF intersects BC at E. Show that $BE = CD$.
13. If P is a point inside a triangle ABC, show that at least one of $\angle PAB$, $\angle PBC$, and $\angle PCA$ is less than or equal to $30°$.
14. In a triangle ABC, suppose the three angle bisectors intersect the circumcircle of ABC again at A', B', and C', respectively. Show that

$$S_{\Delta A'B'C'} \geq S_{\Delta ABC}.$$

15. In a triangle ABC, suppose the three medians intersect the circumcircle of ABC again at A', B', and C', respectively. Show that

$$S_{\Delta A'B'C'} \geq S_{\Delta ABC}.$$

Chapter 18

Concepts and Operations of Vectors

18.1 Key Points of Knowledge and Basic Methods

A. Concepts and basic attributes of vectors

(1) Vectors

A vector is an object with both a magnitude and a direction. The vector starting at A and ending at B is denoted as $\overrightarrow{AB}$. If where the vector starts or ends is irrelevant, we can also denote it as $\vec{a}$, or sometimes $\mathbf{a}$ (in boldface lower case). If the starting point of a vector is specified, then it is called a position vector. Otherwise it is a free vector.

(2) Norms

The magnitude of a vector $\overrightarrow{AB}$, or equivalently the length of the line segment AB, is called the norm of $\overrightarrow{AB}$, denoted as $|\overrightarrow{AB}|$ or $|\vec{a}|$.

(3) Equal vectors

If two vectors have the same norm and direction (here the starting points and ending points need not be the same), then they are equal vectors. (Hence, the norm and the direction are the fundamental attributes of a vector.)

(4) Special vectors

Vectors with norm 1 are called unit vectors. The vector with norm 0 is called the zero vector, denoted as $\vec{0}$. The zero vector is the only vector that does not have a unique direction.

B. Operations of vectors

(1) Addition and subtraction

The addition of vectors follows the parallelogram law. We draw a parallelogram with $\vec{a}$ and $\vec{b}$ as adjacent sides, and then the vector corresponding to the diagonal starting from the common starting point of $\vec{a}$ and $\vec{b}$ is the sum of the two vectors, denoted as $\vec{a}+\vec{b}$. The subtraction of vectors is the inverse operation of addition. If $\vec{b}+\vec{c}=\vec{a}$, then $\vec{c}$ is called the difference of $\vec{a}$ and $\vec{b}$, or $\vec{c}=\vec{a}-\vec{b}$. The subtraction of vectors also follows the triangle rule. If we translate $\vec{a}$ and $\vec{b}$ to make them start at the same point, then the vector from the ending point of $\vec{b}$ to the ending point of is the difference vector $\vec{a}-\vec{b}$.

(2) Scalar multiplication

The product of a real number m and a vector $\vec{a}$ is also a vector, denoted as $m\vec{a}$. Its norm is $|m\vec{a}|=|m||\vec{a}|$, while its direction is the same as $\vec{a}$ when $m>0$ and opposite to $\vec{a}$ when $m<0$. When $m=0$, the direction of $m\vec{a}(=\vec{0})$ is uncertain. Clearly, $(mn)\vec{a}=m(n\vec{a})$ for any two real numbers m and n. Specifically, if $\vec{a}\neq\vec{0}$, then $\frac{1}{|\vec{a}|}\vec{a}$ is the unit vector with the same direction as $\vec{a}$.

(3) Inner product of vectors

The inner product (a.k.a. dot product or scalar product) of two vectors $\vec{a}$ and $\vec{b}$ is defined as the product of their norms times the cosine of the angle between their directions. (If we translate the vectors to make them start at the same point and rotate one vector counterclockwise around their common starting point until it has the same direction as the second vector, then the rotation angle is the angle between their directions.) The inner product of $\vec{a}$ and $\vec{b}$ is denoted as $\vec{a}\cdot\vec{b}$, so by definition $\vec{a}\cdot\vec{b}=|\vec{a}|\cdot|\vec{b}|\cdot\cos(\vec{a},\vec{b})$, where $(\vec{a},\vec{b})$ is the angle between the directions of the two vectors. In particular, we have $|\vec{a}|^2=\vec{a}\cdot\vec{a}$.

C. Vectors in the coordinate form

In the coordinate plane xOy, if a vector $\vec{a}$ starts at the origin, then $\vec{a}$ is uniquely determined by its ending point. Hence, we can write this kind of position vector in the coordinate form. The operations of vectors can also be expressed with coordinates. For example, if $\vec{a}=(x_1,y_1)$ and $\vec{b}=(x_2,y_2)$, then by the distance formula between two points together with the law of cosines,

$$\vec{a} \cdot \vec{b} = x_1x_2 + y_1y_2.$$

D. Basic rules in the operations of vectors

(1) Addition follows the commutative law and the associative law:

$$\vec{a} + \vec{b} = \vec{b} + \vec{a}, (\vec{a} + \vec{b}) + \vec{c} = \vec{a} + (\vec{b} + \vec{c}).$$

(2) Scalar multiplication follows the distributive law:

$$\lambda(\vec{a} + \vec{b}) = \lambda\vec{a} + \lambda\vec{b}, (\lambda + \mu)\vec{a} = \lambda\vec{a} + \mu\vec{a},$$

where λ and μ are arbitrary real numbers.

(3) Inner product follows the commutative law and the distributive law with respect to addition:

$$\vec{a} \cdot \vec{b} = \vec{b} \cdot \vec{a}, (\vec{b} + \vec{c}) \cdot \vec{a} = \vec{b} \cdot \vec{a} + \vec{c} \cdot \vec{a},$$
$$\vec{a} \cdot (\vec{b} + \vec{c}) = \vec{a} \cdot \vec{b} + \vec{a} \cdot \vec{c}.$$

E. Applications of vectors in geometry

Some geometric problems can be reduced to computational problems of vectors, where we can apply algebraic methods to solve them. The following statements are frequently used in analytic geometry:

(1) A quadrilateral $ABCD$ is a parallelogram if and only if $\overrightarrow{AB} = -\overrightarrow{CD}$.
(2) Two vectors $\vec{a}$ and $\vec{b}$ are collinear (or linearly dependent) if and only if there exist real numbers m and n (not both 0) such that $m\vec{a} + n\vec{b} = \vec{0}$. This statement can be used to prove that three points are collinear.
(3) Two vectors $\vec{a}$ and $\vec{b}$ are perpendicular (or orthogonal) if and only if $\vec{a} \cdot \vec{b} = 0$.

18.2 Illustrative Examples

Example 1. Let $\vec{a}$ and $\vec{b}$ be the position vectors of A and B, respectively (with the same starting point). Let P be a point on the line AB such that P divides the directed line segment AB with proportion λ (i.e., $\overrightarrow{AP} = \lambda\overrightarrow{BP}$), where $\lambda \neq -1$. Find the position vector of P.

Solution. It follows from the condition that $\overrightarrow{AP} = \frac{\lambda}{1+\lambda}\overrightarrow{AB}$ and $\overrightarrow{BP} = \frac{1}{1+\lambda}\overrightarrow{BA}$.

Therefore, $\overrightarrow{AP} + \lambda\overrightarrow{BP} = \vec{0}$. Also, since $\overrightarrow{AP} = \vec{p} - \vec{a}$ and $\overrightarrow{BP} = \vec{p} - \vec{b}$,

$$(\vec{p} - \vec{a}) + \lambda(\vec{p} - \vec{b}) = \vec{0},$$

or equivalently $(1 + \lambda)\vec{p} = \vec{a} + \lambda\vec{b}$, hence $\vec{p} = \frac{\vec{a}+\lambda\vec{b}}{1+\lambda}$.

Remark. This is the formula for a definite proportion in the vector form. In particular, if $\lambda = 1$, then we get the midpoint formula.

Example 2. Let O be a point inside a triangle ABC. Show that there exist positive real numbers α, β, and γ with $\alpha + \beta + \gamma = 1$ such that

$$\alpha\overrightarrow{OA} + \beta\overrightarrow{OB} + \gamma\overrightarrow{OC} = \vec{0}.$$

Solution. As shown in Figure 18.1, we extend AO to intersect BC at D. Suppose O divides AD with proportion λ and D divides BC with proportion μ. Then since P lies inside the triangle, λ and μ are positive real numbers. From the result of Example 1, $\overrightarrow{OD} = \frac{\overrightarrow{OB}+\mu\overrightarrow{OC}}{1+\mu}$, and since

$$\overrightarrow{OD} = \frac{1}{\lambda}\overrightarrow{AO} = -\frac{1}{\lambda}\overrightarrow{OA},$$

we get $-\frac{1}{\lambda}\overrightarrow{OA} = \frac{\overrightarrow{OB}+\mu\overrightarrow{OC}}{1+\mu}$. Hence,

$$\frac{1}{\lambda}\overrightarrow{OA} + \frac{1}{1+\mu}\overrightarrow{OB} + \frac{\mu}{1+\mu}\overrightarrow{OC} = \vec{0}.$$

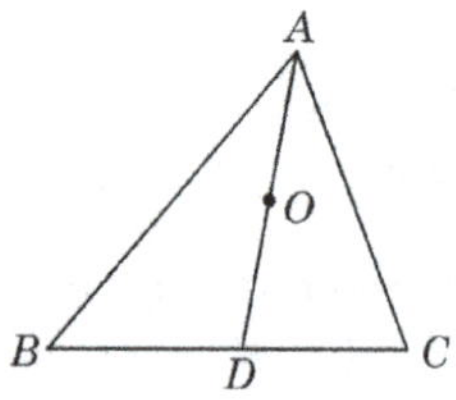

Figure 18.1

Dividing both sides by $1+\frac{1}{\lambda}$, we get

$$\frac{1}{1+\lambda}\overrightarrow{OA}+\frac{\lambda}{(1+\lambda)(1+\mu)}\overrightarrow{OB}+\frac{\lambda\mu}{(1+\lambda)(1+\mu)}\overrightarrow{OC}=\vec{0}.$$

Therefore, we can let $\alpha=\frac{1}{1+\lambda}$, $\beta=\frac{\lambda}{(1+\lambda)(1+\mu)}$, and $\gamma=\frac{\lambda\mu}{(1+\lambda)(1+\mu)}$ (note that $\alpha+\beta+\gamma=1$), and the proposition holds.

Remark. By the assertion in this problem, we can express the position of any point inside a triangle ABC in terms of the position vectors of A, B, and C. In particular, if $\alpha=\beta=\gamma=\frac{1}{3}$, then the corresponding point is the barycenter of the triangle.

Alternatively, we construct an oblique coordinate system with B as the origin, $\overrightarrow{BC}$ as the x-axis, and $\overrightarrow{BA}$ as the y-axis, and let $|\overrightarrow{BC}|=a$ and $|\overrightarrow{BA}|=c$ (so that A has coordinates $(0,c)$ and C has coordinates $(a,0)$). Suppose O has the coordinates (x,y). Then solving the equation we can take $(\alpha,\beta,\gamma)=(1-\frac{x}{a}-\frac{y}{c},\frac{y}{c},\frac{x}{a})$, which also proves the proposition. From this point of view, we see that for any point O on the plane, there always exist α, β, and γ such that $\alpha+\beta+\gamma=1$ and $\alpha\cdot\overrightarrow{OA}+\beta\cdot\overrightarrow{OB}+\gamma\cdot\overrightarrow{OC}=\vec{0}$ (while the price is that they may not be all positive).

Example 3. In a quadrilateral *ABCD*, let E be the midpoint of AB and K be the midpoint of *CD*. Show that the midpoints of *AK*, *CE*, *BK*, and *DE* are the vertices of a parallelogram.

Solution. This is a well-known result in plane geometry, which is easy to prove by using vectors. Let Y, Y_1, X_1, and X be the midpoints of *AK*, *CE*, *BK*, and *DE*, respectively. Then

$$\overrightarrow{EX}=\frac{1}{2}(\overrightarrow{EA}+\overrightarrow{AD}),$$

$$\overrightarrow{EY}=\overrightarrow{EA}+\frac{1}{2}(\overrightarrow{AD}+\overrightarrow{DK}).$$

Hence, $\overrightarrow{XY}=\overrightarrow{EY}-\overrightarrow{EX}=\frac{1}{2}(\overrightarrow{EA}+\overrightarrow{DK})=\frac{1}{4}(\overrightarrow{BA}+\overrightarrow{DC})$, and similarly $\overrightarrow{X_1Y_1}=\frac{1}{4}(\overrightarrow{BA}+\overrightarrow{DC})$.

Therefore, $\overrightarrow{XY}=\overrightarrow{X_1Y_1}$, so XYY_1X_1 is a parallelogram.

Example 4. Let A_1, B_1, and C_1 be points on the sides *BC*, *CA*, and *AB* of ΔABC, respectively. Let G be the barycenter of ΔABC, and G_a, G_b, and G_c be the barycenters of ΔAB_1C_1, ΔBA_1C_1, and ΔCA_1B_1, respectively.

Also, let G_1 and G_2 be the barycenters of $\Delta A_1B_1C_1$ and $\Delta G_aG_bG_c$, respectively. Show that G, G_1, and G_2 are collinear.

Solution. Choose any point O in the plane as the origin. By the conclusion in Example 1,

$$\overrightarrow{OA_1} = \alpha\overrightarrow{OB} + (1-\alpha)\overrightarrow{OC},$$
$$\overrightarrow{OB_1} = \beta\overrightarrow{OC} + (1-\beta)\overrightarrow{OA},$$
$$\overrightarrow{OC_1} = \gamma\overrightarrow{OA} + (1-\gamma)\overrightarrow{OB},$$

where α, β, $\gamma \in (0,1)$. Then

$$\overrightarrow{OG} = \frac{1}{3}(\overrightarrow{OA} + \overrightarrow{OB} + \overrightarrow{OC}),$$
$$\overrightarrow{OG_1} = \frac{1}{3}(\overrightarrow{OA_1} + \overrightarrow{OB_1} + \overrightarrow{OC_1})$$
$$= \frac{1}{3}[(\gamma+1-\beta)\overrightarrow{OA} + (\alpha+1-\gamma)\overrightarrow{OB} + (\beta+1-\alpha)\overrightarrow{OC}],$$
$$\overrightarrow{OG_a} = \frac{1}{3}[(2-\beta+\gamma)\overrightarrow{OA} + (1-\gamma)\overrightarrow{OB} + \beta\overrightarrow{OC}],$$
$$\overrightarrow{OG_b} = \frac{1}{3}[(2-\gamma+\alpha)\overrightarrow{OB} + (1-\alpha)\overrightarrow{OC} + \gamma\overrightarrow{OA}],$$
$$\overrightarrow{OG_c} = \frac{1}{3}[(2-\alpha+\beta)\overrightarrow{OC} + (1-\beta)\overrightarrow{OA} + \alpha\overrightarrow{OB}].$$

Consequently,

$$\overrightarrow{OG_2} = \frac{1}{3}(\overrightarrow{OG_a} + \overrightarrow{OG_b} + \overrightarrow{OG_c})$$
$$= \frac{1}{9}[(3-2\beta+2\gamma)\overrightarrow{OA} + (3-2\gamma+2\alpha)\overrightarrow{OB} + (3-2\alpha+2\beta)\overrightarrow{OC}].$$

Further,

$$\overrightarrow{GG_1} = \overrightarrow{OG_1} - \overrightarrow{OG} = \frac{1}{3}[(\gamma-\beta)\overrightarrow{OA} + (\alpha-\gamma)\overrightarrow{OB} + (\beta-\alpha)\overrightarrow{OC}],$$
$$\overrightarrow{GG_2} = \overrightarrow{OG_2} - \overrightarrow{OG} = \frac{2}{9}[(\gamma-\beta)\overrightarrow{OA} + (\alpha-\gamma)\overrightarrow{OB} + (\beta-\alpha)\overrightarrow{OC}].$$

Therefore, $\overrightarrow{GG_1} = \frac{3}{2}\overrightarrow{GG_2}$, so G, G_1, and G_2 are collinear.

Example 5. In a triangle ABC, suppose $|\overrightarrow{BA} - t\overrightarrow{BC}| \geq |\overrightarrow{AC}|$ for all real numbers t. Show that ABC is a right triangle.

Solution. Let $\angle ABC = \beta$. Then for all $t \in \mathbf{R}$,

$$|\overrightarrow{BA} - t\overrightarrow{BC}|^2 \geq |\overrightarrow{BA} - \overrightarrow{BC}|^2,$$
$$|\overrightarrow{BA}|^2 - 2t\overrightarrow{BA} \cdot \overrightarrow{BC} + t^2|\overrightarrow{BC}|^2 \geq |BA|^2 - 2\overrightarrow{BA} \cdot \overrightarrow{BC} + |\overrightarrow{BC}|^2,$$

so

$$(t^2 - 1)|\overrightarrow{BC}|^2 \geq 2(t-1)\overrightarrow{BA} \cdot \overrightarrow{BC} = 2(t-1)|\overrightarrow{BA}| \cdot |\overrightarrow{BC}| \cos\beta.$$

Hence, for all $t \in \mathbf{R}$,

$$(t-1)\left(t + 1 - \frac{2|\overrightarrow{BA}| \cdot \cos\beta}{|\overrightarrow{BC}|}\right) \geq 0.$$

Let $\frac{|\overrightarrow{BA}| \cdot \cos\beta}{|\overrightarrow{BC}|} = u$. Then the above condition is equivalent to $(t-1)(t+1-2u) \geq 0$ for all $t \in \mathbf{R}$. In other words, $t^2 - 2ut + 2u - 1 \geq 0$, and this inequality always holds if and only if

$$\Delta = 4u^2 - 4(2u-1) \leq 0,$$

namely $4u^2 - 8u + 4 \leq 0$. That is, $4(u-1)^2 \leq 0$. Therefore, $u = 1$.

Thus, we see that $|\overrightarrow{BA}| \cdot \cos\beta = |\overrightarrow{BC}|$, which means that $\angle C = 90°$, so the proposition is true.

Example 6. In a rectangle $ABCD$, we have $AB = 2$ and $AD = 1$. Let P be a moving point on the side DC (the endpoints are included) and Q be a moving point on the extension of CB (the point B is included) such that $|\overrightarrow{DP}| = |\overrightarrow{BQ}|$. Find the minimum value of the inner product $\overrightarrow{PA} \cdot \overrightarrow{PQ}$.

Solution. Without loss of generality we denote $A(0, 0)$, $B(2, 0)$, and $C(0, 1)$. Let P have the coordinates $(t, 1)$ $(0 \leq t \leq 2)$. By $|\overrightarrow{DP}| = |\overrightarrow{BQ}|$, the coordinates of Q are $(2, -t)$. Hence,

$$\overrightarrow{PA} = (-t, -1), \overrightarrow{PQ} = (2-t, -t-1),$$

so

$$\begin{aligned}\overrightarrow{PA}\cdot\overrightarrow{PQ} &= (-t)\cdot(2-t)+(-1)\cdot(-t-1)\\ &= t^2-t+1=\left(t-\frac{1}{2}\right)^2+\frac{3}{4}\geq\frac{3}{4}.\end{aligned}$$

When $t=\frac{1}{2}$, we obtain that $(\overrightarrow{PA}\cdot\overrightarrow{PQ})_{\min}=\frac{3}{4}$.

Example 7. In a triangle ABC, suppose $\overrightarrow{AB}\cdot\overrightarrow{AC}+2\overrightarrow{BA}\cdot\overrightarrow{BC}=3\overrightarrow{CA}\cdot\overrightarrow{CB}$.

(1) Let a, b, and c be the side lengths opposite to A, B, and C, respectively. Show that $a^2+2b^2=3c^2$.
(2) Find the minimum value of $\cos C$.

Solution. (1) By the definition of inner products and the law of cosines,

$$\overrightarrow{AB}\cdot\overrightarrow{AC}=cb\cos A=\frac{b^2+c^2-a^2}{2}.$$

Similarly, $\overrightarrow{BA}\cdot\overrightarrow{BC}=\frac{a^2+c^2-b^2}{2}$ and $\overrightarrow{CA}\cdot\overrightarrow{CB}=\frac{a^2+b^2-c^2}{2}$. Therefore, the condition becomes

$$b^2+c^2-a^2+2(a^2+c^2-b^2)=3(a^2+b^2-c^2),$$

or equivalently $a^2+2b^2=3c^2$.

(2) By the law of cosines and basic inequalities,

$$\begin{aligned}\cos C &= \frac{a^2+b^2-c^2}{2ab}=\frac{a^2+b^2-\frac{1}{3}(a^2+2b^2)}{2ab}=\frac{a}{3b}+\frac{b}{6a}\\ &\geq 2\sqrt{\frac{a}{3b}\cdot\frac{b}{6a}}=\frac{\sqrt{2}}{3}.\end{aligned}$$

The equality here is valid if and only if $a:b:c=\sqrt{3}:\sqrt{6}:\sqrt{5}$. Therefore, the minimum value of $\cos C$ is $\frac{\sqrt{2}}{3}$.

Example 8. Let $A_1, A_2, \ldots, A_n$ and $B_1, B_2, \ldots, B_n$ be $2n$ distinct points in the plane. Show that we can arrange $B_1, B_2, \ldots, B_n$ in some order $B_1', \ldots, B_n'$ such that for all $1\leq i<j\leq n$, the vectors $\overrightarrow{A_iA_j}$ and $\overrightarrow{B_i'B_j'}$ form an acute or right angle.

Solution. We choose an arbitrary point O in the plane as the origin. Consider the following function:

$$S=\sum_{i=1}^{n}\overrightarrow{OA_i}\cdot\overrightarrow{OB_i},$$

where $A_1, \ldots, A_n$ are fixed and $B_1, \ldots, B_n$ can be arranged in any order (so that S is a function of the permutations of B_i).

Since there are finitely many permutations of $B_1, \ldots, B_n$, there exists a permutation $B'_1, \ldots, B'_n$ such that S attains its maximum value S'.

Next, we show that the permutation $B'_1, \ldots, B'_n$ satisfies the target condition.

In fact, if there exist $1 \leq i < j \leq n$ such that the angle between $\overrightarrow{A_iA_j}$ and $\overrightarrow{B'_iB'_j}$ is obtuse, then $\overrightarrow{A_iA_j} \cdot \overrightarrow{B'_iB'_j} < 0$. If we interchange B'_i and B'_j in $B'_1, \ldots, B'_n$ and denote the resulting value of S as S'', then

$$\begin{aligned}
S' - S'' &= \overrightarrow{OA_i} \cdot \overrightarrow{OB'_i} + \overrightarrow{OA_j} \cdot \overrightarrow{OB'_j} - \overrightarrow{OA_i} \cdot \overrightarrow{OB'_j} - \overrightarrow{OA_j} \cdot \overrightarrow{OB'_i} \\
&= \overrightarrow{OA_i} \cdot (\overrightarrow{OB'_i} - \overrightarrow{OB'_j}) - \overrightarrow{OA_j} \cdot (\overrightarrow{OB'_i} - \overrightarrow{OB'_j}) \\
&= (\overrightarrow{OA_i} - \overrightarrow{OA_j}) \cdot (\overrightarrow{OB'_i} - \overrightarrow{OB'_j}) \\
&= \overrightarrow{A_jA_i} \cdot \overrightarrow{B'_jB'_i} = \overrightarrow{A_iA_j} \cdot \overrightarrow{B'_iB'_j} < 0,
\end{aligned}$$

which is contradictory to the assumption that S' is maximal.

Therefore, the proposition is true.

18.3 Exercises

Group A

1. Let $\vec{m}$ and $\vec{n}$ be two unit vectors and the angle between them is $60°$. Find the angle between $2\vec{m} + \vec{n}$ and $2\vec{n} - 3\vec{m}$.
2. Two nonzero vectors $\vec{a}$ and $\vec{b}$ satisfy that $\vec{a} + 3\vec{b}$ is perpendicular to $7\vec{a} - 5\vec{b}$ and $\vec{a} - 4\vec{b}$ is perpendicular to $7\vec{a} - \vec{b}2$. Find the angle between $\vec{a}$ and $\vec{b}$.
3. In a triangle ABC, given that $AB = 7$, $BC = 5$, and $CA = 6$. Find the value of $\overrightarrow{BA} \cdot \overrightarrow{BC}$.
4. Suppose four points A, B, C, and D in the space satisfy the following conditions: $|\overrightarrow{AB}| = 3$, $|\overrightarrow{BC}| = 7$, $|\overrightarrow{CD}| = 11$, and $|\overrightarrow{DA}| = 9$. Find the value of $\overrightarrow{AC} \cdot \overrightarrow{BD}$.
5. Let O be a point inside a triangle ABC and let $\alpha = \frac{S_{\Delta BOC}}{S_{\Delta ABC}}$, $\beta = \frac{S_{\Delta COA}}{S_{\Delta ABC}}$, and $\gamma = \frac{S_{\Delta AOB}}{S_{\Delta ABC}}$. Show that
$$\alpha\overrightarrow{OA} + \beta\overrightarrow{OB} + \gamma\overrightarrow{OC} = \vec{0}.$$
6. Suppose the position vectors of a convex quadrilateral $ABCD$ are $\vec{a}$, $\vec{b}$, $\vec{c}$, and $\vec{d}$, respectively, and $|\vec{a}| = |\vec{b}| = |\vec{c}| = |\vec{d}|$ with $\vec{a} + \vec{b} + \vec{c} + \vec{d} = \vec{0}$. Describe the shape of $ABCD$.

7. Let O be a point inside a triangle ABC such that $\overrightarrow{OA}+2\overrightarrow{OB}+3\overrightarrow{OC} = 3\overrightarrow{AB}+2\overrightarrow{BC}+\overrightarrow{CA}$. Find the value of $\frac{S_{\triangle AOB}+2S_{\triangle BOC}+3S_{\triangle COA}}{S_{\triangle ABC}}$.
8. In a triangle ABC, suppose $\overrightarrow{AB}\cdot\overrightarrow{AC}+2\overrightarrow{BA}\cdot\overrightarrow{BC} = 3\overrightarrow{CA}\cdot\overrightarrow{CB}$. Find the maximum value of $\sin C$.
9. Let F be the midpoint of the side CD of a parallelogram $ABCD$ and let E be the intersection point of AF and BD. Show that E is a point of trisection on BD.
10. In a triangle ABC, suppose $AB = AC$, D is the midpoint of the side BC, and E is a point on the side AC such that $DE \perp AC$. Let F be the midpoint of DE. Show that $AF \perp BE$.

Group B

11. Let P be a point inside a given convex n-gon $A_1A_2\ldots A_n$ or on its boundary and let $f(P) = \sum_{i=1}^{n}|PA_i|$. Show that $f(P)$ attains the maximum value when P is some vertex A_i of the polygon.
12. We fill the integers $1, 2, \ldots, n^2$ in an $n \times n$ table such that each cell has a different number. Now, we draw a vector between the centers of each pair of the cells, with each vector starting at the cell of the smaller number and ending at the cell of the larger one. Suppose the sums of numbers in each column and row are the same. Show that the sum of all these vectors is the zero vector.
13. We fill the cells in an $n \times n$ table with 1 and -1 (one number in each cell). Suppose for each pair of rows, their corresponding vectors have inner product 0. (Namely if the i-th row has numbers $x_1, \ldots, x_n$, and the j-th row has numbers $y_1, \ldots, y_n$, then $\sum_{k=1}^{n} x_k y_k = 0$.) If there is an $a \times b$ "subtable" (that is the intersection of a rows with b columns) such that all the numbers in the "subtable" are 1, show that $ab \leq n$.

Chapter 19

"Angles" and "Distances" in Spaces

19.1 Key Points of Knowledge and Basic Methods

Problems regarding "angles" and "distances" are common in solid geometry. Definitions and basic theorems are the starting point when solving these problems, and it is important to have a sufficient ability in spatial visualization and deductive reasoning.

A. Basic concepts related to "angles"

In solid geometry, the following notions of "angle" are involved frequently: the angle between two lines (especially when they are skew lines), the angle between a line and a plane, and dihedral angles.

B. Concepts related to "distances"

"Distances" that we consider in solid geometry primarily include the distance from a point to a line, the distance between two skew lines, the distance from a point to a plane, the distance between a line and a plane (when they are parallel), and the distance between two parallel planes.

C. Commonly used theorems

Apart from the axioms of solid geometry, the theorems that we use frequently include the theorem of three perpendiculars, the property and criterion of line-plane perpendicularity, the property and criterion of plane-plane perpendicularity, etc.

D. Basic methods

When solving for an angle, we usually need to put it in a plane.

When solving for a distance, we usually reduce it to the distance from a point to a line, or further to the length of a line segment (such as the common perpendicular line segment).

In addition, most problems regarding angles and distances can be solved with space vectors.

19.2 Illustrative Examples

Example 1. As shown in Figure 19.1, A, B, C, and D are four points in the space, and E and F are points on the line segments BC and AD, respectively. Suppose $AB = 4$, $CD = 20$, $EF = 7$, and $\frac{AF}{FD} = \frac{BE}{EC} = \frac{1}{3}$. Find the angle between the skew lines AB and CD.

Solution. We choose the point G on BD such that $\frac{BG}{GD} = \frac{1}{3}$, and form the line segments EG and FG.

In the triangle BCD, since $\frac{BE}{EC} = \frac{BG}{GD}$, so $EG \| CD$, and by $\frac{EG}{CD} = \frac{BE}{BC} = \frac{1}{4}$ we have $EG = 5$. Similarly, $FG \| AB$ and $\frac{FG}{AB} = \frac{DF}{AD} = \frac{3}{4}$, so $FG = 3$. In the triangle EFG, we apply the law of cosines to get

$$\cos \angle FGE = \frac{EG^2 + GF^2 - EF^2}{2 \cdot EG \cdot GF} = \frac{3^2 + 5^2 - 7^2}{2 \times 3 \times 5} = -\frac{1}{2},$$

which implies that $\angle FGE = 120°$.

On the other hand, by $EG \| CD$ and $FG \| AB$, the angle between EG and FG is equal to the angle between AB and CD, so the angle between AB and CD is $60°$.

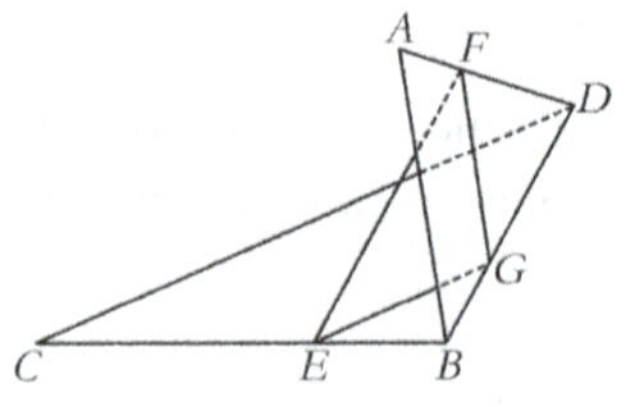

Figure 19.1

Remark. In general, when solving for the angle between skew lines, we can translate one or both of them to the same plane and find the angle between the intersecting lines in the plane.

Example 2. Suppose there is a plane such that the angles between all 12 edges of a cube and this plane are equal to α. Find the value of α.

Solution. The difficulty of this problem lies in how to determine the position of the plane relative to the cube.

As shown in Figure 19.2, it follows from the condition that AB, AC, and AD form equal angles with the plane, so the plane in the problem is parallel to the plane BCD. Since the plane BCD satisfies the condition (the angle between all the edges and the plane BCD are the same), we only need to find the angle between AD and the plane BCD.

Let H be the projection of A onto the plane BCD, so that AH is perpendicular to the plane BCD. Then $\angle ADH = \alpha$.

Note that BCD is an equilateral triangle, and we can prove that H is the circumcenter of ΔBCD (using $AD = AB = AC$), so H is the barycenter of ΔBCD. Let a be the edge length of the cube. Then $DH = \frac{2}{3} CD \cdot \sin 60° = \frac{\sqrt{6}}{3} a$. Since $AH \perp$ plane BCD, we get that $AH \perp DH$ and $\angle AHD = 90°$. Hence,

$$\cos\alpha = \cos\angle ADH = \frac{DH}{AD} = \frac{\sqrt{6}}{3},$$

and $\alpha = \arccos\frac{\sqrt{6}}{3}$.

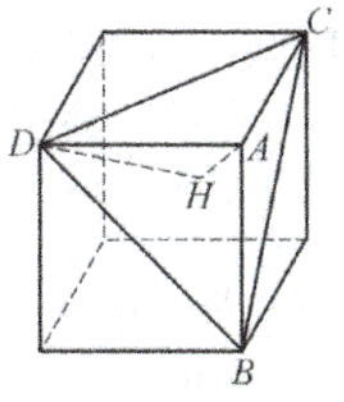

Figure 19.2

Remark. We can also solve this problem by space vectors. As in Figure 19.3, we construct a space coordinate system with $A_1(1, 0, 0)$.

Suppose a plane β forms an angle α with all the edges of the cube. Then we consider its normal vector $\vec{n} = (x_0, y_0, z_0)$ $(x_0 > 0)$. From

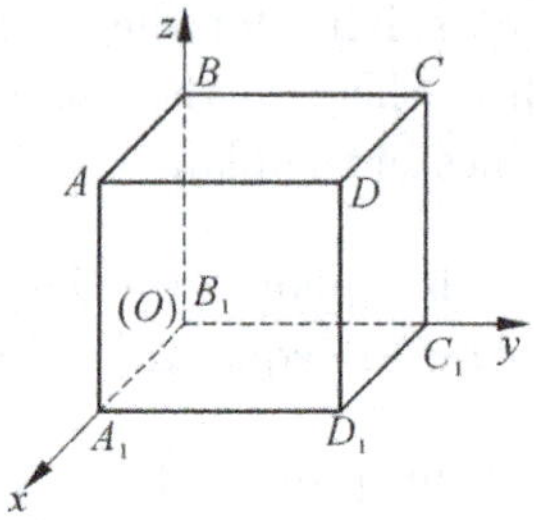

Figure 19.3

$\overrightarrow{DA} = (0, -1, 0)$, $\overrightarrow{DD_1} = (0, 0, -1)$, and $\overrightarrow{DC} = (-1, 0, 0)$,

$$|\cos(90^\circ - \alpha)| = \left|\frac{\overrightarrow{DA} \cdot \vec{n}}{|\overrightarrow{DA}| \cdot |\vec{n}|}\right| = \left|\frac{|\overrightarrow{DD_1} \cdot \vec{n}|}{|\overrightarrow{DD_1}| \cdot |\vec{n}|}\right| = \left|\frac{\overrightarrow{DC} \cdot \vec{n}}{|\overrightarrow{DC}| \cdot |\vec{n}|}\right|,$$

so

$$\sin\alpha = \frac{x_0}{\sqrt{x_0^2 + y_0^2 + z_0^2}} = \frac{y_0}{\sqrt{x_0^2 + y_0^2 + z_0^2}} = \frac{z_0}{\sqrt{x_0^2 + y_0^2 + z_0^2}}.$$

Therefore, $x_0 = y_0 = z_0$ and $\sin\alpha = \frac{\sqrt{3}}{3}$, hence $\alpha = \arcsin\frac{\sqrt{3}}{3}$.

Example 3. Suppose the base of a triangular pyramid S-ABC is an equilateral triangle and the projection of A onto the face SBC is the orthocenter of ΔSBC, denoted as H. If $SA = 2\sqrt{3}$ and $V_{SABC} = \frac{9}{4}\sqrt{3}$, find the magnitude of the dihedral angle H-AB-C.

Solution. As in Figure 19.4, let BH intersect SC at E and form the line segment AE. Let O be the projection of S onto the face ABC and F be the intersection point of CO and AB.

By $AH\perp$ plane SBC, we have $AH\perp SC$, and since H is the orthocenter of ΔSBC, we also have $BE\perp SC$, so $SC\perp$ plane ABE, and consequently $SC\perp AB$. By the inverse theorem of the theorem of three perpendiculars, $CF\perp AB$, and similarly $BO\perp$AC. Therefore, O is the orthocenter of ΔABC. Hence, O is also the circumcenter of ΔABC (since it is equilateral) and $SB = SC = SA = 2\sqrt{3}$.

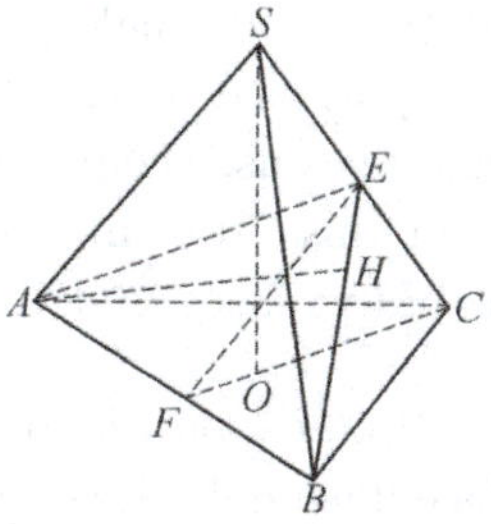

Figure 19.4

Further, since EF is the projection of CF onto the plane ABE and $CF \perp AB$, again by the inverse theorem of the theorem of three perpendiculars, $EF \perp AB$, so the magnitude of the dihedral angle H-AB-C equals the magnitude of the dihedral angle E-AB-C, which is exactly $\angle EFC$.

Let $\angle EFC = \alpha$. Then in the right triangle EFC we have $\angle ECF = 90° - \alpha$ and

$$SO = SC \cdot \sin \angle ECO = SC \cdot \cos \alpha,$$

$$CO = SC \cdot \cos \angle ECO = SC \cdot \sin \alpha.$$

Thus, by $CO = \frac{2}{3} \times \frac{\sqrt{3}}{2} \times AB$ and $S_{\Delta_{ABC}} = \frac{\sqrt{3}}{4} \times AB^2$,

$$S_{\Delta_{ABC}} = \frac{\sqrt{3}}{4} \times (\sqrt{3}CO)^2 = \frac{3\sqrt{3}}{4}(SC \cdot \sin \alpha)^2,$$

and further,

$$\begin{aligned} V_{S\text{-}ABC} &= \frac{1}{3} \times SO \times S_{\Delta_{ABC}} \\ &= \frac{1}{3} \times \frac{3\sqrt{3}}{4} \times SC^3 \times \sin^2 \alpha \cdot \cos \alpha \\ &= 18 \sin^2 \alpha \cos \alpha. \end{aligned}$$

Since $V_{S\text{-}ABC} = \frac{9}{4}\sqrt{3}$,

$$\sin^2 \alpha \cos \alpha = \frac{\sqrt{3}}{8},$$

so $\cos \alpha = \frac{\sqrt{3}}{2}$ or $\frac{\sqrt{7}-\sqrt{3}}{4}$. Therefore, the magnitude of the dihedral angle H-AB-C is $30°$ or $\arccos \frac{\sqrt{7}-\sqrt{3}}{4}$.

Remark. In order to find the plane angle of a dihedral angle, we can either use the definition or apply the inverse theorem of the theorem of three perpendiculars, or sometimes use the perpendicular plane to an edge.

Vectors are also useful in this problem, since we can find the angle between the normal vectors of two faces to determine their dihedral angle.

Example 4. Let $ABCD\text{-}A_1B_1C_1D_1$ be a cube with edge length 1. Find the distance between the face diagonals A_1B and B_1D_1.

Analysis 1. As shown in Figure 19.5, the theorem of three perpendiculars implies that $AC_1 \perp B_1D_1$ and $AC_1 \perp A_1B$. Hence, if we can find a line perpendicular to AC_1 that intersects both B_1D_1 and A_1B, then we can find the distance between the two lines by calculating the length of the common perpendicular line segment.

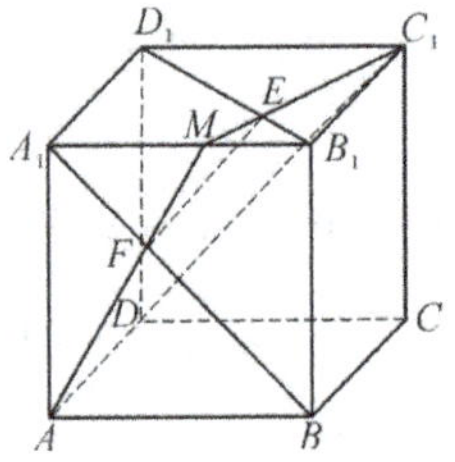

Figure 19.5

Solution 1. Let M be the midpoint of A_1B_1 and E be the intersection point of C_1M and B_1D_1. Then since $AC_1 \| A_1B$, we have $C_1E : EM = 2 : 1$. Let AM intersect A_1B at F. Then similarly $AF : FM = 2 : 1$. This implies that $EF \| C_1A$, so the length of EF is exactly the distance between B_1D_1 and A_1B.

By $AC_1 = \sqrt{3}$, we have $EF = \frac{\sqrt{3}}{3}$, so the distance between B_1D_1 and A_1B is $\frac{\sqrt{3}}{3}$.

Analysis 2. Since the distance between two skew lines is the minimal distance between two points lying on either line, we can find this distance by calculating the minimum value of a function.

Solution 2. Choose a point E on D_1B_1 and let G be a point on A_1B_1 such that $EG \perp A_1B_1$. Let F be a point on A_1B. Then $FG \perp GE$. Let $B_1G = x$. Then $GE = x$ and $EF^2 = GE^2 + GF^2$. Note that when GE is fixed, GF is minimal when $GF \perp A_1B$, so in this case $GF = \frac{\sqrt{2}}{2}(1-x)$. It follows that

$$EF^2 = GE^2 + GF^2 = x^2 + \frac{1}{2}(1-x)^2 = \frac{3}{2}\left(x - \frac{1}{3}\right)^2 + \frac{1}{3}.$$

Therefore, EF attains its minimum value $\frac{\sqrt{3}}{3}$ when $x = \frac{1}{3}$.

Analysis 3. We form the line segments A_1D and BD_1. By $D_1B_1 \| DB$ we have $D_1B_1 \|$ plane A_1DB. Then we can transform the problem of "finding the distance between skew lines D_1B_1 and A_1B" to the problem of "finding the distance between D_1B_1 and plane A_1DB," and it suffices to find the distance h from B_1 to the plane A_1DB (using the volume).

Solution 3. Since $V_{B_1A_1DB} = V_{DA_1B_1B}$, we have $\frac{h}{3} \cdot \frac{\sqrt{3}}{4}(\sqrt{2})^2 = \frac{1}{3} \times 1 \times \frac{1}{2}$, which immediately shows that $h = \frac{\sqrt{3}}{3}$.

Analysis 4. Apparently the plane D_1B_1C is parallel to the plane A_1BD, so the distance that we desire to find is equal to the distance between the planes D_1B_1C and A_1BD. It is simple to see that AC_1 is perpendicular to both planes, so we can calculate the distance from A to the plane A_1DB to find the result. (The details are left for the readers to check.)

Remark. From the solutions above, we can see that there are multiple ways to find the distance between two lines, and the crucial point is to realize it as the length of a line segment.

Example 5. Suppose the base of a triangular pyramid S-ABC is an isosceles right triangle with hypotenuse AB, with $SA = SB = SC = 2$ and $AB = 2$. If S, A, B, and C lie on a sphere centered at O, find the distance from O to the plane ABC.

Analysis. By $SA = SB = SC = 2$, the projection of S onto the plane ABC is the circumcenter of ΔABC. Since ABC is a right triangle, its circumcenter is the midpoint of AB. Then the distance from O to the plane ABC can be found easily.

Solution. As in Figure 19.6, let D be the projection of S onto the plane ABC. Then by $SA = SB = SC$, we see that D is the circumcenter of ΔABC, so D is the midpoint of AB. Since O is the center of the circumscribed sphere of S-ABC, the projection of O onto the plane ABC is also D. Therefore, OD is the distance from O to the base plane.

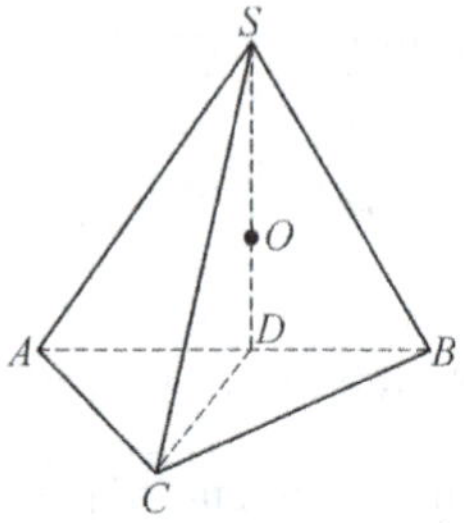

Figure 19.6

Let $OD = x$. Then by $OA = OB = OC = OS$ we have $\sqrt{1^2 + x^2} = SD - x = \sqrt{2^2 - 1^2} - x = \sqrt{3} - x$.

Solving the equation gives $x = \frac{\sqrt{3}}{3}$.

Remark. The circumcenter of a right triangle is the midpoint of its hypotenuse. By showing that S, O, and D are collinear, we determined that OD is the desired distance.

Example 6. In a triangular pyramid S-ABC, it is assumed that $SC = AB$, and the angle between SC and the plane ABC is $60°$. Suppose A, B, and C and the midpoints of each side edge (SA, SB, and SC) lie on the same sphere with radius 1. Show that the center of this sphere lies on AB and find the altitude of the triangular pyramid.

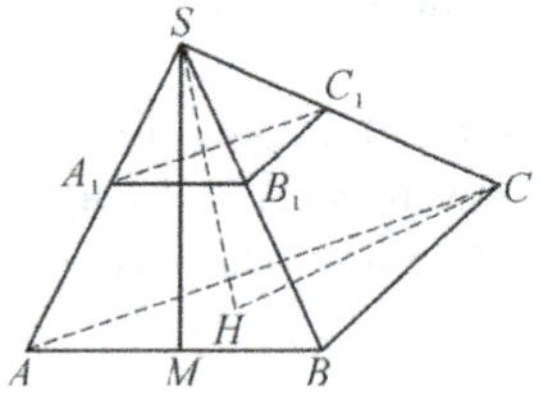

Figure 19.7

Analysis. We use the same method. Let M be the midpoint of AB, and we will show that SM is the altitude of S-ABC.

Solution. Let H be the projection of S onto the plane ABC. Then $\angle SCH = 60°$. In the triangle CHS, we have $SH = \frac{\sqrt{3}}{2}a$ and $HC = \frac{a}{2}$ (here a is the length of SC). Since A, B, C, and the midpoints A_1, B_1, and C_1 of each side lie on the same sphere, we deduce that A_1, A, B_1, and B are concyclic.

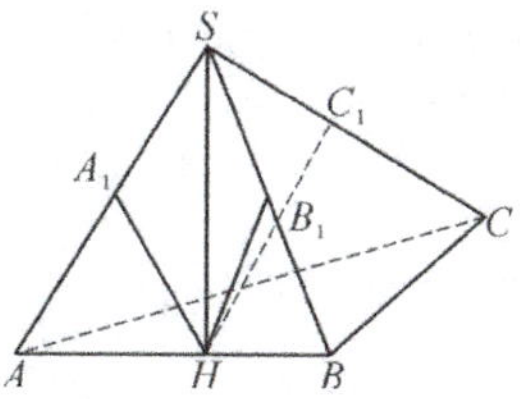

Figure 19.8

Since $A_1B_1 \| AB$, we see that AA_1B_1B is an isosceles trapezoid, and

$$AA_1 = BB_1 = \frac{1}{2}SA = \frac{1}{2}SB.$$

Similarly, $SA = SB = SC = AB = a$, so ASB is an equilateral triangle. Hence, $SM = \frac{\sqrt{3}}{2}a$, and since $SH = \frac{\sqrt{3}}{2}a$, we have $SM = SH$. This implies that $M = H$.

Now, we use Figure 19.8 instead. Since $\angle SAH = \angle SBH = \angle SCH = 60°$, we have $HA = HB = HC = \frac{a}{2}$ and $HA_1 = HB_1 = HC_1 = \frac{a}{2}$. Therefore, H is the center of the sphere and lies on AB, hence $a = 2$ and $SH = \sqrt{3}$.

Example 7. Suppose a cube $ABCD$-$A'B'C'D'$ has edge length 1. Let P be a point of AC, denote as a the angle between the plane $PA'B'$ and the bottom face, and denote as β the angle between the plane $PB'C'$ and the bottom face. Find the minimum value of $\alpha + \beta$.

Analysis. We can first determine the plane angles of the two dihedral angles, and then put them in the same triangle.

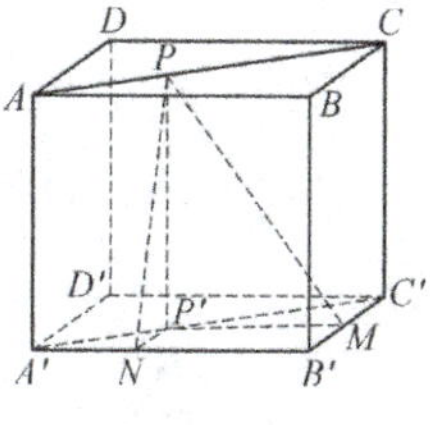

Figure 19.9

Solution. As in Figure 19.9, let P' be the projection of P onto the bottom face. Then P' lies on $A'C'$. Let N and M be the points on $A'B'$ and $B'C'$ respectively such that $P'N \perp A'B'$ and $P'M \perp B'C'$. Next, we rotate the triangles $PP'N$ and $PP'M$ around the axis PP' until they lie in the plane

$PA'P'$. Then we obtain a new triangle $PN'M'$, which has α and β as its inner angles.

Let $AP = x$. Then $N'P' = NP' = \frac{x}{\sqrt{2}}$ and $P'M' = P'M = \frac{\sqrt{2}-x}{\sqrt{2}}$, so

$$N'M' = N'P' + P'M' = 1.$$

Since $\begin{cases} S_\Delta = \frac{1}{2}ab\sin\alpha, \\ \cos\alpha = \frac{a^2+b^2-c^2}{2ab}, \end{cases}$ when the side $N'M'$ and the altitude PP' are fixed, $\angle N'PM'$ is maximal when the other two sides are equal, and $\alpha+\beta = \pi - \angle N'PM'$ is minimal. In this case $N'P' = \frac{1}{2}$ and $\tan\alpha = 2$. Also, since P is the midpoint of AC, we have $\alpha = \beta$, so $\tan(\alpha+\beta) = \tan 2\alpha = -\frac{4}{3}$. Therefore, $(\alpha+\beta)_{\min} = \pi - \arctan\frac{4}{3}$.

Remark. We have simplified the problem by rotating space objects onto a plane.

Example 8. Determine whether there exist a plane π and a regular tetrahedron T such that the cross section of T in π is a triangle with an angle greater than 120°.

Solution. As in Figure 19.10, we may assume that π intersects the edges AB, AC, and AD at P, Q, and R, respectively, and $\angle PQR > 120°$.

For convenience we assume that $AP = x$, $AQ = 1$, and $AR = y$. Then by the law of cosines,

$$PQ^2 = x^2 - x + 1, QR^2 = y^2 - y + 1, RP^2 = x^2 + y^2 - xy.$$

Here we have used the fact that each face of a regular tetrahedron is an equilateral triangle. By $\angle PQR > 120°$,

$$\begin{aligned} RP^2 &= PQ^2 + QR^2 - 2PQ\cdot QR\cdot\cos\angle PQR \\ &> PQ^2 + QR^2 + PQ\cdot QR. \end{aligned}$$

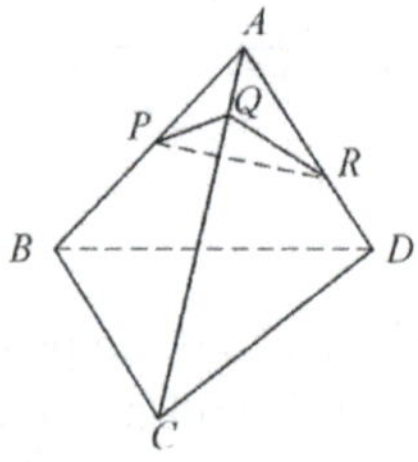

Figure 19.10

Hence,

$$\begin{aligned} x^2+y^2-xy &> x^2+y^2-x-y+2+\sqrt{(x^2-x+1)(y^2-y+1)} \\ &\Leftrightarrow x+y-xy-2>\sqrt{(x^2-x+1)(y^2-y+1)} \\ &\Leftrightarrow (x+y-xy-2)^2>(x^2-x+1)(y^2-y+1) \\ &\Leftrightarrow 3+5xy>x^2y+y^2x+3(x+y) \\ &\Leftrightarrow 3+5xy>(x+y)(xy+3). \text{①} \end{aligned}$$

Here we have used $x+y>xy+2$ (it may be obtained from $\angle PQR>90°$). Further, we can use this inequality again in ① and get

$$3+5xy>(xy+2)(xy+3) \Leftrightarrow (xy)^2+3<0,$$

which gives a contradiction. Therefore, such π and T do not exist.

Remark. The angles in a cross section are plane angles, which means that we need to use methods from plane geometry.

19.3 Exercises

Group A

1. Suppose a regular triangular pyramid S-ABC has equal side edge and base edge. If E and F are the midpoints of SC and AB, respectively, then the angle between EF and SA is ____________.
2. In a rectangle $ABCD$, we have $AB=3$ and $BC=4$. Let $PA\perp$ plane $ABCD$ and $PA=1$. Then the distance from P to the diagonal BD is ____________.
3. Suppose the angle between skew lines a and b is $80°$. Let P be a fixed point in the space and a line passing through P forms an angle of $50°$ with both a and b. The number of such lines is ____________.
4. The number of the planes that are equidistant to four given points in the space (not in the same plane) is ____________.
5. Let $ABCD$ be a square and E be the midpoint of AB. Suppose we fold ΔDAE and ΔCBE along DE and CE respectively so that AE coincides with BE. Let P be the point where A and B coincide. Then the magnitude of the dihedral angle between the face PCD and the face ECD is ____________.
6. In a cube $ABCD$-$A_1B_1C_1D_1$, the magnitude of the dihedral angle B-A_1C-D is ____________.

7. Suppose a cube $ABCD$-$A_1B_1C_1D_1$ has edge length 1, and M and N are the midpoints of BB_1 and B_1C_1, respectively. Find the distance between the lines MN and AC_1.
8. Suppose all the edges of a triangular prism ABC-$A_1B_1C_1$ have length 1 and $\angle A_1AB = \angle A_1AC = \angle BAC$. Let P be a point lying on the line segment A_1B such that $A_1P = \frac{\sqrt{3}}{3}$ and form the line segment PC_1. Find the angle between the lines PC_1 and AC.
9. In a cube $ABCD$-$A_1B_1C_1D_1$, we know that E is the midpoint of the edge BC, and let F be a point on AA_1 with $\frac{A_1F}{FA} = \frac{1}{2}$. Find the magnitude of the dihedral angle between the plane B_1EF and the top face $A_1B_1C_1D_1$.
10. Let the base of a triangular pyramid S-ABC be an equilateral triangle with the side length $4\sqrt{2}$. Suppose $SC \perp$ plane ABC and $SC = 2$. Let E and D be the midpoints of BC and BA, respectively. Find the distance between the lines CD and SE.
11. In a regular triangular prism ABC-$A_1B_1C_1$, assume that $AC = 6$ and $CC_1 = 8$, and D is the midpoint of AC. Find the distance between the lines BD and AC_1.

Group B

12. A tetrahedron A-BCD has equal edge lengths, and E and F are the midpoints of AD and BC, respectively. Form the line segments AF and CE.

    (1) Find the magnitude of the angle between the skew lines AF and CE.
    (2) Find the magnitude of the angle between CE and the face BCD.

13. Let l and m be skew lines. There are three points A, B, and C lying on l with $AB = BC$ such that AD, BE, and CF are perpendicular to m, where D, E, and F are the feet of perpendicular. Suppose $AD = \sqrt{15}$, $BE = \frac{7}{2}$, and $CF = \sqrt{10}$. Find the distance between l and m.
14. Let P be a point on the unit sphere, and three chords PA, PB, and PC are pairwise perpendicular.

    (1) Find the maximum value of the volume of the tetrahedron P-ABC.
    (2) Find the maximum value of the distance from P to the plane ABC.

15. Let α be a given plane, and A and B be two fixed points outside α. Find a point P in α, such that the angle $\angle APB$ is maximal.

Chapter 20

Cross Sections, Folding, and Unfolding

20.1 Key Points of Knowledge and Basic Methods

Cross sections, folding, and unfolding are three typical topics in solid geometry. Solving these problems involves much spatial imagination.

A. Cross sections

Cross section problems often involve construction and calculation. We should understand that lines or circles that we draw in the space are plane objects, and constructing the cross section correctly is crucial to solving relevant problems.

B. Folding plane objects

The important part in folding problems is to find which quantities are invariant and which quantities will change after folding.

C. Unfolding space objects

In some problems, we are required to find quantities such as the distance that a point has travelled along the surface of a polyhedron, in which case we may unfold the faces of this polyhedron onto a plane and solve the problem for the plane object. This is a common idea of reducing a solid geometry problem to a plane geometry problem.

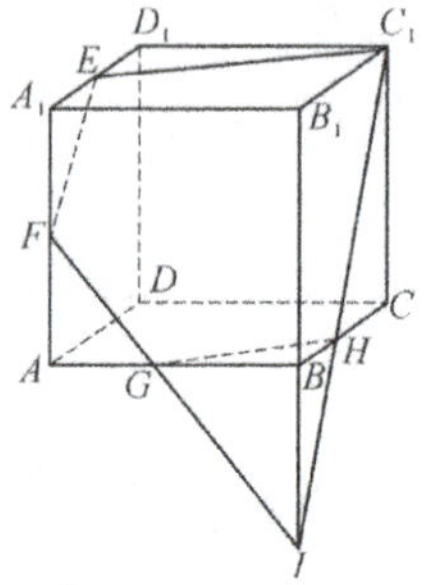

Figure 20.1

20.2 Illustrative Examples

Example 1. Is it possible to cut a cube with a plane such that the cross section is a pentagon? Further, is it possible that the cross section is a regular pentagon?

Solution. As in Figure 20.1, let I be a point on the extension of B_1B such that $IB = \frac{1}{2}BB_1$. Let E be the midpoint of A_1D_1 and F be a point on A_1A such that $\frac{AF}{A_1F} = \frac{1}{3}$. Then the pentagon C_1EFGH is the cross section of the cube cut by the plane containing EF and C_1I (here $EF \| C_1I$).

The cross section cannot be a regular pentagon. In fact, if the cross section is a regular pentagon, then its five sides belong to different faces, and by the pigeonhole principle, two of these faces must be parallel. Hence, the intersection lines of these two faces with the cutting plane should also be parallel, which is a contradiction, since no two sides of a regular pentagon are parallel.

Remark. In fact, we can easily construct a cross section that is a regular triangle, square, or regular hexagon, and Example 1 arised from a further exploration in this topic. We hope that the readers can keep thinking and exploring, and make progress constantly.

Example 2. Suppose the cube in Figure 20.1 has edge length 1, and I, E, and F are as described in Example 1. Find the perimeter of the cross section C_1EFGH.

Analysis. We can find the length of each side with proportions of line segments.

Solution. By the assumption in Example 1, $\frac{AF}{AA_1} = \frac{1}{4}$ and $\frac{BI}{BB_1} = \frac{1}{2}$. Combining them with $AF \| BI$ and $BH \| B_1C_1$, we have

$$\frac{AG}{GB} = \frac{AF}{BI} = \frac{1}{2}, \frac{BH}{B_1C_1} = \frac{BI}{B_1I} = \frac{1}{3},$$

so $BG = \frac{2}{3}$ and $BH = \frac{1}{3}$. Hence,

$$GH = \sqrt{\left(\frac{2}{3}\right)^2 + \left(\frac{1}{3}\right)^2} = \frac{\sqrt{5}}{3},$$

$$C_1H = \sqrt{1 + \left(\frac{2}{3}\right)^2} = \frac{\sqrt{13}}{3},$$

$$FG = \sqrt{\left(\frac{1}{4}\right)^2 + \left(\frac{1}{3}\right)^2} = \frac{5}{12},$$

$$EF = \sqrt{\left(\frac{1}{2}\right)^2 + \left(\frac{3}{4}\right)^2} = \frac{\sqrt{13}}{4},$$

$$C_1E = \sqrt{\left(\frac{1}{2}\right)^2 + 1} = \frac{\sqrt{5}}{2}.$$

Therefore, the perimeter of the cross section C_1EFGH is $\frac{5}{12} + \frac{5\sqrt{5}}{6} + \frac{7\sqrt{13}}{12}$.

Remark. The quantitative condition on I, E, and F is not necessary in Example 1 to find a pentagon cross section. It is only used in Example 2. In addition, if we extend C_1E and IF and let them intersect at J, then by similar triangles and Heron's formula we can find the area of the cross section.

Example 3. Let $ABCD$ be a rectangle with side lengths 3 and 4 such that $AD = 4$. We fold the rectangle along the diagonal AC such that the two parts form a right dihedral angle B_1-AC-D.

(1) Find the magnitude of the dihedral angle B_1-CD-A.
(2) Find the distance between the skew lines AB_1 and CD.

Analysis.(1) We can use the property of perpendicular planes to construct a perpendicular line, and then apply the theorem of three perpendiculars to determine the plane angle of the dihedral angle.

(2) Observe that the distance between the two lines is equal to the distance from CD to the plane ABB_1, so we may use the volume to find it.

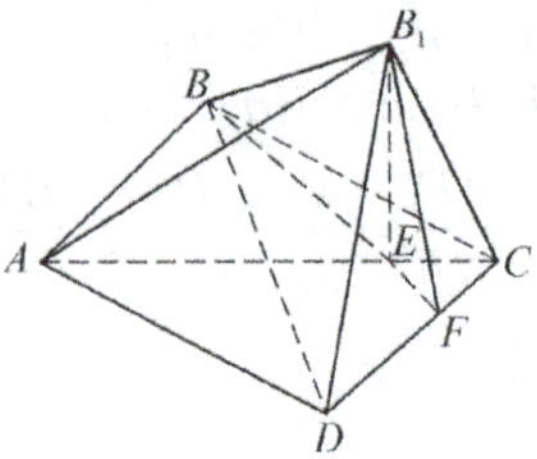

Figure 20.2

Solution. (1) Let E be a point on AC such that $B_1E \perp AC$ and F be a point on CD such that $EF \perp CD$. Form the line segment B_1F, and since B_1-AC-D is a right dihedral angle, $B_1E \perp$ plane ACD. By the theorem of three perpendiculars, $B_1F \perp CD$, so $\angle B_1FE$ is the plane angle of the dihedral angle B_1-CD-A.

By the area formula for triangles,

$$\frac{1}{2}B_1E \cdot AC = \frac{1}{2}AB_1 \cdot B_1C.$$

Hence, $B_1E = \frac{12}{5}$, and consequently

$$CE = \sqrt{4^2 - \left(\frac{12}{5}\right)^2} = \frac{16}{5}.$$

Note that $EF \| AD$, thus $\Delta CEF \backsim \Delta CAD$, from which

$$\frac{EF}{AD} = \frac{CE}{CA} = \frac{16}{25}.$$

Hence, $EF = \frac{64}{25}$ and

$$\tan \angle B_1FE = \frac{B_1E}{EF} = \frac{12}{5} \times \frac{25}{64} = \frac{15}{16}.$$

Therefore, the dihedral angle $B_1 - CD - A = \arctan \frac{15}{16}$.

(2) Note that $AB \| CD$, so the distance between the lines AB_1 and CD is equal to the distance h from D to the plane AB_1B. Since

$$\begin{aligned} V_{B_1\text{-}ABD} &= \frac{1}{3} \times B_1E \times \frac{1}{2}AB \times AD \\ &= \frac{1}{3} \times \frac{12}{5} \times \frac{1}{2} \times 4 \times 3 = \frac{24}{5}, \end{aligned}$$

and also

$$\begin{aligned} BB_1^2 &= BE^2 + B_1E^2 = AB^2 + AE^2 - 2AB \times AE \times \cos\angle BAC + B_1E^2 \\ &= 9 + \left(5 - \frac{16}{5}\right)^2 - 2 \times 3 \times \left(5 - \frac{16}{5}\right) \times \frac{3}{5} + \left(\frac{12}{5}\right)^2, \end{aligned}$$

we see that $BB_1 = \frac{12\sqrt{2}}{5}$, and the distance h_1 from A to BB_1 is

$$h_1 = \sqrt{AB^2 - \left(\frac{BB_1}{2}\right)^2} = \sqrt{9 - \left(\frac{6\sqrt{2}}{5}\right)^2} = \frac{3\sqrt{17}}{5}.$$

Consequently,

$$V_{D\text{-}AB_1B} = \frac{1}{3}h \times \frac{1}{2} \times h_1 \times BB_1 = \frac{6\sqrt{34}}{25}h,$$

and since $V_{D\text{-}AB_1B} = V_{B_1\text{-}ABD}$, we get $h = \frac{10\sqrt{34}}{17}$.

Therefore, the distance between the skew lines AB_1 and CD is $\frac{10\sqrt{34}}{17}$.

Remark. The problem of finding the distance between skew lines was already involved in the previous chapter. Such problems are usually transformed in the following way:

Line-line distance $\xrightarrow{\text{line-plane parallel}}$ line-plane distance $\longrightarrow$ point-plane distance, and then solve with volume, etc.

Example 4. In a circular frustum (as shown in Figure 20.3), the top face has radius 5, the bottom face has radius 10, and the generatrix $A_1A_2 = 20$. Suppose an ant starts at the midpoint M of A_1A_2 and travels one cycle on

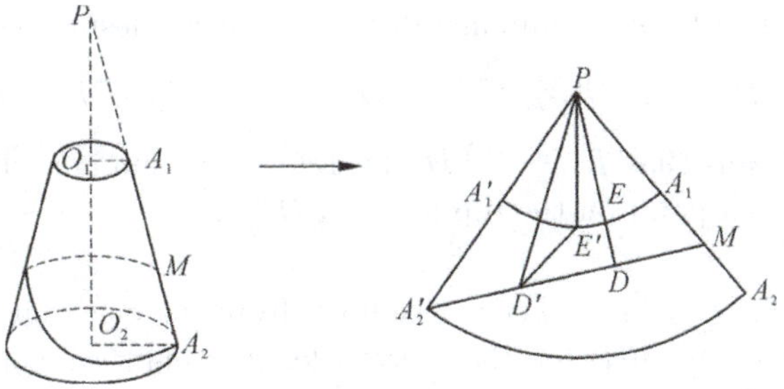

Figure 20.3

the side of the circular frustum until it reaches A_2 (so that going directly from M to A_2 is prohibited).

(1) Find the minimal distance that the ant travels.
(2) If the ant travels along the shortest path, find the minimal distance from the ant to some point on the top face (during travelling).

Solution. We cut the side surface of the circular frustum along the generatrix A_1A_2, getting a sectored annulus.

Let the extensions of A_2A_1 and $A'_2A'_1$ intersect at P. Since $PA_1O_1 \backsim \Delta PA_2O_2$,

$$\frac{PA_1}{PA_2} = \frac{O_1A_1}{O_2A_2} = \frac{1}{2},$$

so $PA_1 = \frac{1}{2}PA_2 = \frac{1}{2}(PA_1 + A_1A_2) = \frac{1}{2}(PA_1 + 20)$, hence $PA_1 = 20$.

Therefore, $\angle A'_1PA_1 = \frac{10\pi}{20} = \frac{\pi}{2}$.

(1) Form the line segment A'_2M. If A'_2M has no intersection point with the sector A'_1PA_1, then A'_2M is the minimal distance. Since $\angle A'_2PM = \frac{\pi}{2}$,

$$A'_2M^2 = A'_2P^2 + PM^2 = 40^2 + 30^2 = 50^2.$$

Hence, $A'_2M = 50$. Let D be a point on A'_2M with $PD \perp A'_2M$. Then

$$PD = \frac{PA'_2 \cdot PM}{A'_2M} = \frac{30 \times 40}{50} = 24 > PA_1,$$

which means that A'_2M has no intersection point with the sector A'_1PA_1. Therefore, the minimal distance that the ant travels is 50.

(2) Let E be the intersection point of PD with the arc A'_1A_1. Then we show that ED is the minimal distance from the ant to a point on the top face (during travelling).

In fact, if $D'E'$ is the minimal distance that we desire to obtain, then

$$PE' + E'D' \geq PD' \geq PD = PE + ED = PE' + ED,$$

which means that $E'D' \geq ED$. Therefore, the minimal distance from the ant to a point on the top face is $ED = 4$.

Remark. The first inequality here follows from the fact that the line segment between two points has the shortest length, and the second inequality follows since the perpendicular line segment between a point and a line has the shortest length.

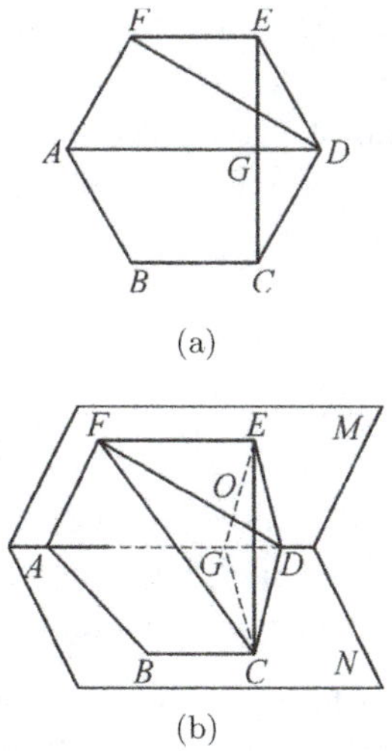

Figure 20.4

Example 5. Let $ABCDEF$ be a regular hexagon with side length a. Fold the hexagon along AD to form a dihedral angle M-AD-N. If the angle between the lines FC and AD is 45°, what is the magnitude of the dihedral angle M-AD-N? And find the volume of the triangular pyramid F-ECD in this case.

Analysis. As in Figure. 20.4, we compare the space object with the plane object. If G is the intersection point of EC and AD, then after folding we still have $EG \perp AD$ and $CG \perp AD$. Hence, $\angle EGC$ is the plane angle of the dihedral angle M-AD-N. Since $EF \| AD$, the problem can be reduced to finding $\angle EGC$ and V_{FECD} when $\angle CFE = 45°$.

Solution. Since $EG \perp AD$ and $CG \perp AD$, we have $AD \perp$ plane EGC. Hence, $\angle EGC$ is the plane angle of the dihedral angle M-AD-N. Also, since $FE \| AD$, we see that $FE \perp$ plane EGC, so

$$FE \perp EC, \angle EFC = 45°, EC = EF = a.$$

Consequently, $EG = CG = \frac{\sqrt{3}}{2}a$.

Applying the law of cosines to ΔEGC, we get

$$\cos \angle EGC = \frac{EG^2 + CG^2 - EC^2}{2EG \cdot CG} = \frac{1}{3},$$

so $\angle EGC = \arccos \frac{1}{3}$.

By $FE = ED = a$ and $\angle FED = 120°$, we find that $FD = \sqrt{3}a$, and since $FC = \sqrt{2}a$ and $CD = a$, we deduce that ΔFDC is a right triangle.

Let O be the projection of E onto the plane FDC. Since $EF = ED = EC = a$, we see that O is the circumcenter of ΔFDC, or in other words the midpoint of FD since it is a right triangle. Then $EO = \sqrt{EF^2 - FO^2} = \sqrt{a^2 - \left(\frac{\sqrt{3}}{2}a\right)^2} = \frac{a}{2}$, and

$$V_{FEDC} = V_{EFDC} = \frac{1}{3}EO \cdot S_{\Delta FDC} = \frac{1}{3} \cdot \frac{1}{2}a\left(\frac{1}{2}a \cdot \sqrt{2}a\right) = \frac{\sqrt{2}}{12}a^3.$$

Remark. The facts $EG \perp AD$ and $CG \perp AD$ do not change after folding, so they can help us determine the plane angle of the dihedral angle and reduce the problem into a problem of solving triangles.

Example 6. Suppose each face of a tetrahedron $ABCD$ has area 1. Let E, F, and G be points on the edges AB, AC, and AD, respectively. Show that $S_{\Delta EFG} \leq 1$.

Solution. It is equivalent to say that the area of the cross section does not exceed the area of any face (when all the faces have the same area). As shown in Figure 20.5, let π be the plane containing ΔEFG and assume that C has the smallest distance to π among B, C, and D. Let the plane passing through C and parallel to π intersect AB and AD at E' and G', respectively. Then $S_{\Delta EFG} \leq S_{\Delta CE'G'}$.

Let the distances from A, G', and D to the line CE' be d_1, d_2, and d_3, respectively. We will show that $d_2 \leq \max\{d_1, d_3\}$.

As in Figure 20.6, let XY be the common perpendicular line segment between AD and CE'. We may assume without loss of generality that D is farther from X than G'. Then by Pythagorean theorem $d_2 \leq d_3$, so $d_2 \leq \max\{d_1, d_3\}$.

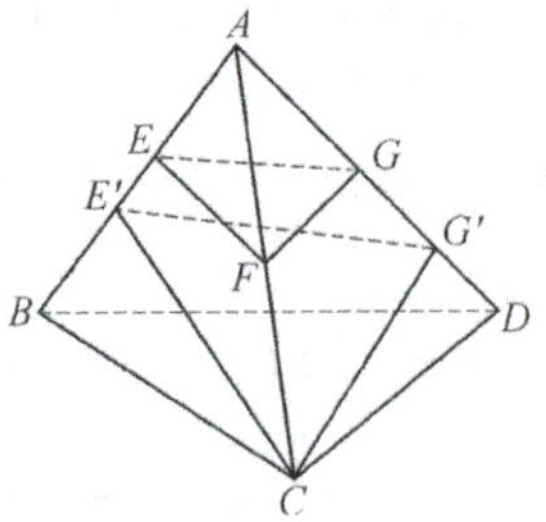

Figure 20.5

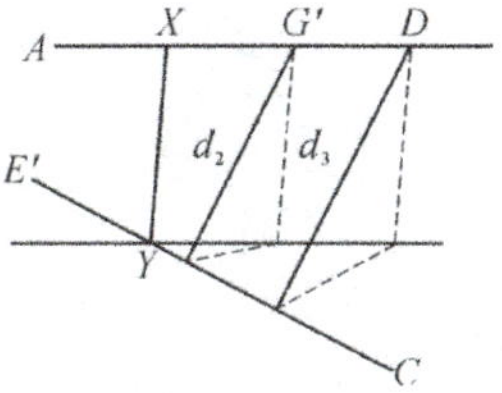

Figure 20.6

Therefore, $S_{\Delta CE'G'} \leq \max\{S_{\Delta ACE'}, S_{\Delta DCE'}\}$, and since $S_{\Delta ACE'} \leq S_{\Delta ABC}$, by the same method we get $S_{\Delta DCE'} \leq \max\{S_{\Delta ACD}, S_{\Delta BCD}\}$. Hence, $S_{\Delta EFG} \leq \max\{S_{\Delta ABC}, S_{\Delta ACD}, S_{\Delta BCD}\}$. If we drop the assumption that the distance from C to π is the smallest, then the general result will be $S_{\Delta EFG} \leq \max\{S_{\Delta ABC}, S_{\Delta ACD}, S_{\Delta BCD}, S_{\Delta ABD}\}$. Therefore, $S_{\Delta EFG} \leq 1$.

Remark. The result $d_2 \leq \max\{d_1, d_3\}$ is frequently used and we leave for the readers to extend it to general cases. Essentially what we have proven in this problem is that the area of the cross section in a tetrahedron cut by any plane does not exceed the greatest area of its faces.

Example 7. As shown in Figure 20.7, let H be the orthocenter of ΔABC and D, E, and F be the midpoints of BC, CA, and AB, respectively. A circle centered at H intersects DE at P and Q, intersects EF at R and S, and intersects FD at T and V. Show that $CP = CQ = AR = AS = BT = BV$.

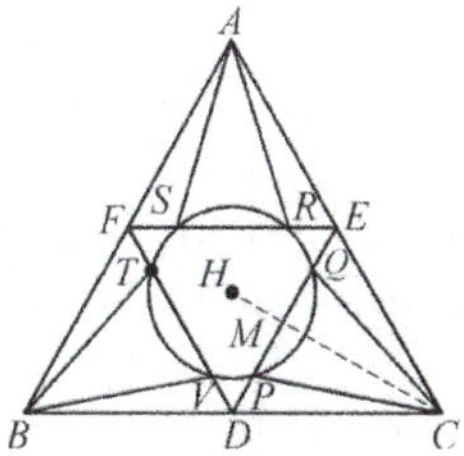

Figure 20.7

Analysis. This is a plane geometry problem that is hard to solve with plane geometry methods. Note that since D, E, and F are the midpoints of the three sides, we may fold the triangle along EF, DE, and DF, and make the three vertices coincide. Thus, we get a triangular pyramid and a

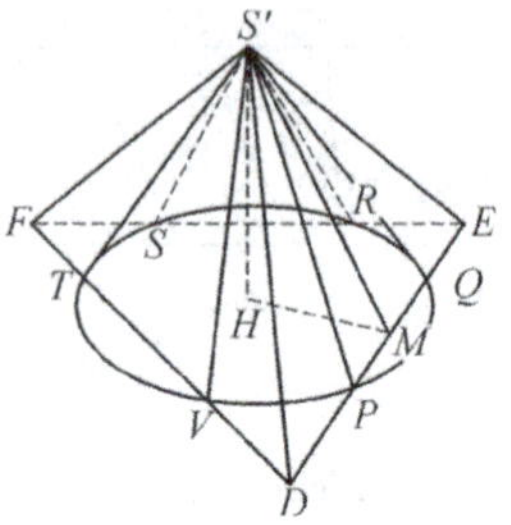

Figure 20.8

cone. Since AS, AR, CQ, CP, BV, and BT are all generatrices of the cone, they are necessarily equal.

Solution. As in Figure 20.7, since H lies inside ΔABC, the triangle is necessarily an acute one. We fold triangles AEF, BFD, and CED along EF, FD, and DE, respectively to get a tetrahedron S'-EFD. As shown in Figure 20.8, the vertex S' is where A, B, and C coincide after folding.

In Figure 20.7, let CH intersect DE at M. Then $CH \perp DE$. In Figure 20.8, it becomes that $S'M \perp DE$ and $HM \perp DE$. Hence, $DE \perp$ plane $S'MH$, so $S'H \perp DE$. Similarly, $DF \perp S'H$, so $S'H \perp$ plane DEF.

This implies that S' and H are the apex and the center of the base of a right circular cone. Since the generatrix has a fixed length, we obtain that

$$S'P = S'Q = S'R = S'S = S'T = S'V.$$

Since these lengths are invariant under folding,

$$CP = CQ = AR = AS = BT = BV.$$

Remark. Here folding is a tool to transform the plane geometry problem into a solid geometry problem, which is the key to this solution.

Example 8. Let O, A, B, and C be points lying in the same plane such that $OA = 4$, $OB = 2\sqrt{3}$, and $OC = \sqrt{22}$. Find the maximum value of the area of ΔABC.

Solution. We consider a tetrahedron M-NPQ (its existence will be proven later) such that MN, MP, and MQ are pairwise perpendicular. Let H be the projection of M onto the plane NPQ such that $HN = 4$, $HP = 2\sqrt{3}$, and $HQ = \sqrt{22}$.

Now, we assume that the plane NPQ is exactly the one containing O, A, B, and C, and O is located at H. Then

$$MA = \sqrt{MO^2 + OA^2} = \sqrt{MH^2 + HN^2} = MN.$$

Similarly, $MB = MP$ and $MC = MQ$, so

$$\begin{aligned} V_{M\text{-}ABC} &\leq \frac{1}{3} \cdot S_{\Delta ABM} \cdot CM \\ &\leq \frac{1}{3} \cdot \left(\frac{1}{2} AM \cdot BM\right) \cdot CM \\ &= \frac{1}{6} MN \cdot MP \cdot MQ = V_{M\text{-}NPQ}. \end{aligned}$$

This implies that $S_{\Delta ABC} \leq S_{\Delta NPQ}$. (Note that we have calculated the volume of the tetrahedron in two different ways, which is a common step when using the volume method.) Then it suffices to show the existence of this tetrahedron and find the area of ΔNPQ.

The key to the proof is to find necessary conditions that the length of MH needs to satisfy. Let $MH = x$ and let NH intersect PQ at R (Figure 20.9). Then

$$MN = \sqrt{x^2 + 16}, \quad MP = \sqrt{x^2 + 12}, \quad MQ = \sqrt{x^2 + 22}.$$

Note that $\Delta NMR \backsim \Delta NHM$ as right triangles. Thus,

$$MR = HM \cdot \frac{NM}{NH} = \frac{x}{4} \cdot \sqrt{x^2 + 6}.$$

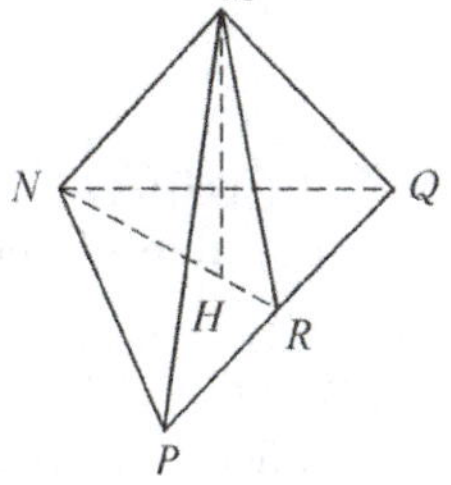

Figure 20.9

Further, $MN \perp$ plane MPQ, so $MN \perp PQ$. Also, $MH \perp$ plane NPQ, hence $MH \perp PQ$ and $PQ \perp$ plane NMR, which implies that $PQ \perp MR$. By calculating the area of ΔPMQ in different ways, $MR \cdot PQ = MP \cdot MQ$, namely

$$\frac{x}{4} \cdot \sqrt{x^2+16} \cdot \sqrt{(x^2+12)+(x^2+22)} = \sqrt{x^2+12} \cdot \sqrt{x^2+22}.$$

Squaring both sides, we have

$$x^6 + 25x^4 - 3 \times 11 \times 64 = 0,$$

which can be viewed as a cubic equation for x^2. This equation has a unique positive real solution $x^2 = 8$, from which $x = 2\sqrt{2}$.

Thus, we can determine the lengths of MN, MP, and MQ as well as the sides of ΔNPQ. Now, we first draw ΔNPQ and find its orthocenter, and then draw a line through H and perpendicular to the plane NPQ, and find a point M on the line such that $MH = 2\sqrt{2}$. This shows the existence of the tetrahedron $M - NPQ$.

Finally,

$$\begin{aligned} S_{\Delta NPQ} &= \frac{3 \cdot V_{M\text{-}NPQ}}{MH} = \frac{MN \cdot MP \cdot MQ}{2MH} \\ &= \frac{\sqrt{(x^2+16)(x^2+12)(x^2+22)}}{2x} \\ &= 15\sqrt{2}. \end{aligned}$$

Therefore, the maximum value of the area of ΔABC is $15\sqrt{2}$.

Remark. This is an interesting and creative solution, where we used solid geometry methods to solve a plane geometry problem. This method is simpler than directly showing that $S_{\Delta ABC}$ is maximal when O is the orthocenter of ΔABC.

20.3 Exercises

Group A

1. Suppose a unit cube $ABCD\text{-}A_1B_1C_1D_1$ is cut by a plane containing BD_1. Then the minimum value of the area of the cross section is __________.
2. In a rectangle $EFGH$, it is given that $EF = 2\sqrt{3}$ and $FG = 2$. If we fold it along EG to form a right dihedral angle, then the distance between F and H (after folding) is __________.

3. In a square $ABCD$ with side length 2, let E be the midpoint of AB. Now, we fold the square along EC and ED, and make EA and EB coincide. Then we obtain a tetrahedron $CDEA$ (here B and A are the same point now). The volume of this tetrahedron is ____________.
4. If we cut a cube of edge length a with a plane such that the cross section is a quadrilateral, then the value range of the area of this quadrilateral is ____________.
5. Suppose the generatrix of a circular cone has length l and its apex angle is α. Then the maximal area of a cross section passing through the apex is ____________.
6. Let $ABCD$ be a rhombus with side length a and $\angle BAD = 60°$. If we fold $ABCD$ along BD to form a dihedral angel A-BD-C that is equal to $120°$, then the distance between the lines AC and BD is ____________.
7. In a triangular pyramid P-ABC, suppose $PA \perp$ plane ABC and $AB \perp AC$ with $AC = a$, $AB = 2a$, and $PA = 3a$. If a cross section containing AB intersects PC at D, find the minimal area of the cross section (i.e., triangle ABD).
8. In a truncated triangular pyramid ABC-DEF, the top face ΔDEF has area a^2 and the bottom face ΔABC has area b^2 $(0 < a < b)$. Suppose the distance between BC and the plane AEF is equal to the altitude h of the truncated triangular pyramid, find the area of the cross section ΔAEF.
9. A square is folded as the side faces of a right square prism, so that a diagonal of the square is divided into four line segments. Find the angles between any two of these line segments.
10. Let n be a positive integer such that there exists a circular cone, whose surface area is n times the surface area of its inscribed sphere. Find all possible values of n.

Group B

11. Suppose a sphere S and a polyhedron P in the space have the following property: for every edge AB of P, its line segment XY cut by S satisfies $|AX| = |XY| = |YB| = \frac{1}{3}|AB|$. Show that there exists a sphere that is tangent to every edge of P.
12. Let PO be the altitude of a regular triangular pyramid P-ABC and M be the midpoint of PO. We draw a plane passing through AM and

parallel to BC such that the plane cuts the triangular pyramid into two parts. Find the ratio between the volumes of the two parts.

13. For an acute triangle with side lengths $2a$, $2b$, and $2c$, we connect the midpoints of each side and cut the triangle into four smaller triangles. This is the unfolding picture of a tetrahedron. Find the volume of this tetrahedron.
14. Let ABC be an equilateral triangle with side length a. A line parallel to BC intersects AB at E and intersects AC at F. Suppose we fold ΔAEF along EF to $\Delta A'EF$ such that plane $A'EF \perp$ plane $BCFE$. Find the distance between EF and BC such that $A'B$ is minimal.

Chapter 21

Projections and the Area Projection Theorem

21.1 Key Points of Knowledge and Basic Methods

In a right triangle ABC, if $\angle C = 90°$, then $AC = AB \cdot \cos \angle A$. Here AC can be understood as the projection of AB onto the line AC, so the formula above shows the relationship between the length of a line segment and the length of its projection onto some line.

Projection is an important notion in geometry, and the result above can be generalized to the space in the following way:

If a line segment in the space has length l and its projection onto a plane π has length d, then $d = l \cos \alpha$, where α is the angle between π and the line containing the segment.

Further, we may ask what the relationship is between the area of a polygon (or arbitrary convex plane object) and the area of its projection onto a plane. The area projection theorem to be expounded in this chapter is an important result of thinking about this problem.

The area projection theorem

In a dihedral angle with magnitude α, if a polygon in one half plane has area S, and its projection onto the second half plane has area S', then

$$S' = S \cdot \cos \alpha.$$

Note that if $\alpha > 90°$, then the projection no longer belongs to the second half plane, but instead lies on its extension. In this case, we may either apply the absolute value on the right side, or understand it as some sort of "directed area".

Further, the theorem above holds for not only polygons, but also circles. In fact, it holds for all convex objects in a plane.

21.2 Illustrative Examples

Example 1. Suppose each face of a tetrahedron has an equal area. Show that each pair of opposite edges in the tetrahedron has an equal length.

Analysis. We can use the area projection theorem to convert the condition into the relationship among dihedral angles, which may be related to the edges.

Solution. As shown in Figure 21.1, let α, β, and γ denote the magnitudes of the dihedral angles $A-DC-B$, $A-DB-C$, and $A-BC-D$, respectively, and let x, y, and z denote the magnitudes of the dihedral angles $C-AB-D$, $B-AC-D$, and $C-AD-B$, respectively. By the area projection theorem,

$$S_{\Delta BCD} = S_{\Delta ACD} \cdot \cos\alpha + S_{\Delta ADB} \cdot \cos\beta + S_{\Delta ABC} \cdot \cos\gamma,$$

so from $S_{\Delta BCD} = S_{\Delta ACD} = S_{\Delta ADB} = S_{\Delta ABC}$ we have $\cos\alpha + \cos\beta + \cos\gamma = 1$.

Similarly,

$$\cos x + \cos y + \cos\gamma = 1,$$

$$\cos x + \cos\beta + \cos z = 1,$$

$$\cos\alpha + \cos y + \cos z = 1.$$

Note that α, β, γ, x, y, and z all lie in $(0, \pi)$, so the above equalities imply that $\cos\alpha = \cos x$, $\cos\beta = \cos y$, and $\cos\gamma = \cos z$, which means that $\alpha = x$, $\beta = y$, and $\gamma = z$.

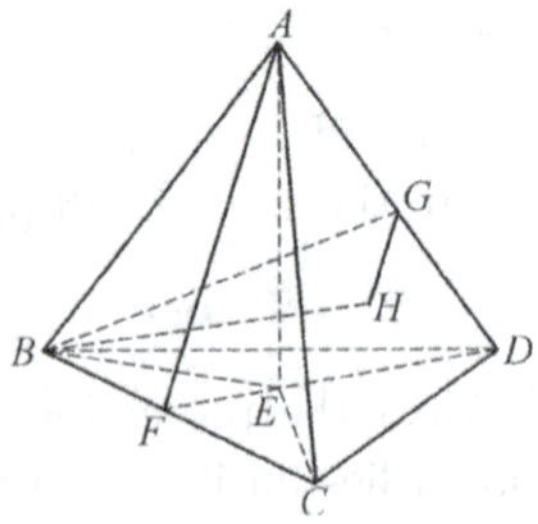

Figure 21.1

Now, let E be the projection of A onto the plane BCD and let F lie on BC such that $EF \perp BC$. Then $\angle AFE = \gamma$. Similarly, we construct $\angle BGH = z$. Since

$$V_{A\text{-}BCD} = \frac{1}{3} \times AE \times S_{\triangle BCD},$$

$$V_{B\text{-}ACD} = \frac{1}{3} \times BH \times S_{\triangle ACD},$$

while the volumes on the left side are equal, and $S_{\triangle BCD} = S_{\triangle ACD}$, we get $AE = BH$. By $AE = AF \cdot \sin\gamma$, $BH = BG \sin z$, and $\gamma = z$, we have $AF = BG$. Further, since

$$S_{\triangle ABC} = \frac{1}{2} AF \cdot BC,$$

$$S_{\triangle BAD} = \frac{1}{2} BG \cdot AD,$$

$$S_{\triangle ABC} = S_{\triangle BAD},$$

we get $BC = AD$. Similarly, we can prove that $AB = CD$ and $AC = BD$. Therefore, the proposition holds.

Remark. Applications of the area projection theorem are usually combined with area formulas and dihedral angles.

Example 2. Let $ABCD$ be a regular tetrahedron and let π be an arbitrary plane in the space. Show that the squared sum of the projection-lengths of all the edges of $ABCD$ onto π is a fixed number (irrelevant of the choice of π).

Solution. As shown in Figure 21.2, we draw a plane through each edge that is parallel to the opposite edge, which results in a cube.

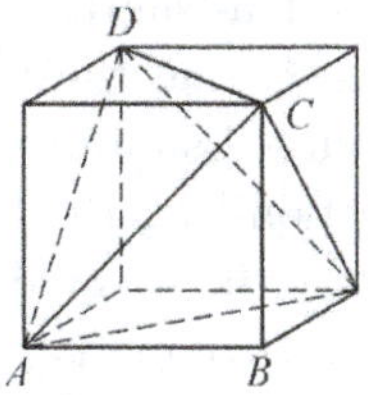

Figure 21.2

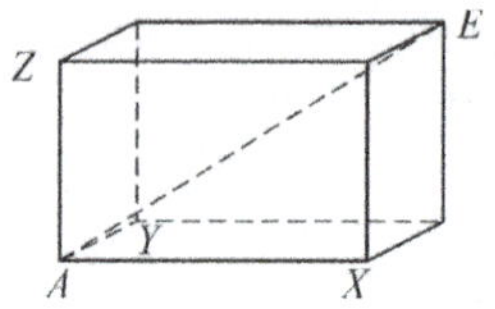

Figure 21.3

Suppose the three edges of the cube starting from A (in fact the lines containing them) form angles α, β, and γ with the normal of π. Then

$$\cos^2\alpha + \cos^2\beta + \cos^2\gamma = 2. \text{①}$$

Here ① is a well-known fact, which can be shown in the following way. Let E be the projection of A onto π, and we consider the cuboid with AE as its diagonal, whose edges starting from A share the same lines with the edges of the cube (as is Figure 21.3). Then $\angle XAE = \alpha$, $\angle YAE = \beta$, and $\angle ZAE = \gamma$. If $AX = x$, $AY = y$, $AZ = z$, and $AE = 1$, then

$$\cos\alpha = \sqrt{y^2 + z^2},$$
$$\cos\beta = \sqrt{z^2 + x^2},$$
$$\cos\gamma = \sqrt{x^2 + y^2}.$$

Since we also have $AE^2 = AX^2 + XE^2 = x^2 + y^2 + z^2$, the equality ① holds.

From this fact we see that the squared sum of the projection-lengths of all the twelve edges on the cube onto π is equal to $4b^2(\sin^2\alpha + \sin^2\beta + \sin^2\gamma) = 4b^2$, which is a constant. (Here b is the edge length of the cube obtained from the tetrahedron, and if the edge length of $ABCD$ is a, then $a^2 = 2b^2$.)

Note that the projection of each face of the cube onto π is a parallelogram, so the squared sum of the projection-lengths of its diagonals equals the squared sum of the projection-lengths of all its sides. (Here we have applied a result in plane geometry that the squared sum of the diagonals of a parallelogram equals the squared sum of all its sides.) This means that the squared sum of the projection-lengths of all the face-diagonals (twelve in total) of the cube equals twice the squared sum of the projection-lengths of all the edges of the cube. By symmetry, we conclude that the squared sum of the projection-lengths of all the edges of $ABCD$ onto π is equal to $4b^2$, namely $2a^2$, which is a constant.

Remark. Here the idea of placing the tetrahedron in a cube (or cuboid) deserves some attention.

Example 3. As shown in Figure 21.4, in a triangular pyramid $S-ABC$, $SA \perp$ planeABC, $AB \perp BC$, E is the midpoint of SC, and D lies on AC. Suppose $DE \perp SC$, and $SA = AB$ and $SB = BC$. Find the magnitude of the dihedral angle E-BD-C.

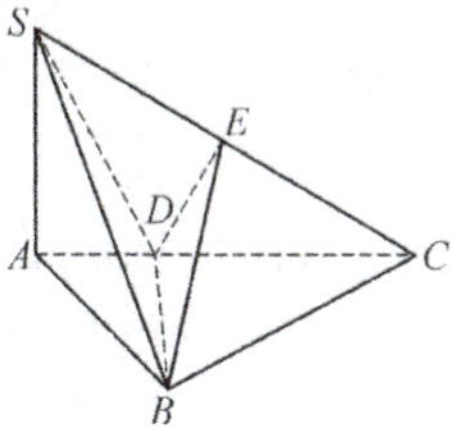

Figure 21.4

Analysis. We can apply the area projection theorem to find the dihedral angle.

Solution. Since E is the midpoint of SC and $SB = BC$, we have $BE \perp SC$. Combining the fact with $DE \perp SC$, we see that $SC \perp$ plane BDE. Hence, ΔBDE is the projection of ΔBDC onto the plane BDE. Let α be the dihedral angle that we desire to find. Then by the area projection theorem,

$$S_{\Delta BDE} = S_{\Delta BDC} \cdot \cos\alpha.$$

Next, we find α by calculating the ratio between $S_{\Delta BDE}$ and $S_{\Delta BDC}$.

Since $SC \perp$ plane BDE and $SE = EC$, we have $V_{SBDE} = V_{CBDE}$, namely

$$V_{CBDE} = \frac{1}{3} \times SE \times S_{\Delta BDE}.$$

Note that $SA \perp$ plane ABC and E is the midpoint of SC, so the distance from E to the plane BCD equals $\frac{1}{2}SA$. Hence,

$$V_{E\text{-}BCD} = \frac{1}{3} \times \left(\frac{1}{2}SA\right) \times S_{\Delta BDC}.$$

Since $V_{C\text{-}BDE} = V_{E\text{-}BCD}$, it follows that

$$\frac{1}{3} \times SE \times S_{\Delta BDE} = \frac{1}{6} \times SA \times S_{\Delta BDC},$$

and $\cos\alpha = \frac{S_{\Delta BDE}}{S_{\Delta BDC}} = \frac{SA}{2SE} = \frac{SA}{SC}$.

The condition that $SA\perp$ plane ABC ensures that $SA \perp BC$, and since $BC\perp AB$, we have $BC\perp$ plane SAB, so $BC\perp AB$. Hence,

$$SC = \sqrt{2}SB = \sqrt{2}(\sqrt{2}SA) = 2SA.$$

Therefore, $\cos\alpha = \frac{1}{2}$ and $\alpha = \frac{\pi}{3}$, and the dihedral angle is equal to $\frac{\pi}{3}$.

Remark. The area projection theorem is frequently used to find the magnitude of a dihedral angle. Compared to regular methods, this approach does not require the construction of the plane angle of a dihedral angle, so that the picture of the space object looks simpler.

Example 4. Let A and B be points in half planes π_1 and π_2, respectively, where the two half planes form a dihedral angle of magnitude θ. Let α be the angle between AB and π_1 and β be the angle between AB and π_2. Suppose the distances from A and B to the edge of the dihedral angle are b and a, respectively, and $AB = c$. Show that

$$\frac{a}{\sin\alpha} = \frac{b}{\sin\beta} = \frac{c}{\sin\theta}.$$

Analysis. This is a generalization of the law of sines to the space. We can construct right triangles to prove the proposition.

Solution. As in Figure 21.5, let AC be perpendicular to the edge of the dihedral angle, and $AO\perp\pi_2$. Since CO is the projection of AC onto π_2, by the inverse theorem of the theorem of three perpendiculars, $\angle ACO = \theta$. Since BO is the projection of AB onto π_2, we also have $\angle ABO = \beta$. Now, observe that

$$AO = AC\cdot\sin\theta = b\sin\theta,$$

$$AO = AB\sin\beta = c\sin\beta,$$

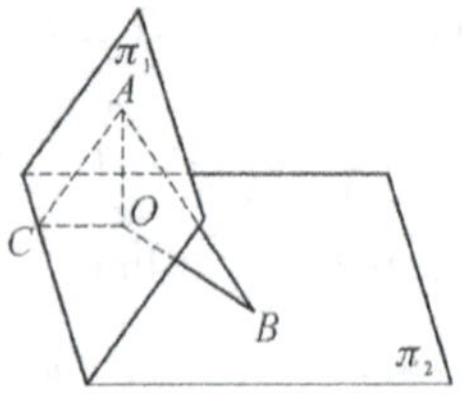

Figure 21.5

so $\frac{b}{\sin\beta} = \frac{c}{\sin\theta}$. Similarly, we can show that $\frac{a}{\sin\alpha} = \frac{c}{\sin\theta}$. Therefore, the proposition is true.

Remark. If BC is also perpendicular to the edge in Figure 21.5, then the proposition is exactly the law of sines.

Example 5. Let $ABCD$ be a parallelogram and $\alpha - MN - \beta$ be a dihedral angle with edge MN such that A lies on MN. Suppose B, C, and D all lie on α with $AB = 2AD$, $\angle DAN = 45°$, and $\angle BAD = 60°$ (shown in Figure 21.6). If the plane angle of the dihedral angle $\alpha - MN - \beta$ is θ, find $\cos\theta$ when the projection of $ABCD$ onto β is a (1) rhombus; (2) rectangle.

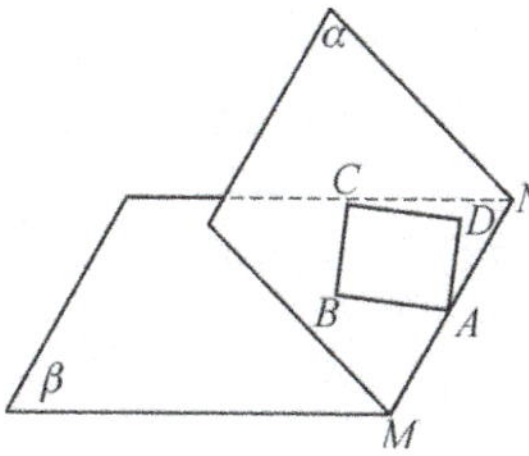

Figure 21.6

Analysis. Note that the projections of AD and BC onto β have equal lengths and the projections of AB and CD onto β also have equal lengths. So the projection of $ABCD$ onto β is always a parallelogram (or a line segment).

Solution. Let x denote the angle between AD and β. Then

$$\frac{AD}{\sin\theta} = \frac{AD\sin 45°}{\sin x}, \qquad \sin x = \frac{\sqrt{2}}{2}\sin\theta.$$

Here we have used the fact that the angle between AD and MN is 45° (so that the distance from D to MN is $AD \cdot \sin 45°$).

Hence, if we denote the projections of B, C, and D onto β as B', C', and D', respectively, then

$$AD' = \sqrt{1 - \frac{1}{2}\sin^2\theta} \cdot AD,$$

and similarly $AB' = \sqrt{1 - \sin^2 75° \sin^2\theta} \cdot AB$.

(1) If $AB'C'D'$ is a rhombus, then $AB' = AD'$, so

$$\sqrt{1-\frac{1}{2}\sin^2\theta}\cdot AD = \sqrt{1-\sin^2 75^\circ\cdot\sin^2\theta}\cdot AB,$$

$$1-\frac{1}{2}\sin^2\theta = 4(1-\sin^2 75^\circ\sin^2\theta) = 4-2(1-\cos 150^\circ)\sin^2\theta.$$

Hence, $\sin^2\theta = 4\sqrt{3}-6$ and

$$\cos\theta = \sqrt{1-(4\sqrt{3}-6)} = 2-\sqrt{3}.$$

(2) If is $AB'C'D'$ is a rectangle, then

$$S_{AB'C'D'} = AB'\times AD',$$

so $S_{ABCD}\cdot\cos\theta = AB'\times AD'$, or equivalently

$$\frac{\sqrt{3}}{2}\cos\theta\cdot AB\cdot AD = \sqrt{1-\frac{1}{2}\sin^2\theta}\cdot\sqrt{1-\sin^2 75^\circ\sin^2\theta}\cdot AB\cdot AD.$$

This shows that

$$\frac{3}{4}\cos^2\theta = \left(1-\frac{1}{2}\sin^2\theta\right)\left(1-\frac{1+\frac{\sqrt{3}}{2}}{2}\sin^2\theta\right).$$

Let $t=\sin^2\theta$. Then

$$(2+\sqrt{3})t^2-2(1+\sqrt{3})t+2=0,$$

so $t=\frac{2(1+\sqrt{3})}{2(2+\sqrt{3})}=\sqrt{3}-1$, hence

$$\cos\theta=\sqrt{2-\sqrt{3}}=\frac{\sqrt{6}-\sqrt{2}}{2}.$$

Therefore, if the projection of $ABCD$ onto β is a rhombus, then $\cos\theta = 2-\sqrt{3}$; if it is a rectangle, then $\cos\theta=\frac{\sqrt{6}-\sqrt{2}}{2}$.

Remark. We can obtain that $\sin x = \frac{\sqrt{2}}{2}\sin\theta$ from the generalized law of sines.

Example 6. Let P-$ABCD$ be a regular square pyramid. Suppose the dihedral angle between its side face and bottom face equals α and the dihedral angle between two adjacent side faces equals β. Show that

$$\cos\beta = -\cos^2\alpha.$$

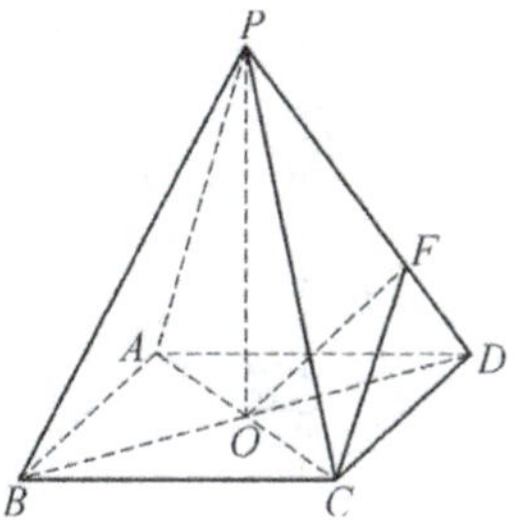

Figure 21.7

Solution. As shown in Figure 21.7, let O be the projection of P onto the plane $ABCD$ and let F be a point on DP such that $CF \perp DP$. Suppose the side length of the square is a and $PC = b$.

Note that in ΔCPF and ΔAPF, $PC = PA, PF = PF$, and $\angle FPA = \angle CPF$, so $\Delta CPF \cong \Delta APF$. Combining them with $CF \perp PD$ we have $AF \perp DP$, thus $\angle AFC = \beta$.

Hence,

$$\cos\beta = \frac{AF^2 + CF^2 - AC^2}{2AF \times CF}$$
$$= 1 - \frac{2OC^2}{CF^2} = 1 - \frac{a^2}{CF^2}. \text{①}$$

On the other hand, by the area projection theorem,

$$\cos\alpha = \frac{S_{\Delta OCD}}{S_{\Delta PCD}} = \frac{a^2}{4S_{\Delta PCD}} = \frac{a^2}{2b \cdot CF}. \text{②}$$

Calculating the area of ΔPCD in different ways, we get $\frac{1}{2}b \times CF = \frac{1}{2}a \times \sqrt{b^2 - \left(\frac{a}{2}\right)^2}$, so

$$CF = \frac{a\sqrt{4b^2 - a^2}}{2b}.$$

With ① and ②, we have

$$\cos\beta = 1 - \frac{4b^2}{4b^2 - a^2} = \frac{-a^2}{4b^2 - a^2},$$
$$-\cos^2\alpha = \frac{-a^2}{4b^2 - a^2}.$$

Therefore, $\cos\beta = -\cos^2\alpha$.

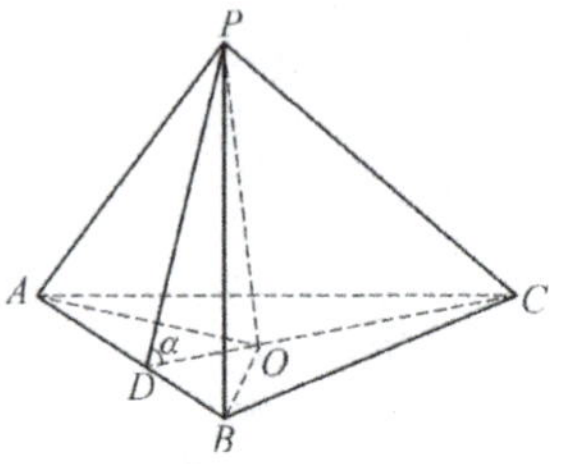

Figure 21.8

Example 7. Let P be a point outside a plane containing ΔABC such that PA, PB, and PC are pairwise perpendicular. Let the areas of ΔPAB, ΔPBC, and ΔPAC be S_1, S_2, and S_3, respectively, and let S denote the area of ΔABC. Show that $S^2 = S_1^2 + S_2^2 + S_3^2$.

Analysis. This example is a generalization of Pythagorean theorem.

Solution. Let the dihedral angles $P-AB-C$, $P-BC-A$, and $P-AC-B$ be equal to α, β, and γ, respectively. Let O be the projection of P onto the plane ABC, and the lines CO and AB intersect at D.

Since $PC \perp PA$ and $PC \perp PB$, we have $PC \perp$ plane PAB, so $PC \perp AB$, $CD \perp AB$, and $PD \perp AB$. Hence, $\angle PDC = \alpha$, and

$$S = \frac{1}{2} AB \cdot CD = \frac{AB \cdot PD}{2\cos\alpha} = \frac{S_1}{\cos\alpha},$$

$$S_{\Delta OAB} = \frac{1}{2} AB \cdot DO = \frac{1}{2} AB \cdot PD \cdot \cos\alpha = S_1 \cdot \cos\alpha.$$

This means that $S \cdot S_{\Delta AOB} = {S_1}^2$. Similarly, we can prove that $S \cdot S_{\Delta BOC} = S_2^2$ and $S \cdot S_{\Delta COA} = {S_3}^2$.

Therefore, $S(S_{\Delta AOB} + S_{\Delta BOC} + S_{\Delta COA}) = S^2 = {S_1}^2 + {S_2}^2 + {S_3}^2$.

Remark. Many results in plane geometry can be generalized to the space, but not all such statements are true, so we need to verify carefully.

Example 8. If all the six dihedral angles in a tetrahedron are equal, does it have to be a regular tetrahedron? What if at least five of them are equal?

Solution. Let $ABCD$ be a tetrahedron such that all six of its dihedral angles have the same magnitude θ. Then by the method in Example 1,

$$S_{\Delta BCD} = (S_{\Delta ABC} + S_{\Delta ACD} + S_{\Delta ADB}) \cos\theta.$$

Summing cyclically, we get $\cos\theta = \frac{1}{3}$.

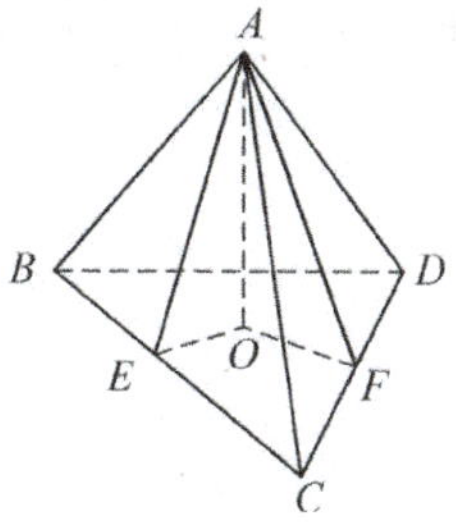

Figure 21.9

Let O be the projection of O onto the plane BCD and let E and F lie on BC and CD, respectively such that $OE \perp BC$ and $OF \perp CD$ (Figure 21.9). Then $\angle AEO = \angle AFO = \theta$, hence $AE = AF$ ($= AO \cdot \csc\theta$), so Rt $\Delta AEC \cong$ RtAFC, from which $\angle ACB = \angle ACD$.

Similarly, we can prove that $\angle BCD = \angle ACD$. Therefore, $\angle ACD = \angle ACB = \angle BCD$, in other words, the three face angles at C are equal. Similar results hold for other vertices as well.

Let the face angle at A, B, C, and D be α, β, γ, and δ, respectively. Then in ΔABC and ΔABD,

$$\alpha + \beta + \gamma = \alpha + \beta + \delta = 180°,$$

which implies that $\gamma = \delta$. Similarly, $\alpha = \beta = \gamma = \delta$, and hence they are all $60°$. Therefore, each face of the tetrahedron is an equilateral triangle, and it is thus a regular tetrahedron.

If only five dihedral angles are equal, then the tetrahedron is not necessarily regular. In fact, we can construct a tetrahedron such that

$$\angle ABC = \angle CBD = \angle DBA = \angle ACB = \angle BCD = \angle DCA = 40°,$$

$$\angle BAD = \angle CAD = \angle CDA = \angle BDA = 70°,$$

$$\angle BAC = \angle BDC = 100°.$$

In this example, we may first construct ΔABC and ΔDBC, and then fold along BC until it reaches the desired position.

All the dihedral angles in this tetrahedron are equal except for $B - AD - C$. Therefore, it is not a regular tetrahedron.

Remark. The converse of this proposition (i.e., any regular tetrahedron has six equal dihedral angles) is obviously true, so whether a tetrahedron has six equal dihedral angles can be used as a criterion for whether it is regular.

21.3 Exercises

Group A

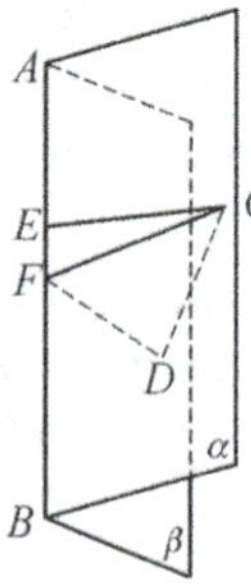

Exercise 1

1. As shown in the figure, suppose a dihedral angle $\alpha - AB - \beta$ is acute and C lies in α (but not on the edge AB of the dihedral angle). Let D be the projection of C onto β and let E be any point on the edge AB such that $\angle CEB$ is an acute angle. Which of $\angle CEB$ and $\angle DEB$ is greater?
2. Suppose each side edge of a triangular pyramid is perpendicular to its opposite edge on the bottom face. Then the projection of the apex onto the bottom face is the ____________ of the bottom triangle.
3. Suppose ΔABC is a right triangle with hypotenuse BC lying in a plane α. Let D be the projection of A onto α. If $\angle ABD = 30°$ and $\angle ACD = 45°$, then the dihedral angle between the plane ABC and α equals ____________.
4. In a right dihedral angle $\alpha - l - \beta$, let $A \in \alpha$ and $B \in \beta$ such that neither of A and B lies on l. Let x be the angle between AB and α, let y be the angle between AB and β, and let z be the angle between AB and l. Then $\cos^2 x + \cos^2 y - \cos^2 z =$____________.
5. For a cube with edge length 1, the greatest possible area of its projection onto some plane is ____________.
6. In a regular tetrahedron $ABCD$, points E and F lie on the edges AB and AC respectively such that $BE = 3$ and $EF = 4$, and EF is parallel to the plane BCD. Then the area of DEF is ____________.

7. In a triangular pyramid $P - ABC$, suppose the dihedral angle between each side face and the bottom face has an equal magnitude. Show that the projection of P onto the plane ABC is either the incenter or an escenter of ΔABC.
8. Suppose a cube $ABCD - A_1B_1C_1D_1$ has edge length a, and M, N, and P are the midpoints of AB, AD, and CC_1, respectively. Find the area of the cross section passing through M, N, and P.
9. For a regular triangular frustum, suppose its top face has side length 6, its bottom face has side length 10, and the dihedral angle between the side face and the bottom face equals 45°. Find its total surface area.
10. In a cube $ABCD - A_1B_1C_1D_1$ with edge length 1, let P be a point on AC, let α be the magnitude of the (acute) dihedral angle between the planes PA_1B_1 and $A_1B_1C_1D_1$, and let β be the magnitude of the (acute) dihedral angle between the planes PB_1C_1 and $A_1B_1C_1D_1$. Find the position of P when $\tan(\alpha + \beta)$ is minimal.

Group B

11. In a tetrahedron $ABCD$, suppose $BD \perp CD$ and the projection of D onto the plane ABC is the orthocenter of ABC. Show that

$$(AB + BC + CA)^2 \leq 6(AD^2 + BD^2 + CD^2).$$

12. Suppose each side face of a triangular pyramid $P - ABC$ forms an equal dihedral angle of 45° with the bottom faceand the bottom face is an obtuse triangle whose sides are consecutive even numbers. Find the volume of this triangular pyramid.
13. In a regular square pyramid $S - ABCD$, suppose the bottom face has side length 2 and its altitude is h. Let π be a plane parallel to AC, and suppose the dihedral angle between π and the bottom face equals α. What is the maximal area of the polyhedron's projection onto π? And what is the value of α when this maximum value is attained?
14. Suppose in a regular pyramid whose bottom face is a regular n – gon, the dihedral angle between each side face and the bottom face equals α, and the angle between each side edge and the bottom face is β. Show that $\sin^2 \alpha - \sin^2 \beta \leq \tan^2 \left(\frac{\pi}{2n}\right)$.

Chapter 22

Partitions of Sets

22.1 Key Points of Knowledge and Basic Methods

In this chapter we discuss problems regarding partitions of sets. The methods we use include mappings, constructions, induction, etc.

If $A_1, A_2, \ldots, A_n$ are a collection of nonempty subsets of X such that

(1) for all $1 \leq i < j \leq n$, there hold $A_i \cap A_j = \emptyset$,
(2) $X = A_1 \cup A_2 \cup \cdots \cup A_n$,

then the collection $A_1, A_2, \ldots, A_n$ is called an n-partition of X. In particular, if $A_1, A_2, \ldots, A_n$ satisfy (2) (while (1) is not necessarily satisfied), then we call the collection $A_1, A_2, \ldots, A_n$ a covering of X.

22.2 Illustrative Examples

Example 1. Find the smallest and second smallest positive integers n such that the set $\{1, 2, \ldots, 3n-1, 3n\}$ can be partitioned into n three-element subsets $\{x, y, z\}$ with $x + y = 3z$.

Solution. Consider the sum of numbers in all these subsets:

$$\sum_{i=1}^{n}(x_i + y_i + z_i) = 4(z_1 + z_2 + \cdots + z_n) = \frac{3n(3n+1)}{2}.$$

If n is even, then $8|n$;
if n is odd, then $8|3n+1$, so $n = 5, 13, \ldots$;

On the other hand, when $n = 5$, the set $\{1, 2, \ldots, 3n-1, 3n\}$ can be partitioned in the following way: $\{4, 1, 11\}$, $\{5, 2, 13\}$, $\{6, 3, 15\}$, $\{7, 9, 12\}$, $\{8, 10, 14\}$.

When $n = 8$, the set $\{1, 2, \ldots, 3n-1, 3n\}$ can be partitioned in the following way: $\{5, 1, 14\}$, $\{7, 2, 19\}$, $\{8, 3, 21\}$, $\{9, 4, 23\}$, $\{10, 6, 24\}$, $\{11, 15, 18\}$, $\{12, 16, 20\}$, $\{13, 17, 22\}$.

Therefore, 5 and 8 are the smallest and second smallest positive integers n with the desired property.

Example 2. Find all positive integers n such that the set $\{1, 2, \ldots, n\}$ can be partitioned into five subsets, with the sums of numbers in each subset the same.

Solution. We first find a necessary condition. If $\{1, 2, \ldots, n\}$ can be partitioned into five subsets with the same sums of numbers of each subset, then

$$1 + 2 + \cdots + n = \frac{n(n+1)}{2}$$

has to be a multiple of 5, so $n = 5k$ or $n = 5k - 1$.

Apparently when $k = 1$ the condition above is not sufficient. However, we can prove with induction that for $k \geq 2$, the above condition is also sufficient.

When $k = 2$, we have $n = 9$ or 10. We partition the corresponding sets in the following way:

$$\{1, 8\},\ \{2, 7\},\ \{3, 6\},\ \{4, 5\},\ \{9\};$$

$$\{1, 10\},\ \{2, 9\},\ \{3, 8\},\ \{4, 7\},\ \{5, 6\}.$$

When $k = 3$, we have $n = 14$ or 15, and there are the respective partitions

$$\{1, 2, 3, 4, 5, 6\},\ \{7, 14\},\ \{8, 13\},\ \{9, 12\},\ \{10, 11\};$$

$$\{1, 2, 3, 5, 6, 7\},\ \{4, 8, 12\},\ \{9, 15\},\ \{10, 14\},\ \{11, 13\}.$$

Next, we show that if $\{1, 2, \ldots, n\}$ can be partitioned into 5 subsets with equal sums of numbers of each subset, then $\{1, 2, \ldots, n, n+1, \ldots, n+10\}$ also satisfies this condition. In fact, let

$$\{1, 2, \ldots, n\} = A_1 \cup A_2 \cup A_3 \cup A_4 \cup A_5$$

be a partition satisfying the condition. Then we may choose

$$B_1 = A_1 \cup \{n+1, n+10\},\ B_2 = A_2 \cup \{n+2, n+9\},$$

$$B_3 = A_3 \cup \{n+3, n+8\},\ B_4 = A_4 \cup \{n+4, n+7\},$$

$$B_5 = A_5 \cup \{n+5, n+6\},$$

and it is easy to see that B_1, B_2, B_3, B_4, B_5 constitute a desired partition of $\{1, 2, \ldots, n, n+1, \ldots, n+10\}$.

Hence, by induction, for all $n = 5k - 1$ or $n = 5k$ (here $k \geq 2$), the set $\{1, 2, \ldots, n\}$ can be partitioned into five subsets with equal sums of numbers of each subset.

Example 3. (1) Show that the set of positive integers $\mathbf{N}^*$ can be partitioned into three nonintersecting subsets such that for all m, $n \in \mathbf{N}^*$, if $|m - n| = 2$ or 5, then m and n belong to different subsets.

(2) Show that the set of positive integers $\mathbf{N}^*$ can be partitioned into four nonintersecting subsets such that for all m, $n \in \mathbf{N}^*$, if $|m - n| = 2$, 3, or 5, then m and n belong to different subsets. Also, show that if we only partition $\mathbf{N}^*$ into three subsets, then the above conclusion cannot hold.

Solution. (1) Let

$$A = \{3k | k \in \mathbf{N}^*\}, \ B = \{3k - 2 | k \in \mathbf{N}^*\}, \ C = \{3k - 1 | k \in \mathbf{N}^*\}.$$

Then A, B, and C are pairwise disjoint, and $A \cup B \cup C = \mathbf{N}^*$. Since the difference between any two numbers in the same subset is divisible by 3, it cannot be 2 or 5. Therefore, such a partition satisfies the condition.

(2) Let $A = \{4k | k \in \mathbf{N}^*\}$, $B = \{4k - 1 | k \in \mathbf{N}^*\}$, $C = \{4k - 2 | k \in \mathbf{N}^*\}$, and $D = \{4k - 3 | k \in \mathbf{N}^*\}$. By similar arguments as in (1) we see that such a partition satisfies the condition.

Suppose we can partition $\mathbf{N}^*$ into three subsets A, B, and C with the same property, then for all m, $n \in \mathbf{N}^*$, whenever $|m - n| = 2, 3$, or 5, it holds that m and n belong to different subsets. Without loss of generality we assume that $1 \in A$. Then $3 \notin A$, so $3 \in B$ or $3 \in C$. Again assume without loss of generality that $3 \in B$. Since $6 \notin A$ and $6 \notin B$, we have $6 \in C$. Continue this argument and we have $4 \in B$, $8 \in A$, $5 \in C$, and $7 \in A$. Now, since 9 cannot belong to any of the three subsets, there is a contradiction. Therefore, the desired property cannot be satisfied if we partition $\mathbf{N}^*$ into three subsets.

Example 4. Let S be a set consisting of n positive real numbers. For each nonempty subset A of S, let $f(A)$ denote the sum of numbers in A. Show that the set $\{f(A) | A \subseteq S, A \neq \emptyset\}$ can be partitioned into n subsets such that in each subset, the ratio of the largest and smallest numbers is less than 2.

Solution. Suppose the numbers in S are $(0 <) u_1 < u_2 < \cdots < u_n$, and let

$$T_1 = \{f(A) | A = \{u_1\}\},$$

$$T_k = \{f(A) | u_1 + u_2 + \cdots + u_{k-1} < f(A) \\ \leq u_1 + u_2 + \cdots + u_k, A \subseteq S, A \neq \emptyset\}, \quad k = 2, 3, \ldots, n.$$

Then for $1 \leq i < j \leq n$, we have $T_i \cap T_j = \emptyset$, and

$$T_1 \cup T_2 \cup \cdots \cup T_n = \{f(A) | A \subseteq S, A \neq \emptyset\}.$$

Hence, T_1, T_2, ..., T_n constitute an n–partition of $\{f(A) | A \subseteq S, A \neq \emptyset\}$.

In T_1, the largest and smallest numbers are the same, so their ratio is $1 (< 2)$.

For $k \geq 2$, the largest number in T_k is $u_1 + u_2 + \cdots + u_k$, and we consider the following two cases to show that T_k satisfies the condition.

If $u_k \leq u_1 + u_2 + \cdots + u_{k-1}$, then the ratio M_k of the largest and smallest numbers in T_k satisfies

$$M_k < \frac{u_1 + u_2 + \cdots + u_k}{u_1 + u_2 + \cdots + u_{k-1}} = 1 + \frac{u_k}{u_1 + \cdots + u_{k-1}} \leq 2.$$

If $u_k > u_1 + u_2 + \cdots + u_{k-1}$, then since for each element $f(A)$ of T_k,

$$f(A) > u_1 + u_2 + \cdots + u_{k-1},$$

it follows that A contains some u_i with $i \geq k$. Hence, u_k is the smallest number in T_k, and

$$M_k = \frac{u_1 + \cdots + u_k}{u_k} = \frac{u_1 + \cdots + u_{k-1}}{u_k} + 1 < 2.$$

Therefore, the partition $T_1, T_2, \ldots, T_n$ satisfies all the conditions.

Remark. Direct construction is a common method for existence problems. In this problem we defined T_k with inequalities and compared u_k with $u_1 + \cdots + u_{k-1}$ to determine the ratio of the largest and smallest numbers in T_k.

Example 5. Find all positive real numbers a with the following property: there exist a positive integer n and n pairwise disjoint infinite sets $A_1, A_2, \ldots, A_n$ with $A_1 \cup A_2 \cup \cdots \cup A_n = \mathbf{N}^*$ (i.e., an n–partition of $\mathbf{N}^*$ where all the subsets are infinite) such that for every two numbers b and $c (b > c)$ in some A_i, we have $b - c \geq a^i$.

Solution. If $0 < a < 2$, then $2^{n-1} > a^n$ for sufficiently large n. Now, let

$$A_i = \{2^{i-1}m|m \text{ is odd}\}, i = 1, 2, \ldots, n-1,$$

$$A_n = \{\text{multiples of } 2^{n-1}\}.$$

Then this partition satisfies the desired condition.

If $a \geq 2$, then suppose $A_1, A_2, \ldots, A_n$ satisfy the condition. Let $M = \{1, 2, \ldots, 2^n\}$, and we will show that

$$|A_i \cap M| \leq 2^{n-i}.$$

Let $A_i \cap M = \{x_1, x_2, \ldots, x_m\}$ with $x_1 < x_2 < \cdots < x_m$. Then

$$2^n > x_m - x_1 = (x_m - x_{m-1}) + (x_{m-1} - x_{m-2}) + \cdots + (x_2 - x_1) \geq (m-1)2^i.$$

Hence, $m - 1 < 2^{n-i}$, namely $m < 2^{n-i} + 1$, so $m \leq 2^{n-i}$.

On the other hand, $A_i \cap M$ with $i = 1, 2, \ldots, n$ form a partition of M, so

$$2^n = |M| = \sum_{i=1}^{n} |A_i \cap M| \leq \sum_{i=1}^{n} 2^{n-i} = 2^n - 1,$$

which is a contradiction.

Therefore, a can be any positive real number less than 2.

Example 6. Find all positive integers n such that there exists a set S with the following properties:

(1) S consists of n positive integers that are less than 2^{n-1};
(2) for every two (distinct) nonempty subsets A and B of S, the sum of numbers in A is not equal to the sum of numbers in B.

Solution. It is simple to see that such S does not exist for $n = 2$ and 3. When $n = 4$, we choose $S = \{3, 5, 6, 7\}$. Then S satisfies (1) and (2).

For $n \geq 5$, let

$$S = \{3, 2^3, 2^4, \ldots, 2^{n-2}, 2^{n-1} - 3, 2^{n-1} - 2, 2^{n-1} - 1\},$$

and we show that such S satisfies the conditions.

Let A and B be two subsets of S and let $f(X)$ be the sum of numbers in X. Then we need to show that $f(A) \neq f(B)$. Here we may assume that $A \cap B = \emptyset$.

Note that $1 + 2 + \cdots + 2^{m-1} = 2^m - 1 < 2^m$ for each positive integer m. Let

$$a = 2^{n-1} - 3, b = 2^{n-1} - 2, c = 2^{n-1} - 1.$$

Then we can see that if none of a, b, and c belong to $A \cup B$, then $f(A) \neq f(B)$.

Also, $3 + 2^3 + 2^4 + \cdots + 2^{n-2} = 2^{n-1} - 5$, so if exactly one of a, b, and c belongs to $A \cup B$, say $a \in A$, then $f(A) > f(B)$, so $f(A) \neq f(B)$. Similarly, if two or three of a, b, and c belong $A \cup B$, then we can still derive that $f(A) \neq f(B)$.

Therefore, such S exists for all positive integers $n \geq 4$.

Remark. There is also an interesting construction by induction. Let S_n denote the set satisfying the condition for n, with $S_4 = \{3, 5, 6, 7\}$. If $S_k (k \geq 4)$ is chosen, then $S_{k+1} = \{2a | a \in S_k\} \cup \{1\}$ also satisfies the condition. Therefore, n can be any integer at least 4.

Example 7. Let $A_1, A_2, \ldots, A_n$ and $B_1, B_2, \ldots, B_n$ be two n-partitions of a set M. Suppose for each pair of disjoint sets A_i and B_j $(1 \leq i, j \leq n)$, we have $|A_i \cup B_j| \geq n$. Show that $|M| \geq \frac{n^2}{2}$.

Solution. Let $k = \min\{|A_i|, |B_j|, 1 \leq i, j \leq n\}$. Without loss of generality we assume that $|A_1| = k$.

If $k \geq \frac{n}{2}$, then by

$$|M| = \sum_{i=1}^{n} |A_i| \geq nk \geq \frac{n^2}{2},$$

the result holds immediately.

If $k < \frac{n}{2}$, then since B_1, B_2, ..., B_n are pairwise disjoint, there are at most k sets in B_1, B_2, ..., B_n that have a nonempty intersection with A_1. Hence, at least $n - k$ of them are disjoint with A_1, so the number of elements in each of these sets is at least $n - k$. Now, by $n > 2k$ we have $n - k > k$, and

$$|M| = \sum_{i=1}^{n} |B_i| \geq k \cdot k + (n-k) \cdot (n-k) = k^2 + (n-k)^2$$

$$\geq 2\left(\frac{k + (n-k)}{2}\right)^2 = \frac{n^2}{2}.$$

Therefore, the proposition holds.

Example 8. For a set A of real numbers, define $d(A) = \{|x - y| | x, y \in A, x \neq y\}$. Does there exist a partition of $\mathbf{N}^*$ into finitely many subsets $A_1, \ldots, A_m (m \geq 2)$ such that each $|A_i| \geq 2$ and $d(A_i) \cap d(A_j) = \emptyset$ for all $1 \leq i < j \leq m$?

Solution. Such a partition does not exist.

Suppose it exists, Let $A_1, \ldots, A_m$ be such a partition. Then at least one of them must be an infinite set. Assume that A_1 is infinite. Then choose a number n in $d(A_2)$. For m distinct numbers $a_1, \ldots, a_m$ of A_1, consider the following numbers:

$$a_1 + n, a_2 + n, \ldots, a_m + n. \text{①}$$

By $d(A_1) \cap d(A_2) = \emptyset$, we have $n \notin d(A_1)$, so none of the numbers in ① belong to A_1, so they belong to $A_2 \cup \cdots \cup A_m$. Hence, by the pigeonhole principle two of these numbers belong to the same $A_k (2 \le k \le m)$. Suppose $a_i + n, a_j + n \in A_k$. Then $|a_i - a_j| \in d(A_k)$, while a_i and a_j belong to A_1, so $|a_i - a_j| \in d(A_1)$, contradictory to $d(A_1) \cap d(A_k) = \emptyset$.

Therefore, such a partition does not exist.

Remark. Partition of sets is a part of combinatorics, and some basic methods such as the least number principle and the pigeonhole principle are used frequently in such problems.

Example 9. For two sets X and Y consisting of rational numbers, define $XY = \{xy | x \in X, y \in Y\}$. A 3-partition A, B, C of $\mathbf{Q}_+$ (the set of positive rational numbers) is called a "good partition" if it satisfies $BA = B$, $BB = C$, and $BC = A$. Find the maximal positive integer N, such that there exists a "good partition" A, B, C of $\mathbf{Q}_+$ satisfying the property that for every $n \in \{1, 2, \ldots, N-1\}$, the numbers n and $n+1$ belong to different subsets.

Solution. The maximum value of N is 8.

We first construct an example with $N = 8$. Let p_k be the kth smallest prime number. Then every positive rational number x can be written as

$$x = 2^{\alpha_2} \cdot 3^{\alpha_3} \cdots p_k^{\alpha_{pk}}, \alpha_2, \alpha_3, \ldots, \alpha_{p_k} \in \mathbf{Z}.$$

Let $A = \{x \in \mathbf{Q}_+ | \alpha_2 - \alpha_7 \equiv 0 (\text{mod} 3)\}$, $B = \{x \in \mathbf{Q}_+ || \alpha_2 - \alpha_7 \equiv 1 (\text{mod} 3)\}$, and $C = \{x \in \mathbf{Q}_+ | \alpha_2 - \alpha_7 \equiv 2 (\text{mod} 3)\}$. It is easy to see that A, B, and C are pairwise disjoint and compose a 3-partition of $\mathbf{Q}_+$. Further, for $a = 2^{\alpha_2} \cdot 7^{\alpha_7} \cdots \in A$ and $b = 2^{\beta_2} \cdot 7^{\beta_7} \cdots \in B$, we have $ba = 2^{\alpha_2 + \beta_2} \cdot 7^{\alpha_7 + \beta_7} \cdots$, and $(\alpha_2 + \beta_2) - (\alpha_7 + \beta_7) = (\alpha_2 - \alpha_7) + (\beta_2 - \beta_7) \equiv 0 + 1 \equiv 1 (\text{mod} 3)$, so $ba \in B$. Similarly, we get that $BB = C$ and $BC = A$. Hence, A, B, C compose a "good partition" of $\mathbf{Q}_+$.

Next, we show that when $N \ge 9$, such a partition does not exist.

In fact, if A, B, C is such a partition, then

$$B = BA = B(BC) = (BB)C = CC,$$

and since $C = BB = B(BA) = (BB)A = CA$, we see that B and C are "symmetric."

Next, we consider two numbers x and y in B (possibly equal). If $\frac{x}{y} \in B$, then y, $\frac{x}{y}$, and x all belong to B. However, $y \cdot \frac{x}{y} \in BB = C$, so $x \in C$, a contradiction. If $\frac{x}{y} \in C$, then by $y \in B$ and $\frac{x}{y} \in C$, we have $x = y \cdot \frac{x}{y} \in BC = A$, also a contradiction. Therefore, $\frac{x}{y} \in A$, and in particular $1 = \frac{x}{x} \in A$ (note that B and C cannot be empty).

This implies that $2 \notin A$. Since B and C are symmetric, we may assume that $2 \in B$, so $2 \times 2 \in C$, namely $4 \in C$. Since $3 \notin B$ and $3 \notin C$, necessarily $3 \in A$. By $BC = A$, we have $8 = 2 \times 4 \in A$, and since

$$A = BC = (BA)C = A(BC) = AA,$$

we get $9 = 3 \times 3 \in A$, which means that 8 and 9 both belong to A, a contradiction.

Therefore, the maximum value of N is 8.

22.3 Exercises

Group A

1. Let p be an odd prime number, n be a positive integer, and $T = \{1, 2, \ldots, n\}$. We call n "p-subdivisible" if there exist nonempty subsets $T_1, T_2, \ldots, T_p$ of T such that

   (1) $T = T_1 \cup T_2 \cup \cdots \cup T_p$;
   (2) $T_1, T_2, \ldots, T_p$ are pairwise disjoint;
   (3) $T_i (1 \leq i \leq p)$ have the same sums of numbers.

   Prove the following propositions :

   (1) if n is "$p-$subdivisible", then p divides either n or $n + 1$;
   (2) if n is divisible by $2p$, then n is "p-subdivisible".

2. Prove that there exists an infinite partition of $\mathbf{N}^*$ into $A_1, A_2, \ldots$ such that the statement "if x, y, z, and w belong to the same A_i, then $x - y$ and $z - w$ belong to the same A_k (here i and k are not necessarily equal)" is true if and only if $\frac{x}{y} = \frac{z}{w}$.
3. For a positive integer n, find (with a proof) the smallest positive integer $h(n)$ with the following property: for any n-partition of $\{1, 2, \ldots, h(n)\}$, there exist a nonnegative integer a, and integers x and y with $1 \leq x \leq y \leq h(n)$, such that $a + x$, $a + y$, and $a + x + y$ belong to the same subset in the partition.

4. Find the smallest positive integer n with the following property: for any 2-partition of $\{1, 2, \ldots, n\}$, one of the subsets contains three distinct numbers x, y, and z such that $xy = z$.
5. Determine whether it is possible to partition $\mathbf{N}^*$ into two subsets with the following properties:

   (1) no three (distinct) numbers of A form an arithmetic sequence;
   (2) B does not contain an (nonconstant) arithmetic sequence of infinite length.

6. For integers a and b, define $S_{a,b} = \{n^2 + an + b | n \in \mathbf{Z}\}$. How many sets of this type can we choose at most such that they are pairwise disjoint?
7. Suppose we partition $\mathbf{N}^*$ into two subsets A and B with

   (1) $1 \in A$;
   (2) no two distinct numbers of A have a sum of the form $2^k + 2(k = 0, 1, \ldots)$;
   (3) B also has property (2).

   Prove that such a partition is unique.

Group B

8. Let r, s, and n be positive integers such that $r > 1$, $s > 1$, and $r+s = n$. Sets A and B are defined as follows:

$$A = \left\{ \left[\frac{in}{r} \right] | i = 1, 2, \ldots, r - 1 \right\};$$

$$B = \left\{ \left[\frac{in}{s} \right] | i = 1, 2, \ldots, s - 1 \right\}.$$

   Show that A and B compose a 2-partition of $\{1, 2, \ldots, n-2\}$ if and only if $(r, n) = 1$.
9. Let n be a positive integer. Consider 3-partitions A, B, C of the set $\{1, 2, \ldots, n\}$ with the following properties (here A, B, and C are allowed to be empty):

   (1) if we arrange the numbers in each subset in increasing order, then the numbers have alternate parity (so that if the smallest number is odd, then the second smallest is even, then odd, and so on);
   (2) if none of A, B, and C are empty, then exactly one of them has an even number as its smallest number.

   Find the number of such partitions.

10. Let p and q be relatively prime positive integers ($p \neq q$). Suppose we want to partition the set of positive integers into 3 subsets A, B, and C such that for every $z \in \mathbf{N}^*$, the sets A, B, and C each contains one of the three numbers z, $z+p$, and $z+q$. Show that such a partition exists if and only if $3|p+q$.
11. Suppose a rectangle R can be partitioned into n (here n is a given positive integer) pairwise non-overlapping rectangles $R_1, R_2, \ldots, R_n$, where the sides of $R_1, R_2, \ldots, R_n$ are parallel to the sides of R and each of them has at least one side of integer length. Show that R has a side of integer length.

Chapter 23

Synthetical Problems of Quadratic Functions

23.1 Illustrative Examples

Example 1. Suppose $f(x) = x^2 + (a^2 + b^2 - 1)x + a^2 + 2ab\text{-}b^2$ is an even function. Find the maximal possible value of the minimum value of $f(x)$.

Solution. It follows from the condition that $a^2 + b^2 - 1 = 0$ and the minimum value of $f(x)$ is $a^2 + 2ab\text{-}b^2$. Let $a = \cos\theta$ and $b = \sin\theta$. Then

$$\begin{aligned} a^2 + 2ab\text{-}b^2 &= \cos^2\theta + 2\sin\theta\cos\theta - \sin^2\theta \\ &= \cos 2\theta + \sin 2\theta \\ &= \sqrt{2}\sin\left(2\theta + \frac{\pi}{4}\right) \leq \sqrt{2}, \end{aligned}$$

where the equality holds when $a = \cos\frac{\pi}{8}$ and $b = \sin\frac{\pi}{8}$.

Therefore, the maximal possible value of the minimum value of $f(x)$ is $\sqrt{2}$.

Example 2. Suppose the graph of a quadratic function $y = x^2 + bx + c$ has the vertex D, intersects the $x-$axis at A and B (such that A lies on the left), and intersects the $y-$axis at C. If ΔABD and ΔOBC are both isosceles right triangles (here O is the origin), find the value of $b + 2c$.

Solution. First it follows that the coordinates of the points are $C(0, c)$, $A\left(\frac{-b-\sqrt{b^2-4c}}{2}, 0\right)$, $B\left(\frac{-b+\sqrt{b^2-4c}}{2}, 0\right)$, and $D\left(-\frac{b}{2}, -\frac{b^2-4c}{4}\right)$.

Let E be a point on AB with $DE \perp AB$. Then $2DE = AB$, so $2 \times \frac{b^2-4c}{4} = \sqrt{b^2-4c}$, and $\sqrt{b^2-4c} = 0$ or $\sqrt{b^2-4c} = 2$. Since $b^2 - 4c > 0$, we have $\sqrt{b^2-4c} = 2$.

Also, since $OC = OB$, we have $c = \frac{-b+\sqrt{b^2-4c}}{2}$, hence $b + 2c = \sqrt{b^2-4c} = 2$.

Example 3. Let a and b be real constants. Suppose for all real numbers k, the graph of

$$y = (k^2 + k + 1)x^2 - 2(a + k)^2 x + (k^2 + 3ak + b)$$

always passes through $A(1, 0)$.

(1) Find the values of a and b;
(2) if B is the other intersection point of the graph of the function and the x – axis, find the maximum value of $|AB|$ as k varies.

Solution. (1) Since $A(1, 0)$ lies on the graph of the function,

$$(k^2 + k + 1) - 2(a + k)^2 + (k^2 + 3ak + b) = 0,$$

or equivalently $(1 - a)k + (b + 1 - 2a^2) = 0$. Since the equality is valid for all real k, necessarily

$$1 - a = 0, b + 1 - 2a^2 = 0.$$

Solving the equations we get $a = 1$ and $b = 1$, and the function is

$$y = (k^2 + k + 1)x^2 - 2(k + 1)^2 x + (k^2 + 3k + 1).$$

(2) Let $B(x_2, 0)$. Then $|AB| = |x_2 - 1|$. Since 1 and x_2 are the two roots of the quadratic equation

$$(k^2 + k + 1)x^2 - 2(k + 1)^2 x + k^2 + 3k + 1 = 0,$$

we have $1 \cdot x_2 = \frac{k^2+3k+1}{k^2+k+1}$, and hence $(1 - x_2)k^2 + (3 - x_2)k + (1 - x_2) = 0$.

Note that in this quadratic equation for k, the discriminant $\Delta = (3 - x_2)^2 - 4(1 - x_2)^2 \geq 0$, so

$$3{x_2}^2 - 2x_2 - 5 \leq 0,$$

which means that $-1 \leq x_2 \leq \frac{5}{3}$, namely $-2 \leq x_2 - 1 \leq \frac{2}{3}$.

Therefore, $|AB| = |x_2 - 1| \leq 2$, where the equality is satisfied when $k = -1$, and the maximum value of $|AB|$ is 2.

Example 4. Let a function $f(x) = ax^2 + bx + c$. Suppose $|f(x)| \leq 1$ for all $x \in [-1, 1]$. Show that when $x \in [-1, 1]$, the following inequality always holds:

$$|ax + b| + |c| \leq 3.$$

Solution. From $f(-1) = a - b + c$, $f(1) = a + b + c$, and $f(0) = c$,

$$a = \frac{1}{2}f(-1) + \frac{1}{2}f(1) - f(0), \quad b = \frac{1}{2}f(1) - \frac{1}{2}f(-1), \quad c = f(0).$$

Hence, $|c| = |f(0)| \leq 1$, and

$$\begin{aligned} |ax + b| &= \left|\frac{1}{2}f(-1)x + \frac{1}{2}f(1)x - f(0)x + \frac{1}{2}f(1) - \frac{1}{2}f(-1)\right| \\ &= \left|\frac{1}{2}(x-1)f(-1) + \frac{1}{2}(x+1)f(1) - f(0)x\right| \\ &\leq \frac{1}{2}|x - 1| + \frac{1}{2}|x + 1| + |f(0)| \cdot |x| \\ &\leq \frac{1}{2}(1 - x) + \frac{1}{2}(x + 1) + 1 = 2. \end{aligned}$$

Therefore, $|ax + b| + |c| \leq 3$.

Remark. The condition "$|f(x)| \leq 1$ for all $x \in [-1, 1]$" implies that $|f(-1)| \leq 1, |f(0)| \leq 1$, and $|f(1)| \leq 1$ (usually these are all we need). By expressing a, b, and c in terms of $f(-1)$, $f(0)$, and $f(1)$, we can derive specific properties of a, b, and c. Similarly, if the condition says "$|f(x)| \leq 1$ for all $x \in [0, 1]$," then we may use $|f(0)| \leq 1$, $\left|f\left(\frac{1}{2}\right)\right| \leq 1$, and $|f(1)| \leq 1$ instead.

Example 5. For a function $f(x)$, if $f(x) = x$, then x is called a "fixed point" of $f(x)$. If $f(f(x)) = x$, then x is called a "stable point" of $f(x)$. Let A and B be the set of fixed points and the set of stable points of a function $f(x)$, respectively. (In other words, $A = \{x | f(x) = x\}$ and $B = \{x | f(f(x)) = x\}$.)

(1) Show that $A \subseteq B$;
(2) if $f(x) = ax^2 - 1$ ($a \in \mathbf{R}$ and $x \in \mathbf{R}$), and $A = B \neq \emptyset$, find the value range of a.

Solution. (1) If $A = \emptyset$, then the proposition holds obviously. If $A \neq \emptyset$, then $f(x_0) = x_0$ for every $x_0 \in A$, so $f(f(x_0)) = f(x_0) = x_0$, and hence $x_0 \in B$. Therefore, $A \subseteq B$.

(2) By

$$f(f(x)) = af^2(x) - 1 = af^2(x) - ax^2 + ax^2 - 1$$
$$= a(f(x) + x)(f(x) - x) + f(x) = x,$$

we get

$$(f(x) - x)[a(f(x) + x) + 1] = 0,$$
$$(ax^2 - x - 1)(a^2x^2 + ax\text{-}a + 1) = 0.$$

Therefore, B is the set of real roots of the equation above.

Since $A \neq \emptyset$, we have $a = 0$ or $\begin{cases} a \neq 0, \\ \Delta_1 = 1 + 4a \geq 0, \end{cases}$ so $a \geq -\frac{1}{4}$.

On the other hand, since $A = B$, the equation

$$a^2x^2 + ax\text{-}a + 1 = 0 \text{ ①}$$

has either no real solutions or only solutions that are also solutions to

$$ax^2 - x - 1 = 0. \text{ ②}$$

If ① has no real solutions, then $\Delta_2 = a^2 - 4a^2(1 - a) < 0$, so $a < \frac{3}{4}$.

If ① has real solutions, but both solutions are also solutions of ②, then $a^2x^2 = ax + a$. Combining them with ① gives

$$2ax + 1 = 0, x = -\frac{1}{2a}.$$

Plugging into ②, we obtain that

$$\frac{1}{4a} + \frac{1}{2a} - 1 = 0, a = \frac{3}{4}.$$

Therefore, the value range of a is $\left[-\frac{1}{4}, \frac{3}{4}\right]$.

Example 6. Let $f(x) = ax^2 + (b + 1)x + (b - 1)(a \neq 0)$.

(1) If $a = 1$ and $b = -2$, find the fixed points of $f(x)$;
(2) if for all real numbers b, the function $f(x)$ always has two distinct fixed points, find the value range of a;
(3) suppose the condition in (2) holds, and A and B are the points on the graph of $y = f(x)$ that correspond to the fixed points such that A and B are symmetric with respect to the line $y = kx + \frac{1}{2a^2+1}$ (k is a real number). Find the minimum value of b.

Solution. (1) If $a = 1$ and $b = -2$, then $f(x) = x^2 - x - 3$. Solving $x^2 - x - 3 = x$ we get $x_1 = -1$, and $x_2 = 3$. Therefore, the fixed points of $f(x)$ are -1 and 3.

(2) If for all real numbers b, the function $f(x)$ always has two distinct fixed points, then the equation

$$ax^2 + (b+1)x + (b-1) = x$$

always has two distinct real roots (regardless of the choice of b). Hence,

$$\Delta = b^2 - 4a(b-1) > 0$$

for all real b, which means that $b^2 - 4ab + a > 0$ for all b, so

$$\Delta' = (4a)^2 - 16a < 0.$$

Solving the inequality we have $0 < a < 1$. Therefore, the value range of a is (0, 1).

(3) Since A and B correspond to the fixed points of $f(x)$, both of them lie on the line $y = x$. Denote $A(x_1, x_1)$ and $B(x_2, x_2)$, and since they are symmetric with respect to $y = kx + \frac{1}{2a^2+1}$, necessarily $k = -1$.

Let the midpoint of AB be $M(x', y')$. Since x_1 and x_2 are the roots of $ax^2 + bx + b - 1 = 0$, we have $x' = y' = \frac{x_1+x_2}{2} = -\frac{b}{2a}$.

Since M lies on the line $y = -x + \frac{1}{2a^2+1}$,

$$-\frac{b}{2a} = \frac{b}{2a} + \frac{1}{2a^2+1},$$

$$b = -\frac{a}{2a^2+1},$$

$$2ba^2 + a + b = 0.$$

We view the above equality as an equation for a. Then $\Delta = 1 - 8b^2 \geq 0$. Hence, $-\frac{1}{2\sqrt{2}} \leq b \leq \frac{1}{2\sqrt{2}}$. When $a = \frac{\sqrt{2}}{2} \in (0, 1)$, we have $b = -\frac{1}{2\sqrt{2}}$, so the minimum value of b is $-\frac{1}{2\sqrt{2}} = -\frac{\sqrt{2}}{4}$.

Example 7. Suppose $f(x) = ax^2 + bx + c$ satisfies $|f(x)| \leq 1$ for all $x \in [0, 1]$. Find the maximum value of $|a| + |b| + |c|$.

Analysis. Compare with the proof in Example 4. We first express a, b, and c in terms of $f(0), f\left(\frac{1}{2}\right)$, and $f(1)$. Then we use $|f(0)| \leq 1, |f\left(\frac{1}{2}\right)| \leq 1$, and $|f(1)| \leq 1$ to estimate an upper bound for $|a| + |b| + |c|$. Finally we find an instance to show that the bound is reachable.

Solution. Since

$$\begin{cases} f(0) = c, \\ f\left(\frac{1}{2}\right) = \frac{1}{4}a + \frac{1}{2}b + c, \\ f(1) = a + b + c, \end{cases}$$

$$\begin{cases} a = 2f(1) + 2f(0) - 4f\left(\frac{1}{2}\right), \\ b = 4f\left(\frac{1}{2}\right) - f(1) - 3f(0), \\ c = f(0). \end{cases}$$

Hence,

$$\begin{aligned} |a| &= \left|2f(1) + 2f(0) - 4f\left(\frac{1}{2}\right)\right| \\ &\leq 2|f(1)| + 2|f(0)| + 4\left|f\left(\frac{1}{2}\right)\right| \leq 8, \\ |b| &= \left|4f\left(\frac{1}{2}\right) - f(1) - 3f(0)\right| \\ &\leq 4\left|f\left(\frac{1}{2}\right)\right| + |f(1)| + 3|f(0)| \leq 8, \\ |c| &= |f(0)| \leq 1, \end{aligned}$$

$$|a| + |b| + |c| \leq 8 + 8 + 1 = 17.$$

Next, consider $f(x) = 8x^2 - 8x + 1$. For $x \in [0,1]$, we have $8(x^2 - x) \leq 0$ and $8(x - x^2) = 8(1-x)x \leq 8\left(\frac{1-x+x}{2}\right)^2 = 2$, so $-1 \leq f(x) \leq 1$. Since $|a| + |b| + |c| = 17$, we conclude that the maximum value of $|a| + |b| + |c|$ is 17.

Example 8. Let $M(a)$ denote the maximum value of $f(x) = |x^2 - a|$ for $x \in [-1,1]$. Find the minimum value of $M(a)$.

Solution 1. Since $f(x) = |x^2 - a|$ is an even function on $[-1,1]$, it suffices to consider $M(a)$ as the maximum value of $f(x)$ on $[0,1]$.

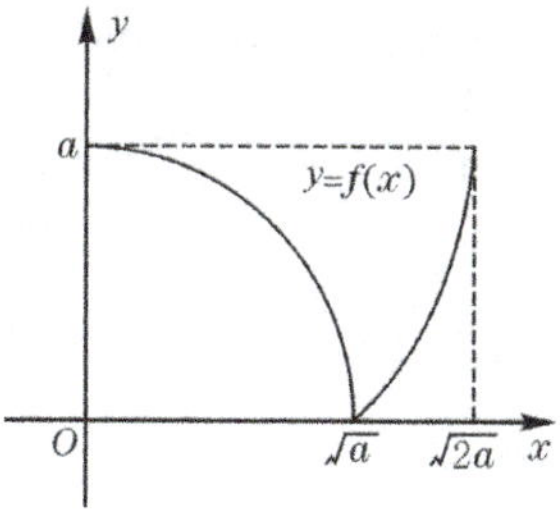

Figure 23.1

If $a \le 0$, then $f(x) = x^2 - a$, which is increasing on $[0, 1]$, so $M(a) = 1 - a$.

If $a > 0$, then we consider the graph in Figure 23.1. If $\sqrt{2a} \ge 1$, namely $a \ge \frac{1}{2}$, then $M(a) = a$; if $\sqrt{2a} < 1$, that is $a < \frac{1}{2}$, then $M(a) = \max\{a, f(1)\} = \max\{a, 1 - a\} = 1 - a$. Hence,

$$M(a) = \begin{cases} 1 - a, & a < \dfrac{1}{2}, \\ a, & a \ge \dfrac{1}{2}. \end{cases}$$

It is simple to observe that this function takes its minimum value at $a = \frac{1}{2}$, and the minimum value of $M(a)$ is $\frac{1}{2}$.

Solution 2. Consider $M(a)$ as the maximum value of $f(x)$ on $[0, 1]$. Then

$$M(a) \ge \max\{|a|, |1 - a|\}.$$

Hence, $M(a) \ge \frac{1}{2}(|a| + |1 - a|) \ge \frac{1}{2}|a + 1 - a| = \frac{1}{2}$.

On the other hand, when $a = \frac{1}{2}$, we have $f(x) = \left|x^2 - \frac{1}{2}\right| \le \frac{1}{2}$, while $f(0) = \frac{1}{2}$, so $M(a) = \frac{1}{2}$. Therefore, the minimum value of $M(a)$ is $\frac{1}{2}$.

Example 9. Determine whether there exists a quadratic function $f(x)$ such that for each positive integer k, if $x = \underbrace{55\cdots5}_{k \text{ copies}}$, then $f(x) = \underbrace{55\cdots5}_{2k \text{ copies}}$. Explain the reason.

Solution. Let $f(x) = ax^2 + bx + c$. Then for $k = 1$, 2, and 3,

$$f(5) = 25a + 5b + c = 55, \text{ ①}$$

$$f(55) = 3025a + 55b + c = 5555, \text{ ②}$$

$$f(555) = 308025a + 555b + c = 555555. \text{ ③}$$

Solving the system of equations ①, ②, and ③ we get $a = \frac{9}{5}, b = 2$, and $c = 0$. Hence $f(x) = \frac{9}{5}x^2 + 2x$.

Next, we show that $f(x) = \frac{9}{5}x^2 + 2x$ satisfies the desired condition. Since

$$\underbrace{55\cdots5}_{k \text{ copies}} = 5(1 + 10 + 100 + \cdots + 10^{k-1}) = \frac{5}{9}(10^k - 1),$$

$$\underbrace{55\cdots5}_{k \text{ copies}} = \frac{5}{9}(10^{2k} - 1),$$

we have

$$\begin{aligned} f(\underbrace{55\cdots5}_{k \text{ copies}}) &= f\left(\frac{5}{9}(10^k - 1)\right) \\ &= \frac{9}{5}\left[\frac{5}{9}(10^k - 1)\right]^2 + 2 \times \frac{5}{9}(10^k - 1) \\ &= \frac{5}{9}(10^k - 1)^2 + 2 \times \frac{5}{9}(10^k - 1) \\ &= \frac{5}{9}(10^k - 1)(10^k + 1) \\ &= \frac{5}{9}(10^{2k} - 1) = \underbrace{55\cdots5}_{2k \text{ copies}}. \end{aligned}$$

Therefore, $f(x) = \frac{9}{5}x^2 + 2x$ is the desired function.

Example 10. Suppose there are two points $A(m_1, f(m_1))$ and $B(m_2, f(m_2))$ on the graph of $f(x) = ax^2 + bx + c(a > b > c)$ such that

$$a^2 + [f(m_1) + f(m_2)]a + f(m_1)f(m_2) = 0,$$

$$f(1) = 0.$$

(1) Show that $b \geq 0$;

(2) show that the value range of the distance between the two intersection points of the graph of $f(x)$ and the $x-$axis is $[2, 3)$;

(3) is it true that at least one of $f(m_1 + 3)$ and $f(m_2 + 3)$ is positive? Prove your result.

Solution. (1) Since $f(m_1)$ and $f(m_2)$ satisfy

$$a^2 + [f(m_1) + f(m_2)]a + f(m_1)f(m_2) = 0,$$

we have $[a + f(m_1)][a + f(m_2)] = 0$, so $f(m_1) = -a$ or $f(m_2) = -a$. In other words, one of m_1 and m_2 is a root of $f(x) = -a$.

Hence, $\Delta \geq 0$ in this equation, namely $b^2 \geq 4a(a+c)$. Since $f(1) = 0$, we have $b = -(a+c)$, thus

$$(a+c)^2 - 4a(a+c) = -3a^2 - 2ac + c^2 \geq 0,$$

or equivalently $(3a-c)(a+c) \leq 0$.

Also, by $a+b+c=0$ and $a > b > c$, we get $a > 0, c < 0$, and $3a - c < 0$, so $a + c \leq 0$, that is $b \geq 0$.

(2) Let x_1 and x_2 be the roots of $f(x) = ax^2 + bx + c = 0$. Since $f(1) = a + b + c = 0$, one root is 1, and the other is equal to $\frac{c}{a}$.

Since $a > 0$ and $c < 0$, we have $\frac{c}{a} < 0$. By $a > b > c$ and $b = -a - c \geq 0$, we see that

$$a > -a - c \geq 0.$$

Therefore, $-2 < \frac{c}{a} \leq -1$, which shows that $2 \leq |x_1 - x_2| < 3$.

(3) Let $f(x) = a(x - x_1)(x - x_2) = a(x-1)\left(x - \frac{c}{a}\right)$.

Since $f(m_1) = -a$ or $f(m_2) = -a$, we assume without loss of generality that $f(m_1) = -a$. Then,

$$a(m_1 - 1)\left(m_1 - \frac{c}{a}\right) = -a < 0,$$

and hence $\frac{c}{a} < m_1 < 1$ and $m_1 + 3 > \frac{c}{a} + 3 > 1$.

Since the axis of symmetry for $f(x) = ax^2 + bx + c$ is $x = -\frac{b}{2a}$ and $a > b \geq 0$, so

$$-\frac{1}{2} < -\frac{b}{2a} \leq 0.$$

Hence, $f(x)$ is increasing on $[1, +\infty)$ and $f(m_1 + 3) > f(1) = 0$. Similarly, if $f(m_2) = -a$, then $f(m_2 + 3) > 0$.

Therefore, at least one of $f(m_1 + 3)$ and $f(m_2 + 3)$ is positive.

23.2 Exercises

Group A

1. As shown in the figure, a parabola $y = -\frac{1}{7}x^2 + bx + c$ intersects the $x-$axis at A and B, and $AB = 4$. Also P is a point on the parabola and C lies on the $x-$axis with $PC \perp x-$axis. Suppose the $x-$coordinate of C is -1, and $\angle PAC = 45°$ and $\tan \angle PBC = \frac{3}{7}$. Find the value of $b + c$.

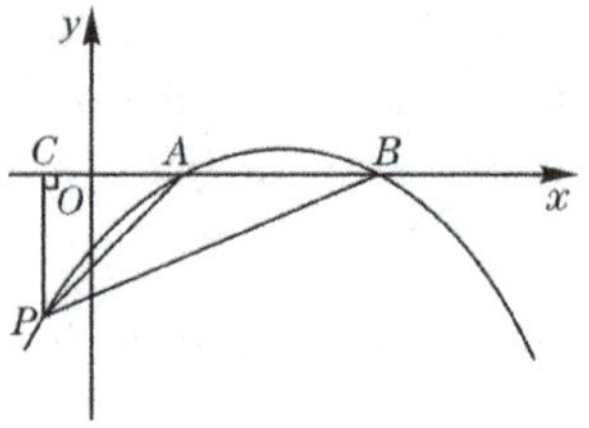

Exercise 1

2. Suppose the graph of $y = x^2 - |x| - 12$ intersects the x-axis at two distinct points A and B, and another parabola $y = ax^2 + bx + c$ passing through A and B has vertex P, such that ΔAPB is an isosceles right triangle. Find the values of a, b, and c.
3. Show that for all real values of p, the parabola $y = x^2 + (p+1)x + \frac{1}{2}p + \frac{1}{4}$ passes through a fixed point and the vertices of these parabolas lie on a fixed parabola.
4. Let x_1 be a root of $ax^2 + bx + c = 0$ and x_2 be a root of $-ax^2 + bx + c = 0$ such that both of them are nonzero and $x_1 \neq x_2$. Show that the equation $\frac{a}{2}x^2 + bx + c = 0$ has a real root between x_1 and x_2.
5. Suppose the equation $4\cos^2 x + (a-5)\cos x + 1 = 0$ $\left(x \in \left(0, \frac{\pi}{2}\right)\right)$ has two different real solutions. Find the value range of a.
6. Suppose real numbers a, b, and c satisfy

$$\begin{cases} a^2 - bc - 8a + 7 = 0, \\ b^2 + c^2 + bc - 6a + 6 = 0. \end{cases}$$

   Find the value range of a.
7. Let a and b be real constants. Suppose $f(x) = x^2 + 2bx + 1$ and $g(x) = 2a(x+b)$. We view each pair of real numbers (a, b) as a point in the $a-b$ coordinate plane, and let S be the set of points (a, b) such that the graphs of $y = f(x)$ and $y = g(x)$ have no common point. Find the area of S.
8. Suppose the equation $x^2 + (2m-1)x + (m-6) = 0$ has one root at most -1 and one root at least 1.

   (1) Find the value range of m;
   (2) find the maximum and minimum values of the sum of squares of the two roots.
9. Let $f(x) = ax^2 + bx + c$ be such that $f(x) \in [-1, 1]$ for all $x \in [-1, 1]$. Show that $f(x) \in [-7, 7]$ for all $x \in [-2, 2]$.

Group B

10. Suppose the graph of a quadratic function $y = mx^2 - \left(3m + \frac{4}{3}\right)x + 4$ intersects the $x-$axis at A and B and intersects the $y-$axis at C. If ΔABC is an isosceles triangle, find the formula of the quadratic function.
11. In the coordinate plane, the points whose coordinates are both integers are called grid points. Find all the grid points on the graph of $y = \frac{x^2}{10} - \frac{x}{10} + \frac{9}{5}$ such that $y \leq |x|$ and explain why there are no more of them.
12. Let $f(x)$ be a function defined on $(-\infty, +\infty)$, which is periodic with period 2. For $k \in \mathbf{Z}$, let I_k denote the interval $(2k-1, 2k+1]$. Suppose $f(x) = x^2$ for $x \in I_0$.

    (1) Find the formula of $f(x)$ on I_k;
    (2) for a positive integer k, find the set (explicitly)

    $$M_k = \{a | \text{the equation } f(x) = ax \text{ has two distinct real roots on } I_k\}.$$

13. Let a, b, and c be positive integers, and A and B be the (distinct) intersection points of the graph of $y = ax^2 + bx + c$ with the $x-$axis. If the distances from A and B to the origin are both less than 1, find the minimum value of $a + b + c$.
14. Let $f(x) = ax^2 + bx + c$ be such that $|f(x)| \leq 1$ for all $|x| \leq 1$. Show that $|2ax + b| \leq 4$ when $|x| \leq 1$.
15. Suppose $|ax^2 + bx + c| \leq 1$ for all $x \in [-1, 1]$. Show that $|cx^2 \pm bx + a| \leq 2$ for all $x \in [-1, 1]$.

Chapter 24

Maximum and Minimum Values of Discrete Quantities

24.1 Key Points of Knowledge and Basic Methods

Extreme value problems are popular topics in math competitions. In previous chapters, we have discussed extreme value problems of functions, and tried to find the maximum or minimum values of functions when the argument takes all real values in the domain.

Another type of extreme value problems involves arguments that are integers, positive integers, or sets and subsets. In these problems we are required to find the maximum or minimum values of a target function when the arguments vary with such restrictions. This kind of extreme value problems is called the extreme value problem of discrete quantities.

Similar to the continuous case, solving extreme value problems of discrete quantities usually involves two parts, namely "proof" and "construction": On one hand we need to find an upper bound or lower bound for the target function with a proof, and on the other hand we need to construct an example to show that this bound can be reached under the specified restriction.

Since this kind of problems requires thinking from both sides, and sometimes the proof and the construction involve quite different ideas, such problems tend to reflect the contestant's ability of overall thinking and rigorous reasoning.

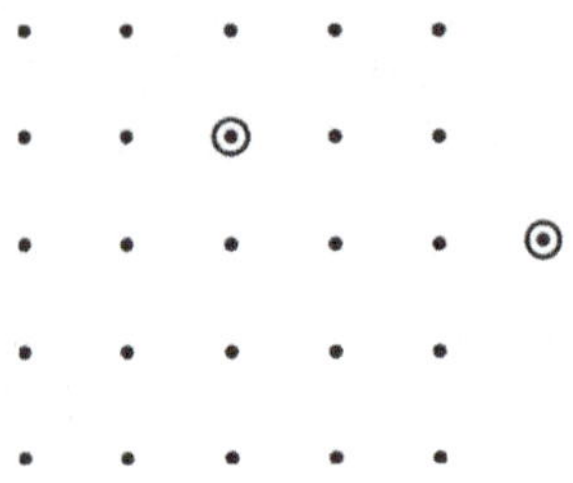

Figure 24.1

24.2 Illustrative Examples

Example 1. Let $S = \{(a, b)|1 \leq a, b \leq 5, a, b \in \mathbf{Z}\}$ be a set of grid points in the coordinate plane. Suppose T is another set of grid points with the property that for each point P in S, there exists a point Q in T $(P \neq Q)$ such that the line segment PQ does not contain any grid point other than P and Q. How many elements does T have at least?

Solution. The minimum value of the number of elements in T is 2. We first show that T cannot contain only one point.

Suppose the converse is true. Then let $T = \{Q(x_0, y_0)\}$, and there exists a point $P(x_1, y_1)$ in S, such that $(x_1, y_1) \neq (x_0, y_0)$, while x_1 has the same parity with x_0 and y_1 has the same parity with y_0. (This is because there are four classes of grid points depending on the parity of the coordinates, and S contains at least two points in each class.) Then the midpoint of PQ is also a grid point, a contradiction.

Next, we construct an example of T containing two points. Consider the following case:

We can verify that the condition is satisfied. Therefore, the minimum value of $|T|$ is 2.

Example 2. Let n be a given positive integer and let A be a subset of $\{1, 2, \ldots, 2n - 1\}$ such that no two distinct elements of A sum to $2n - 1$ or $2n$. Find the maximum value of $|A|$.

Solution. Note that if $A = \{n, n + 1, \ldots, 2n - 1\}$, then the sum of the smallest and second smallest numbers in A is $2n + 1$, so no two distinct numbers of A sum to $2n - 1$ or $2n$. Hence, the maximum value of $|A|$ is at least n.

On the other hand, consider the following sequence (a permutation of 1, 2, ..., $2n-1$):

$$2n-1, 1; 2n-2, 2; \ldots; n+1, n-1; n.$$

The sum of any two consecutive numbers is either $2n-1$ or $2n$, so A cannot contain two consecutive numbers in this sequence. If $|A| \geq n+1$, then by the pigeonhole principle A has to contain two consecutive numbers, which is a contradiction. Hence, $|A| \leq n$.

Therefore, the maximum value of $|A|$ is n.

Remark. In fact, the argument above shows that if $|A| = n$, then necessarily $A = \{n, n+1, \ldots, 2n-1\}$.

Example 3. Let n be a positive integer. Find the smallest positive integer k such that there exist k sequences of length $2n+2$, whose terms are either 0 or 1, that satisfy the following property: for any sequence A of 0 and 1 with length $2n+2$, there is some sequence B in the k sequences that has at least $n+2$ terms in common with A.

Solution. The smallest integer is $k = 4$. (Finding this answer is where the difficulty of this problem lies.)

We consider the following 4 sequences:

$$A = (0, 0, \ldots, 0), \; B = (1, 1, \ldots, 1),$$
$$C = (1, 0, \ldots, 0), \; D = (0, 1, \ldots, 1).$$

Let X be a sequence of 0 and 1 with length $2n+2$. If the number of 0 (resp. 1) in X is at least $n+2$, then X has at least $n+2$ terms in common with A (resp. B). If the numbers of 0 and 1 in X are both $n+1$, then consider the first term of X. If the first term is 1 (resp. 0), then X has exactly $n+2$ terms in common with D (resp. C). Hence, $k \leq 4$.

On the other hand, if we have only three sequences, let them be

$$A = (a_1, a_2; a_3, a_4; \cdots; a_{2n+1}, a_{2n+2}),$$
$$B = (b_1, b_2; b_3, b_4; \cdots; b_{2n+1}, b_{2n+2}),$$
$$C = (c_1, c_2; c_3, c_4; \cdots; c_{2n+1}, c_{2n+2}).$$

For $m = 1, 2, \ldots, n+1$, there exists a tuple (d_{2m-1}, d_{2m}) different from any of $(a_{2m-1}, a_{2m}), (b_{2m-1}, b_{2m})$, or (c_{2m-1}, c_{2m}). (Note that since each term is 0 or 1, such a tuple has four options.) Now, we choose

$$D = (d_1, d_2; d_3, d_4; \cdots ; d_{2n+1}, d_{2n+2}).$$

Then D has at least $n+1$ terms different from each of A, B, and C, so these three sequences cannot satisfy the condition.

Therefore, the smallest integer is $k = 4$.

Example 4. In a country, some pairs of cities have direct flights between them (in both directions). Suppose for each pair of cities, it is possible to travel from one to the other through at most 100 flights, and it is also possible to travel from one to the other through an even number of flights. Find the smallest integer d such that for each pair of cities, it is possible to go from one to the other through an even number of flights that is at most d. (Here it is allowed to pass through a city multiple times or take the same flight multiple times.)

Solution. The smallest integer is $d = 200$.

First we show that d is at least 200. Consider the following example: there are 201 cities in the country such that the flights among them form a cycle. Apparently the conditions are satisfied, and for two adjacent cities on the cycle, if we travel from one to the other through an even number of flights, then we need to take at least 200 flights.

Then we show that $d = 200$ is valid, i.e., for any two cities A and B, it is possible to travel from A to B through an even number of flights that is at most d.

In fact, if we need to take at least $2k$ flights from A to B (if an even number of flights is required) with $k > 100$, then suppose we first travel from A to C through k flights. We know that it is possible to travel from A to C through m flights with $m \leq 100$. If $m \equiv k(\text{mod}2)$, then there is a shorter path from A to B with an even number of flights, which is contradictory to the minimality of k. Similarly, we can travel from C to B through n flights with $n \leq 100$. If $n \equiv k(\text{mod}2)$, then it also results in a contradiction. Thus we have found a path from A to B through C, where the number of flights is $m + n(\leq 200)$, which is even since m and n have the same parity.

Therefore, $d = 200$.

Example 5. Suppose M is a set of 100 points in a coordinate plane. Find the maximal possible number of rectangles with the following property: the vertices of these rectangles are all points in M, and their sides are parallel to the coordinate axes.

Solution. For $M = \{(x, y)|1 \leq x, y \leq 10, x, y \in \mathbf{N}^*\}$, the number of rectangles with the given property is $(C_{10^2})^2 = 45^2 = 2025$. (In this case the points in M form a 10×10 grid, and choosing any two vertical lines and two horizontal lines gives a rectangle.) Hence, the maximal number of such rectangles is at least 2025.

Next, we show that the number of such rectangles cannot exceed 2025. Choose any point O in M, and without loss of generality assume that O is the origin of the coordinate plane. Suppose now there are x points of M on the $x-$axis and y points of M on the $y-$axis (here O itself is not counted in these two numbers). Then the number T of valid rectangles with O as one vertex satisfies:

(1) $T \leq xy$. This is because such a rectangle needs to have one vertex on the $x-$axis and one vertex on the $y-$axis (excluding O), while every such pair corresponds to at most one such a rectangle.
(2) $T \leq 100 - (x + y + 1)$. This is because the opposite vertex of O in the rectangle does not lie on the coordinate axes, while the rectangle is uniquely determined by O and its opposite vertex.

Hence, $T \leq \min\{xy, 99 - x - y\}$. If $x + y \leq 18$, then $T \leq xy \leq \left(\frac{x+y}{2}\right)^2 = 81$, and if $x + y > 18$, then $T \leq 99 - x - y < 81$. Thus, $T \leq 81$, so each point in M is a vertex of at most 81 valid rectangles. Therefore, the total number of valid rectangles does not exceed $\frac{81 \times 100}{4} = 2025$ (here we divided by 4 since each rectangle has 4 vertices).

To summarize, the maximal possible number of the valid rectangles is 2025.

Remark. In this proof we solved the global problem by considering the local behavior, which is a common method when the local parts have certain kind of symmetry.

Example 6. Suppose sixteen students attend a math contest, where every problem is a multiple choice with four options and students choose one of the options as the answer. After the contest, it turns out that any two students have the same answer in at most one problem. How many problems are there in the contest at most? Explain the reason.

Solution. We first estimate an upper bound for the number of problems.

Suppose there are k problems in total and each problem has options 1, 2, 3, 4. For a particular problem S, let a_i be the number of students who choose option i in problem S for $i = 1, 2, 3, 4$. Then $a_1 + a_2 + a_3 + a_4 = 16$.

If two students have one answer in common, then we call them a "pair." Now, consider the number A of "pairs" who share the same answer in S. We have

$$A = \sum_{i=1}^{4} C_{a_i}^2 = \frac{1}{2}\sum_{i=1}^{4}(a_i{}^2 - a_i) = \frac{1}{2}\sum_{i=1}^{4} a_i{}^2 - \frac{1}{2}\sum_{i=1}^{4} a_i$$

$$= \frac{1}{2}\sum_{i=1}^{4} a_i^2 - 8 \geq \frac{1}{2}\cdot\frac{1}{4}\left(\sum_{i=1}^{4} a_i\right)^2 - 8 = 24.$$

Hence, k problems will generate at least $24k$ different "pairs." However, every two students correspond to at most one such "pair" by assumption, so $24k \leq C_{16}^2$, that is $k \leq 5$.

Next, we find an example to show that $k = 5$ is possible.

$$P_1 : 11111,\quad P_2 : 12222,\quad P_3 : 13333,\quad P_4 : 14444,$$
$$P_5 : 21234,\quad P_6 : 22143,\quad P_7 : 23412,\quad P_8 : 24321,$$
$$P_9 : 31342,\quad P_{10} : 32431,\quad P_{11} : 33124,\quad P_{12} : 34213,$$
$$P_{13} : 41423,\quad P_{14} : 42314,\quad P_{15} : 43241,\quad P_{16} : 44132.$$

Here, P_1, $P_2, \ldots, P_{16}$ stand for the sixteen students and each student has a sequence representing the answers that he/she chooses on problem 1 through 5.

Therefore, there are at most five problems.

Example 7. Let n be an integer greater than 10 and let A be a set with $2n$ elements. We call a collection of subsets of A "good" if the collection $\{A_i | i = 1, 2, \ldots, m\}$ satisfies the following properties:

(1) every A_i consists of n elements in A;
(2) for all indices i, j, and k with $1 \leq i < j < k \leq m$, we have $|A_i \cap A_j \cap A_k| \leq 1$.

For each n, find the maximal positive integer m such that there exists a "good" collection that consists of m subsets.

Solution. The maximum value of m is 4.

On one hand, we choose two arbitrary different subsets A_1 and A_2 of A with n elements that are not complements to each other, and let $A_3 = A \backslash A_1$ and $A_4 = A \backslash A_2$. Then the collection $\{A_1, A_2, A_3, A_4\}$ is "good." Hence, the maximum value of m is at least 4.

On the other hand, suppose there exists a "good" collection of five subsets $\{A_1, A_2, \ldots, A_5\}$. We let I_t denote the sum of $|A_{i_1} \cap \cdots \cap A_{i_t}|$, where $(i_1, i_2, \cdots, i_t)$ can be any t indices with $1 \leq i_1 \leq \ldots \leq i_t \leq m$, and let U_t denote the sum of $|A_{i_1} \cup \cdots \cup A_{i_t}|$, where $(i_1, i_2, \ldots, i_t)$ can be any t indices with $1 \leq i_1 \leq \cdots \leq i_t \leq m$. Then by the inclusion-exclusion principle,

$$U_4 = 4I_1 - 3I_2 + 2I_3 - I_4,$$

$$U_5 = I_1 - I_2 + I_3 - I_4 + I_5.$$

Since the union of any four sets in $\{A_1, A_2, \ldots, A_5\}$ is a subset of $A_1 \cup \cdots \cup A_5$, we have $5U_5 \geq U_4$. From the two equalities above, $5(I_1 - I_2 + I_3 - I_4 + I_5) \geq 4I_1 - 3I_2 + 2I_3 - I_4$, so $I_1 - 2I_2 + 3I_3 - 4I_4 + 5I_5 \geq 0$.

Note that $I_1 = 5n$ (as $|A_i| = n$ for every i) and $A_1 \cup \cdots \cup A_5 \subseteq A$, hence $U_5 \leq 2n$. Comparing it with the previous results, we get that

$$\begin{cases} 5n - 2I_2 + 3I_3 - 4I_4 + 5I_5 \geq 0, & \text{①} \\ 2n \geq 5n - I_2 + I_3 - I_4 + I_5. & \text{②} \end{cases}$$

By ①+②×2, we have $9n - 2I_2 + 3I_3 - 4I_4 + 5I_5 \geq 10n - 2I_2 + 2I_3 - 2I_4 + 2I_5$, which implies that $I_3 - 2I_4 + 3I_5 \geq n$. However, $I_3 \leq C_5^3 = 10 < n$ by condition (2), which requires that $3I_5 > 2I_4$. Since $A_1 \cap \cdots \cap A_5$ is a subset of the intersections of any 4 of $A_1, \ldots, A_5$, we get $2I_4 \geq 10I_5 \geq 3I_5$, which is a contradiction. Therefore, the maximum value of m is 4.

Example 8. There are eight students attending a contest with n True-or-False problems. Suppose for every ordered pair of problems (A, B), two of the students have answers (T, T), two have answers (T, F), two have answers (F, T), and two have answers (F, F). Find the maximum value of n and explain why.

Solution. Consider a table with eight rows and n columns. The cell of row i and column j has number 0 (resp. 1) if the ith student has answer T (resp. F) for problem j.

It follows from the assumption that for every two columns, their eight (ordered) pairs of cells consist of two 00, two 01, two 10 and two 11.

If $n \geq 8$, then we consider only the first eight columns, which compose an 8×8 table. Note that if we switch all the 0's and 1's in one column, then the desired condition still holds, so we may assume that the numbers in the first row are all 0. Now, let a_i denote the number of 0's in the ith row. Then by considering the number of 0's contained in each pair of columns, we easily have $\sum_{i=1}^{8} a_i = \frac{1}{2} \times 8 \times 8 = 32$, and the number of pairs "00" in the same row is $\sum_{i=1}^{8} C_{a_i}^2$.

Since $a_1 = 8$, we have $\sum_{i=1}^{8} a_i = 24$, so

$$\begin{aligned}\sum_{i=1}^{8} C_{a_i}^2 &= C_8^2 + \sum_{i=1}^{8} C_{a_i}^2 = 28 + \frac{1}{2}\sum_{i=1}^{8} a_i^2 - \frac{1}{2}\sum_{i=1}^{8} a_i \\ &= 28 - 12 + \frac{1}{2}\sum_{i=1}^{8} {a_i}^2 \\ &\geq 16 + \frac{1}{2} \times \frac{1}{7} \times \left(\sum_{i=2}^{8} a_i\right)^2 = 57\frac{1}{7}.\end{aligned}$$

Hence, the number of pairs "00" in the same row is at least 58.

On the other hand, considering the pairs "00" contained in each pair of columns, we see that the number of such pairs is at most $2C_8^2 = 56$, which means that $56 \geq 58$, a contradiction.

Consequently, $n \leq 7$.

Next, we show that $n = 7$ is possible by an example:

$$\begin{matrix} 0 & 0 & 0 & 0 & 0 & 0 & 0 \\ 0 & 1 & 1 & 1 & 1 & 0 & 0 \\ 0 & 1 & 1 & 0 & 0 & 1 & 1 \\ 0 & 0 & 0 & 1 & 1 & 1 & 1 \\ 1 & 0 & 1 & 0 & 1 & 0 & 1 \\ 1 & 0 & 1 & 1 & 0 & 1 & 0 \\ 1 & 1 & 0 & 0 & 1 & 1 & 0 \\ 1 & 1 & 0 & 1 & 0 & 0 & 1. \end{matrix}$$

We can verify that this example satisfies the required condition. Therefore, the maximum value of n is 7.

Remark. The methods in Examples 6 and 8 have much in common, since they both involve "counting twice." In these problems we counted the same quantity in two different ways to give an estimation of the desired variable, and then showed by an example that the bound is reachable.

Example 9. Let ten students attend n extracurricular groups (here each student may attend more than one group) such that every group contains at most five people, every two students attend at least one common group, and for every two groups, there are at least two students who attend neither of them. Prove that the minimum value of n is 6.

Solution. Let the ten students be S_1, $S_2, \ldots, S_{10}$ and let the n groups be G_1, $G_2, \ldots, G_n$ (they are subsets of $\{S_1, S_2, \ldots, S_{10}\}$).

First, every student attends at least two groups. If not, suppose student S_1 attends only one group. Then S_1 needs to meet every other student in this group, so this group contains ten students, a contradiction.

If some student attends only two groups, suppose S_1 attends only G_1 and G_2. Then by assumption at least two students attend neither of G_1 and G_2, so S_1 never meets them in the same group, a contradiction.

Therefore, each student attends at least three groups and the sum of the numbers of elements in $G_1, G_2, \ldots, G_n$ is at least $3 \times 10 = 30$.

On the other hand, each group contains at most five students, so $5n \geq 30$, namely $n \geq 6$.

Next, we construct an example to show that $n = 6$ is possible. Let

$$G_1 = \{S_1, S_2, S_3, S_4, S_5\}, G_2 = \{S_1, S_2, S_6, S_7, S_8\},$$
$$G_3 = \{S_1, S_3, S_6, S_9, S_{10}\}, G_4 = \{S_2, S_4, S_7, S_9, S_{10}\},$$
$$G_5 = \{S_3, S_5, S_7, S_8, S_9\}, G_6 = \{S_4, S_5, S_6, S_8, S_{10}\}.$$

It is easy to verify that the conditions are satisfied. Therefore, the minimum of value n is 6.

24.3 Exercises

Group A

1. Suppose positive numbers a, b, and c satisfy $4a + b = abc$. Find the minimum value of $a + b + c$.
2. Suppose among any k numbers in 1, 2, ..., 100, there are two numbers that are relatively prime. Find the minimum value of k.
3. The sum of several different positive integers is 200. What is the maximum value of their product?
4. For a finite set X, let $|X|$ denote the number of its elements and let $n(X)$ denote the number of subsets of X (including the empty set and X itself). Suppose sets A, B, and C satisfy

$$n(A) + n(B) + n(C) = n(A \cup B \cup C),$$
$$|A| = |B| = 100.$$

Find the minimum value of $|A \cap B \cap C|$.

5. Let a, b, and c be different positive integers. Suppose

$$\{a+b, b+c, c+a\} = \{n^2, (n+1)^2, (n+2)^2\},$$

where n is a positive integer. Find the minimum value of $a^2+b^2+c^2$.

6. We fill the cells in a 9×2000 table with numbers $1, 2, \ldots, 2000$ such that each number appears nine times and the difference between any two numbers in the same column does not exceed three. Find the maximum of the minimum values of all the column-sums (2000 in total).

7. Let $M = \{1, 2, \ldots, 19\}$ and $A = \{a_1, \ldots, a_k\}$ be a subset of M. Find the minimal integer k such that for any $b \in M$, there exist $a_i, a_j \in A$ satisfying $b \in \{a_i, a_i + a_j, a_i - a_j\}$. (Here a_i and a_j are allowed to be equal.)

8. Suppose in a sequence of real numbers, the sum of any seven consecutive terms is negative, while the sum of any eleven consecutive terms is positive. How many terms does this sequence have at most?

9. There are n high schools in a city. Suppose one day C_i students from the ith high school come to a stadium to watch a soccer game, such that $1 \le C_i \le 40 (1 \le i \le n)$ and $\sum_{i=1}^{n} C_i = 2000$. There are 200 seats in every row of the stadium and the students of the same school must sit in the same row. How many rows of seats does the stadium need at least to ensure that all students can be seated?

Group B

10. Ann and Bob are playing the following game. Ann first chooses a number from $1, 2, \ldots, 144$, but does not tell Bob what the number is. Then in every turn Bob chooses a subset of $\{1, 2, \ldots, 144\}$ and Ann tells whether the chosen number belongs to this subset. Suppose in every turn, if Ann's answer is "Yes", then Bob needs to pay two coins, and if Ann's answer is "No", then Bob needs to pay one coin. How many coins does Bob need to prepare at least to ensure that the chosen number can be found?

11. Initially there are 128 numbers on a blackboard, all of which are 1. We do the following operation repeatedly: choose two numbers a and b on the blackboard, erase them, and write $ab + 1$. After 127 operations, there is only one number left, and let A be the maximum value of this number. Find the last digit of A (under the decimal representation).

12. Let A be a 3×9 table, and each cell is filled with a positive integer. An $m \times n$ $(1 \le m \le 3$ and $1 \le n \le 9)$ subtable of A (here a subtable

is defined to be the intersection of several consecutive columns and several consecutive rows of A) is called a "good rectangle" if the sum of all the numbers contained in it is divisible by 10. A cell in A is called a "bad cell" if it is not contained in any "good rectangles". Find the maximal possible number of "bad cells" in A.

13. Let n be a given positive integer. Suppose a set of positive integers $M = \{a_1, a_2, \ldots, a_n\}$ has the following property: any two disjoint subsets of M have different sums of elements. Find the maximum value of $\frac{1}{a_1} + \frac{1}{a_2} + \cdots + \frac{1}{a_n}$.
14. Let n be a given positive integer ($n \geq 2$). Suppose a, b, c, and d are positive integers such that $\frac{b}{a} + \frac{d}{c} < 1$ and $b + d \leq n$. Find the maximum value of $\frac{b}{a} + \frac{d}{c}$.
15. Let n be a given positive integer. For n positive integers $a_1, a_2, \ldots, a_n$ such that

$$\frac{1}{a_1} + \frac{1}{a_2} + \cdots + \frac{1}{a_n} < 1,$$

find the maximum value of $\frac{1}{a_1} + \frac{1}{a_2} + \cdots + \frac{1}{a_n}$.

Chapter 25

Simple Function Itearation and Functional Equations

25.1 Key Points of Knowledge and Basic Methods

Let $f : D \to D$ be a function. For $x \in D$, denote $f^{(0)}(x) = x, f^{(1)}(x) = f(x), f^{(2)}(x) = f(f(x)), \ldots, f^{(n+1)}(x) = f(f^{(n)}(x))$ for $n \in \mathbf{N}$. We call the function $f^{(n)}(x)$ determined in this way the nth iteration of $f(x)$.

The theory and methods of function iteration have important applications in computational mathematics. However, since some of its methods and results are quite elementary, such problems also appear in high school contests.

Equations containing unknown functions are called functional equations. For example, $f(x + 2\pi) = f(x)$ and $f(xy) = f(x) + f(y)$ are functional equations, where $f(x)$ is the unknown function.

If a function $f(x)$ satisfies the functional equation for all values of x in its domain, then we call $f(x)$ a solution to the functional equation. Solving a functional equation is the process of finding solutions to the equation or showing that it has no solution.

There are few general methods for function iteration or functional equation problems, and we usually find methods based on specific problems. However, some relatively common methods include the fixed point method, the bridge function method, plugging special values, induction, Cauchy's method, inequalities, guess and prove, etc.

25.2 Illustrative Examples

Example 1. Let the domain of a function $f(x)$ be the set of nonzero real numbers. Suppose for all x in the domain,

$$3f(x) + 2f\left(\frac{1}{x}\right) = 4x. \text{ ①}$$

Find $f(x)$.

Solution. Substituting $\frac{1}{x}$ for x in ①, we get

$$3f\left(\frac{1}{x}\right) + 2f(x) = \frac{4}{x}. \text{ ②}$$

Eliminating $f\left(\frac{1}{x}\right)$ in ① and ②, we have

$$f(x) = \frac{4}{5}\left(3x - \frac{2}{x}\right)$$

for all x in the domain. Plugging the function into ①, we can verify that this function satisfies the functional equation. Therefore, the solution to the equation is

$$f(x) = \frac{4}{5}\left(3x - \frac{2}{x}\right) \ (x \neq 0).$$

Remark. Eliminating unknowns in a system of equations is one of the most common methods when solving functional equations. Also, verifying the answer is an important part of solving functional equations, because in general we find $f(x)$ based on the necessary conditions that it needs to satisfy, which may not be sufficient. Verification can be simple, but is always needed.

Example 2. Find all functions $f : \mathbf{R} \to \mathbf{R}$ such that for any x and $y \in \mathbf{R}$,

$$f(x) + f(y) = f(f(x)f(y)). \text{ ①}$$

Solution. We crack the problem by finding the value of $f(0)$.

Suppose $f(0) = a$. Then let $x = y = 0$ in ①, and we have

$$f(a^2) = 2a. \text{ ②}$$

Further, letting $x = y = a^2$ in ①, we get

$$f(4a^2) = 2f(a^2) = 4a.$$

Now, letting $x = 4a^2$ and $y = 0$ in ① gives

$$f(4a^2) + f(0) = f(4a^2).$$

Hence, $f(0) = 0$. Next, let $y = 0$ in ①, and we obtain

$$f(x) + f(0) = f(0),$$

so $f(x) = 0$ for all $x \in \mathbf{R}$. It is easy to verify that $f(x) = 0$ satisfies the functional equation.

Therefore, the solution to the equation is $f(x) = 0 (x \in \mathbf{R})$.

Remark. Finding the value of $f(0)$ was a crucial step in the solution. Once $f(0)$ was found, the remaining steps followed easily.

Example 3. Let $n(\geq 2)$ be a given positive integer. Suppose a function $f : \mathbf{R} \to \mathbf{R}$ satisfies that for all $x, y \in \mathbf{R}$,

$$f(x + y^n) = f(x) + f(y)^n. \text{ ①}$$

Show that $f(x^2) = f(1)^{n-2} f(x)^2$ for all $x \in \mathbf{R}$.

Solution. Let $x = y = 0$ in ①, and we have $f(0) = 0$. Hence, for all $y \in \mathbf{R}$,

$$f(y^n) = f(y)^n.$$

For $z > 0$, let $y = \sqrt[n]{z}$. Then

$$f(x + z) = f(x + y^n) = f(x) + f(y)^n = f(x) + f(y^n) = f(x) + f(z).$$

Let $x = -z$ in the equation above. Then $f(-z) = -f(z)$, so f is an odd function. Consequently, for all $x, y \in \mathbf{R}$,

$$f(x + y) = f(x) + f(y). \text{ ②}$$

The functional equation ② is called Cauchy's equation. By induction we see that $f(kx) = kf(x)$ for all $k \in \mathbf{N}^*$, and since $f(-x) = -f(x)$, we have $f(kx) = kf(x)$ for all $k \in \mathbf{Z}$. Further, if $k \in \mathbf{Q}$, then by letting $k = \frac{p}{q}$ for $p, q \in \mathbf{Z}$, we obtain that $f(q(kx)) = qf(kx)$. Since $q(kx) = px$, we have $f(q(kx)) = f(px) = pf(x)$, and hence $f(kx) = f(\frac{p}{q}x) = kf(x)$.

Now, for $x \in \mathbf{R}$, by the above result we can see that for all $k \in \mathbf{Q}$,

$$f((x+k)^n) = f(x^n + \mathrm{C}_n^1 x^{n-1} k + \cdots + \mathrm{C}_n^n \cdots k^n).$$

By ② and $f(kx) = kf(x)$,

$$\begin{aligned} f((x+k)^n) &= f(x^n) + f(\mathrm{C}_n^1 x^{n-1} \cdot k) + \cdots + f(\mathrm{C}_n^n \cdot k^n) \\ &= f(x^n) + k\mathrm{C}_n^1 f(x^{n-1}) + k^2 \mathrm{C}_{n^2} f(x^{n-2}) + \cdots + \mathrm{C}_n^n \cdot k^n f(1). \end{aligned}$$

On the other hand,

$$\begin{aligned} f((x+k)^n) &= f(x+k)^n = (f(x) + f(k))^n = (f(x) + kf(1))^n \\ &= f(x)^n + \mathrm{C}_n^1 f(x)^{n-1} \cdot k \cdot f(1) + \cdots + \mathrm{C}_n^n k^n \cdot f(1)^n. \end{aligned}$$

Comparing the two equalities above,we get the equality

$$\begin{aligned} &f(x^n) + k\mathrm{C}_n^1 f(x^{n-1}) + \cdots + k^n \mathrm{C}_n^n f(1) \\ &\quad = f(x)^n + k\mathrm{C}_n^1 f(1) f(x)^{n-1} + \cdots + k^n \mathrm{C}_n^n \cdot f(1)^n. \end{aligned}$$

Let x be a constant. The both sides of the equality above can be viewed as polynomials in k. Since this equation holds for all $k \in \mathbf{Q}$, the corresponding coefficients of both sides must be equal. Comparing the coefficients for k^{n-2} gives that

$$\mathrm{C}_n^{n-2} f(x^2) = \mathrm{C}_n^{n-2} f(1)^{n-2} f(x)^2.$$

Therefore, $f(x^2) = f(1)^{n-2} f(x)^2$, and the proposition follows.

Remark. We dealt with ② by Cauchy's method, which proceeded from positive integers to all integers and then to rational numbers.

Example 4. Find all functions $f : \mathbf{R} \to \mathbf{R}$ such that for all $x, y \in \mathbf{R}$,

$$f(x^2 + y + f(y)) = 2y + f(x)^2. \text{①}$$

Solution. First it is easy to see that the function $f(x) = x$ is a solution. Then we show that this is the only solution by proving the following claims.

Claim 1. $f(a) = 0$ if and only if $a = 0$.

Let $y = -\frac{1}{2}f(x)^2$ in ①, and we can see that there exists $a \in \mathbf{R}$ such that $f(a) = 0$. For this a, let $(x, y) = (0, a)$ in ① and we get $0 = 2a + f(0)^2$. Next, let $y = 0$ in ① and we have

$$f(x)^2 = f(x^2 + f(0)).$$

Substituting $-x$ for x in the equation above, we also have

$$f(-x)^2 = f((-x)^2 + f(0)) = f(x^2 + f(0)) = f(x)^2.$$

Hence, $f(-a) = 0$. Now, letting $(x, y) = (0, -a)$ in ① gives $0 = -2a + f(0)^2$. Combining the above with $0 = 2a + f(0)^2$, we conclude that $a = 0$, and claim 1 is proven.

Claim 2. The function $f(x)$ is odd and $f(x) > 0$ when $x > 0$.

In fact, by letting $y = 0$ in ① and combining the equality with claim 1, we get $f(x^2) = f(x)^2$. Hence, $f(x) > 0$ for all $x > 0$. On the other hand, choose $(x, y) = \left(\sqrt{\alpha}, -\frac{1}{2}f(\alpha)\right)$ (here α can be any positive real number) in ①. Then

$$f\left(\alpha - \frac{f(\alpha)}{2} + f\left(-\frac{f(\alpha)}{2}\right)\right) = -f(\alpha) + f(\sqrt{\alpha})^2 = -f(\alpha) + f((\sqrt{\alpha})^2) = 0.$$

By claim 1, it follows that $\alpha = \frac{f(\alpha)}{2} - f\left(-\frac{f(\alpha)}{2}\right)$. Hence,

$$f(-\alpha) = f\left(-\frac{f(\alpha)}{2} + f\left(-\frac{f(\alpha)}{2}\right)\right). \quad ②$$

Choose $(x, y) = \left(0, -\frac{f(\alpha)}{2}\right)$, and we get

$$f\left(-\frac{f(\alpha)}{2} + f\left(-\frac{f(\alpha)}{2}\right)\right) = -f(\alpha).$$

Comparing the above with ② we see that $f(x)$ is an odd function, and claim 2 holds.

Claim 3. $f(x) = x$ for all $x > 0$.

For any $\alpha > 0$, choose $(x, y) = (\sqrt{\alpha}, -\alpha)$. Then

$$f(f(-\alpha)) = -2\alpha + f(\sqrt{\alpha})^2 = -2\alpha + f(\alpha).$$

Hence,

$$\begin{aligned} f(2\alpha) &= f(f(\alpha) + f(f(\alpha))) = f(0^2 + f(\alpha) + f(f(\alpha))) \\ &= 2f(\alpha) + f(0)^2 = 2f(\alpha). \end{aligned}$$

On the other hand, choose $(x, y) = (\sqrt{2\alpha}, -\alpha)$, and we get

$$f(2\alpha - \alpha + f(-\alpha)) = -2\alpha + f(\sqrt{2\alpha})^2 = -2\alpha + f(2\alpha),$$

so $f(\alpha - f(\alpha)) = -2\alpha + 2f(\alpha) = -2(\alpha - f(\alpha))$.

Let $\beta = \alpha - f(\alpha)$. Then $f(\beta) = -2\beta$. If $\beta > 0$, then $-2\beta < 0$, but $f(\beta) > 0$ (claim 2), a contradiction; if $\beta < 0$, then $-2\beta > 0$, but $f(\beta) = -f(-\beta) < 0$, also a contradiction. Therefore, the only possibility is $\beta = 0$, and hence claim 3 is valid.

Combining claims 1, 2, and 3, we conclude that the functional equation has exactly one solution, which is $f(x) = x (x \in \mathbf{R})$.

Example 5. Determine whether there exists a function $f : \mathbf{R}_+ \to \mathbf{R}_+$ such that

$$f(x + y) \geq f(x) + yf(f(x)) \text{ ①}$$

for all $x, y \in \mathbf{R}_+$.

Solution. We prove that such a function does not exist.

In fact, if there exists such a function $f(x)$, then by ① we see that $f(x)$ is increasing on $\mathbf{R}_+$ and is not bounded above (this can be seen by letting $x = 1$ and $y \to +\infty$).

Now, for $x > y > 0$, substitute (x, y) for $(x + y, x)$ in ①, and we get

$$f(x) \geq f(y) + (x - y)f(f(x)),$$

so

$$f(x) - f(y) \geq (x - y)f(f(x)). \text{ ②}$$

If $f(y) > y$, then we choose $x = f(y)$ in ②, obtaining that

$$f(f(y)) - f(y) \geq (f(y) - y)f(f(y)).$$

Further,

$$1 + y - f(y) \geq \frac{f(y)}{f(f(y))} > 0,$$

Thus $f(y) < y + 1$. Note that this shows that $f(x) < x + 1$ for all $x \in \mathbf{R}_+$. Hence, for all $x, y \in \mathbf{R}_+$,

$$1 + x + y > f(x + y) \geq f(x) + yf(f(x)).$$

We fix x and let $y \to +\infty$ in this inequality. Then $f(f(x)) \leq 1$.

Since $f(x)$ is an increasing function without an upper bound, there exists a sequence of positive numbers $\{C_n\}$ (not bounded above) such that $f(C_{n+1}) \geq C_n$ for each $n \in \mathbf{N}^*$. However, we have

$$1 \geq f(f(C_{n+2})) \geq f(C_{n+1}) > C_n,$$

which gives a contradiction. Therefore, such a function does not exist.

Example 6. Find all real numbers a for which there exists a function $f{:}\mathbf{R} \to \mathbf{R}$ such that for all $x, y \in \mathbf{R}$,

$$\frac{f(x) + f(y)}{2} \geq f\left(\frac{x+y}{2}\right) + a|x - y|. \text{①}$$

Solution. If $a \leq 0$, then we let $f(x) = x^2$. By $\frac{x^2+y^2}{2} \geq \left(\frac{x+y}{2}\right)^2$ we see that the inequality holds.

Next, we show that if $a > 0$, such a function does not exist.

In fact, if such a function f exists for some $a > 0$, then for all $n \in \mathbf{N}^*$ and $i \in \mathbf{N}$,

$$\frac{1}{2}\left(f\left(\frac{i+2}{n}\right) + f\left(\frac{i}{n}\right)\right) \geq f\left(\frac{i+1}{n}\right) + \frac{2a}{n}.$$

Hence,

$$f\left(\frac{i+2}{n}\right) - f\left(\frac{i+1}{n}\right) \geq f\left(\frac{i+1}{n}\right) - f\left(\frac{i}{n}\right) + \frac{4a}{n}. \text{②}$$

Summing ② for $i = 0, 1, 2, \ldots, n-1$, we get

$$f\left(\frac{n+1}{n}\right) - f\left(\frac{1}{n}\right) \geq f(1) - f(0) + 4a;$$

summing ② for $i = 1, 2, \ldots, n$, we get

$$f\left(\frac{n+2}{n}\right) - f\left(\frac{2}{n}\right) \geq f\left(\frac{n+1}{n}\right) - f\left(\frac{1}{n}\right) + 4a;$$

$$\cdots\cdots$$

summing ② for $i = n-1, n, \ldots, 2n-2$, we get

$$f(2) - f(1) \geq f\left(\frac{2n-1}{n}\right) - f\left(\frac{n-1}{n}\right) + 4a.$$

Now, we take the sum of all these n inequalities, getting

$$f(2) - f(1) \geq f(1) - f(0) + 4na.$$

Since $a > 0$, the above inequality cannot hold when $n \to +\infty$. Therefore, such a function f does not exist.

In summary, the desired function exists if and only if $a \leq 0$.

Remark. Inequality methods are widely used in all branches of mathematics. The combination of equalities and inequalities is common in mathematical proofs.

The following problems concern iteration of functions.

Example 7. Let $n \in \mathbf{N}^*$. Find the nth iteration $f^{(n)}(x)$ for the following functions.

(1) $f(x) = ax + b$, where a and b are constants.
(2) $f(x) = 2x^2 - 1$ for $|x| \leq 1$.

Solution. (1) If $a = 1$, then $f^{(n)}(x) = x + nb$.

If $a \neq 1$, then we write

$$f(x) = a\left(x - \frac{b}{1-a}\right) + \frac{b}{1-a}.$$

Here $\frac{b}{1-a}$ is the solution to the equation $f(x) = x$, i.e., $ax + b = x$, which is the fixed point of $f(x)$. Then,

$$\begin{aligned}
f^{(2)}(x) &= a\left(f(x) - \frac{b}{1-a}\right) + \frac{b}{1-a} \\
&= a\left(a\left(x - \frac{b}{1-a}\right)\right) + \frac{b}{1-a} \\
&= a^2\left(x - \frac{b}{1-a}\right) + \frac{b}{1-a}.
\end{aligned}$$

By induction, $f^{(n)}(x) = a^n\left(x - \frac{b}{1-a}\right) + \frac{b}{1-a}$.

Therefore, $f^{(n)}(x) = \begin{cases} x + nb, a = 1; \\ a^n\left(x - \frac{b}{1-a}\right) + \frac{b}{1-a}, a \neq 1. \end{cases}$

(2) In order to find $f^{(n)}(x)$, we introduce the notion of "bridge functions." If there exists a function $\varphi(x)$ with the inverse function $\varphi^{-1}(x)$ such that $f(x) = \varphi^{-1}(g(\varphi(x)))$, then we say that $f(x)$ and $g(x)$ are similar through a bridge function $\varphi(x)$.

If $f(x)$ and $g(x)$ are similar through $\varphi(x)$, in other words

$$f(x) = \varphi^{-1}(g(\varphi(x))),$$

then

$$f^{(2)}(x) = \varphi^{-1}(g(\varphi(f(x)))) = \varphi^{-1}(g(\varphi(\varphi^{-1}(g(\varphi(x)))))) = \varphi^{-1}(g^{(2)}(\varphi(x))).$$

By induction, $f^{(n)}(x) = \varphi^{-1}(g^{(n)}(\varphi(x)))$.

In the original problem, we take $g(x) = 2x$ and $\varphi(x) = \arccos x$. Then $\varphi^{-1}(x) = \cos x$. Since

$$f(x) = 2x^2 - 1 = 2\cos^2(\arccos x) - 1 = \cos(2\arccos x) = \varphi^{-1}(g(\varphi(x))),$$

we see that $f(x)$ and $g(x)$ are similar through $\varphi(x) = \arccos x$. Therefore, by $g^{(n)}(x) = 2^n x$ (using the result in (1)) we have $f^{(n)}(x) = \varphi^{-1}(g^{(n)}(\varphi(x))) = \cos(2^n \arccos x)$.

Remark. These two problems used the fixed point method and the bridge function method, respectively, which are common approaches to function iteration problems.

The result in (2) can be generalized to show that $f(x) = \cos(n \arccos x)$ is a polynomial of x with degree n. Such polynomials are called Chebyshev's polynomials, which have wide applications.

Example 8. Let $f(x) = \frac{1}{x+1}$ with $x \in \mathbf{R}_+$. For each $n \in \mathbf{N}^*$, define

$$g_n(x) = x + f^{(1)}(x) + f^{(2)}(x) + \cdots + f^{(n)}(x).$$

Show that (1) $g_n(x) > g_n(y)$ for all x and y such that $x > y > 0$;

(2) $g_n(1) = \frac{F_1}{F_2} + \frac{F_2}{F_3} + \cdots + \frac{F_{n+1}}{F_{n+2}}$, where $F_1 = F_2 = 1$ and $F_{n+2} = F_n + F_{n+1}$ for $n = 1, 2, \ldots$ (this is the famous Fibonacci sequence).

Solution. (1) It is well known that $h(x) = x + \frac{1}{x}$ is increasing for $x > 1$, so $x + f(x) = \frac{1}{x+1} + (x+1) - 1$ is increasing for $x > 0$. Note that $f(x)$ is decreasing for $x > 0$, so $f^{(2)}(x)$ is increasing for $x > 0$, and hence $f^{(2)}(x) + f^{(3)}(x)$ is increasing. (Here we substituted $f^{(2)}(x)$ for x in $x + f(x)$.) Further we can prove that $f^{(2^m)}(x)$ is increasing for $x > 0$ and $f^{(2m)}(x) + f^{(2m+1)}(x)$ is also increasing for $x > 0$. Hence, when $n = 2m$, the function

$$g_n(x) = (x + f(x)) + (f^{(2)}(x) + f^{(3)}(x)) + \cdots + (f^{(2m-2)}(x) \\ + f^{(2m-1)}(x)) + f^{(2m)}(x)$$

is increasing, and when $n = 2m+1$, the function

$$g_n(x) = \sum_{k=0}^{m} (f^{(2k)}(x) + f^{(2k+1)}(x))$$

is also increasing. Therefore, (1) holds.

(2) Since $f\left(\frac{F_n}{F_{n+1}}\right) = \frac{1}{\frac{F_n}{F_{n+1}} + 1} = \frac{F_{n+1}}{F_n + F_{n+1}} = \frac{F_{n+1}}{F_{n+2}}$,

$$f^{(j)}\left(\frac{F_1}{F_2}\right) = f^{(j-1)}\left(\frac{F_2}{F_3}\right) = \cdots = \frac{F_{j+1}}{F_{j+2}}.$$

Hence, $g_n(1) = g_n\left(\frac{F_1}{F_2}\right) = \sum_{j=0}^{n} f^{(j)}\left(\frac{F_1}{F_2}\right) = \frac{F_{j+1}}{F_{j+2}}$, and (2) follows.

Example 9. Let a function $f{:}\mathbf{R} \to \mathbf{R}$ satisfy the following conditions:

(1) for all $x, y \in \mathbf{R}$, it is true that $|f(x) - f(y)| \leq |x - y|$.
(2) there exists $k \in \mathbf{N}^*$ such that $f^{(k)}(0) = 0$.

Show that $f(0) = 0$ or $f(f(0)) = 0$.

Solution. By (2), we let k be the smallest integer such that $f^{(k)}(0) = 0$, and we are required to show that $k = 1$ or $k = 2$. Suppose $k \geq 3$. Then

$$\begin{aligned} |f(0)| &= |f(0) - 0| \\ &\geq |f^{(2)}(0) - f(0)| \\ &\geq \cdots \\ &\geq |f^{(k)}(0) - f^{(k-1)}(0)| \end{aligned}$$

$$\begin{aligned}
&= |f^{(k-1)}(0)|, \\
|f^{(k-1)}(0)| &= |f^{(k-1)}(0) - 0| \\
&\geq |f^{(k)}(0) - f(0)| \\
&= |f(0).
\end{aligned}$$

Hence, $|f^{(k-1)}(0)| = |f(0)|$.

If $f^{(k-1)}(0) = f(0)$, then $f(f(0)) = f^{(k)}(0) = 0$, a contradiction. Thus, $f^{(k-1)}(0) = -f(0)$, and

$$\begin{aligned}
|f(0)| &= |f(0) + 0| \\
&= |f^{(k)}(0) - f^{(k-1)}(0)| \\
&\leq |f^{(k-1)}(0) - f^{(k-2)}(0)| \\
&\leq \cdots \\
&\leq |f^{(2)}(0) - f(0)| \\
&\leq |f(0) - 0| \\
&= |f(0)|.
\end{aligned}$$

Now, each inequality above has to be an equality, so for $2 \leq j \leq k-1$,

$$|f^{(j)}(0) - f^{(j-1)}(0)| = |f(0)|,$$

or equivalently $f^{(j)}(0) - f^{(j-1)}(0) = \pm f(0)$. By the minimality of k, none of $f(0)$, $f^{(2)}(0), \ldots, f^{(k-1)}(0)$ are 0, so combining the fact with the previous results we have $f^{(2)}(0) = 2f(0), f^{(3)}(0) \in \{f(0), 3f(0)\}$, and $f^{(4)}(0) \in \{2f(0), 4f(0)\}$. By induction, we see that $f^{(k-1)}(0)$ is a positive integer multiple of $f(0)$, where $f(0)$ is nonzero, but this is contradictory to $f^{(k-1)}(0) = -f(0)$.

Therefore, $k \leq 2$ and the proposition holds.

25.3 Exercises

Group A

1. Let $f(x) = x^2 + ax + b\cos x$. Find all pairs of real numbers (a, b) such that the equations $f(x) = 0$ and $f(f(x)) = 0$ have the same nonempty set of real solutions.
2. Let $f(x) = 3x + 2$. Show that there exists $m \in \mathbf{N}^*$ such that $f^{(100)}(m)$ is divisible by 2000.

3. Let $f(x) = \sqrt{20x^2 + 10}$. Find the formula for $f^{(n)}(x)$.
4. Find a function $g(x)$ such that $g^{(8)}(x) = x^2 + 2x$.
5. Let $P_n(x, y)$ be polynomial functions such that $P_1(x, y) = 1$ and for every positive integer n,

$$P_{n+1}(x, y) = (x + y - 1)(y + 1)P_n(x, y + 2) + (y - y^2)P_n(x, y).$$

Show that $P_n(x, y) = P_n(y, x)$ for all x and y and every positive integer n.
6. Suppose $f(x) + f\left(\frac{x-1}{x}\right) = 1 + x$ for every real number $x(\neq 0$ or $1)$. Find $f(x)$.
7. Find all functions $f: \mathbf{N}^* \to \mathbf{N}^*$ such that for all $x, y \in \mathbf{N}^*$,

$$f(x + f(y)) = f(x) + y.$$

8. Find all nonzero functions $f: \mathbf{R} \to \mathbf{R}$ such that for all real numbers x and y,

$$f(x)f(y) = f(x - y).$$

9. Find all functions $f(x)$ defined on the set of nonzero real numbers such that

(1) $f(x) = xf\left(\frac{1}{x}\right)$ for all nonzero $x \in \mathbf{R}$;
(2) for all $x, y \in \mathbf{R}$ with $x \neq -y$, we have $f(x) + f(y) = 1 + f(x + y)$.

Group B

10. Let $S = \{1, 2, 3, 4, 5\}$. How many functions $f: S \to S$ are there such that $f^{(50)}(x) = x$ for every $x \in S$?
11. Find all functions f: $\mathbf{Z} \to \mathbf{Z}$ such that for every $x \in \mathbf{Z}$,

$$3f(x) - 2f(f(x)) = x.$$

12. Find all functions $f: \mathbf{R} \to \mathbf{R}$ such that for all $x, y \in \mathbf{R}$,

$$f(x - f(y)) = f(f(y)) + xf(y) + f(x) - 1.$$

13. Find all functions $f: [1, +\infty) \to [1, +\infty)$ such that

(1) $f(x + 1) = \frac{f(x)^2 - 1}{x}$ for $x \in [1, +\infty)$;
(2) $g(x) = \frac{f(x)}{x}$ is a bounded function.

14. Find all functions $f\colon \mathbf{R}_+ \to \mathbf{R}_+$ such that for all $x,\, y \in \mathbf{R}_+$,

$$f(x+y) + f(x)f(y) = f(xy) + f(x) + f(y).$$

15. Prove that there is no function $f\colon \mathbf{R}_+ \to \mathbf{R}_+$ such that for all $x, y \in \mathbf{R}_+$,

$$f(x) - f(x+y) \geq \frac{f(x)y}{f(x)+y}.$$

Chapter 26

Constructing Functions to Solve Problems

26.1 Key Points of Knowledge and Basic Methods

When dealing with some equations or inequalities, we may construct a function, and use the properties of the function to solve the problem. Such constructions can be creative.

26.2 Illustrative Examples

The following problems involve constructing linear functions.

Example 1. Let a, b, and c be real numbers with their absolute values less than 1. Prove that

$$ab + bc + ca + 1 > 0.$$

Solution. We construct a linear function

$$f(x) = (b + c)x + bc + 1, -1 < x < 1.$$

The graph of this function is a line segment that does not include the endpoints $(-1, f(-1))$ and $(1, f(1))$. By the property of linear functions, if we can show that $f(-1)$ and $f(1)$ are both greater than 0, then $f(x) > 0$ for each x in the domain. Since

$$f(-1) = -(b + c) + bc + 1 = (b - 1)(c - 1) > 0,$$
$$f(1) = (b + c) + bc + 1 = (b + 1)(c + 1) > 0,$$

we conclude that $f(x) > 0$ for all $-1 < x < 1$, so

$$(a) = a(b+c) + bc + 1 = ab + bc + ca + 1 > 0.$$

Example 2. Let $x, y, z \in (0,1)$. Show that

$$x(1-y) + y(1-z) + z(1-x) < 1.$$

Solution. Let

$$f(x) = (1-y-z)x + y + z - yz - 1, 0 < x < 1$$

and view y and z as constants. Then $f(x)$ is a linear function of x. Since

$$f(0) = y + z - yz - 1 = -(y-1)(z-1) < 0,$$
$$f(1) = (1-y-z) + y + z - yz - 1 = -yz > 0,$$

we have $f(x) < 0$ for all $0 < x < 1$, so

$$(1-y-z)x + y + z - yz - 1 < 0,$$

or equivalently $x(1-y) + y(1-z) + z(1-x) < 1$.

Remark. Proving inequalities by monotonicity of linear functions is a common method. The key is to construct a good linear function.

The following problems involve constructing quadratic functions.

Example 3. Let $a_1, a_2, \ldots, a_n, b_1, b_2, \ldots, b_n$ be real numbers. Show that

$$({a_1}^2 + {a_2}^2 + \cdots + {a_n}^2)({b_1}^2 + {b_2}^2 + \cdots + {b_n}^2) \geq (a_1b_1 + a_2b_2 + \cdots + a_nb_n)^2.$$

Solution. If ${a_1}^2 + a_2^2 + \cdots + {a_n}^2 = 0$, then $a_1 = a_2 = \cdots = a_n = 0$, and the proposition holds obviously. If ${a_1}^2 + {a_2}^2 + \cdots + {a_n}^2 \neq 0$, then we construct a quadratic function

$$\begin{aligned} f(x) &= ({a_1}^2 + {a_2}^2 + \cdots + {a_n}^2)x^2 - 2(a_1b_1 + a_2b_2 \\ &\quad + \cdots + a_nb_n)x + ({b_1}^2 + {b_2}^2 + \cdots + {b_n}^2) \\ &= (a_1x - b_1)^2 + (a_2x - b_2)^2 + \cdots + (a_nx - b_n)^2. \end{aligned}$$

Since ${a_1}^2 + {a_2}^2 + \cdots + {a_n}^2 > 0$, we see that the graph of $f(x)$ is a parabola opening upwards, and satisfies $f(x) \geq 0$ for all real numbers x.

This implies that

$$\Delta = 4(a_1b_1+a_2b_2+\cdots+a_nb_n)^2-4(a_1^2+a_2^2+\cdots+a_n{}^2)(b_1{}^2+b_2{}^2+\cdots+b_n{}^2) \leq 0,$$

which is equivalent to

$$(a_1{}^2+a_2{}^2+\cdots+a_n{}^2)(b_1{}^2+b_2{}^2+\cdots+b_n{}^2) \geq (a_1b_1+a_2b_2+\cdots+a_nb_n)^2.$$

Here the equality holds if and only if $b_i = ka_i$ for $i = 1, 2, \ldots, n$, where k is some real constant.

Remark. This is famous Cauchy's inequality. From this proof we can see that if we need to prove $AC \leq$ (or $\geq$)B^2, we may first write it as $4AC \leq$ (or $\geq$)$(2B)^2$, and then construct a quadratic function

$$f(x) = Ax^2 - (2B)x + C$$

and determine the sign of its discriminant.

Example 4. Let a, b, c, and d be real numbers such that

$$(a+b+c)^2 \geq 2(a^2+b^2+c^2)+4d.$$

Show that $ab+bc+ca \geq 3d$.

Analysis. If we can prove $ab \geq d$, and similarly $bc \geq d$ and $ca \geq d$, then the proposition follows. The original inequality can be written as

$$c^2 - 2(a+b)c + [(a^2+b^2) - 2ab + 4d] \leq 0,$$

so we may construct a function $f(x) = x^2 - 2(a+b)x + a^2 + b^2 - 2ab + 4d$.

Solution. We construct the function

$$f(x) = x^2 - 2(a+b)x + a^2 + b^2 - 2ab + 4d.$$

Then the graph of $f(x)$ is an upward-opening parabola, and $f(c) \leq 0$. Hence, the parabola has intersection points with the x-axis, so

$$\Delta = 4(a+b)^2 - 4(a^2+b^2-2ab+4d) \geq 0.$$

This is equivalent to $ab \geq d$. Similarly, we can prove that $bc \geq d$ and $ca \geq d$. Adding the three inequalities, we get $ab+bc+ca \geq 3d$.

Example 5. Let a, b, c, d, e, and f be positive integers and $S = a + b + c + d + e + f$. Suppose $abc + def$ and $ab + bc + ca - de - ef - fd$ are both divisible by S. Prove that S is a composite number.

Solution. We construct the function

$$f(x) = (x + a)(x + b)(x + c) - (x - d)(x - e)(x - f).$$

Then $f(x)$ is a quadratic function, and by expanding the formula we get

$$f(x) = Sx^2 + (ab + bc + ca - de - ef - fd)x + (abc + def).$$

By the assumption we see that every coefficient of $f(x)$ is divisible by S, so S $|f(x)$ for every integer x. In particular, choose $x = d$, and we have

$$S|(d + a)(d + b)(d + c).$$

Note that $d+a$, $d+b$, and $d+c$ are all positive integers smaller than S. Therefore, S must be a composite number.

The following problems involve constructing other functions.

Example 6. Suppose $x, y \in \left[-\frac{\pi}{4}, \frac{\pi}{4}\right]$ and $a \in \mathbf{R}$, and

$$\begin{cases} x^3 + \sin x - 2a = 0, \\ 4y^3 + \frac{1}{2}\sin 2y + a = 0. \end{cases}$$

Find the value of $\cos(x + 2y)$.

Solution. The system of equations can be written as

$$\begin{cases} x^3 + \sin x = 2a, \\ (-2y)^3 + \sin(-2y) = 2a. \end{cases}$$

Let $f(t) = t^3 + \sin t$. Then $f(t)$ is increasing on $\left[-\frac{\pi}{2}, \frac{\pi}{2}\right]$. Since $x, -2y \in \left[-\frac{\pi}{2}, \frac{\pi}{2}\right]$ and $f(x) = f(-2y)$, we have $x = -2y$ and $x + 2y = 0$. Therefore, $\cos(x + 2y) = 1$.

Example 7. Let $f(x)$ be a polynomial of degree 98 such that

$$f(k) = \frac{1}{k}, \quad k = 1, 2, \ldots, 99.$$

Find the value of $f(100)$.

Solution. Construct the function $g(x) = xf(x) - 1$. Then

$$g(1) = g(2) = \cdots = g(99) = 0$$

and $g(x)$ is a polynomial of degree 99. It follows that

$$g(x) = a(x-1)(x-2)\cdots(x-99),$$

where the leading coefficient a is to be determined. Now, since

$$f(x) = \frac{g(x)+1}{x} = \frac{a(x-1)(x-2)\cdots(x-99)+1}{x}$$

is a polynomial of degree 98, the constant term of $a(x-1)(x-2)\cdots(x-99)+1$ has to be 0, so $-99!a+1=0$, from which $a = \frac{1}{99!}$.

Therefore, $f(x) = \frac{\frac{1}{99!}(x-1)(x-2)\cdots(x-99)+1}{x}$ and $f(100) = \frac{1+1}{100} = \frac{1}{50}$.

Remark. For a quadratic polynomial $f(x) = ax^2 + bx + c$, if $f(x) = 0$ has two roots x_1 and x_2, then $f(x) = a(x-x_1)(x-x_2)$. In general, if

$$f(x) = a_n x^n + a_{n-1}x^{n-1} + \cdots + a_1 x + a_0$$

and $x_1, x_2, \ldots, x_n$ are its n roots, then

$$f(x) = a_n(x-x_1)(x-x_2)\cdots(x-x_n).$$

Example 8. Let n be a positive integer and θ be a real number. Show that

$$|\sin\theta \sin 2\theta \cdots \sin 2^n\theta| \le \left(\frac{\sqrt{3}}{2}\right)^n. \quad ①$$

Solution. We first find the maximum value of the function

$$f(x) = |\sin x| \cdot |\sin 2x|^{\frac{1}{2}}.$$

Note that

$$f(x) = \sqrt{2} \cdot |\sin x|^{\frac{3}{2}} \cdot |\cos x|^{\frac{1}{2}},$$

so

$$\begin{aligned} f(x)^4 &= 2^2 \cdot (\sin x)^6 \cdot (\cos x)^2 \\ &= \frac{4}{3} \cdot \sin^2 x \cdot \sin^2 x \cdot \sin^2 x \cdot (3\cos^2 x) \\ &\le \frac{4}{3} \cdot \left(\frac{3\sin^2 x + 3\cos^2 x}{4}\right)^4 = \left(\frac{3}{4}\right)^3. \end{aligned}$$

Hence, $f(x) \le \left(\frac{\sqrt{3}}{2}\right)^{\frac{3}{2}}$, where the equality is satisfied when $|\tan x| = \sqrt{3}$.

Now, let A denote the left side of ①. Then

$$A \le \prod_{i=0}^{n-1} |\sin 2^i\theta|,$$

$$A^{\frac{1}{2}} \le |\prod_{i=0}^{n-1} \sin 2^{i+1}\theta|^{\frac{1}{2}}.$$

Further,

$$\begin{aligned} A^{\frac{3}{2}} &\le \left|\prod_{i=0}^{n-1} \sin 2^i\theta\right| \cdot |\sin 2^{i+1}\theta|^{\frac{1}{2}} \\ &= \prod_{i=0}^{n-1} f(2^i\theta) \le \prod_{i=0}^{n-1} \left(\frac{\sqrt{3}}{2}\right)^{\frac{3}{2}}. \end{aligned}$$

Therefore, $A \le \left(\frac{\sqrt{3}}{2}\right)^n$, and ① is valid.

Example 9. Let a, b, and c be the side lengths of ΔABC such that $a+b+c=1$. Show that

$$5(a^2+b^2+c^2) + 18abc \ge \frac{7}{3}.$$

Solution. By the condition $a+b+c=1$,

$$a^2+b^2+c^2 = 1 - 2(ab+bc+ca),$$

so the inequality to be proved is equivalent to

$$\frac{5}{9}(ab+bc+ca) - abc \le \frac{4}{27}.$$

We construct the cubic polynomial

$$\begin{aligned} f(x) &= (x-a)(x-b)(x-c) \\ &= x^3 - (a+b+c)x^2 + (ab+bc+ca)x - abc \\ &= x^3 - x^2 + (ab+bc+ca)x - abc. \end{aligned}$$

Since $a+b+c=1$, and a, b, and c are the side lengths of a triangle, $0 < a, b, c < \frac{1}{2}$, so $0 < a, b, c < \frac{5}{9}$. Note that $f\left(\frac{5}{9}\right) = \left(\frac{5}{9}\right)^3 - \left(\frac{5}{9}\right)^2 + \frac{5}{9}(ab+bc+ca) - abc$. On the other hand,

$$\begin{aligned} f\left(\frac{5}{9}\right) &= \left(\frac{5}{9}-a\right)\left(\frac{5}{9}-b\right)\left(\frac{5}{9}-c\right) \\ &\leq \left(\frac{\left(\frac{5}{9}-a\right)+\left(\frac{5}{9}-b\right)+\left(\frac{5}{9}-c\right)}{3}\right)^3 = \frac{8}{729}. \end{aligned}$$

Hence, $\frac{125}{729} - \frac{25}{81} + \frac{5}{9}(ab+bc+ca) - abc \leq \frac{8}{729}$, thus $\frac{5}{9}(ab+bc+ca) - abc \leq \frac{4}{27}$. Therefore, the desired inequality is true.

26.3 Exercises

Group A

1. Suppose the equation $ax^2+bx+c=0$ has two different real roots. Prove that the equation $ax^2+bx+c+k\left(x+\frac{b}{2a}\right)=0$ (here k is a nonzero real constant) has at least one root that lies between the two roots of the previous equation.
2. Suppose $|a|<1$ and $|b|<1$. Show that

$$\left|\frac{a+b}{1+ab}\right| < 1.$$

3. Let a and b be real numbers such that one of the roots of the equation $x^2-ax+b=0$ lies in the interval $[-1,1]$ and the other root lies in $[1,2]$. Find the value range of $a-2b$.
4. Let a, b, and c be real numbers such that

$$a+b+c=2, \frac{1}{a}+\frac{1}{b}+\frac{1}{c}=\frac{1}{2}.$$

Show that one of a, b, and c equals 2.

5. Suppose a, b, and c are three real numbers such that $a < b < c$. Show that the equation

$$\frac{1}{x-a} + \frac{1}{x-b} + \frac{1}{x-c} = 0$$

has two real roots, one of which lies between $\frac{2a+b}{3}$ and $\frac{a+2b}{3}$ and the other lies between $\frac{2b+c}{3}$ and $\frac{b+2c}{3}$.

6. Let x and y be real numbers such that

$$(3x+y)^5 + x^5 + 4x + y = 0.$$

Find the value of $4x + y$.

7. For the following equation of the unknown x:

$$(ax+1)^2 = a^2(1-x^2), a > 1,$$

prove that the equation has a positive root less than 1 and a negative root greater than -1.

8. Find the maximal real number k such that for all positive real numbers u, v, and w satisfying $u^2 > 4vw$, the following inequality holds:

$$(u^2 - 4vw)^2 > k(2v^2 - uw)(2w^2 - uv).$$

9. Let x_1, x_2, x_3, y_1, y_2, and y_3 be real numbers such that ${x_1}^2 + {x_2}^2 + {x_3}^2 \le 1$. Prove that

$$(x_1y_1 + x_2y_2 + x_3y_3 - 1)^2 \ge (x_1^2 + x_2^2 + x_3^2 - 1)(y_1^2 + y_2^2 + y_3^2 - 1).$$

10. Let ABC be a triangle of side lengths a, b, and c with perimeter 2. Prove that

$$a^2 + b^2 + c^2 + 2abc < 2.$$

Group B

11. Determine whether there exists a function f from $\mathbf{R}$ to $\mathbf{R}$ such that $f(x) \not\equiv x$ (i.e., $f(x)$ does not always equal x) and $f(f(f(x))) = x$.

12. Let a, b, and c be real numbers such that $a+b+c > 0$, $ab+bc+ca > 0$, and $abc > 0$. Show that $a > 0, b > 0$, and $c > 0$.

13. Let $a_1, a_2, \ldots, a_n$ be positive real numbers such that

$$a_1 + a_2 + \cdots + a_n = 1.$$

Also, let $\lambda_1, \lambda_2, \ldots, \lambda_n$ be real numbers with $0 < \lambda_1 \le \lambda_2 \le \cdots \le \lambda_n$. Show that

$$\left(\sum_{i=1}^{n} \frac{a_i}{\lambda_i}\right)\left(\sum_{i=1}^{n} a_i\lambda_i\right) \le \frac{(\lambda_1 + \lambda_n)^2}{4\lambda_1\lambda_n}.$$

14. Suppose $x_1 \geq x_2 \geq x_3 \geq x_4 \geq 2$ and $x_2 + x_3 + x_4 \geq x_1$. Show that

$$(x_1 + x_2 + x_3 + x_4)^2 \leq 4x_1x_2x_3x_4.$$

15. Suppose $\begin{cases} a^2 + b^2 - kab = 1 \\ c^2 + d^2 - kcd = 1 \end{cases}$, where a, b, c, d, and k are real numbers with $|k| < 2$. Prove that

$$|ac - bd| \leq \frac{2}{\sqrt{4 - k^2}}.$$

Chapter 27

Vectors and Geometry

27.1 Key Points of Knowledge and Basic Methods

Vectors contain information about not only quantities but also positions, so they can be used to solve geometric problems. We may interpret geometric problems in an algebraic way with vectors and solve them with algebraic computations. As a tool for geometric problems, the vector method is geometrically intuitive, simple to express, and applicable to a variety of problems.

A. The first step of using the vector method is to specify base vectors, which should be linearly independent. Once the base vectors are specified, other vectors may be expressed in terms of them, and the problem turns into calculations of vectors based on these relations.

B. In order to determine the position of a point, we usually need to use linear relations of vectors (this is an important property of vectors throughout the vector method).

C. Problems with angles usually involve the inner product of vectors.

D. Rotations in geometry correspond to the outer product (cross product) of vectors, and the vector obtained by rotating a given vector can be expressed by the outer product (cross product).

27.2 Illustrative Examples

Example 1. Let O be the circumcenter of ΔABC and H be its orthocenter. Show that $\overrightarrow{OH} = \overrightarrow{OA} + \overrightarrow{OB} + \overrightarrow{OC}$.

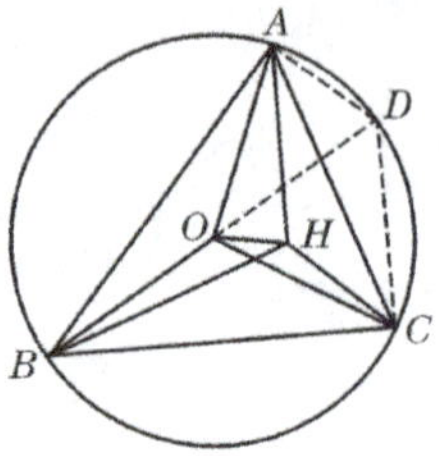

Figure 27.1

Proof. As shown in Figure 27.1, let BD be a diameter of the circumcircle, and form the line segments DA and DC. Then

$$DA \perp AB, DC \perp BC, AH \perp BC, CH \perp AB.$$

Hence, $CH \parallel DA$ and $AH \parallel DC$, and $AHCD$ is a parallelogram, so $\overrightarrow{AH} = \overrightarrow{CD}$. Since $\overrightarrow{DC} = \overrightarrow{OC} - \overrightarrow{OD} = \overrightarrow{OC} + \overrightarrow{OB}$,

$$\begin{aligned}\overrightarrow{OH} &= \overrightarrow{OA} + \overrightarrow{AH} = \overrightarrow{OA} + \overrightarrow{DC} \\ &= \overrightarrow{OA} + \overrightarrow{OB} + \overrightarrow{OC}.\end{aligned}$$

Example 2. Show that for a triangle ΔABC, its circumcenter O, barycenter G, and orthocenter H are collinear and $OG{:}GH=1{:}2$.

Proof. From Example 1, we know that $\overrightarrow{OH} = \overrightarrow{OA} + \overrightarrow{OB} + \overrightarrow{OC}$, and by Example 2 in Chapter 18, we also have

$$\overrightarrow{OG} = \frac{1}{3}(\overrightarrow{OA} + \overrightarrow{OB} + \overrightarrow{OC}).$$

Hence, $\overrightarrow{OG} = \frac{1}{3}\overrightarrow{OH} = \frac{1}{3}(\overrightarrow{OG} + \overrightarrow{GH})$, so $\overrightarrow{OG} = \frac{1}{2}\overrightarrow{GH}$. This implies that O, G, and H are collinear and $|\overrightarrow{OG}| : |\overrightarrow{GH}| = 1 : 2$.

Remark. The straight line that O, G, and H lie on is the Euler line of the triangle. This example shows a simple and clean proof for this result using vectors.

Example 3. As shown in Figure 27.2, let ΔABC and ΔAD_0E_0 be isosceles right triangles with $\angle B = \angle D_0 = 90°$. Suppose E_0 lies on AC with $AE_0 \neq AC$, and rotate ΔAD_0E_0 around A by an arbitrary angle θ to get ΔADE. Prove that there exists a point M on EC such that ΔBMD is also an isosceles right triangle.

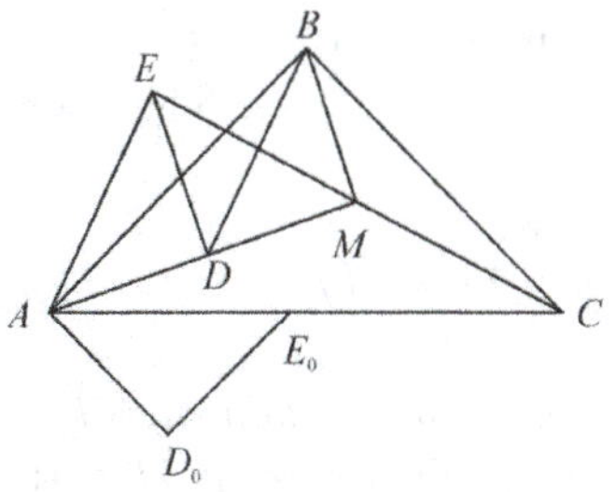

Figure 27.2

Proof. Choose base vectors $\overrightarrow{AC} = \vec{a}$ and $\overrightarrow{AE} = \vec{b}$. Then $|\vec{a}| \neq \left|\vec{b}\right|$. Since $\overrightarrow{AB}$ is obtained by rotating $\frac{\sqrt{2}}{2}\vec{a}$ counterclockwise by 45°,

$$\begin{aligned}\overrightarrow{AB} &= \frac{\sqrt{2}}{2}\left(\vec{a}\cos\frac{\pi}{4} + \vec{k}\times\vec{a}\cdot\sin\frac{\pi}{4}\right)\\ &= \frac{1}{2}(\vec{a}+\vec{k}\times\vec{a}),\end{aligned}$$

where $\vec{k}\times\vec{u}$ is the vector $\vec{u}$ rotated counterclockwise by 90°. Similarly, $\overrightarrow{AD} = \frac{1}{2}(\vec{b}-\vec{k}\times\vec{b})$.

Now, choose M on EC and write $\overrightarrow{AM}$ as

$$\begin{aligned}\overrightarrow{AM} &= m\overrightarrow{AC} + (1-m)\overrightarrow{AE}\\ &= m\vec{a} + (1-m)\vec{b},\end{aligned}$$

where $0 \leq m \leq 1$. Then

$$\begin{aligned}\overrightarrow{MB} &= \overrightarrow{AB} - \overrightarrow{AM}\\ &= \frac{1}{2}(\vec{a}+\vec{k}\times\vec{a}) - m\vec{a} - (1-m)\vec{b}\\ &= \left(\frac{1}{2}-m\right)\vec{a} + \frac{1}{2}\vec{k}\times\vec{a} + (m-1)\vec{b}.\end{aligned}$$

On the other hand, $\overrightarrow{MD} = \overrightarrow{AD} - \overrightarrow{AM} = -m\vec{a} + \left(m-\frac{1}{2}\right)\vec{b} - \frac{1}{2}\vec{k}\times\vec{b}$. In order to make ΔBMD an isosceles right triangle, we only need $\overrightarrow{MD} = \vec{k}\times\overrightarrow{MB}$, which requires that

$$-m\vec{a} + \left(m-\frac{1}{2}\right)\vec{b} - \frac{1}{2}k\times\vec{b} = \left(\frac{1}{2}-m\right)\vec{k}\times\vec{a} - \frac{1}{2}\vec{a} + (m-1)\vec{k}\times\vec{b}.$$

Here, we have used the relation $\vec{k} \times (\vec{k} \times \vec{a}) = -\vec{a}$, since rotating $\vec{a}$ counterclockwise by 90° twice gives $-\vec{a}$. The above equation is equivalent to

$$\left(m - \frac{1}{2}\right)[(\vec{a} - \vec{b}) - \vec{k} \times (\vec{a} - \vec{b})] = \vec{0}.$$

Since $|\vec{a}| \neq |\vec{b}|$, we have $\vec{a} - \vec{b} \neq 0$, and since $\vec{k} \times (\vec{a} - \vec{b})$ is perpendicular to $\vec{a} - \vec{b}$, necessarily $m = \frac{1}{2}$. Therefore, if M is the midpoint of EC, then ΔBMD is an isosceles triangle.

Example 4. As shown in Figure 27.3, the point O is the circumcenter of ΔABC, and the three altitudes AD, BE, and CF intersect at H. Let the lines ED and AB intersect at M, and FD and AC intersect at N. Prove that (1) $OB \perp DF$ and $OC \perp DE$; (2) $OH \perp MN$.

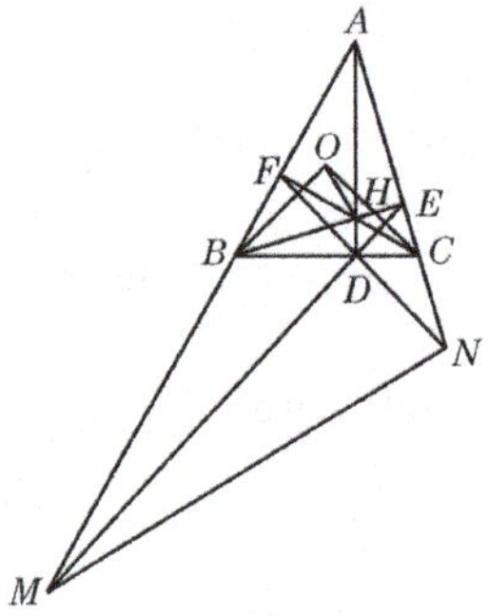

Figure 27.3

Proof. (1) It follows that A, F, C, and D are concyclic, so $\angle BDF = \angle BAC$. Since O is the circumcenter,

$$\angle BOC = 2\angle BAC, \angle OBC = \angle OCB,$$

so $\angle OBC = \frac{1}{2}(180° - \angle BOC) = 90° - \angle BAC$ and $\angle OBC + \angle BDF = 90°$. Hence, $OB \perp DF$, and similarly $OC \perp DE$.

(2) By $MF \perp CH$, $FN \perp OB$, $MD \perp OC$, $AN \perp BH$, $DF \perp OB$, $FA \perp CH$, and $DA \perp BC$, we have

$$\overrightarrow{MF} \cdot \overrightarrow{CH} = 0, \overrightarrow{FN} \cdot \overrightarrow{OB} = 0,$$
$$\overrightarrow{MD} \cdot \overrightarrow{OC} = 0, \overrightarrow{AN} \cdot \overrightarrow{BH} = 0,$$

$$\overrightarrow{DF}\cdot\overrightarrow{OB}=0, \overrightarrow{FA}\cdot\overrightarrow{CH}=0,$$
$$\overrightarrow{DA}\cdot\overrightarrow{BC}=0.$$

Hence,

$$\begin{aligned}\overrightarrow{MN}\cdot\overrightarrow{OH}&=(\overrightarrow{MF}+\overrightarrow{FN})\cdot\overrightarrow{OH}\\&=\overrightarrow{MF}\cdot\overrightarrow{OH}+\overrightarrow{FN}\cdot\overrightarrow{OH}\\&=\overrightarrow{MF}\cdot(\overrightarrow{OC}+\overrightarrow{CH})+\overrightarrow{MF}\cdot(\overrightarrow{OB}+\overrightarrow{BH})\\&=\overrightarrow{MF}\cdot\overrightarrow{OC}+\overrightarrow{FN}\cdot\overrightarrow{BH}\\&=(\overrightarrow{MD}+\overrightarrow{DF})\cdot\overrightarrow{OC}+(\overrightarrow{FA}+\overrightarrow{AN})\cdot\overrightarrow{BH}\\&=\overrightarrow{DF}\cdot\overrightarrow{OC}+\overrightarrow{FA}\cdot\overrightarrow{BH}\\&=\overrightarrow{DF}\cdot(\overrightarrow{OB}+\overrightarrow{BC})+\overrightarrow{MN}\cdot(\overrightarrow{BC}+\overrightarrow{CH})\\&=\overrightarrow{DF}\cdot\overrightarrow{BC}+\overrightarrow{FA}\cdot\overrightarrow{BC}\\&=\overrightarrow{DA}\cdot\overrightarrow{BC}=0.\end{aligned}$$

Therefore, $MN\perp OH$.

Example 5. In a tetrahedron $ABCD$, we know that $AB = AC = AD$. Let O be the center of its circumscribed sphere, G be the barycenter of ΔACD, E be the midpoint of BG, and F be the midpoint of AE. Prove that $OF\perp BG$ if and only if $OD\perp AC$.

Proof. Solid geometry is also an important field for applications of vectors. In this problem we will see the power of vectors when dealing with the perpendicular relationship.

Choose O as the origin, and we use position vectors to rewrite this problem. The conditions imply that

$$\begin{cases}|\overrightarrow{AB}|=|\overrightarrow{AC}|=|\overrightarrow{AD}|, & ①\\ \overrightarrow{OA}^2=\overrightarrow{OB}^2=\overrightarrow{OC}^2=\overrightarrow{OD}^2. & ②\end{cases}$$

Taking the square of ① and combining the result with ② and equalities like $\overrightarrow{OA}-\overrightarrow{OB}=\overrightarrow{AB}$, we get

$$\overrightarrow{OA}\cdot\overrightarrow{OB}=\overrightarrow{OA}\cdot\overrightarrow{OC}=\overrightarrow{OA}\cdot\overrightarrow{OD}. \quad ③$$

Now,

$$\begin{aligned}\overrightarrow{OF}\cdot\overrightarrow{GB} &= \left(\frac{\overrightarrow{OA}+\overrightarrow{OE}}{2}\right)\cdot(\overrightarrow{OB}-\overrightarrow{OG})\\ &= \frac{1}{4}(2\overrightarrow{OA}+\overrightarrow{OB}+\overrightarrow{OG})\cdot\left(\overrightarrow{OB}-\frac{1}{3}(\overrightarrow{OA}+\overrightarrow{OC}+\overrightarrow{OD})\right)\\ &= \frac{1}{36}(18\overrightarrow{OA}\cdot\overrightarrow{OB}-6\overrightarrow{OA}\cdot(\overrightarrow{OA}+\overrightarrow{OC}+\overrightarrow{OD})+9\overrightarrow{OB}^2\\ &\quad-(\overrightarrow{OA}+\overrightarrow{OC}+\overrightarrow{OD})^2).\end{aligned}$$

By ② and ③, we see that

$$\overrightarrow{OF}\cdot\overrightarrow{GB}=\frac{1}{36}(2\overrightarrow{OA}\cdot\overrightarrow{OD}-2\overrightarrow{OC}\cdot\overrightarrow{OD})=\frac{1}{18}\overrightarrow{CA}\cdot\overrightarrow{OD}.$$

Therefore, $OF\perp BG\Leftrightarrow OD\perp AC$, and the proposition is proven.

Example 6. Let $A_1A_2\dots A_n$ be a regular n-gon inscribed to a unit circle Γ. Let P be a point in the plane of Γ. Prove that ① $\sum_{i=1}^n PA_i\ge n$; ② $\sum_{i=1}^n PA_i^2\ge n$.

Proof. Let O be the center of Γ. Then

$$|\overrightarrow{OA_i}|=1, 1\le i\le n$$

and $\sum_{i=1}^n\overrightarrow{OA_i}=\vec{0}$.

(1) We have

$$\begin{aligned}\sum_{i=1}^n PA_i &= \sum_{i=1}^n|\overrightarrow{PA_i}|\cdot|\overrightarrow{OA_i}|\ge\sum_{i=1}^n|\overrightarrow{PA_i}\cdot\overrightarrow{OA_i}|\\ &=\sum_{i=1}^n|\overrightarrow{OP}\cdot\overrightarrow{OA_i}-\overrightarrow{OA_i}\cdot\overrightarrow{OA_i}|=\sum_{i=1}^n|\overrightarrow{OP}\cdot\overrightarrow{OA_i}-1|\\ &\ge\left|\sum_{i=1}^n(\overrightarrow{OP}\cdot\overrightarrow{OA_i}-1)\right|=\left|\sum_{i=1}^n\left(\overrightarrow{OA_i}\right)\cdot\overrightarrow{OP}-n\right|\\ &=|-n|=n.\end{aligned}$$

Therefore, the inequality is valid.

(2) We have

$$PA_i^2 = \overrightarrow{PA_i} \cdot \overrightarrow{PA_i} = (\overrightarrow{OP} - \overrightarrow{OA_i}) \cdot (\overrightarrow{OP} - \overrightarrow{OA_i})$$
$$= \overrightarrow{OP}^2 - 2\overrightarrow{OA_i} \cdot \overrightarrow{OP} + \overrightarrow{OA_i}^2 = \overrightarrow{OP}^2 - 2\overrightarrow{OA_i} \cdot \overrightarrow{OP} + 1.$$

Hence,

$$\sum PA_i^2 = \sum_{i=1}^{n} (\overrightarrow{OP}^2 - 2\overrightarrow{OA_i} \cdot \overrightarrow{OP} + 1)$$
$$= n\overrightarrow{OP}^2 - 2\left(\sum_{i=1}^{n} \overrightarrow{OA_i}\right) \cdot \overrightarrow{OP} + n$$
$$= n + n\overrightarrow{OP}^2 \geq n.$$

Therefore, the inequality is true.

Example 7. In a triangle ABC, suppose $\angle A = 60°$ and P is a point in the plane such that $PA = 6$, $PB = 7$, and $PC = 10$. Find the maximum value of the area of ΔABC.

Solution. As shown in Figure 27.4, we draw the parallelograms $APEC$ and $ABDC$. Then $BPED$ is also a parallelogram. We have

$$PD^2 - PC^2 + PA^2 - PB^2$$
$$= |\overrightarrow{PD}|^2 - |\overrightarrow{PC}|^2 + |\overrightarrow{PA}|^2 - |\overrightarrow{PB}|^2$$
$$= |\overrightarrow{PA} + \overrightarrow{AD}|^2 - |\overrightarrow{PA} + \overrightarrow{AC}|^2 + |\overrightarrow{PA}|^2 - |\overrightarrow{PA} + \overrightarrow{AB}|^2$$
$$= |\overrightarrow{AD}|^2 + 2\overrightarrow{PA} \cdot \overrightarrow{AD} - |\overrightarrow{AC}|^2 - 2\overrightarrow{PA} \cdot \overrightarrow{AC} - |\overrightarrow{AB}|^2 - 2\overrightarrow{PA} \cdot \overrightarrow{AB}.$$

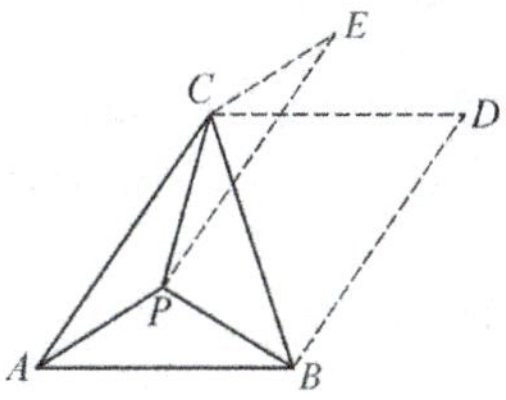

Figure 27.4

Note that $\overrightarrow{AD} = \overrightarrow{AC} + \overrightarrow{AB}$, so the expression above equals

$$\begin{aligned}
&|\overrightarrow{AD}|^2 - |\overrightarrow{AC}|^2 - |\overrightarrow{AB}|^2 \\
&\quad = |\overrightarrow{AD}|^2 - |\overrightarrow{AC}|^2 - |\overrightarrow{CD}|^2 \\
&\quad = -2 \cdot |\overrightarrow{AC}| \cdot |\overrightarrow{CD}| \cos \angle ACD \\
&\quad = 2AB \cdot AC \cos A = AB \cdot AC \\
&\quad = CD \cdot PE.
\end{aligned}$$

Hence,

$$PD^2 - 100 + 36 - 49 = CD \cdot PE.$$

By Ptolemy's theorem,

$$CD \cdot PE \leq PD \cdot CE + PC \cdot ED = PD \cdot PA + PC \cdot PB = 6PD + 70,$$

so $PD^2 - 113 \leq 6PD + 70$, and solving the inequality gives $PD \leq 3 + 8\sqrt{3}$, where the equality holds if and only if P, C, E, and D are concyclic. Consequently,

$$\begin{aligned}
S_{\Delta_{ABC}} &= \frac{1}{2} AB \cdot AC \sin A = \frac{\sqrt{3}}{4} AB \cdot AC \\
&= \frac{\sqrt{3}}{4} CD \cdot PE \leq \frac{\sqrt{3}}{4}(6PD + 70) \\
&\leq 36 + 22\sqrt{3},
\end{aligned}$$

where the equality is true when P, C, E, and D are concyclic, or equivalently $\angle CPE = \angle CDE$, namely $\angle ACP = \angle ABP$. Therefore, the maximum value of the area of ABC is $36 + 22\sqrt{3}$.

Example 8. Let n be a positive integer that is at least 5 and let $a_1, b_1, \ldots, a_n, b_n$ be integers such that

(1) the pairs $(a_i, b_i)(i = 1, 2, \ldots, n)$ are pairwise distinct;
(2) for each $1 \leq i \leq n$, we have $|a_i b_{i+1} - a_{i+1} b_i| = 1$, where $(a_{n+1}, b_{n+1}) = (a_1, b_1)$.

Prove that there exist indices i and j with $1 \leq |i, j| \leq n$ such that $1 < |i - j| < n - 1$ and

$$|a_i b_j - a_j b_i| = 1.$$

Proof. This problem looks algebraic, but is in essence geometric. Using vectors helps us see its geometric background.

Let $O(0, 0)$ be the origin and let $V_i(a_i, b_i)$ be points in the plane. Then $\overrightarrow{OV_i}$ is the position vector of V_i. Without loss of generality we may assume that $|\overrightarrow{OV_1}|$ is maximal, and $\overrightarrow{OV_2}$ and $\overrightarrow{OV_n}$ lie on the different sides of OV_1 (otherwise we replace V_2 by $(-a_2, -b_2)$, in which case $|a_1b_2 - a_2b_1|$ and $|a_3b_2 - a_2b_3|$ are invariant). Further we may assume that the intersection point P of OV_1 and V_2V_n lies on the ray OV_1 (otherwise we replace V_1 by $(-a_1, -b_1)$).

Now, since $|\overrightarrow{OP}| \le \max\{|\overrightarrow{OV_2}|, |\overrightarrow{OV_n}|\} \le |\overrightarrow{OV_1}|$, we see that P lies on the line segment OV_1.

If P lies between O and V_1 (not including the endpoints), then $OV_nV_1V_2$ is a convex quadrilateral, so

$$\begin{aligned} S_{\Delta OV_2V_n} + S_{\Delta V_1V_2V_n} &= S_{\Delta OV_1V_2} + S_{\Delta OV_1V_n} \\ &= \frac{1}{2}\left\|\begin{matrix} 1 & 0 & 0 \\ 1 & a_1 & b_1 \\ 1 & a_2 & b_2 \end{matrix}\right\| + \frac{1}{2}\left\|\begin{matrix} 1 & 0 & 0 \\ 1 & a_1 & b_1 \\ 1 & a_n & b_n \end{matrix}\right\| \\ &= \frac{1}{2}|a_1b_2 - a_2b_1| + \frac{1}{2}|a_1b_n - a_nb_1| = 1. \end{aligned}$$

(Here the notation means the absolute value of the determinant of the matrix.)

Hence, $S_{\Delta OV_2V_n} = \frac{1}{2}$ (since the area of a triangle with grid point vertices is at least $\frac{1}{2}$) and the pair $(i, j) = (2, n)$ is a valid pair of indices (using $n \ge 5$).

If $P = V_1$, then one of $\angle OV_1V_2$ and $\angle OV_1V_n$ is at least 90°, so

$$|\overrightarrow{OV_1}| < \max\left\{|\overrightarrow{OV_2}|, |\overrightarrow{OV_n}|\right\},$$

which is contradictory to the maximality of $|\overrightarrow{OV_1}|$.

If $P = O$, then by condition (2) we have $S_{\Delta OV_1V_2} = S_{\Delta OV_1V_n}$, from which

$$\begin{aligned} &\frac{1}{2}|\overrightarrow{OV_1}| \cdot |\overrightarrow{OV_2}| \cdot \sin\alpha \\ &\quad = \frac{1}{2}|\overrightarrow{OV_1}| \cdot |\overrightarrow{OV_n}| \cdot \sin(\pi - \alpha) \text{ (where } \alpha = \angle V_1OV_2), \end{aligned}$$

and hence $|\overrightarrow{OV_2}| = |\overrightarrow{OV_n}|$. This implies that $(a_2, b_2) = (n, -b_n)$ (since V_2, O, and V_n are collinear). Therefore, $|a_3b_2 - a_2b_3| = |a_3b_n - a_nb_3| = 1$ and $(i, j) = (3, n)$ is a valid pair (using $n \geq 5$ again). To summarize, we conclude that the proposition is true.

27.3 Exercises

Group A

1. Let G be the barycenter of ΔABC. Show that
$$AB^2 + BC^2 + CA^2 = 3(GA^2 + GB^2 + GC^2).$$
2. Suppose a convex n-gon $A_1A_2\cdots A_n$ has a circumcircle, and let G be its barycenter (i.e., the mean position of points $A_1, A_2, \ldots, A_n$). Show that
$$\sum_{i=1}^{n} \frac{A_iG}{GB_i} = n.$$
3. In a triangle ABC, let A_1, B_1, and C_1 be points on BC, CA, and AB, respectively, G be the barycenter of ΔABC, and G_a, G_b, and G_c be the barycenters of ΔAB_1C_1, ΔBC_1A_1, and ΔCA_1B_1, respectively. Finally let G_1 and G_2 be the barycenters of $\Delta A_1B_1C_1$ and $\Delta G_aG_bG_c$, respectively. Show that G, G_1, and G_2 are collinear.
4. A quadrilateral $A_1A_2A_3A_4$ is inscribed to $\odot O$, and points H_1, H_2, H_3, and H_4 are the orthocenters of $\Delta A_2A_3A_4, \Delta A_3A_4A_1, \Delta A_4A_1A_2$, and $\Delta A_1A_2A_3$, respectively. Prove that H_1, H_2, H_3, and H_4 lie on the same circle, and determine the center of this circle.
5. Let R denote the radius of the circumcircle of ΔABC and let P be a point lying inside ΔABC or on its boundary. Show that
$$|PA| + |PB| + |PC| \leq 4R.$$
6. Let O and G be the circumcenter and orthocenter of ΔABC, respectively, and let R and r be the radii of its circumcircle and incircle, respectively. Prove that $OG \leq \sqrt{R(R-2r)}$.
7. Prove that three times the sum of squares of the side lengths of a convex pentagon is greater than the sum of squares of its diagonals.

Group B

8. Determine whether there exist four vectors in a plane such that no two of them are collinear and the sum of any two of them is perpendicular to the sum of the other two.

9. Let M, N, P, Q, and R be the midpoints of the sides AB, BC, CD, DE, and EA of a convex pentagon $ABCDE$, respectively. Show that if AP, BQ, CR, and DM are concurrent, then EN also passes through their intersection point.

10. Let O be the center of a regular n-gon $A_1A_2\cdots A_n$, and let $\alpha_1 > \alpha_2 > \cdots > \alpha_n > 0$ be real numbers. Show that

$$\alpha_1\overrightarrow{OA_1} + \alpha_2\overrightarrow{OA_2} + \cdots + \alpha_n\overrightarrow{OA_n} \neq \vec{0}.$$

11. Let $A_1A_2\cdots A_n$ be a regular n-gon inscribed to a circle Γ. Suppose P is a point inside Γ and the line PA_i intersects Γ at another point $B_i (i = 1, 2, \ldots, n)$. Show that

(1) $\sum\limits_{i=1}^{n} PA_i^2 \geq \sum\limits_{i=1}^{n} PB_i^2$;

(2) $\sum\limits_{i=1}^{n} PA_i \geq \sum\limits_{i=1}^{n} PB_i$.

12. Let n be a given positive integer. We fill the integers $1, 2, \ldots, n^2$ in an $n \times n$ table such that each cell has a different number. Now, we draw a vector between the centers of each pair of the cells such that it starts at the cell with a smaller number and ends at the cell with a larger number. Suppose the sums of numbers in each column and row are the same. Show that the sum of all these vectors is the zero vector.

Chapter 28

Tetrahedrons

28.1 Key Points of Knowledge and Basic Methods

Tetrahedrons are the most basic and important geometric objects in solid geometry. They play a role that triangles play in plane geometry. Tetrahedrons and triangles have many common properties, for example:

In a tetrahedron, the line segments between the midpoints of the three pairs of opposite edges are concurrent and the intersection point bisects each line segment.

The line segment between each vertex and the barycenter of its opposite face passes through a common point G, and G divides each of these line segments into two parts with ratio 3:1. The point G is called the barycenter of the tetrahedron.

Every tetrahedron has a circumsphere, and the center O of this sphere is the intersection point of the bisector vertical planes of all edges. The distance from O to each vertex is equal to the radius of the circumsphere.

Every tetrahedron has an inscribed sphere, and the center I of this sphere is the intersection point of the bisector half planes of all dihedral angles between faces. The point I is equidistant to each face and the distance from I to each face is equal to the radius of the inscribed sphere.

28.2 Illustrative Examples

Example 1. Prove that for each tetrahedron, there is a vertex such that the lengths of the three edges starting from this vertex can be those of a triangle.

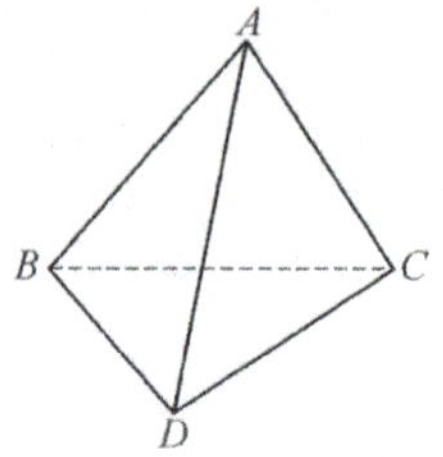

Figure 28.1

Analysis. We consider the longest edge.

Proof. Without loss of generality, we assume that AD is the longest edge (Figure 28.1). Then by the triangle inequality,

$$AC + CD > AD, \quad AB + BD > AD.$$

Hence,

$$(AB + AC) + (CD + BD) > 2AD.$$

This implies that at least one of $AB + AC$ and $CD + BD$ is greater than AD. Therefore, either A or D is a vertex that satisfies the requirement.

Remark. Here we applied the extreme principle and the pigeonhole principle to solve the problem with a clean elaboration.

Example 2. Let R be the radius of the circumsphere of a tetrahedron $ABCD$, and let r be the radius of its inscribed sphere. Prove that $R \geq 3r$.

Analysis. We can prove the result by similar objects.

Proof. Suppose the barycenters of ΔABC, ΔBCD, ΔCDA, and ΔDAB are D', A', B', and C', respectively. Let R_1 be the radius of the circumsphere of the tetrahedron $A'B'C'D'$. Then since $A'B'C'D'$ and $ABCD$ are similar tetrahedrons with similarity ratio $\frac{1}{3}$, we have $R = 3R_1$.

On the other hand, since the circumsphere of $A'B'C'D'$ intersects every face of $ABCD$, it follows that $R_1 \geq r$. (We can construct a tetrahedron, which is similar to $ABCD$ and contains $ABCD$, such that its inscribed sphere is the circumsphere of $A'B'C'D'$.)

Therefore, $R \geq 3r$.

Remark. Similarly, the same method can be used to prove that$R \geq 2r$ in a triangle ABC, where R and r are the radii of its circumcircle and incircle, respectively. This is famous Euler's theorem.

Example 3. Suppose exactly one edge of a tetrahedron has length greater than 1. Show that the volume of this tetrahedron cannot exceed $\frac{1}{8}$.

Proof. Let $ABCD$ be such a tetrahedron and let $AB > 1$, while the other edges have lengths at most 1. Further, let h be the distance from A to the face BCD and h_1 be the distance from A to CD in the triangle ACD. Then $h_1 \geq h$.

Let AE be an altitude of ΔACD. Then at least one of CE and DE has length $\geq \frac{1}{2}a$ (here a denotes the length of CD). Hence, $h_1 \leq \sqrt{1 - \left(\frac{a}{2}\right)^2}$, so $h \leq \frac{1}{2}\sqrt{4 - a^2}$.

Similarly, in the triangle BCD we see that the distance from B to $CD \leq \frac{1}{2}\sqrt{4 - a^2}$, so

$$S_{\Delta BCD} \leq \frac{a}{4}\sqrt{4 - a^2}.$$

Therefore,

$$\begin{aligned} V_{ABCD} &= \frac{1}{3}h \cdot S_{\Delta BCD} \leq \frac{1}{24}a(4 - a^2) \\ &= \frac{1}{24}a(2 - a)(2 + a) \leq \frac{1}{24}\left(\frac{a + 2 - a}{2}\right)^2 (2 + a) \\ &= \frac{1}{24}(2 + a) \leq \frac{1}{24}(2 + 1) = \frac{1}{8}. \end{aligned}$$

Note that the volume of the tetrahedron can be equal to $\frac{1}{8}$, and this happens if and only if ΔACD and ΔBCD are both equilateral triangles with side length 1, and plane $ACD \perp$ plane BCD.

Example 4. For any tetrahedron, prove that the product of the lengths of one pair of opposite edges is less than the sum of the products of those of the other two pairs.

Analysis. Without loss of generality, we only need to prove the following inequality:

$$AB \times CD < BC \times AD + AC \times BD.$$

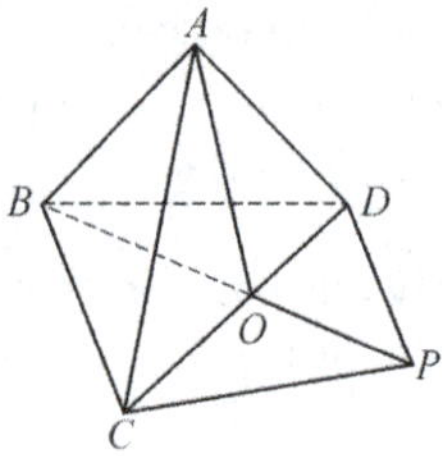

Figure 28.2

Proof. As shown in Figure 28.2, we rotate the face ACD around the edge CD to make it lie on the plane BCD, and let P be the corresponding point of A, so that $AD = DP$ and $AC = CP$.

Since $\Delta ACD \backsimeq \Delta PCD$, we have $AO = OP$, so $BP = BO + OA > AB$. By Ptolemy's theorem,

$$BP \times CD \leq DP \times BC + BD \times CP$$
$$= AD \times BC + BD \times AC. \text{ ①}$$

Also, note that $AB \times CD < BP \times CD$, and we conclude that

$$AB \times CD < AD \times BC + BD \times AC.$$

Remark. We cannot guarantee that $BDPC$ is a convex quadrilateral after the above rotation, but we can still use generalized Ptolemy's theorem to prove ①.

Example 5. Suppose O is a point inside a tetrahedron $ABCD$ such that the lines AO, BO, CO, and DO intersect their opposite faces at A', B', C', and D', respectively. Show that

$$(1)\frac{OA'}{AA'} + \frac{OB'}{BB'} + \frac{OC'}{CC'} + \frac{OD'}{DD'} = 1;$$
$$(2)\frac{AO}{AA'} + \frac{BO}{BB'} + \frac{CO}{CC'} + \frac{DO}{DD'} = 3.$$

Proof. As shown in Figure 28.3, let H be the projection of A onto the plane BCD, and form the line segment $A'H$. Then we draw $OH' \perp A'H$ with H' lying on $A'H$, and we have

$$\frac{A'O}{A'A} = \frac{OH'}{AH}.$$

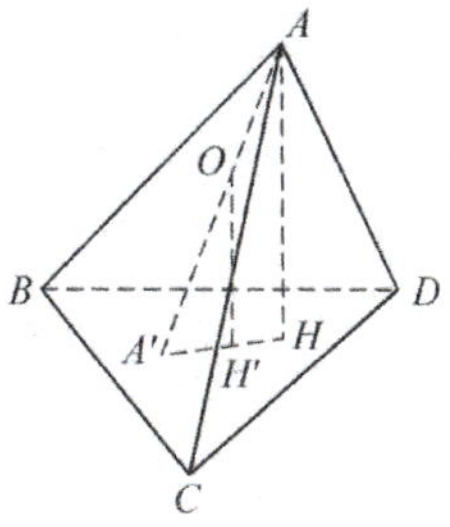

Figure 28.3

Since OH' and AH are the altitudes of the triangular pyramids O-BCD and A-BCD, respectively, and the two triangular pyramids have the same bottom face,

$$\frac{OA'}{AA'} = \frac{OH'}{AH} = \frac{V_{OBCD}}{V_{ABCD}}.$$

Similarly,

$$\frac{OB'}{BB'} = \frac{V_{OACD}}{V_{BACD}}, \frac{OC'}{CC'} = \frac{V_{OABD}}{V_{CABD}}, \frac{OD'}{DD'} = \frac{V_{OABC}}{V_{DABC}}.$$

Since the four equalities above have the same denominator, we can take the sum and get

$$\frac{OA'}{AA'} + \frac{OB'}{BB'} + \frac{OC'}{CC'} + \frac{OD'}{DD'} = 1.$$

Here, we used the fact that $V_{OABC} + V_{OBCD} + V_{OCDA} + V_{ODAB} = V_{ABCD}$. Therefore, ① holds.

Further, by $\frac{AO}{AA'} = 1 - \frac{OA'}{AA'}$ and other similar equalities, together with the result (1) we can get (2) immediately.

Remark. Similar results are valid in a triangle, where the proof uses the method of area. Here we used the method of volume.

Example 6. In a tetrahedron $ABCD$, let E and F be the midpoints of AB and CD, respectively. Let M be an arbitrary plane that passes through E and F. Prove that M divides the tetrahedron into two parts with an equal volume.

Analysis. We cut the volume into pieces, and prove the result through objects with equal volumes.

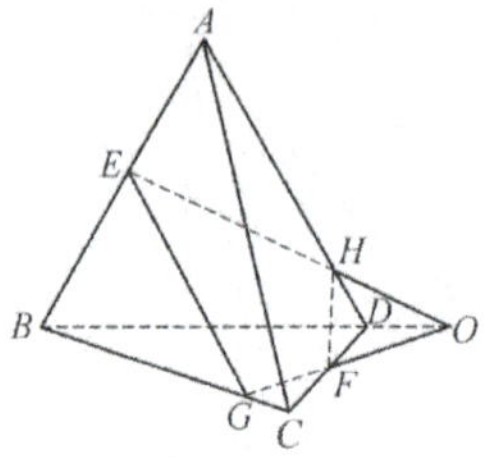

Figure 28.4

Proof. Let M intersect BC at G and intersect AD at H, and let $EH \cap GF = O$. Then form the line segments ED, EF, and EC (Figure 28.4). Since O belongs to both plane ABD and plane BCD, it must belong to their intersection, so $O \in BD$.

Since E is the midpoint of AB,

$$V_{AECD} = V_{BECD} = \frac{1}{2} V_{ABCD}.$$

Since F is the midpoint of CD,

$$\frac{V_{E\text{-}GCF}}{V_{ABCD}} = \frac{EB \cdot GC \cdot CF}{AB \cdot BC \cdot CD} = \frac{1}{4} \cdot \frac{CG}{BC}.$$

Similarly, $V_{E\text{-}FDH} = \frac{1}{4}\frac{DH}{AD} \cdot V_{ABCD}$.

Now, we apply the Menelaus theorem to the plane ABO, getting

$$\frac{DH}{HA} \cdot \frac{AE}{EB} \cdot \frac{BO}{OD} = 1,$$

so $\frac{DH}{AH} = \frac{OD}{OB}$. Also, in the plane BCO, we have $\frac{CF}{FD} \cdot \frac{DO}{OB} \cdot \frac{BG}{GC} = 1$, from which $\frac{CG}{BG} = \frac{OD}{OB}$, and hence

$$\frac{DH}{AH} = \frac{OD}{OB} = \frac{CG}{BG}.$$

Consequently, $\frac{DH}{AH+DH} = \frac{CG}{BG+CG}$ and $\frac{DH}{AD} = \frac{CG}{BC}$.

Therefore, $V_{E\text{-}GCF} = \frac{1}{4}\frac{CG}{BC} \cdot V_{ABCD} = \frac{1}{4}\frac{DH}{AD} \cdot V_{ABCD} = V_{E\text{-}FDH}$ and $V_{AEGCFH} = V_{EBGFDH}$.

If $EH \| GF$, then EH, GF, and BD are pairwise parallel, as in Figure 28.5. Since

$$V_{E\text{-}GCF} = \frac{1}{4}\frac{CG}{BC} \cdot V_{ABCD}, V_{E\text{-}FDH} = \frac{1}{4}\frac{DH}{AD} \cdot V_{ABCD},$$

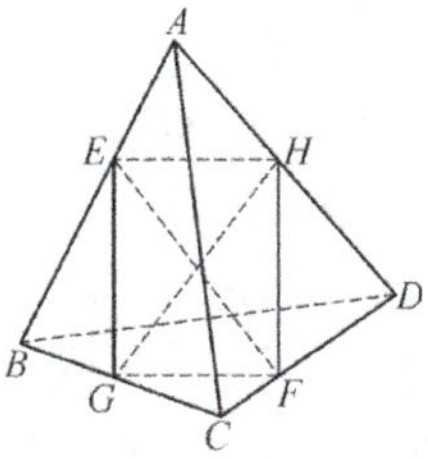

Figure 28.5

while $\frac{CG}{BC} = \frac{1}{2} = \frac{DH}{AD}$, we also have $V_{EGCF} = V_{EFOH}$, hence $V_{AEGCFH} = V_{EBGFDH}$.

Remark. This is a common method when dealing with the volume of a geometric object, where we find an object of reference, and find the relationship between their volumes (either equal or with a certain multiple).

Example 7. Let O be a point on the edge AB of a tetrahedron $ABCD$. Suppose the circumsphere of $AOCD$ intersects the lines BC and BD at M and N (different from C and D) respectively, and the circumsphere of $BOCD$ intersects the lines AC and AD at P and Q (different from C and D) respectively. Show that $\Delta OMN \backsim \Delta OQP$.

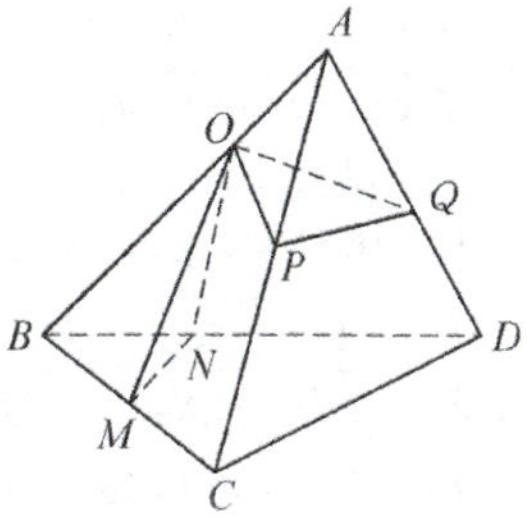

Figure 28.6

Analysis. We need to use the fact that the intersection of a sphere and a plane is a circle (like cutting a watermelon).

Proof. As shown in Figure 28.6, it follows that in each of the following groups:

$$CDNM,\ CDPQ,\ CPOB,\ CAOM, DBOQ,\ DAON,$$

the four points are concyclic. Hence,

$$\angle AOQ = \angle ADB = \angle BON,\ \angle AQO = \angle ABD.$$

Consequently, $\Delta AOQ \backsim \Delta NOB$, and so

$$\frac{OQ}{OB} = \frac{AO}{NO} = \frac{AQ}{BN}. \quad ①$$

Similarly, $\Delta AOP \backsim \Delta MOB$, and so

$$\frac{OP}{OB} = \frac{AO}{MO} = \frac{AP}{BM}. \quad ②$$

Comparing ① and ②, we get

$$\frac{OQ}{OB} \cdot \frac{AO}{MO} = \frac{AO}{NO} \cdot \frac{OP}{OB} = \frac{AQ \cdot AP}{BN \cdot BM}.$$

Hence,

$$\frac{OQ}{OM} = \frac{OP}{ON} = \frac{BO \cdot AP \cdot AQ}{AO \cdot BM \cdot BN}. \quad ③$$

On the other hand, since C, D, M, and N are concyclic, and C, D, Q, and P are concyclic, $\Delta BMN \backsim \Delta BDC$ and $\Delta APQ \backsim \Delta ADC$. It follows that

$$\frac{MN}{CD} = \frac{BM}{BD}, \frac{PQ}{CD} = \frac{AP}{AD}.$$

We take the ratio of the two equalities on both sides to get

$$\frac{PQ}{MN} = \frac{AP \cdot BD}{BM \cdot AD}. \quad ④$$

Also, since B, D, Q, and O are concyclic, and A, D, N, and O are concyclic, $\Delta AOQ \backsim \Delta ADB$ and $\Delta BON \backsim \Delta BDA$, from which

$$\frac{AO}{AQ} = \frac{AD}{AB}, \frac{BO}{BN} = \frac{BD}{BA}.$$

Similarly, we take the ratio of the equalities on both sides, obtaining

$$\frac{BD}{AD} = \frac{BO \cdot AQ}{AO \cdot BN}, \quad ⑤$$

By ③, ④, and ⑤ together we have

$$\frac{OQ}{OM} = \frac{OP}{ON} = \frac{PQ}{MN}.$$

Therefore, $\Delta OMN \backsim \Delta OQP$, and the proposition holds.

Remark. Problems combining tetrahedrons and spheres usually involve circles, just like in the plane case.

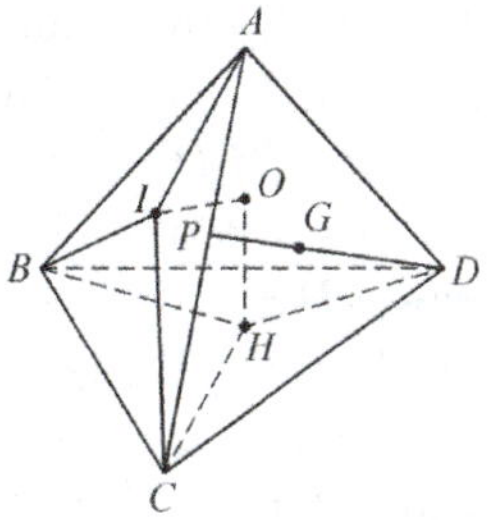

Figure 28.7

Example 8. Suppose there is a sphere that is tangent to three faces of a tetrahedron, and the three points of tangency are the incenter, orthocenter, and barycenter of their respective faces. Prove that this tetrahedron is a regular one.

Analysis. We try to prove that each face is an equilateral triangle.

Proof. As shown in Figure 28.7, let the sphere centered at O be tangent to the face ABC at its incenter I, tangent to the face BCD at its orthocenter H, and tangent to the face ACD at its barycenter G. Then $OI = OG = OH$, and $OI \perp$ plane ABC, $OH \perp$ plane BCD, and $OG \perp$ plane CDA.

Let $\angle IAB = \angle IAC = \alpha$, $\angle IBA = \angle IBC = \beta$, and $\angle ICB = \angle ICA = \gamma$. Since $\text{Rt}\Delta OIA \backsimeq \Delta \text{Rt} OGA$, we have $AI = AG$.

Similarly, $CI = CG$, so $\Delta AIC \backsimeq \Delta AGC$. This implies that $\angle GAC = \angle IAC = \alpha$ and $\angle GCA = \angle ICA = \gamma$. By $\Delta BIC \backsimeq \Delta BHC$, we have $\angle HBC = \angle IBC = \beta$ and $\angle HCB = \angle ICB = \gamma$. Note that we also have $\Delta CHD \backsimeq \Delta CGD$.

Now, since H is the orthocenter of ΔBCD, and using the fact that $\alpha + \beta + \gamma = \frac{\pi}{2}$, we see that

$$\angle HBD = \frac{\pi}{2} - \angle HCB\text{-}\angle HBC = \alpha,$$

and $\angle HBD = \angle HCD = \alpha$ since $\angle HCD = \frac{\pi}{2} - \angle BDC = \angle HBD$. Further we can get $\angle HDC = \beta$ and $\angle HDB = \gamma$. Combining these with the result that $\Delta CHD \backsimeq \Delta CGD$, we obtain $\angle GCD = \alpha$ and $\angle GDC = \beta$.

Let P be the intersection point of DG and AC. Then P is the midpoint of AC (since G is the barycenter), so

$$\angle DPC = \pi - \angle GDC - \angle GCD - \angle GCA = \pi - \beta - \alpha - \gamma = \frac{\pi}{2},$$

or equivalently $DP \perp AC$. Hence, ΔADC is an isosceles triangle, which is symmetric with respect to DP. Further we have $\angle GAC = \angle GCA$, thus $\alpha = \gamma$ and CH is the bisector of $\angle BCD$. Combining them with $CH \perp BD$, we see that ΔBCD is also an isosceles triangle.

Now, the proved equalities $AD = CD$, $BC = CD$, and $AB = BC$ (since $\alpha = \gamma$) impliy $AB = BC = CD = DA$. Hence, $BAC \backsimeq DAC$, and together with $\Delta AIC \backsimeq \Delta AGC$, we know that G is the incenter of ΔADC. Since G is both the barycenter and the incenter, so ΔADC is necessarily an equilateral triangle. Further we can derive that H is the incenter of ΔBCD, so ΔBCD is also equilateral. Finally since $\Delta ABC \backsimeq \Delta ADC$, we see that ΔABD is equilateral.

Therefore, $ABCD$ is a regular tetrahedron.

Remark. A regular tetrahedron is the solid object encompassed by four congruent equilateral triangles, which is a special type of regular triangular pyramid.

Combining the above with Example 8 in Chapter 21, we see that it is not that easy to determine whether a set of conditions is sufficient for a tetrahedron to be regular. Sometimes we need to switch between planes and spaces.

28.3 Exercises

Group A

1. For the eight vertices of a cube, four of them constitute the vertices of a regular tetrahedron. Then the ratio between the surface areas of the cube and the tetrahedron is ____________.
2. In a triangular frustum, let the areas of its top face and bottom face be S_t and S_b, respectively, and let its altitude be h. Consider two edges, one in the top face and the other in the bottom face, such that they are not parallel. Then the tetrahedron determined by the four vertices of these two edges has volume ____________ (in terms of the given quantities).
3. Let O be the center of a regular tetrahedron $ABCD$. Then $\angle AOB =$ ____________.
4. In a regular tetrahedron $ABCD$, let E and F be points on the edges AB and CD respectively such that $\frac{AE}{EB} = \frac{CF}{FD}$. Let α and β denote the angles between the line EF and the lines BD and AC, respectively. Then $\alpha + \beta =$ ____________.

5. In a tetrahedron $ABCD$, suppose the incenter I, the midpoint E of AB, and the midpoint F of CD are collinear. Prove that the circumcenter of the tetrahedron also lies on the line EF.
6. In a regular triangular pyramid P-ABC, let O be the center of the bottom face ΔABC. Suppose π is a plane containing O and intersects the rays PA, PB, and PC at Q, R, and S, respectively. Show that the sum $\frac{1}{PQ} + \frac{1}{PR} + \frac{1}{PS}$ is a constant.
7. In a triangular pyramid S-ABC, suppose the side edges SA, SB, and SC are pairwise perpendicular. Let M be the barycenter of ΔABC, D be the midpoint of AB, and PD be a line passing through D and parallel to SC. Prove that the lines PD and SM intersect at the circumcenter of S-ABC.
8. Let O be the center of a regular tetrahedron $ABCD$, let E be a point on the face ABC, and let F, G, and H be the projection of E onto the other faces. Show that OE passes through the barycenter Q of ΔFGH.
9. In a right-angled tetrahedron P-ABC, suppose ΔABC is not a right triangle, and $PA = a$, $PB = b$, and $PC = c$.

   (1) Let h be the altitude from P to the face ABC. Show that

   $$h^{-2} = a^{-2} + b^{-2} + c^{-2}.$$

   (2) Let M be a point on the face ABC, and let x, y, and z denote the distances from M to the faces PBC, PCA, and PAB, respectively. Find the maximum value of $\frac{xyz}{abc}$.

Group B

10. Let P be a point inside a tetrahedron $ABCD$. Suppose for a point X on the surface of the tetrahedron, the maximal and minimal distances from P to X are H and h, respectively. Show that $H \geq 3h$.
11. Let P and Q be two points inside a regular tetrahedron $ABCD$. Show that $\cos \angle PAQ > \frac{1}{2}$.
12. For each edge of a regular tetrahedron with edge length 1, we draw a sphere whose diameter is the edge. Let S be the intersection of all the six balls (including the interior). Prove that the distance between any two points in S does not exceed $\frac{1}{\sqrt{6}}$.

Chapter 29

The Five Centers of a Triangle

29.1 Key Points of Knowledge and Basic Methods

A. The barycenter

The three medians of a triangle intersect at one point, which is called the barycenter of the triangle.

Property 1. The barycenter divides each median into two parts with ratio 2:1 (the distance to the vertex is greater).

Property 2. The three triangles formed by the barycenter and any two vertices (of the big triangle) have the same area.

Property 3. The barycenter is the point in a triangle such that the sum of the squares of its distances to the three vertices is minimal.

B. The orthocenter

The three altitudes (or the lines containing them) of a triangle intersect at one point, which is called the orthocenter of the triangle.

Property 1. The distance from the orthocenter to a vertex is twice the distance from the circumcenter to the opposite side of the vertex.

Property 2. The circumcenter O, barycenter G, and orthocenter H of a triangle are collinear (and the line is called the Euler line of the triangle), and $OG : GH = 1 : 2$.

C. The circumcenter

The center of the circumcircle of a triangle is called its circumcenter.

Property 1. The circumcenter is the intersection point of the perpendicular bisectors of the three sides.

Property 2. The circumcenter of a triangle is equidistant to the three vertices.

Property 3. The circumcenter of a right triangle is the midpoint of its hypotenuse; the circumcenter of an acute triangle lies inside the triangle; the circumcenter of an obtuse triangle lies outside the triangle.

Property 4. If O is the circumcenter of a triangle ABC, then

$$\angle BOC = 2\angle A, \angle COA = 2\angle B, \angle AOB = 2\angle C.$$

D. The incenter

The center of the inscribed circle of a triangle is called its incenter.

Property 1. The incenter is the intersection point of the three (inner) angle bisectors of a triangle.

Property 2. The incenter is equidistant to the three sides.

Property 3. ("The claw theorem") If I is the incenter of ΔABC and the line AI intersects the circumcircle of ΔABC again at M, then $MI = MB = MC$.

Property 4. If I is the incenter of ABC, then

$$\angle BIC = 90^\circ + \frac{1}{2}\angle A,\ \angle CIA = 90^\circ + \frac{1}{2}\angle B,\ \angle AIB = 90^\circ + \frac{1}{2}\angle C.$$

Property 5. (Euler's formula) If I and O are the incenter and circumcenter of ΔABC, respectively, and r and R denote the radii of the incircle and circumcircle, respectively, then

$$OI^2 = R^2 - 2Rr.$$

E. The escenter

The center of an escircle of a triangle (i.e., a circle that is tangent to one side and the extensions of the other two sides) is called an escenter of the triangle.

Property 1. Every triangle has three escenters.

Property 2. The bisectors of one interior angle and two exterior angles of a triangle intersect at one point, which is an escenter of the triangle.

Property 3. If I_A is the center of the escircle of ΔABC that is tangent to the side BC, and E and F are the projections of I_A onto the lines AB and AC, respectively, then

$$AE = AF = \frac{BC + CA + AB}{2}.$$

29.2 Illustrative Examples

Example 1. As shown in Figure 29.1, $ABCD$ is a parallelogram, G is the barycenter of ΔABD, and P and Q lie on the line BD, such that $GP \perp PC$ and $GQ \perp QC$. Show that AG is the angle bisector of $\angle PAQ$.

Proof. As in Figure 29.2, we form the line segment AC, and let M be the intersection point of AC and BD. Then M is the midpoint of both AC and BD, and hence G lies on AC.

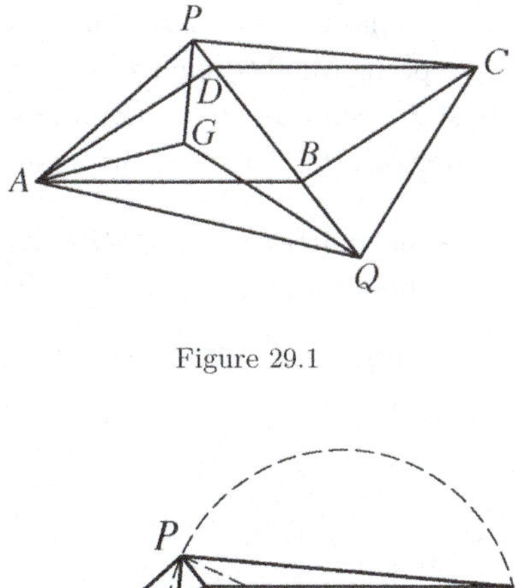

Figure 29.1

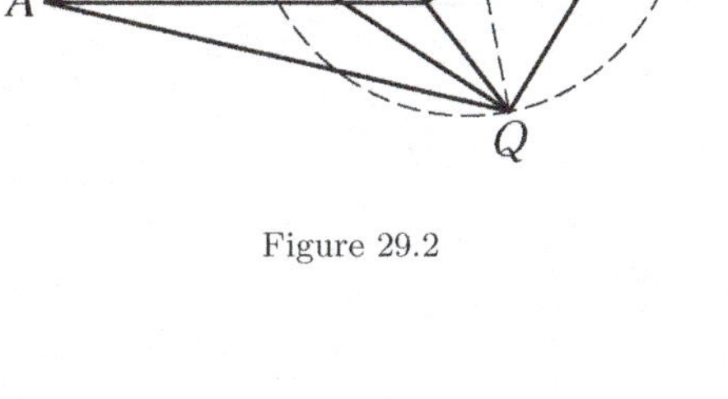

Figure 29.2

Since $\angle GPC = \angle GQC = 90°$, the points P, G, Q, and C are concyclic, and GC is the diameter of this circle. By the intersecting chords theorem,

$$PM \cdot MQ = GM \cdot MC. \text{ ①}$$

Let O be the midpoint of GC. Then since $AG : GM : MC = 2 : 1 : 3$,

$$OC = \frac{1}{2}GC = AG.$$

Hence, G and O are symmetric with respect to M, and

$$GM \cdot MC = AM \cdot MO. \text{ ②}$$

Combining ① and ②, we have $PM \cdot MQ = AM \cdot MO$, so A, P, O, and Q are concyclic. Since $OP = OQ = \frac{1}{2}GC$, we see that $\angle PAO = \angle QAO$, hence AG bisects $\angle PAQ$.

Remark. By $\frac{AG}{GM} = \frac{AC}{CM} = 2$, the circle with diameter GC is in fact an Apollonius circle, which passes through P and Q. This implies that $\frac{PA}{PM} = \frac{QA}{QM} = 2$, and hence AG bisects $\angle PAQ$.

Example 2. In an acute triangle ABC, suppose $AB \neq AC$ and AD is an altitude. Let H be a point on AD such that BH intersects AC at E and CH intersects AB at F. Suppose B, C, E, and F are concyclic. Is H necessarily the orthocenter of ABC? Prove your result.

Solution. The answer is affirmative. The proof is as follows.

As in Figure 29.3, we choose a point G on the ray AD such that

$$AH \cdot AG = AF \cdot AB = AE \cdot AC.$$

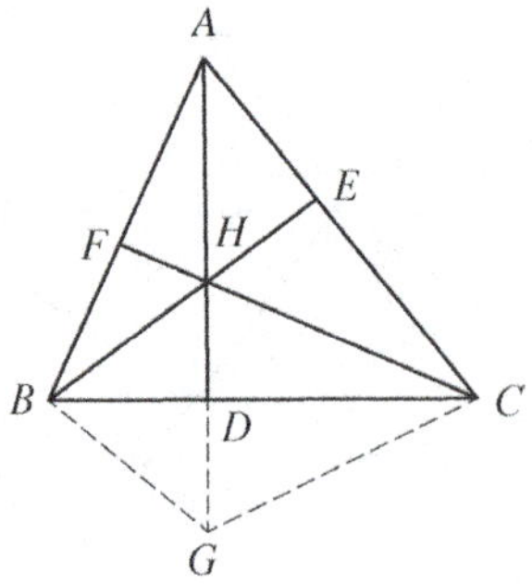

Figure 29.3

(1) If G and D do not coincide, then

$$\angle AFH = \angle AGB, \angle AEH = \angle AGC.$$

Since B, C, E, and F are concyclic, $\angle BFC = \angle CEB$ and $\angle AFH = \angle AEH$. It also follows that $\angle AGB = \angle AGC$, which implies that $BD = BC$, and hence $AB = AC$, a contradiction.

(2) If G and D coincide, then

$$\angle AFH = \angle ADB = 90^\circ, \angle AEH = \angle ADC = 90^\circ,$$

and H has to be the orthocenter of the triangle ABC.

Example 3. As in Figure 29.4, in a triangle ABC, assume that $AB = AC$, I is its incenter, and D is a point inside ΔABC, such that I, B, C, and D are concyclic. Let E be a point on the line AD such that CE is parallel to BD. Show that $CD^2 = BD \cdot CE$.

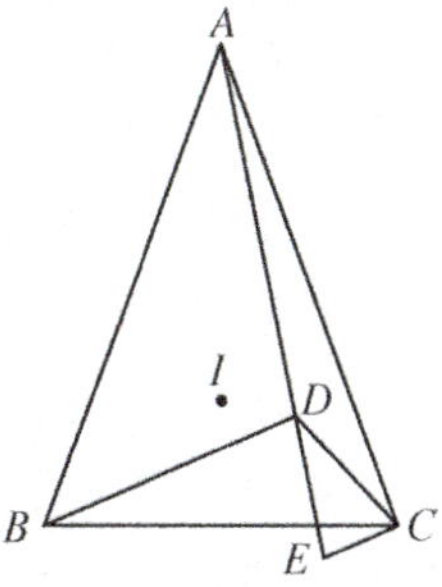

Figure 29.4

Proof. As shown in Figure 29.5, we form the line segments BI and CI. Suppose I, B, C, and D lie on a circle centered at O, and the extension of DE intersects the circle at F. Form the line segments FB and FC.

Since $BD \| CE$,

$$\angle DCE = 180^\circ - \angle BDC = \angle BFC.$$

Also, since $\angle CDE = \angle CDF = \angle CBF$, we get $\Delta BFC \backsim \Delta DCE$, so

$$\frac{DC}{CE} = \frac{BF}{FC}. \text{①}$$

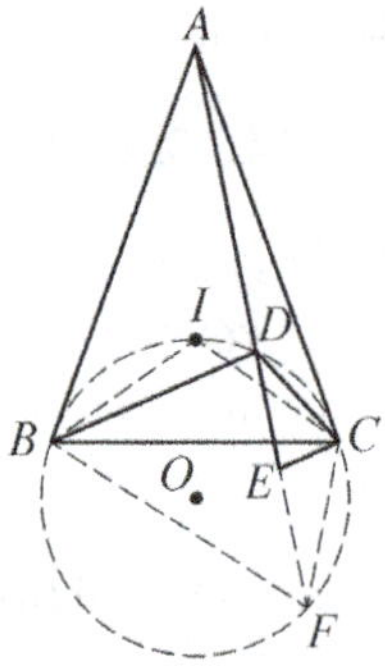

Figure 29.5

Next, we show that AB and AC are tangent to the circle. In fact, $\angle ABI = \frac{1}{2}\angle ABC = \frac{1}{2}\angle ACB = \angle ICB$, so AB is tangent to the circle centered at O, and AC is also tangent to the circle centered at O by the same token.

Hence, $\Delta ABD \backsim \Delta AFB$ and $\Delta ACD \backsim \Delta AFC$, thus

$$\frac{BD}{BF} = \frac{AB}{AF} = \frac{AC}{AF} = \frac{DC}{CF},$$

or equivalently

$$\frac{BF}{FC} = \frac{BD}{DC}. \quad ②$$

Combining ① and ②, we get $\frac{DC}{CE} = \frac{BD}{DC}$, namely $CD^2 = BD \cdot CE$.

Example 4. As shown in Figure 29.6, let ΔABC be an isosceles triangle with $AB = AC$ and I be its incenter. Let Γ_1 be the circle centered at A

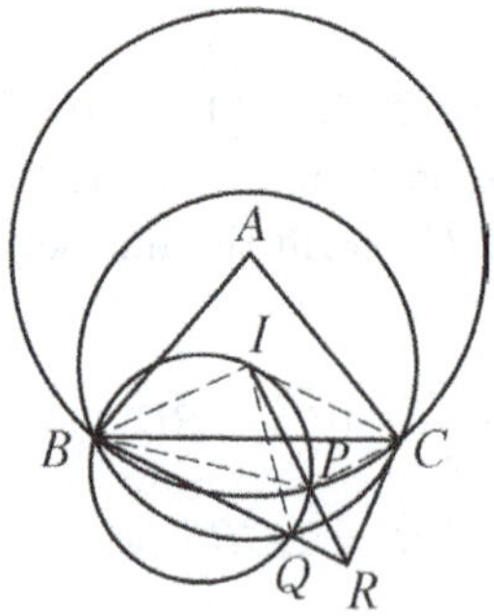

Figure 29.6

with radius AB and Γ_2 be the circle centered at I with radius IB. Suppose Γ_3 is a circle passing through I and B and intersects Γ_1 and Γ_2 at P and Q respectively (different from B), and R is the intersection point of IP and BQ. Prove that $BR \perp CR$.

Proof. We form the line segments IB, IC, IQ, PB, and PC.

Since Q lies on Γ_2, we have $IB = IQ$, so

$$\angle IBQ = \angle IQB.$$

Since B, I, P, and Q are concyclic, $\angle IQB = \angle IPB$, so $\angle IBQ = \angle IPB$ and $\Delta IBP \backsim \Delta IRB$, which shows that $\angle IRB = \angle IBP$ and

$$\frac{IB}{IR} = \frac{IP}{IB}.$$

Note that $AB = AC$ and I is the incenter. It follows that $IB = IC$, from which

$$\frac{IC}{IR} = \frac{IP}{IC}.$$

Then, $\Delta ICP \backsim \Delta IRC$, thus $\angle IRC = \angle ICP$.

Now, P lies on the arc BC on Γ_1 As a result $\angle BPC = 180° - \frac{1}{2}\angle A$, and hence

$$\begin{aligned}\angle BRC &= \angle IRB + \angle IRC = \angle IBP + \angle ICP \\ &= 360° - \angle BIC\text{-}\angle BPC \\ &= 360° - \left(90° + \frac{1}{2}\angle A\right) - \left(180° - \frac{1}{2}\angle A\right) \\ &= 90°.\end{aligned}$$

Therefore, $BR \perp CR$.

Example 5. In ΔABC, let D be the point of tangency of the incircle $\odot I$ and BC. The line DI intersects AC at X, and let XY be a tangent line to $\odot I$ (here XY does not coincide with AC) with Y lying on AB. Let Z be the intersection point of YI and BC. Show that $AB = BZ$.

Proof. Since $\odot I$ is the A-escenter of triangle AXY, we have $\angle XIY = 90° - \frac{1}{2}\angle BAC$, and

$$\angle IZB = 90° - \angle DIZ = 90° - \angle XIY = \frac{1}{2}\angle BAC.$$

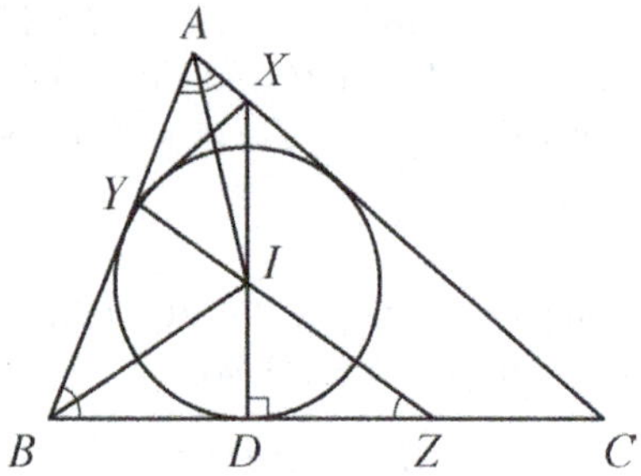

Figure 29.7

With $\angle IZB = \angle IAB$, $\angle IBZ = \angle IBA$, and $IB = IB$, we see that

$$\Delta ABI \cong \Delta ZBI.$$

Hence, $AB = BZ$.

Example 6. As shown in Figure 29.8, let I and O be the incenter and circumcenter of ΔABC, respectively. Prove that $\angle AIO \leq 90°$ if and only if $2BC \leq AB + AC$.

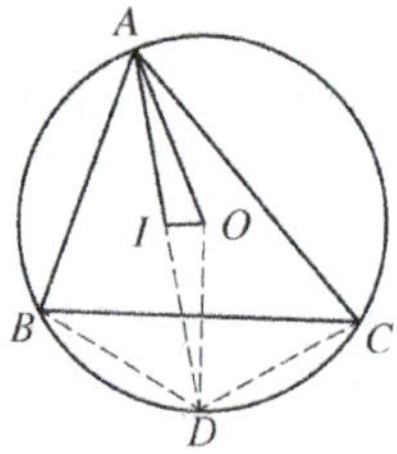

Figure 29.8

Proof. Let AI intersect the circumcircle again at D and draw the line segments BD, CD, and OD. Then

$$\angle AIO \leq 90° \Leftrightarrow AI \geq ID \Leftrightarrow 2 \leq \frac{AD}{DI}.$$

By Property 3 of the incenter, $DI = DB = DC$. By Ptolemy's theorem,

$$\begin{aligned} AD \cdot BC &= AB \cdot CD + AC \cdot BD \\ &= AB \cdot DI + AC \cdot DI. \end{aligned}$$

Hence, $\frac{AD}{DI} = \frac{AB+AC}{BC}$. Therefore, $\angle AIO \leq 90°$ if and only if $2 \leq \frac{AB+AC}{BC}$, in other words $2BC \leq AB + AC$.

Remark. The key point in this proof is to find the equivalence of $\angle AIO \leq 90°$ and $AI \geq ID$. Ptolemy's theorem says that for a quadrilateral that has a circumcircle, the sum of the products of the lengths of the two pairs of opposite sides equals the product of the lengths of the two diagonals.

Example 7. As shown in Figure 29.9, let O be the circumcenter of ΔABC. Let K be the intersection point of CO and the altitude above BC, and P and M are the midpoints of AK and AC, respectively. The lines PO and BC intersect at Y, and the circumcircle of ΔBCM intersects AB at $X(\neq B)$. Prove that B, O, X, and Y are concyclic.

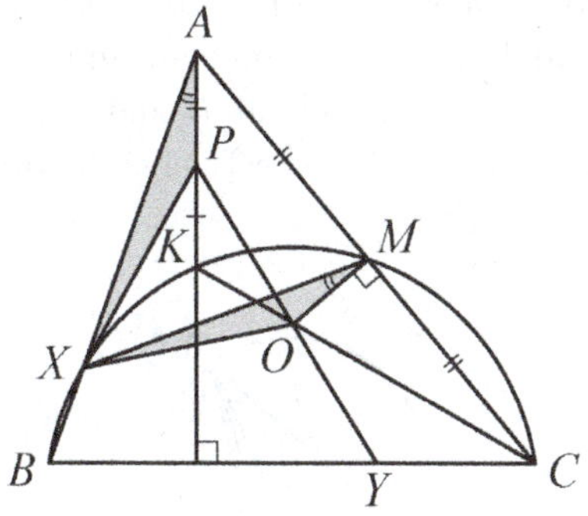

Figure 29.9

Proof. It suffices to show that $\angle XOP = \angle ABC$. First we prove that $\Delta XPA \backsim \Delta XOM$.

Since $\angle ABC = \angle XMA$,

$$\angle XMO = 90° - \angle XMA = 90° - \angle ABC = \angle XAK.$$

Now, we show that $\frac{AX}{XM} = \frac{AP}{OM}$. Since B, X, M, and C are concyclic, $\Delta AXM \backsim \Delta ACB$, from which

$$\frac{AX}{XM} = \frac{AC}{BC}.$$

Since O is the circumcenter of ΔABC,

$$\angle OCA = 90° - \angle ABC, \angle AKC = 180° - \angle BAC.$$

Applying the law of sines to ΔAKC and ΔABC, and using the definition of trigonometric functions for the right triangle OMC, we get

$$\frac{AC}{\sin\angle BAC}=\frac{AK}{\sin(90^\circ-\angle ABC)},\ OC=\frac{OM}{\sin(90^\circ-\angle ABC)}.$$

Dividing the two equalities on both sides we get $\frac{AK}{OM}=\frac{AC}{OC\cdot\sin\angle BAC}$, and since P is the midpoint of AK,

$$\frac{AP}{OM}=\frac{AC}{2OC\cdot\sin\angle BAC}=\frac{AC}{BC}.$$

Hence, $\Delta XPA\backsim\Delta XOM$, and also $\Delta XOP\backsim\Delta XMA$. Therefore,

$$\angle XOP=\angle XMA=\angle ABC.$$

Example 8. As shown in Figure 29.10, let $\odot O$ be the circumcircle of ΔABC and let AM and AT be the median and angle bisector from A, respectively. The tangent lines of $\odot O$ at B and C intersect at P, and AP intersects BC and $\odot O$ at D and E respectively. Prove that T is the incenter of the triangle AME.

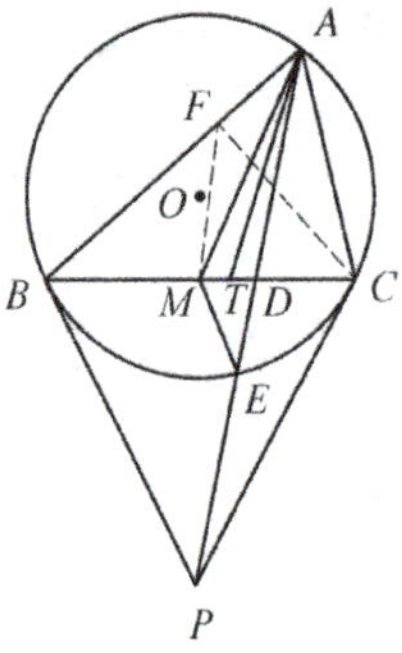

Figure 29.10

Proof. We first show that AT bisects $\angle MAE$, namely $\angle BAM=\angle CAP$.

Let F be a point on AB such that $CF\perp AB$, and form the line segment MF. Then $FM=\frac{1}{2}BC=MC$, and since $\angle BAC=\angle BCP$,

$$\frac{FA}{AC}=\cos\angle BAC=\cos\angle BCP=\frac{CM}{PC}=\frac{FM}{PC},$$

so $\frac{FA}{FM}=\frac{CA}{CP}$.

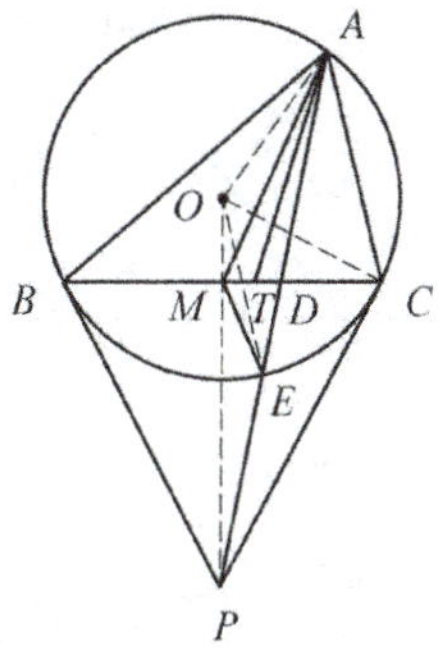

Figure 29.11

Also, since $\angle AFM = 180° - \angle BFM = 180° - \angle FBC = \angle ACP$, it follows that $\Delta AFM \backsim \Delta ACP$, and hence $\angle BAM = \angle CAP$.

Next, we show that MD bisects $\angle AME$.

Since M is the midpoint of BC, the points O, M, and P are collinear, and $OP \perp BC$. Draw the line segments OA, OC, and OE as in Figure 29.11. Then we obtain that

$$PE \cdot PA = PC^2 = PM \cdot PO.$$

Hence, M, O, A, and E are concyclic, and

$$\angle OMA = \angle OEA = \angle OAE = \angle PME.$$

Therefore, $\angle AMD = \angle EMD$, and T is the incenter of the triangle AME.

Example 9. In a triangle ABC, let X and Y be two points on the line BC such that X, B, C, and Y lie in that order on this line (Figure 29.12), and

$$BX \cdot AC = CY \cdot AB.$$

Let the circumcenters of ΔACX and ΔABY be O_1 and O_2, respectively, and the line O_1O_2 intersects AB and AC at U and V, respectively.

Prove that AUV is an isosceles triangle.

Proof 1. As in Figure 29.13, let the angle bisector of $\angle BAC$ intersect BC at P and let the circumcircles of the triangles ACX and ABY be ω_1 and ω_2, respectively. Then by the property of angle bisectors, $\frac{BP}{CP} = \frac{AB}{AC}$. Since

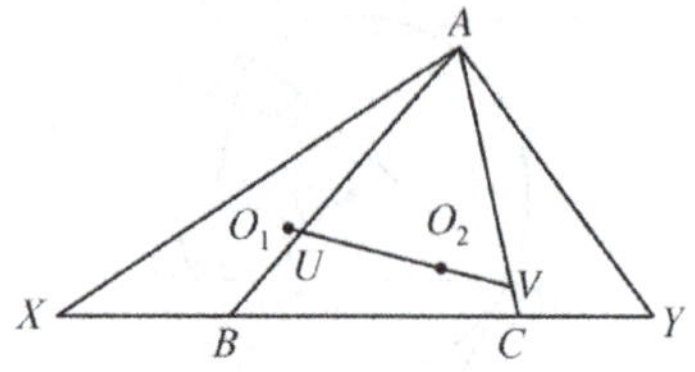

Figure 29.12

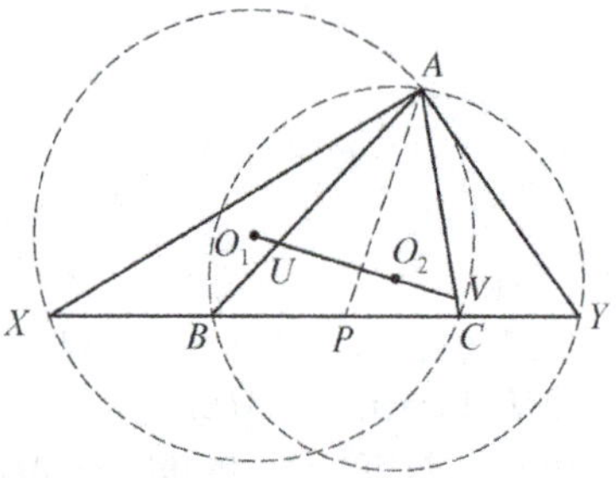

Figure 29.13

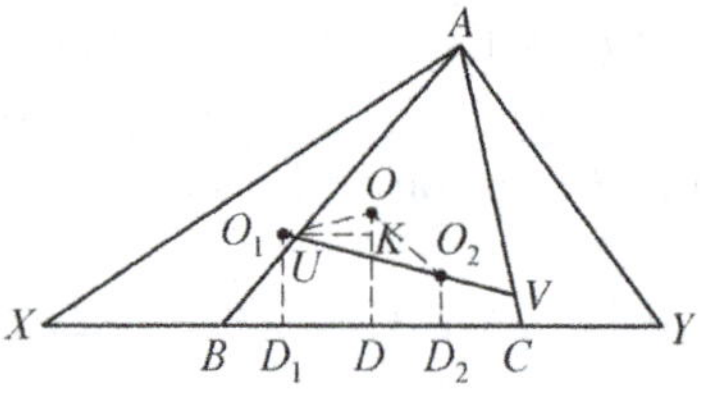

Figure 29.14

$\frac{BX}{CY} = \frac{AB}{AC}$, it follows that

$$\frac{PX}{PY} = \frac{BX + BP}{CY + CP} = \frac{AB}{AC} = \frac{BP}{CP},$$

and $CP \cdot PX = BP \cdot PY$.

Hence, the powers of P with respect to the circles ω_1 and ω_2 are the same, so P lies on their radical axis.

This implies that $AP \perp O_1O_2$, so U, V are symmetric with respect to AP, and thus AUV is an isosceles triangle.

Proof 2. As in Figure 29.14, let O be the circumcenter of ΔABC, and draw the line segments OO_1 and OO_2. We draw a perpendicular line from each of O, O_1, and O_2 to BC, and let D, D_1, and D_2 be the respective feet of perpendicular. Let K be a point on OD such that $O_1K \perp OD$.

We will show that $OO_1 = OO_2$. In the right triangle OKO_1,

$$OO_1 = \frac{O_1K}{\sin\angle O_1OK}.$$

By the property of the circumcenter, $OO_1 \perp AC$, and since $OD \perp BC$, we have $\angle O_1OK = \angle ACB$.

Since D and D_1 are the midpoints of BC and CX, respectively,

$$DD_1 = CD_1 - CD = \frac{1}{2}CX - \frac{1}{2}BC = \frac{1}{2}BX.$$

Hence,

$$OO_1 = \frac{O_1K}{\sin\angle O_1OK} = \frac{DD_1}{\sin\angle ACB} = \frac{\frac{1}{2}BX}{\frac{AB}{2R}} = R\cdot\frac{BX}{AB},$$

where R is the radius of the circumcircle of ABC. Similarly, $OO_2 = R\cdot\frac{CY}{AC}$.

Now, $\frac{BX}{AB} = \frac{CY}{AC}$ by the condition, so it follows that $OO_1 = OO_2$.

Since $OO_1 \perp AC$, we see that $\angle AVU = 90^\circ - \angle OO_1O_2$, and similarly $\angle AUV = 90^\circ - \angle OO_2O_1$. Combining them with $OO_1 = OO_2$, we get $\angle OO_1O_2 = \angle OO_2O_1$, hence $\angle AUV = \angle AVU$. Therefore, $AU = AV$, and ΔAUV is an isosceles triangle.

Example 10. As in Figure 29.15, let BE and CF be the angle bisectors of $\angle B$ and $\angle C$ respectively in a triangle ABC, with E lying on AC and

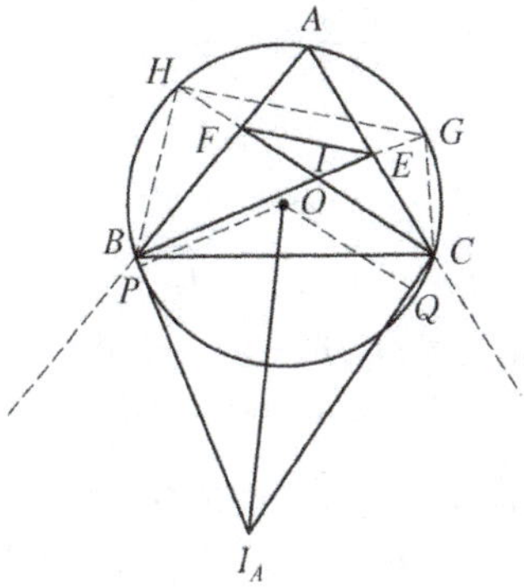

Figure 29.15

F lying on AB. Let O be the circumcenter of ΔABC and let I_A be the escenter of ΔABC inside $\angle A$. Show that $OI_A \perp EF$.

Proof. Let I be the incenter of ΔABC, and let BI and CI intersect the circumcircle of ΔABC again at points G and H, respectively. Let P and Q be the projections of O onto I_AB and I_AC, respectively. Since $OP \perp I_AB$ and $GB \perp I_AB$, we have $OP \| BG$, thus $OP = \frac{1}{2}BG$.

Similarly, $OQ \| CH$ and $OQ = \frac{1}{2}CH$. Hence,

$$\frac{OP}{OQ} = \frac{BG}{CH}. \quad ①$$

On the other hand, from $\Delta ICE \backsim \Delta BGH$, we see that

$$\frac{BG}{BH} = \frac{IC}{IE}. \quad ②$$

Since $\Delta IBF \backsim \Delta CHG$, we obtain

$$\frac{CG}{CH} = \frac{IF}{IB}. \quad ③$$

From ② and ③ we get

$$\frac{BG}{CH} \cdot \frac{CG}{BH} = \frac{IF}{IE} \cdot \frac{IC}{IB}. \quad ④$$

Since $\Delta IHB \backsim \Delta IGC$, we have

$$\frac{CG}{BH} = \frac{IC}{IB}. \quad ⑤$$

From ④ and ⑤ we get

$$\frac{BG}{CH} = \frac{IF}{IE}. \quad ⑥$$

From ① and ⑥ we get

$$\frac{OP}{OQ} = \frac{IF}{IE}.$$

Now, since $OP \| IE$ and $OQ \| IF$, we have $\Delta OPQ \backsim \Delta IFE$, so $\angle OPQ = \angle IFE$. Hence, $\angle OI_AQ = \angle IFE$ and $OI_A \perp EF$.

Example 11. As in Figure 29.16, $ABCD$ is a convex quadrilateral, and the lines AD and BC intersect at P. Let I_1 and I_2 be the incenters of ΔPAB and ΔPDC, respectively. Let O be the circumcenter of ΔPAB and H be the orthocenter of ΔPDC. Prove that the circumcircles of ΔAI_1B

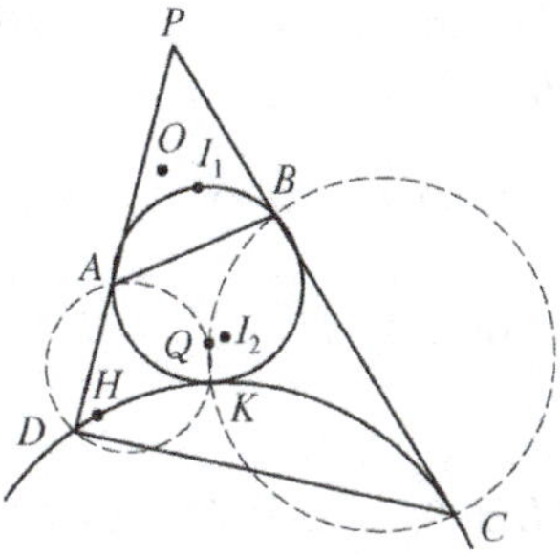

Figure 29.16

and ΔDHC are tangent to each other if and only if the circumcircles of ΔAOB and ΔDI_2C are tangent to each other.

Proof. Suppose the circumcircles of ΔAI_1B and ΔDHC are tangent at K, and let Q be the second intersection point of the circumcircles of ΔAKD and ΔBKC. Then

$$\angle DHC = \angle DKC = 180^\circ - \angle P,$$

$$\angle PDK + \angle PCK = \angle DKC - \angle P = 180^\circ - 2\angle P.$$

Since A, Q, K, and D are concyclic, $\angle AQK = 180^\circ - \angle PDK$, and since B, Q, K, and C are concyclic, $\angle BQK = 180^\circ - \angle PCK$. Hence,

$$\begin{aligned}\angle AQB &= 360^\circ - \angle AQK - \angle BQK \\ &= \angle PDK + \angle PCK \\ &= 180^\circ - 2\angle P = 180^\circ - \angle AOB.\end{aligned}$$

This implies that A, O, B, and Q are concyclic.

Now, $\angle AKD = \angle AQD, \angle BKC = \angle BQC$, and $\angle AQB = \angle DKC - \angle P$, so

$$\begin{aligned}\angle CQD &= 360^\circ - \angle AQB - \angle AQD - \angle BQC \\ &= 360^\circ - (\angle DKC - \angle P) - \angle AKD - \angle BKC \\ &= \angle AKB + \angle P = 180^\circ - \angle AI_1B + \angle P \\ &= 180^\circ - \left(90^\circ + \frac{\angle P}{2}\right) = 90^\circ + \frac{\angle P}{2},\end{aligned}$$

and since $\angle CI_2D = 90^\circ + \frac{\angle P}{2}$, we see that C, D, Q, and I_2 are concyclic.

Next, we prove that the circumcircles of ΔAOB and ΔDI_2C are tangent at Q, and it suffices to show that

$$\angle ABQ + \angle DCQ = \angle AQD.$$

Since the circumcircles of ΔAI_1B and ΔDHC are tangent at K, $\angle ABK + \angle DCK = \angle AKD$, and thus

$$(\angle ABQ + \angle KBQ) + (\angle DCQ - \angle KCQ) = \angle AKD.$$

Since $\angle KBQ = \angle KCQ$ and $\angle AKD = \angle AQD$, we get $\angle ABQ + \angle DCQ = \angle AQD$. Hence, the circumcircles of ΔAOB and ΔDI_2C are tangent at Q.

For the other side of the problem, we may assume that the circumcircles of ΔCI_2D and ΔAOB are tangent at Q, and let K be the second intersection point of the circumcircles of ΔAQD and ΔBQC. By similar arguments, we can prove that the circumcircles of ΔAI_1B and ΔDHC are tangent at K.

Example 12. As in Figure 29.17, let P and Q be two points on the side BC of ΔABC such that they have an equal distance to the midpoint of BC. Let E and F be points on AC and AB, respectively, such that $EP \perp BC$ and $FQ \perp BC$. Let M be the intersection point of PF and EQ and let H_1 and H_2 be the orthocenters of ΔBFP and ΔCEQ, respectively. Show that $AM \perp H_1H_2$.

Proof. First we show that the line AM is independent of the choice of P and Q. In fact it suffices to show that $\frac{\sin \angle MAB}{\sin \angle MAC}$ is a constant. Applying the law of sines to ΔAFM and ΔAEM, we obtain that

$$\frac{\sin \angle MAB}{\sin \angle MAC} = \frac{\sin \angle AFM}{\sin \angle AEM} \cdot \frac{FM}{EM}. \quad ①$$

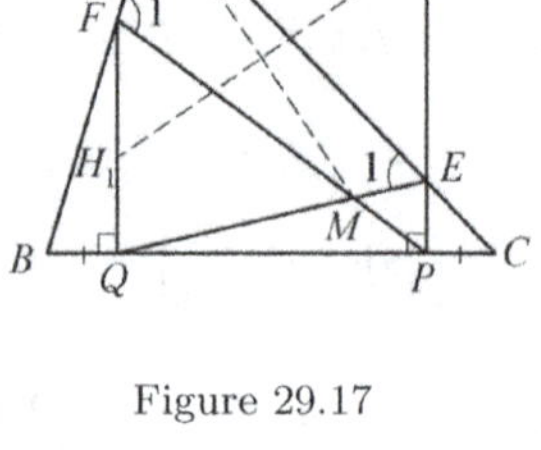

Figure 29.17

In the triangles ΔFBP and ΔCEQ,

$$\sin \angle AFM = \frac{BP}{PF} \sin \angle B, \sin \angle AEM = \frac{CO}{EQ} \cdot \sin \angle C.$$

Since $BP = CQ$,

$$\frac{\sin \angle AFM}{\sin \angle AEM} = \frac{\sin \angle B}{\sin \angle C} \cdot \frac{EQ}{FP}. \text{②}$$

By ① and ②,

$$\frac{\sin \angle MAB}{\sin \angle MAC} = \frac{\sin \angle B}{\sin \angle C} \cdot \frac{EQ}{FP} \cdot \frac{FM}{EM}. \text{③}$$

Since $\Delta FMQ \backsim \Delta EMP$,

$$\frac{FM}{FP} = \frac{FQ}{FQ + EP}, \frac{EQ}{EM} = \frac{FQ + EP}{EP}.$$

Combining the above with ③, we get

$$\frac{\sin \angle MAB}{\sin \angle MAC} = \frac{\sin \angle B}{\sin \angle C} \cdot \frac{FQ}{EQ}. \text{④}$$

On the other hand, since

$$\tan \angle B = \frac{FQ}{BQ}, \tan \angle C = \frac{EP}{CP}, BQ = CP,$$

by ④,

$$\frac{\sin \angle MAB}{\sin \angle MAC} = \frac{\sin \angle B}{\sin \angle C} \cdot \frac{\tan \angle B}{\tan \angle C},$$

which is a constant.

Let α be the angle between H_1H_2 and BC. Then,

$$\tan \alpha = \frac{H_2P - H_1Q}{QP}. \text{⑤}$$

Since H_1 and H_2 are the orthocenters of ΔBFP and ΔCEQ, respectively,

$$QF \cdot H_1Q = BQ \cdot QP, \quad EP \cdot H_2P = CP \cdot PQ,$$

or equivalently $H_1Q = \frac{BQ \cdot QP}{FQ}$ and $H_2P = \frac{CP \cdot PQ}{EP}$. Also, note that $CP = BQ$. Thus

$$H_2P - H_1Q = \frac{PQ \cdot BQ \cdot (FQ - EP)}{EP \cdot FQ}.$$

Combining them with ⑤, we get

$$\tan\alpha = \frac{BQ \cdot (FQ - EP)}{EP \cdot FQ} = \frac{BQ}{EP} - \frac{BQ}{FQ} = \frac{CP}{EP} - \frac{BQ}{FQ} = \cot C - \cot B.$$

Let θ be the angle between AM and BC. We will show that

$$\tan\alpha \cdot \tan\theta = 1. \text{ ⑥}$$

Let X be the intersection point of AM and BC. Then

$$\frac{BX}{CX} = \frac{\sin\angle MAB}{\sin\angle MAC} \cdot \frac{\sin\angle C}{\sin\angle B} = \frac{\tan\angle B}{\tan\angle C}.$$

Let AD be an altitude in ΔABC. Then

$$\frac{BX}{CX} = \frac{\tan\angle B}{\tan\angle C} = \frac{\frac{AD}{BD}}{\frac{AD}{CD}} = \frac{CD}{BD},$$

so $BD = CX$. Hence,

$$\tan\theta = \frac{AD}{DX} = \frac{AD}{CD - CX} = \frac{AD}{CD - BD}$$
$$= \frac{1}{\frac{CD}{AD} - \frac{BD}{AD}} = \frac{1}{\cot\angle C - \cot\angle B}.$$

Therefore, ⑥ holds, and $AM \perp H_1H_2$.

29.3 Exercises

Group A

1. In a right triangle ABC, assume that $\angle A = 90°$ and $\angle C = 30°$. Let Ω be a circle that passes through A and is tangent to BC at the midpoint of BC. Let Ω intersect AC and the circumcircle of ΔABC at N and M, respectively (different from A). Prove that $MN \perp BC$.
2. In a triangle ABC, we have $\angle C = \angle A + 90°$. Let D be a point on the extension of BC such that $AC = AD$. A point E lies on the other side of BC (different from A) such that $\angle EBC = \angle A$ and $\angle EDC = \frac{1}{2}\angle A$. Prove that $\angle CED = \angle ABC$.

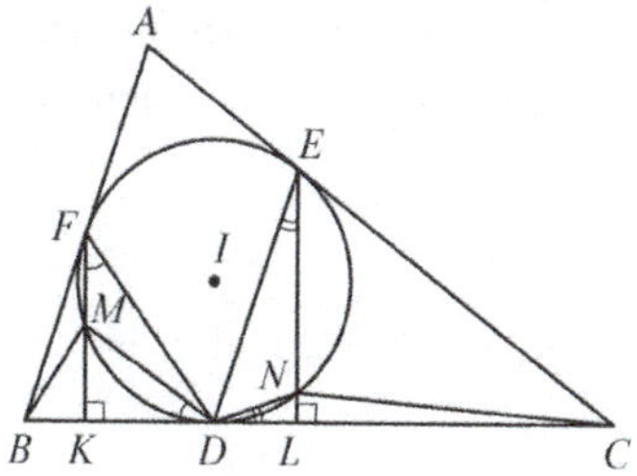

Exercise 3

3. In a triangle ABC, let D, E, and F be the points of tangency of the incircle on the sides AB, AC, and BC, respectively. Let K and L be the projections of F and E onto BC, respectively. Suppose FK and EL intersect the incircle again at M and N, respectively. Show that $\frac{S_{\Delta BMD}}{S_{\Delta CND}} = \frac{DK}{DL}$.
4. Let K and L be points on arcs AB (not including C) and BC (not including A) of the circumcircle of a triangle ABC, respectively, such that KL is parallel to AC. Show that the incenters of ΔABK and ΔCBL have an equal distance to the midpoint of the arc AC (including B).
5. Let ABC be an equilateral triangle with the circumcircle ω and the center O. Let P be a point on the arc BC (not including A), and suppose the tangent line to ω at P intersects the lines AB and AC at K and L, respectively. Show that $\angle KOL > 90°$.

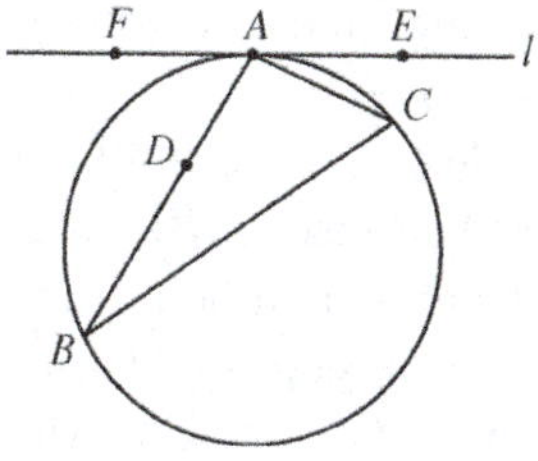

Exercise 6

6. As shown in the above figure, $AB > AC$ in a triangle ABC and l is the tangent line to the circumcircle of ΔABC at A. Suppose the circle centered at A with radius AC intersects AB at D and intersects l at E and F. Prove that one of the lines DE and DF passes through the incenter of ΔABC.
7. Let ΔABC be an acute isosceles triangle with $AB = AC$ and let O be its circumcenter. The ray BO intersects AC at B' and the ray CO

intersects AB at C'. Let l be a line passing through C' and parallel to AC. Show that l is tangent to the circumcircle of $\Delta B'OC$.

Group B

8. Let I be the incenter of ΔABC and let P be a point inside ΔABC such that

$$\angle PBA + \angle PCA = \angle PBC + \angle PCB.$$

Show that $AP \geq AI$ and the equality holds if and only if $P = I$.

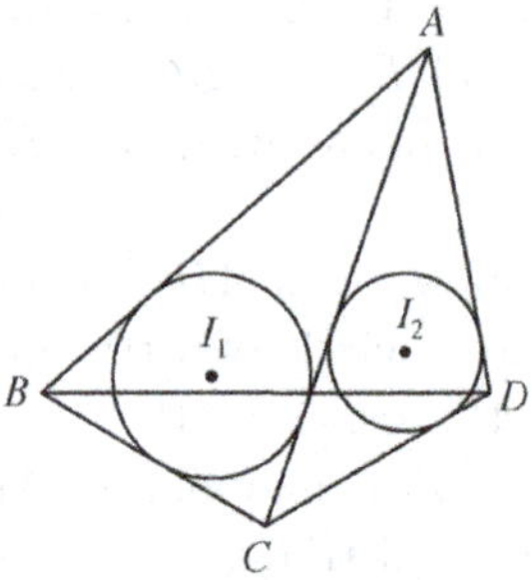

Exercise 9

9. As shown in the above figure, a quadrilateral $ABCD$ has a circumcircle and $\angle BAC = \angle DAC$. Let $\odot I_1$ and $\odot I_2$ be the incircles of ΔABC and ΔADC, respectively. Show that one external common tangent line of $\odot I_1$ and $\odot I_2$ is parallel to BD.
10. Let Ω be the circumcircle of a triangle ABC with its center O. Suppose a circle Γ centered at A intersects BC at D and E, such that B, D, E, and C are different points and lie in that order on the line BC. Let F and G be the intersection points of Γ and Ω, such that A, F, B, C, and G lie in that order on Ω. Let K be the second intersection point of the circumcircle of the triangle BDF and the line segment AB, and let L be the second intersection point of the circumcircle of the triangle CGE and the line segment CA.

    If the lines FK and GL do not coincide, and intersect at X, show that X lies on AO.
11. Let $ABCD$ be a convex quadrilateral and let I_1 and I_2 be the incenters of ΔABC and ΔDBC, respectively. The line I_1I_2 intersects AB and CD at E and F, respectively. Let the lines AB and DC intersect at P. Suppose $PE = PF$. Show that A, B, C, and D are concyclic.

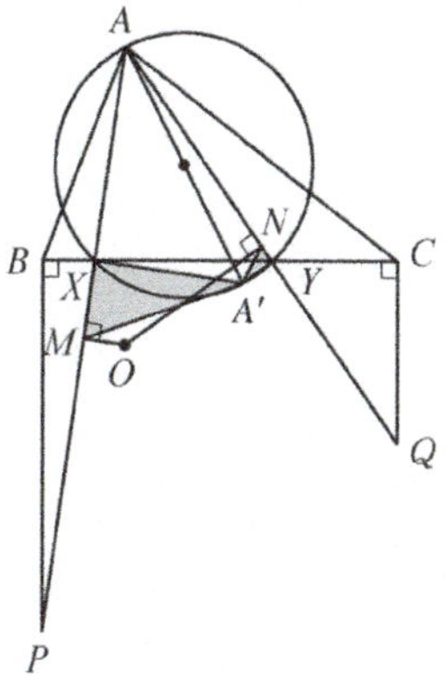

Exercise 12

12. In a triangle ABC, let X and Y be two points on BC (here X lies between B and Y) such that $2XY = BC$. Let AA' be a diameter of the circumcircle of ΔAXY. The points P and Q lie on the lines AZ and AY, respectively, such that $BP \perp BC$ and $CQ \perp BC$. Prove that the tangent line to the circumcircle of ΔAXY at A' passes through the circumcenter of ΔAPQ.
13. Let O and I be the circumcenter and incenter of a scalene triangle ABC, respectively. Let B' be the reflection of B with respect to the line OI, and B' lies inside $\angle ABI$. Show that the tangent lines to the circumcircle of $\Delta BIB'$ at points B' and I, and the line AC are concurrent.
14. Suppose $ABCD$ is a convex quadrilateral, and I_A, I_B, I_C, and I_D be the incenters of ΔDAB, ΔABC, ΔBCD, and ΔCDA, respectively. If $\angle BI_AA + \angle I_CI_AI_D = 180°$, show that $\angle BI_BA + \angle I_CI_BI_D = 180°$.

Chapter 30

Some Famous Theorems in Plane Geometry

30.1 Key Points of Knowledge and Basic Methods

A. Ceva's theorem and its inverse theorem

Ceva's theorem. As shown in Figure 30.1, let P be an arbitrary point in the plane that does not lie on the lines containing the sides of ΔABC. The lines AP, BP, and CP intersect the lines containing their opposite sides at D, E, and F, respectively. Then

$$\frac{BD}{DC} \cdot \frac{CE}{EA} \cdot \frac{AF}{FB} = 1.$$

The inverse theorem of Ceva's theorem. Let D, E, and F be points on lines BC, CA, and AB, respectively (here A, B, and C are the vertices of a triangle), such that either the three points all lie inside the line segments, or two of them lie outside the line segments while the third lies inside the line segment. If

$$\frac{BD}{DC} \cdot \frac{CE}{EA} \cdot \frac{AF}{FB} = 1,$$

then the lines AD, BE, and CF are either concurrent or pairwise parallel.

The angular form of Ceva's theorem. In the same setting of Ceva's theorem,

$$\frac{\sin \angle ABE}{\sin \angle CBE} \cdot \frac{\sin \angle BCF}{\sin \angle ACF} \cdot \frac{\sin \angle CAD}{\sin \angle BAD} = 1.$$

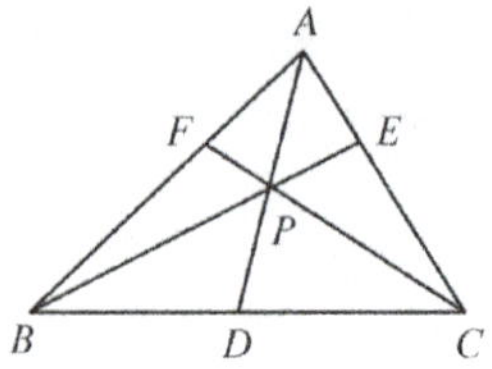

Figure 30.1

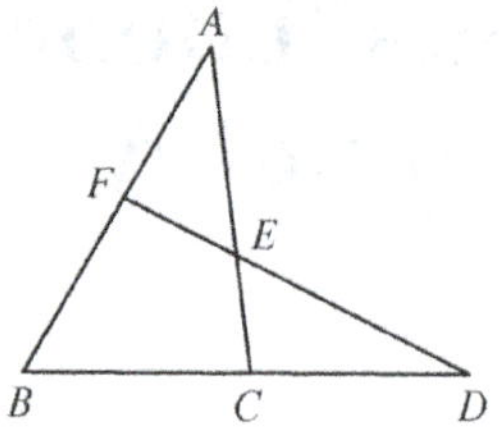

Figure 30.2

The inverse theorem of the angular form of Ceva's theorem. In the same setting of the inverse theorem of Ceva's theorem, if

$$\frac{\sin\angle ABE}{\sin\angle CBE}\cdot\frac{\sin\angle BCF}{\sin\angle ACF}\cdot\frac{\sin\angle CAD}{\sin\angle BAD}=1,$$

then the lines AD, BE, and CF are either concurrent or pairwise parallel.

B. The Menelaus theorem and its inverse theorem

The Menelaus theorem. As in Figure 30.2, for a triangle ABC, if a line DEF intersects the lines BC, CA, and AB at D, E, and F, respectively, then

$$\frac{BD}{DC}\cdot\frac{CE}{EA}\cdot\frac{AF}{FB}=1.$$

The inverse theorem of the Menelaus theorem. For a triangle ABC, if D, E, and F lie on the lines BC, CA, and AB, such that either none of them lie inside the line segment, or two of them lie on the line segments and the third lies outside the line segment, and satisfies

$$\frac{BD}{DC}\cdot\frac{CE}{EA}\cdot\frac{AF}{FB}=1,$$

then D, E, and F are collinear.

The angular form of the Menelaus theorem. Let D, E, and F be points on lines BC, CA, and AB, respectively (here A, B, and C are the vertices of a triangle). Then D, E, and F are collinear if and only if

$$\frac{\sin\angle BAD}{\sin\angle DAC}\cdot\frac{\sin\angle CBE}{\sin\angle EBA}\cdot\frac{\sin\angle ACF}{\sin\angle FCB}=1.$$

C. Ptolemy's theorem (and inequality)

Ptolemy's theorem. If a quadrilateral $ABCD$ has a circumcircle, then

$$AB\cdot CD+AD\cdot BC=AC\cdot BD.$$

In other words, the sum of the products of the lengths of the two pairs of opposite sides equals the product of the lengths of the diagonals.

The inverse theorem of Ptolemy's theorem. In a convex quadrilateral $ABCD$, if

$$AB\cdot CD+AD\cdot BC=AC\cdot BD,$$

then A, B, C, and D are concyclic.

Ptolemy's inequality. In a quadrilateral $ABCD$, the following inequality holds:

$$AB\cdot CD+AD\cdot BC\geq AC\cdot BD,$$

where the equality is satisfied if and only if $ABCD$ has a circumcircle.

D. Simson's theorem and its inverse theorem

Simson's theorem. As in Figure 30.3, P is a point on the circumcircle of ΔABC (different from A, B, and C). Let D, E, and F be the projections of P onto AB, BC, and AC, respectively. Then D, E, and F are collinear. The line DEF is called the Simson line of ΔABC with respect to P.

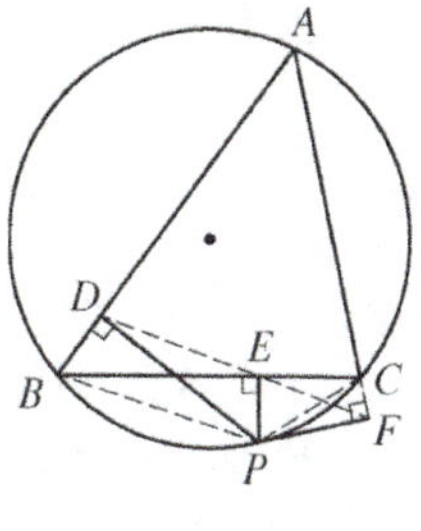

Figure 30.3

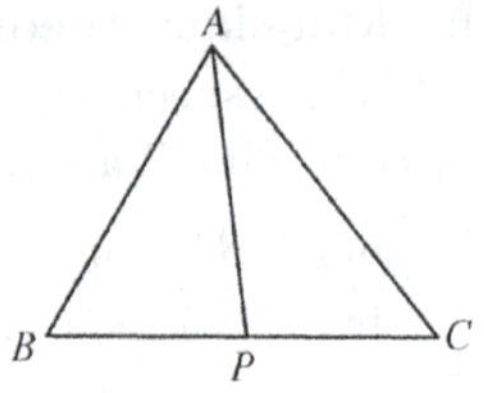

Figure 30.4

The inverse theorem of Simson's theorem. For a point P in the plane of ΔABC, if the projections of P onto the lines AB, BC, and CA are collinear, then P lies on the circumcircle of ΔABC.

E. Stewart's theorem

Stewart's theorem As in Figure 30.4, P is a point on the side BC of a triangle ABC. Then

$$BP \cdot AC^2 + CP \cdot AB^2 - BC \cdot PA^2 = BP \cdot CP \cdot BC.$$

If $\frac{BP}{PC} = \frac{m}{n}$, then this equality can be written alternatively as

$$mAC^2 + nAB^2 = (m+n)AP^2 + \frac{mn}{m+n}BC^2.$$

F. Monge's theorem

Monge's theorem. For three circles in the plane, the radical axes of each pair of them (if they exist) are either concurrent or pairwise parallel.

Monge's theorem is also called **the radical center theorem.**

Corollary to Monge's theorem. If any two of three circles (their centers are not collinear) intersect at two points, then the lines of the three common chords intersect at one point.

30.2 Illustrative Examples

Example 1. As in Figure 30.5, a line PA is tangent to $\odot O$ at A, and a line PBC intersects $\odot O$ at B and C. Let M be the midpoint of PA, and a chord AD and BC intersect at E. A point F lies on the extension of AB such that $\angle FBD = \angle FED$. Prove that P, F, and D are collinear if and only if M, B, and D are collinear.

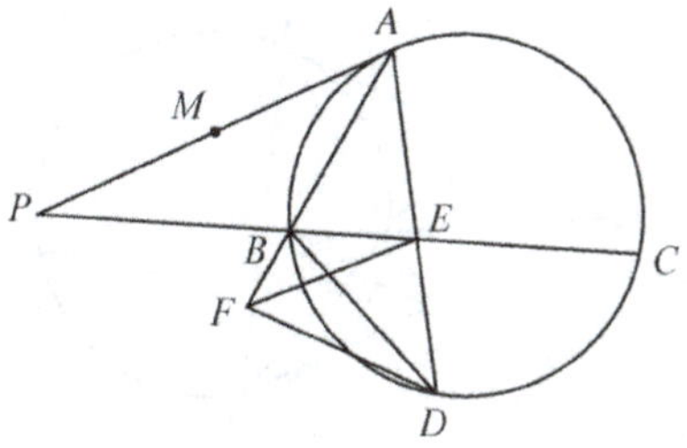

Figure 30.5

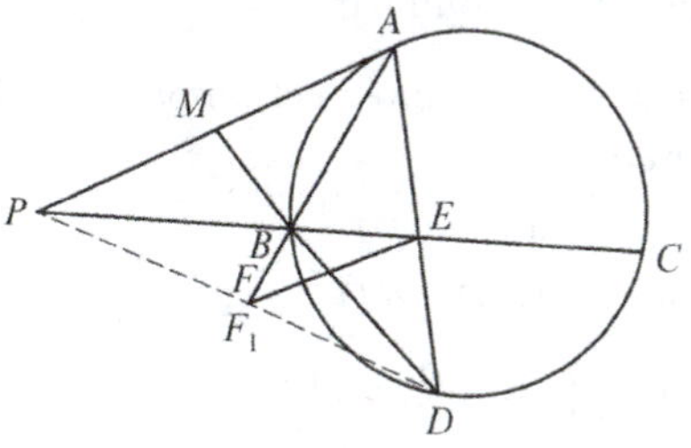

Figure 30.6

Proof. Since PA is tangent to $\odot O$, we have $\angle PAD + \angle ABD = 180°$. Since

$$\angle FBD + \angle ABD = 180°,$$

it follows that $\angle PAD = \angle FBD = \angle FED$. Hence, $EF \parallel AP$.

(1) Suppose M, B, and D are collinear.

As shown in Figure 30.6, let F_1 be the intersection point of AB and DP. By Ceva's theorem,

$$\frac{AM}{MP} \cdot \frac{PF_1}{F_1D} \cdot \frac{DE}{EA} = 1.$$

Since $AM = MP$,

$$\frac{PF_1}{F_1D} = \frac{AE}{ED}, EF_1 \parallel AP.$$

Since F and F_1 both lie on the line AB, they are necessarily the same point. Therefore, P, F, and D are collinear.

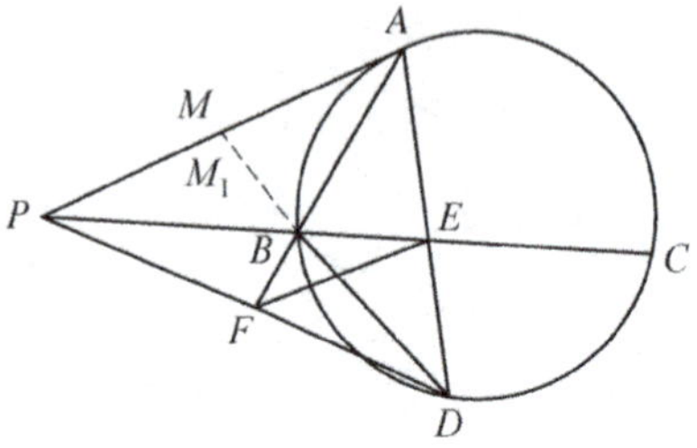

Figure 30.7

(2) Suppose P, F, and D are collinear.

As in Figure 30.7, let M_1 be the intersection point of the lines DB and AP. Then by Ceva's theorem,

$$\frac{AM_1}{M_1P} \cdot \frac{PF}{FD} \cdot \frac{DE}{EA} = 1.$$

Hence, $EF \| AP$ and $\frac{PF}{FD} = \frac{AE}{ED}$, and

$$\frac{AM_1}{M_1P} = 1, AM_1 = M_1P.$$

Therefore, M_1 is the midpoint of PA, which means that $M = M_1$. This shows that M, B, and D are collinear.

From (1) and (2) we conclude that P, F, and D are collinear if and only if M, B, and D are collinear.

Example 2. As in Figure 30.8, in an obtuse triangle ABC, suppose $AB > AC$. Let O be its circumcenter, and D, E, and F be the midpoints of BC, CA, and AB, respectively. Suppose AD intersects the lines OF and OE at M and N, respectively, and the lines BM and CN intersect at P. Show that $OP \perp AP$.

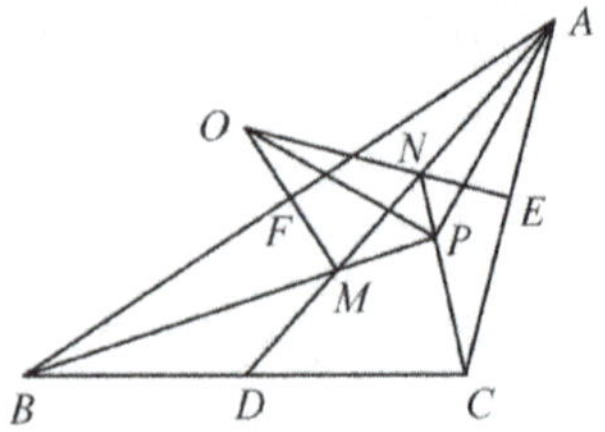

Figure 30.8

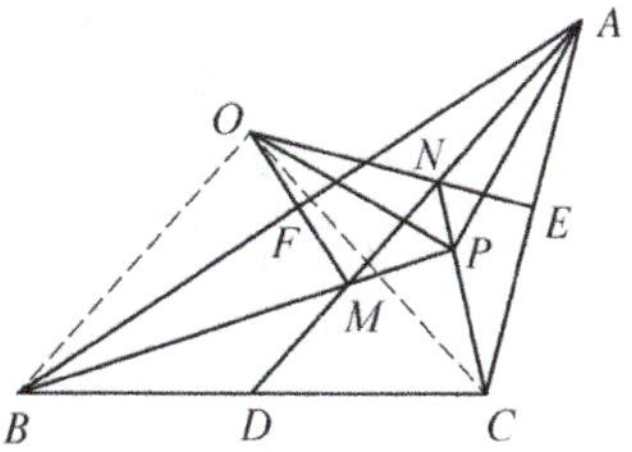

Figure 30.9

Proof. It follows from the conditions that $BM = AM$, $CN = AN$, $\angle AMP = 2\angle BAM$, and $\angle PND = 2\angle CAM$. We form the line segments OB and OC (Figure 30.9). Since O is the circumcenter of ΔABC,

$$\angle BOC = 2\angle BAC = 2\angle BAM + 2\angle CAM$$

$$= \angle AMP + \angle PND = \angle BPC.$$

Hence, B, O, P, and C are concyclic, and

$$\angle BPO = \angle BCO = \angle CBO = \angle OPN.$$

On the other hand, we apply the Menelaus theorem to ΔBCP and the line DMN, so

$$\frac{BD}{DC} \cdot \frac{CN}{NP} \cdot \frac{PM}{MB} = 1.$$

Combining it with $BD = DC$, $BM = AM$, and $CN = AN$ we obtain

$$\frac{PM}{PN} = \frac{MB}{NC} = \frac{AM}{AN},$$

Hence PA is the exterior angle bisector of $\angle MPN$. Since PO is the angle bisector of $\angle MPN$, it follows that $OP \perp AP$.

Example 3 (Pappus' theorem). As shown in Figure 30.10, let A, C, and E be points on one line, and let B, D, and F be points on another line. Let the lines AB and DE intersect at L, the lines CD and FA intersect at M, and the lines EF and BC intersect at N. Prove that L, M, and N are collinear.

Proof. Suppose the lines AB, CD, and EF encompass a triangle UVW (here the lines AB and EF intersect at V, the lines AB and CD intersect at V, and the lines CD and EF intersect at W. Also if two of the three lines are parallel, then let their intersection point be considered at infinity).

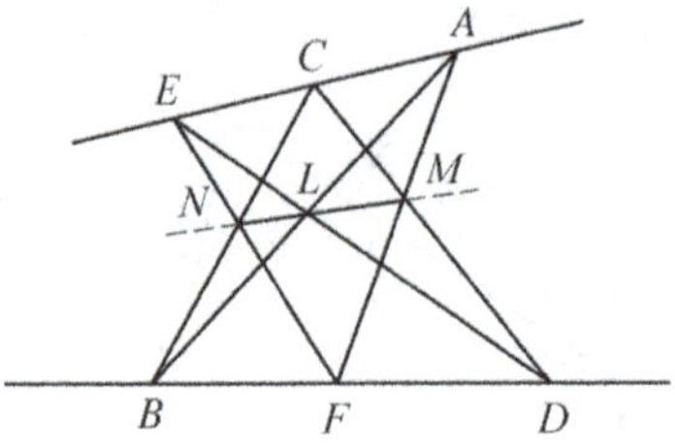

Figure 30.10

Then we apply the Menelaus theorem to ΔUVW and the following five lines: ELD, AMF, BCN, ACE, and BDF. Thus,

$$\frac{VL}{LW}\cdot\frac{WD}{DU}\cdot\frac{UE}{EV}=1,\qquad \frac{VA}{AW}\cdot\frac{WM}{MU}\cdot\frac{UF}{FV}=1,$$
$$\frac{VB}{BW}\cdot\frac{WC}{CU}\cdot\frac{UN}{NV}=1,\qquad \frac{VA}{AW}\cdot\frac{WC}{CU}\cdot\frac{UE}{EV}=1,$$
$$\frac{VB}{BW}\cdot\frac{WD}{DU}\cdot\frac{UF}{FV}=1.$$

Dividing the product of the last two equalities by the product of the first three equalities, we get

$$\frac{NV}{UN}\cdot\frac{MU}{WM}\cdot\frac{LW}{VL}=1.$$

Therefore, by the inverse theorem of the Menelaus theorem, we conclude that L, M, and N are collinear.

Example 4 (Pascal's theorem). As shown in Figure 30.11, let $ABCDEF$ be a hexagon inscribed to a circle. Suppose the lines AB and DE intersect

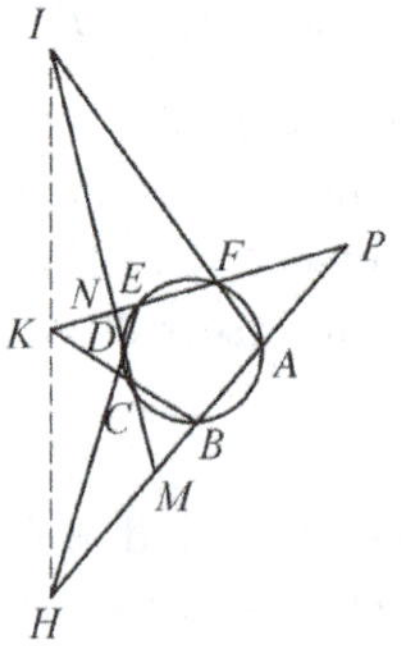

Figure 30.11

at H, the lines BC and EF intersect at K, and the lines CD and FA intersect at I. Show that H, K, and I are collinear.

Proof. Suppose the lines AB, CD, and EF enclose a triangle MNP, and apply the Menelaus theorem to the triangle MNP and the line AFI. Then

$$\frac{MA}{AP} \cdot \frac{PF}{FN} \cdot \frac{NI}{IM} = 1, \text{①}$$

and similarly

$$\frac{MB}{BP} \cdot \frac{PK}{KN} \cdot \frac{NC}{CM} = 1, \text{②}$$

$$\frac{MH}{HP} \cdot \frac{PE}{EN} \cdot \frac{ND}{DM} = 1. \text{③}$$

On the other hand, by the circle-power theorem,

$$MA \cdot MB = MC \cdot MD, \text{④}$$

$$ND \cdot NC = NE \cdot NF, \text{⑤}$$

$$PA \cdot PB = PF \cdot PE. \text{⑥}$$

Taking the product of ①, ②, and ③, and combining the result with ④, ⑤, and ⑥, we have

$$\frac{NI}{IM} \cdot \frac{MH}{HP} \cdot \frac{PK}{KN} = 1.$$

Hence, by the inverse theorem of the Menelaus theorem, we conclude that H, K, and I are collinear.

Example 5. Let a, b, and c be the three sides of ΔABC, and a', b', and c' be the three sides of $\Delta A'B'C'$. Suppose $\angle B = \angle B'$ and $\angle A + \angle A' = 180°$. Show that $aa' = bb' + cc'$.

Analysis. Note that the equality to be proved is similar to the conclusion of Ptolemy's theorem. Combining $\angle B = \angle B'$ and $\angle A + \angle A' = 180°$, we may construct a circle and apply Ptolemy's theorem.

Proof. As in Figure 30.12, we draw the circumcircle of ΔABC and a line CD parallel to AB that intersects the circumcircle at D. Draw the line segments AD and BD.

Since $\angle A + \angle A' = 180° = \angle A + \angle CDB$,

$$\angle B' = \angle B = \angle BCD.$$

Hence, $\angle A' = \angle CDB$ and $\angle B' = \angle BCD$, so $\Delta A'B'C' \backsim \Delta DCB$.

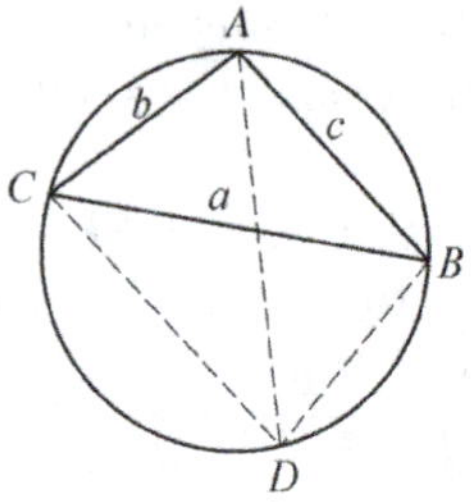

Figure 30.12

This implies that $\frac{A'B'}{DC} = \frac{B'C'}{CB} = \frac{A'C'}{DB}$, namely $\frac{c'}{DC} = \frac{a'}{a} = \frac{b'}{DB}$, and $DC = \frac{c'a}{a'}$ and $DB = \frac{b'a}{a'}$.

We apply Ptolemy's theorem to the quadrilateral $ACDB$. Then

$$AC \cdot BD + AB \cdot CD = AD \cdot BC,$$

so $b \cdot \frac{b'a}{a'} + c \cdot \frac{c'a}{a'} = AD \cdot a$.

Since $AB \| CD$, it follows that $AD = BC = a$. Therefore, $b \cdot \frac{b'a}{a'} + c \cdot \frac{c'a}{a'} = a \cdot a$, and the desired equality follows.

Example 6. As shown in Figure 30.13, let ΔDBC, ΔEAC, and ΔFAB be triangles outside ΔABC such that $\angle DBC = \angle ECA = \angle FAB$ and $\angle DCB = \angle EAC = \angle FBA$. Show that

$$FA + FB + BD + DC + CE + EA \geq AD + BE + CF.$$

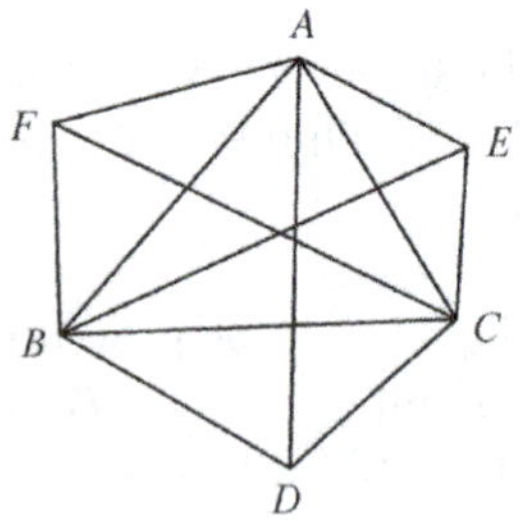

Figure 30.13

Proof. We apply Ptolemy's inequality to the quadrilateral $ABCD$. Then

$$AB \cdot CD + AC \cdot BD \geq AD \cdot BC,$$

so $\frac{AB \cdot CD}{BC} + \frac{AC \cdot BD}{BC} \geq AD$.

The condition $\Delta BCD \backsim \Delta CAE \backsim \Delta ABF$ implies that

$$\frac{AB \cdot CD}{BC} = BF, \quad \frac{AC \cdot BD}{BC} = CE.$$

Hence, $BF + CE \geq AD$. Similarly, $AF + CD \geq BE$ and $AE + BD \geq CF$. Adding the three inequalities we get the desired result.

By the equality condition for Ptolemy's inequality, we see that the equality in this problem holds if and only if A, B, C, and D are concyclic, B, C, E, and A are concyclic, and C, A, F, and B are concyclic. This implies that

$$\angle BAC = 180^\circ - \angle BDC = 180^\circ - \angle CEA = \angle ABC,$$

and similarly $\angle BCA = \angle ABC = \angle CAB$. Hence, ΔABC has to be an equilateral triangle, and also $\angle BDC = \angle CEA = \angle AFB = 180^\circ - \angle ACB = 120^\circ$.

Therefore, the equality holds if and only if ΔABC is an equilateral triangle and $\angle BDC = 120^\circ$.

Example 7. As in Figure 30.14, line segment AB is a chord in a circle ω. Let P be a point on the arc AB, and E and F are points on AB such that $AE = EF = FB$. Suppose the lines PE and PF intersect ω again at C and D, respectively. Show that $EF \cdot CD = AC \cdot BD$.

Proof. As shown in Figure 30.15, we form the line segments AD, BC, CF, and DE. Since $AE = EF = FB$,

$$\frac{BC \cdot \sin \angle BCE}{AC \cdot \sin \angle ACE} = \frac{BE}{AE} = 2. \text{①}$$

Similarly,

$$\frac{AD \cdot \sin \angle ADF}{BD \cdot \sin \angle BDF} = \frac{AF}{BF} = 2. \text{②}$$

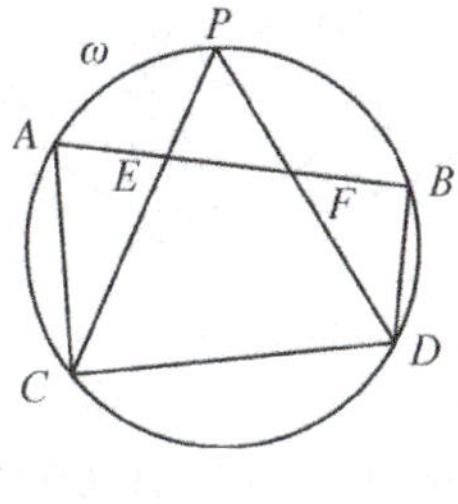

Figure 30.14

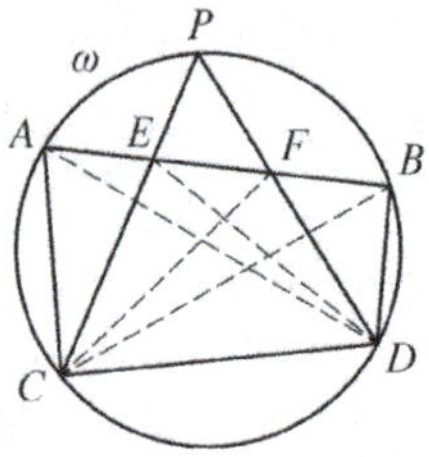

Figure 30.15

On the other hand, since

$$\angle BCE = \angle BCP = \angle BDP = \angle BDF,$$

$$\angle ACE = \angle ACP = \angle ADP = \angle ADF,$$

we can multiply ① and ② to get $\frac{BC \cdot AD}{AC \cdot BD} = 4$, or equivalently

$$BC \cdot AD = 4AC \cdot BD. \text{ ①}$$

From Ptolemy's theorem,

$$AD \cdot BC = AC \cdot BD + AB \cdot CD. \text{ ②}$$

By ③ and ④ we obtain that $AB \cdot CD = 3AC \cdot BD$, so $EF \cdot CD = AC \cdot BD$.

Example 8. As in Figure 30.16, two circles $\odot O_1$ and $\odot O_2$ intersect at P and Q, and AB is an exterior common tangent line of them, where A lies on $\odot O_1$ and B lies on $\odot O_2$. A circle Γ passing through A and B intersects $\odot O_1$ and $\odot O_2$ again at D and C, respectively. Show that $\frac{CP}{CQ} = \frac{DP}{DQ}$.

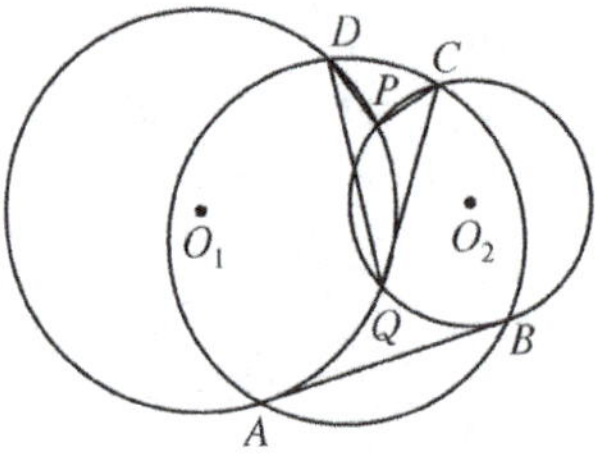

Figure 30.16

Proof. By Monge's theorem, the lines AD, QP, and BC have a common point, denoted as K. We draw the line segments AP, AQ, BP, and BQ.

Since $\Delta KPD \backsim \Delta KAQ$,

$$\frac{DP}{AQ} = \frac{KP}{KA}.$$

Since $\Delta KPA \backsim \Delta KDQ$,

$$\frac{AP}{DQ} = \frac{KA}{KQ}.$$

Multiplying the two equalities gives

$$\frac{AP \cdot DP}{AQ \cdot DQ} = \frac{KP}{KQ},$$

and similarly $\frac{BP \cdot CP}{BQ \cdot CQ} = \frac{KP}{KQ}$. Hence,

$$\frac{AP \cdot DP}{AQ \cdot DQ} = \frac{BP \cdot CP}{BQ \cdot CQ}. \text{①}$$

Let M be the intersection point of PQ and AB. Since $\Delta AQM \backsim \Delta PAM$,

$$\frac{AQ}{AP} = \frac{AM}{PM} = \frac{QM}{AM},$$

so $\left(\frac{AQ}{AP}\right)^2 = \frac{AM}{PM} \cdot \frac{QM}{AM} = \frac{QM}{PM}$. Similarly, $\left(\frac{BQ}{BP}\right)^2 = \frac{QM}{PM}$, hence $\left(\frac{AQ}{AP}\right)^2 = \left(\frac{BQ}{BP}\right)^2$, namely

$$\frac{AQ}{AP} = \frac{BQ}{BP}. \text{②}$$

Putting ① and ② together, we see that $\frac{DP}{DQ} = \frac{CP}{CQ}$.

Example 9. As in Figure 30.17, point H is the orthocenter of a triangle ABC. Let l_1 and l_2 be perpendicular lines that both pass through H.

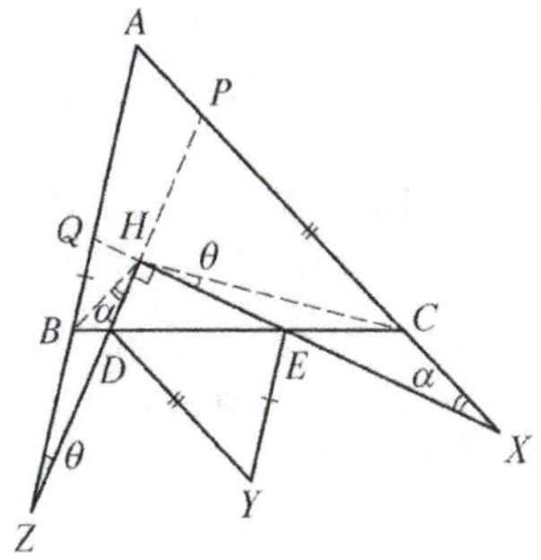

Figure 30.17

The line l_1 intersects BC at D and intersects the extension of AB at Z. The line l_2 intersects BC at E and intersects the extension of AC at X. The point Y satisfies $YD \| AC$ and $YE \| AB$. Show that X, Y, and Z are collinear.

Proof. Let ZH intersect AC at P, let XH intersect AB at Q, and form the line segments BH and CH. Then we apply the Menelaus theorem to ΔAQX and ΔAPZ, getting

$$\frac{CX}{AC} \cdot \frac{AB}{BQ} \cdot \frac{QE}{EX} = 1, \text{ ①}$$

$$\frac{BZ}{AB} \cdot \frac{AC}{PC} \cdot \frac{PD}{DZ} = 1. \text{ ②}$$

Since H is the orthocenter of ABC, we have $BH \perp AC$. Since $\angle DHE = 90°$, we have $\angle HXA = \angle BHZ$ (denoted as α). Similarly, $\angle HZA = \angle CHX$ (denoted as θ). Hence,

$$\frac{PC}{CX} = \frac{PH \sin \angle PHC}{HX \sin\theta} = \frac{PH}{HX} \cdot \frac{\cos\theta}{\sin\theta} = \frac{\tan\alpha}{\tan\theta}.$$

By the same token, $\frac{BZ}{BQ} = \frac{\tan\alpha}{\tan\theta}$. Consequently, $\frac{PC}{CX} = \frac{BZ}{BQ}$, or equivalently

$$\frac{CX \cdot BZ}{BQ \cdot PC} = 1. \text{ ③}$$

By ①× ②× ③ we get $\frac{QE}{EX} \cdot \frac{PD}{DZ} = 1$, so

$$\frac{EX}{QE} = \frac{PD}{DZ}. \text{ ④}$$

Suppose the line passing through E and parallel to AB intersects ZX at Y_1. Then $\frac{Y_1X}{ZY_1} = \frac{EX}{QE}$. Suppose the line passing through D and parallel to AC intersects ZX at Y_2. Then $\frac{Y_2X}{ZY_2} = \frac{PD}{ZD}$.

By ④ we have $\frac{Y_1X}{ZY_1} = \frac{Y_2X}{ZY_2}$, hence we conclude that Y_1 and Y_2 are the same point (it is exactly Y), and X, Y, and Z are collinear.

Example 10. As in Figure 30.18, a triangle ABC has its circumcircle ω. Let M be the midpoint of the arc $\overset{\frown}{BC}$ and N be the midpoint of BC. Let I

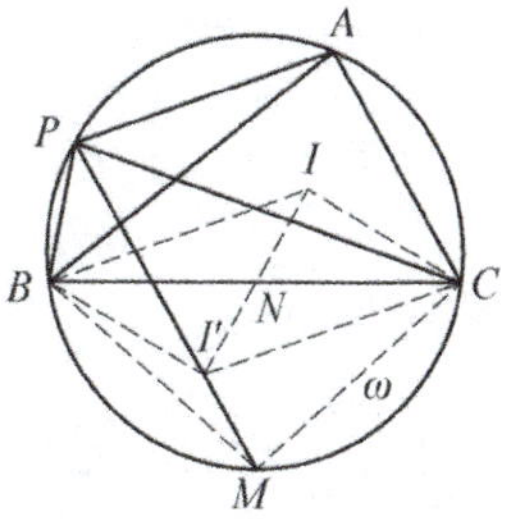

Figure 30.18

be the incenter of ΔABC, and its reflection with respect to the point N is I'. Suppose the ray MI' intersects $\overset{\frown}{AB}$ again at P. Show that $PA+PB=PC$.

Proof. Let $AB=c$, $AC=b$, and $BC=a$. Then by Ptolemy's theorem,

$$PC\cdot c-PB\cdot b=PA\cdot a. \text{①}$$

The equality that we desire can be written as

$$PC\cdot a-PB\cdot a=PA\cdot a. \text{②}$$

Hence, it suffices to show that $PC\cdot c-PB\cdot b=PC\cdot a-PB\cdot a$, or equivalently

$$\frac{PB}{PC}=\frac{a-c}{a-b}. \text{③}$$

Now, we proceed to prove ③. Applying the law of sines to $\Delta BMI'$ and $\Delta CMI'$, we get

$$\frac{\sin\angle BMI'}{BI'}=\frac{\sin\angle MBI'}{MI'},$$
$$\frac{\sin\angle CMI'}{CI'}=\frac{\sin\angle MCI'}{MI'}.$$

Dividing the two equalities, we have

$$\frac{\sin\angle BMI'}{\sin\angle CMI'}=\frac{\sin\angle MBI'}{\sin\angle MCI'}\cdot\frac{BI'}{CI'}.$$

Since $IBI'C$ is a parallelogram,

$$\angle MBI' = \angle MBC - \angle I'BC = \angle MAC - \angle ICB = \frac{\angle A - \angle C}{2},$$

and similarly $\angle MCI' = \frac{A-B}{2}$. Hence,

$$\begin{aligned}\frac{\sin \angle BMI'}{\sin \angle CMI'} &= \frac{\sin \frac{A-C}{2}}{\sin \frac{A-B}{2}} \cdot \frac{IC}{IB} = \frac{\sin \frac{A-C}{2} \cdot \sin \frac{B}{2}}{\sin \frac{A-B}{2} \cdot \sin \frac{C}{2}} \\ &= \frac{\sin \frac{A-C}{2} \cdot \cos \frac{A+C}{2}}{\sin \frac{A-B}{2} \cdot \cos \frac{A+B}{2}} = \frac{\frac{1}{2}(\sin A - \sin C)}{\frac{1}{2}(\sin A - \sin B)} \\ &= \frac{a-c}{a-b}.\end{aligned}$$

Applying the law of sines to ω ensures that

$$\frac{PB}{PC} = \frac{\sin \angle BMI'}{\sin \angle CMI'} = \frac{a-c}{a-b},$$

which completes the proof.

Example 11. Suppose a circle intersects the three sides BC, CA, and AB of ΔABC at D_1, D_2, E_1, E_2, F_1, and F_2 (as in Figure 30.19). Let D_1E_1 and F_2D_2 intersect at L, E_1F_1 and D_2E_2 intersect at M, and F_1D_1 and E_2F_2 intersect at N. Prove that AL, BM, and CN are concurrent.

Proof. By the law of sines,

$$\frac{\sin \angle BAL}{\sin \angle LF_2A} = \frac{F_2L}{AL},$$

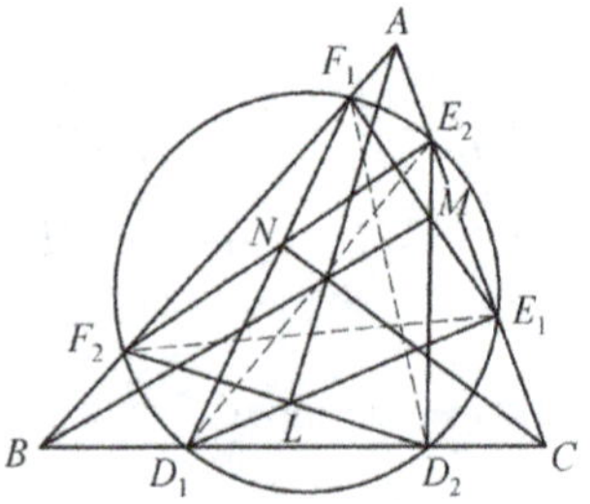

Figure 30.19

$$\frac{\sin \angle LE_1A}{\sin \angle LAC} = \frac{AL}{LE_1},$$
$$\frac{\sin \angle LF_2A}{\sin \angle LE_1A} = \frac{F_1D_2}{D_1E_2}.$$

Hence,

$$\begin{aligned}\frac{\sin \angle BAL}{\sin \angle LAC} &= \frac{\sin \angle BAL}{\sin \angle LF_2A} \cdot \frac{\sin \angle LE_1A}{\sin \angle LAC} \cdot \frac{\sin \angle LF_2A}{\sin \angle LE_1A} \\ &= \frac{F_2L}{AL} \cdot \frac{AL}{LE_1} \cdot \frac{F_1D_2}{D_1E_2} = \frac{F_2L}{LE_1} \cdot \frac{F_1D_2}{D_1E_2}.\end{aligned}$$

Since $\Delta LD_1F_2 \backsim \Delta LD_2E_1$, we have $\frac{F_2L}{LE_1} = \frac{F_2D_1}{D_2E_1}$, so

$$\frac{\sin \angle BAL}{\sin \angle LAC} = \frac{F_2D_1}{D_2E_1} \cdot \frac{F_1D_2}{D_1E_2}.$$

Similarly, $\frac{\sin \angle CBM}{\sin \angle MBA} = \frac{D_2E_1}{E_2F_1} \cdot \frac{D_1E_2}{E_1F_2}$ and $\frac{\sin \angle ACN}{\sin \angle NCB} = \frac{E_2F_1}{F_2D_1} \cdot \frac{E_1F_2}{F_1D_2}$. Therefore,

$$\begin{aligned}&\frac{\sin \angle BAL}{\sin \angle LAC} \cdot \frac{\sin \angle CBM}{\sin \angle MBA} \cdot \frac{\sin \angle ACN}{\sin \angle NCB} \\ &\quad = \frac{F_2D_1}{D_2E_1} \cdot \frac{F_1D_2}{D_1E_2} \cdot \frac{D_2E_1}{E_2F_1} \cdot \frac{D_1E_2}{E_1F_2} \cdot \frac{E_2F_1}{F_2D_1} \cdot \frac{E_1F_2}{F_1D_2} = 1.\end{aligned}$$

Apparently AL and BN intersect, so by the angular form of Ceva's theorem, we conclude that AL, BM, and CN are concurrent.

Example 12. Let ABA_1B_2, BCB_1C_2, and CAC_1A_2 be rectangles outside a triangle ABC. A point C' satisfies $C'A_1 \perp A_1C_2$ and $C'B_2 \perp B_2C_1$, and A' and B' are defined similarly. Show that AA', BB', and CC' are concurrent.

Proof. As in Figure 30.20, let l_A denote the line passing through A and perpendicular to B_2C_1, and define l_B and l_C similarly. Let $CB_1 = BC_2 = x$, $BA_1 = AB_2 = y$, and $AC_1 = CA_2 = z$. Then

$$\frac{y}{z} = \frac{AB_2}{AC_1} = \frac{\sin \angle B_2C_1A}{\sin \angle C_1B_2A} = \frac{\sin \angle A_1}{\sin \angle A_2}.$$

(Here $\angle A_1$ denotes the angle inside $\angle A$ with number 1, and others are similar.)

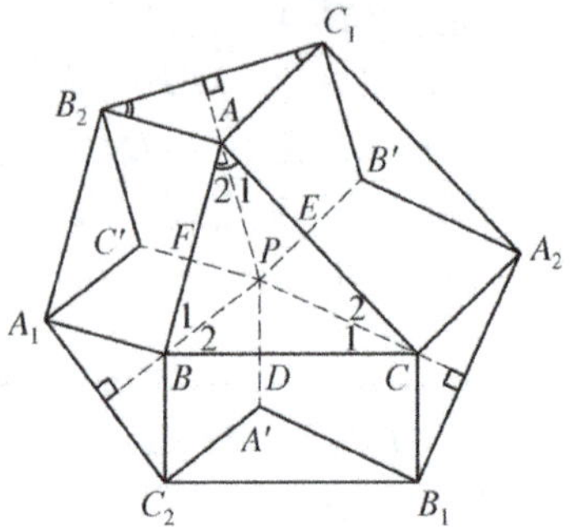

Figure 30.20

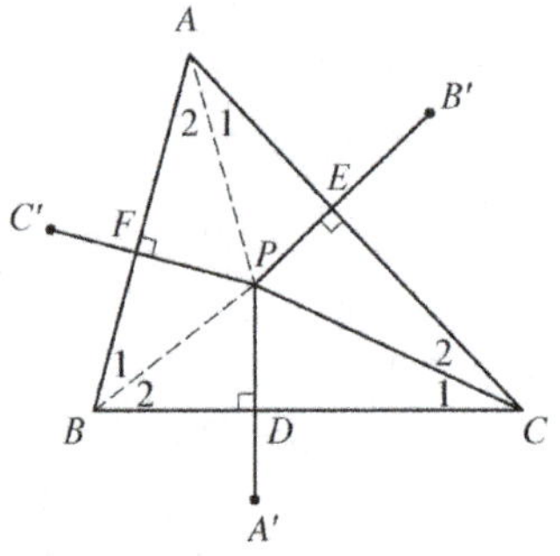

Figure 30.21

Similarly, $\frac{x}{y} = \frac{\sin\angle B_1}{\sin\angle B_2}$ and $\frac{z}{x} = \frac{\sin\angle C_1}{\sin\angle C_2}$. Hence, $\frac{\sin\angle A_1}{\sin\angle A_2} \cdot \frac{\sin\angle B_1}{\sin\angle B_2} \cdot \frac{\sin\angle C_1}{\sin\angle C_2} = 1$.

By the angular form of Ceva's theorem, l_A, l_B, and l_C are concurrent, and suppose they intersect at P. Then it follows that

$$\Delta PBC \backsimeq \Delta A'C_2B_1.$$

Thus, $PB = C_2A'$, and since $PB \| C_2A'$, we see that PBC_2A' is a parallelogram. Hence, $PA' = BC_2 = x$ and $PA' \| BC_2$. Since $BC_2 \perp C_2B_1$, we have $PA' \perp C_2B_1$. Also, $BC \| B_1C_2$, so $PA' \perp BC$. Similarly, $PC' = y$, $PB' = z$, $PC' \perp AB$, and $PB' \perp AC$.

As in Figure 30.21, let PA', PB', and PC' intersect BC, CA, and AB at D, E, and F, respectively. Let $PD = m$, $PE = n$, and $PF = t$. Then $\frac{y}{z} = \frac{\sin\angle A_1}{\sin\angle A_2} = \frac{n}{t}$, $\frac{x}{y} = \frac{\sin\angle B_1}{\sin\angle B_2} = \frac{t}{m}$, and $\frac{z}{x} = \frac{\sin\angle C_1}{\sin\angle C_2} = \frac{m}{n}$.

From the first equality we may assume that $n = ky$, so $t = kz$ and $m = \frac{kyz}{x}$.

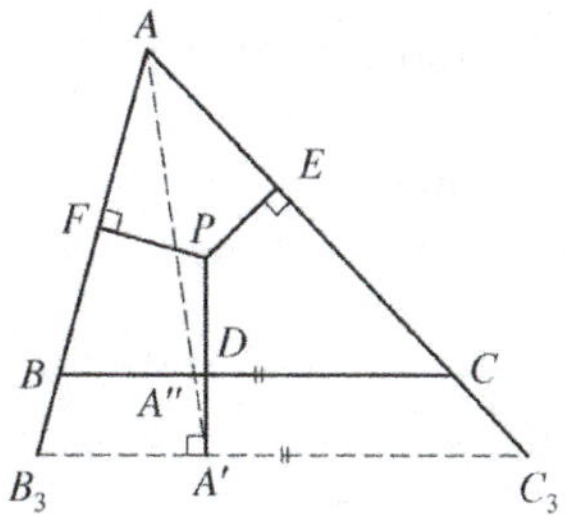

Figure 30.22

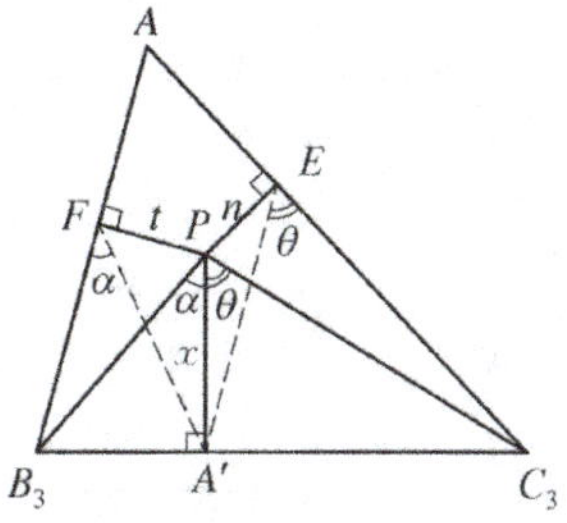

Figure 30.23

Now, we look at Figure 30.22. Let B_3C_3 be a line passing through A' and parallel to BC, with B_3 and C_3 lying on the lines AB and AC, respectively. Let A'' be the intersection point of AA' and BC. Then $\frac{BA''}{CA''} = \frac{B_3A'}{C_3A'}$.

As in Figure 30.23, let $\angle B_3PA' = \alpha$ and $\angle C_3PA' = \theta$. Since P, F, B_3, and A' are concyclic, and P, E, C_3, and A' are concyclic,

$$\begin{aligned}\angle PA'F &= \angle PB_3F = \angle B - \angle PB_3A' \\ &= \angle B - (90^\circ - \alpha) = \angle B + \alpha - 90^\circ, \\ \angle PFA' &= \angle PB_3A' = 90^\circ - \alpha.\end{aligned}$$

Hence, $\angle B_3FA' = \alpha$ and $\angle C_3EA' = \theta$. Applying the law of sines to $\Delta PFA'$, we get

$$\begin{aligned}\frac{t}{x} &= \frac{PF}{PA'} = \frac{\sin\angle PA'F}{\sin\angle PFA'} = \frac{\sin(\angle B + \alpha - 90^\circ)}{\sin(90^\circ - \alpha)} \\ &= \frac{-\cos(\angle B + \alpha)}{\cos\alpha} = \tan\alpha\sin\angle B - \cos\angle B.\end{aligned}$$

Then, $\tan\alpha = \frac{\frac{t}{x} - \cos\angle B}{\sin\angle B}$, and similarly $\tan\theta = \frac{\frac{n}{x} - \cos\angle C}{\sin\angle C}$. Consequently,

$$\frac{BA''}{CA''} = \frac{B_3A'}{C_3A'} = \frac{PA'\tan\alpha}{PA'\tan\theta} = \frac{\tan\alpha}{\tan\theta} = \frac{x\cos\angle B - t}{x\cos\angle C - n} \cdot \frac{\sin\angle C}{\sin\angle B},$$

and we have the other two similar equalities.

By the inverse theorem of Ceva's theorem, it suffices to show the following:

$$\frac{x\cos\angle B - t}{x\cos\angle C - n} \cdot \frac{\sin\angle C}{\sin\angle B} \cdot \frac{z\cos\angle C - m}{z\cos\angle A - t} \cdot \frac{\sin\angle A}{\sin\angle C} \cdot \frac{y\cos\angle A - n}{y\cos\angle B - m} \cdot \frac{\sin\angle B}{\sin\angle A} = 1,$$

or equivalently $\frac{x\cos\angle B - t}{x\cos\angle C - n} \cdot \frac{z\cos\angle C - m}{z\cos\angle A - t} \cdot \frac{y\cos\angle A - n}{y\cos\angle B - m} = 1$.

Since $n = ky$, $t = kz$, and $m = \frac{kyz}{x}$,

$$\begin{aligned} &\frac{x\cos\angle B - t}{x\cos\angle C - n} \cdot \frac{z\cos\angle C - m}{z\cos\angle A - t} \cdot \frac{y\cos\angle A - n}{y\cos\angle B - m} \\ &= \frac{x\cos\angle B - kz}{x\cos\angle C - ky} \cdot \frac{z\cos\angle C - \frac{kyz}{x}}{z\cos\angle A - kz} \cdot \frac{y\cos\angle A - ky}{y\cos\angle B - \frac{kyz}{x}} \\ &= \frac{x\cos\angle B - kz}{x\cos\angle C - ky} \cdot \frac{x\cos\angle C - ky}{x\cos\angle A - kx} \cdot \frac{x\cos\angle A - kx}{x\cos\angle B - kz} = 1. \end{aligned}$$

Therefore, the lines AA', BB', and CC' are concurrent.

30.3 Exercises

Group A

1. As shown in the following figure, ΔABC is inscribed to $\odot O$, and the tangent line of $\odot O$ at A intersects the line BC at P. Let D be the midpoint of AB and DP intersects AC at M. Show that $\frac{PA^2}{PC^2} = \frac{AM}{MC}$.

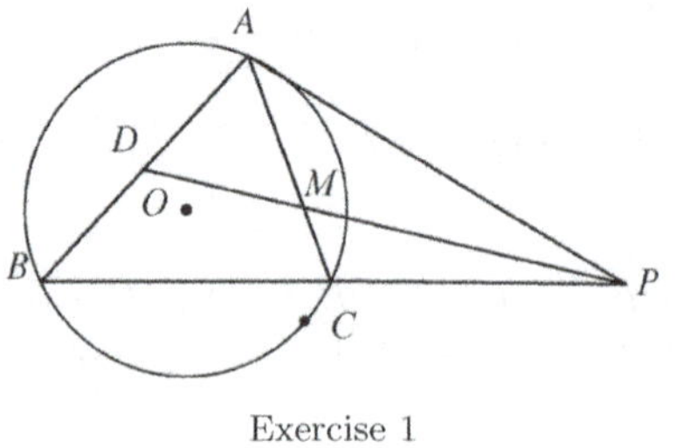

Exercise 1

2. As shown in the following figure, $\angle BAC = 90°$ in ABC. Let G be a given point on AB, which is not the midpoint of AB. Let D be an arbitrary point on CG. Suppose AD intersects BC at E, BD intersects AC at F, and EF intersects AB at H. Show that the position of H is independent of the choice of D.

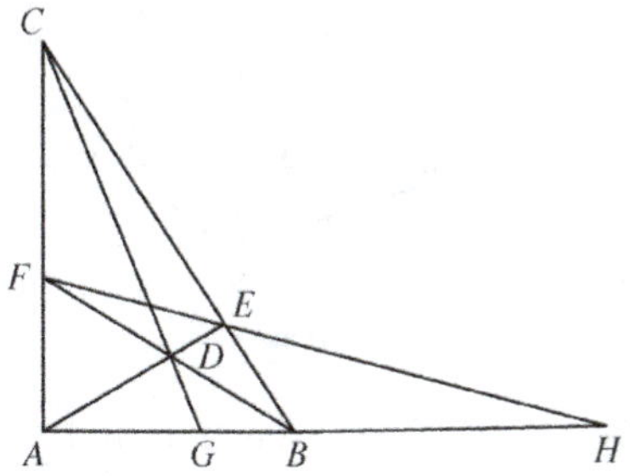

Exercise 2

3. In a triangle ABC, suppose the circle with diameter BC intersects AB and AC again at D and E, respectively. Let F and G be the projections of D and E onto BC, respectively, and the lines DG and EF intersect at M. Show that $AM \perp BC$.
4. As shown in the following figure, suppose AC bisects $\angle BAD$ in a quadrilateral $ABCD$. Let E be an arbitrary point on CD, the line BE intersects AC at F, and DF intersects BC at G. Show that $\angle GAC = \angle EAC$.

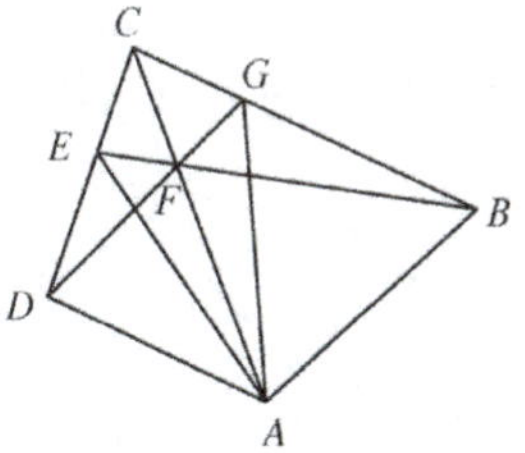

Exercise 4

5. In a scalene triangle ABC, suppose $BC = \frac{1}{2}(AB + AC)$. Let O and I be its circumcenter and incenter, respectively. Let E be the intersection point of the outer angle bisector of $\angle BAC$ and the circumcircle of ΔABC (different from A). Prove that $OI = \frac{1}{2}AE$.

6. As shown in the following figure, a quadrilateral $ABCD$ has a circumcircle and satisfies $\frac{AB}{AD} = \frac{CB}{CD}$. The diagonals AC and BD intersect at P and E is the midpoint of AC. Show that $\frac{BE}{ED} = \frac{BP}{PD}$.

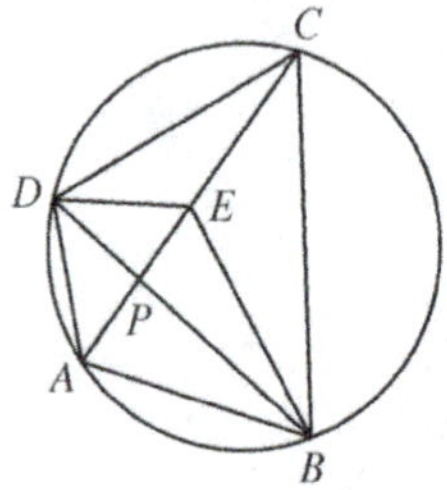

Exercise 6

7. In a triangle ABC, let AM be a median and D is a point inside the triangle such that

$$\angle ABD = \angle BAM, \angle ACD = \angle CAM.$$

Show that $\frac{AB^2}{AC^2} = \frac{BD}{CD}$.

Group B

8. As shown in the following figure, $AB > AC$ in an acute triangle ABC. Let M be the midpoint of BC, let I be the incenter of ΔABC, and MI intersects AC at D. Let E be the second intersection point of BI and the circumcircle of ΔABC. Show that $\frac{ED}{EI} = \frac{IC}{IB}$.

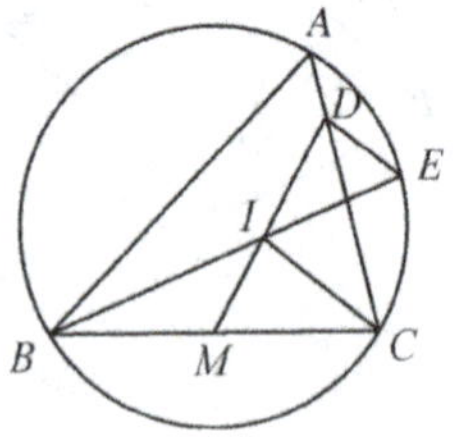

Exercise 8

9. As shown in the following figure, let D be the midpoint of the arc BC (containing A) on the circumcircle ω of ΔABC. The line DA intersects the tangent lines to ω at B and C at P and Q, respectively. Suppose BQ intersects AC at X, CP intersects AB at Y, and BQ intersects CP at T. Show that AT bisects the line segment XY.

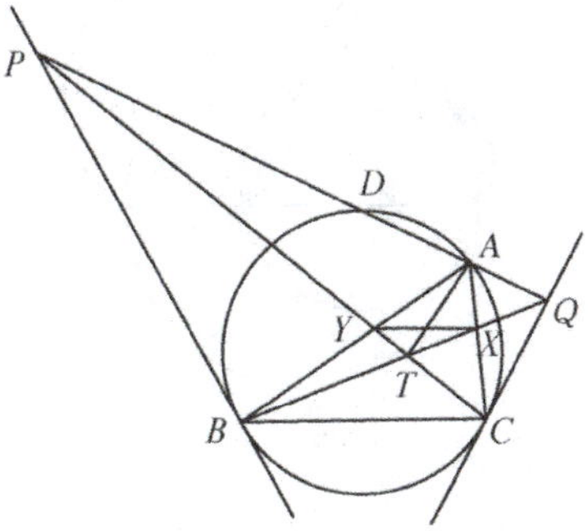

Exercise 9

10. In ΔABC, let AD be the angle bisector of $\angle BAC$, where D lies on BC. Suppose the two exterior common tangent lines of the circumcircles of ΔABD and ΔACD intersect the line AD at P and Q, respectively. Show that $PQ^2 = AB \cdot AC$.

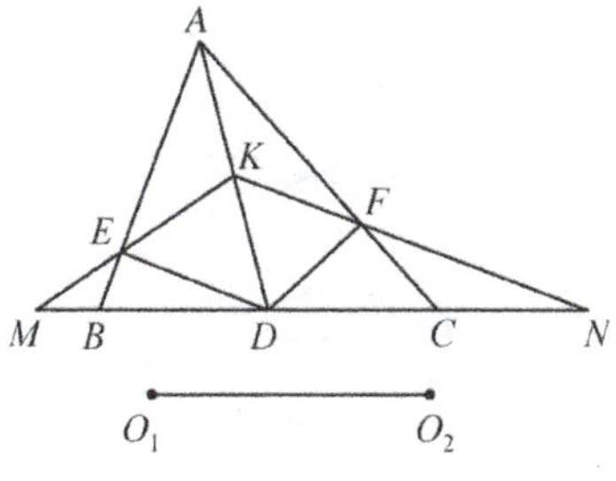

Exercise 11

11. As shown in the following figure, $AB \neq AC$ in a triangle ABC. Let K be the midpoint of the median AD, and let E and F be the projections

of D onto AB and AC, respectively. The lines KE and KF intersect BC at M and N, respectively. Let O_1 and O_2 be the circumcenters of ΔDEM and ΔDFN, respectively. Show that $O_1O_2 \| BC$.

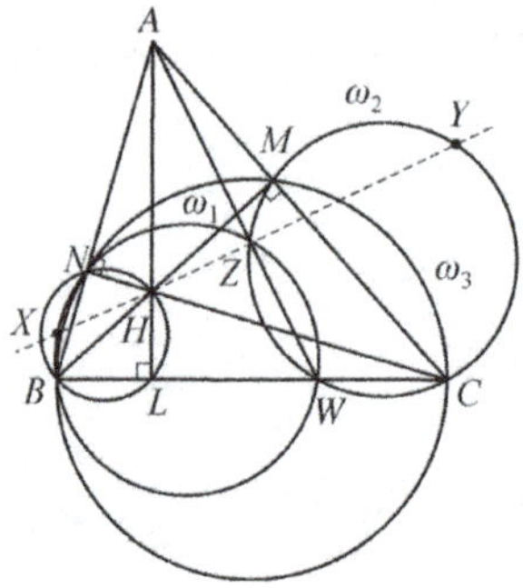

Exercise 12

12. Let ABC be an acute triangle with its orthocenter H. Let W be a point on BC (the endpoints are excluded), and BM and CN are altitudes of the triangle. Let ω_1 be the circumcircle of the triangle BWN and X lies on ω_1 such that WX is a diameter of ω_1. Similarly, let ω_2 be the circumcircle of the triangle CWM and WY is a diameter of ω_2. Show that X, Y, and H are collinear.

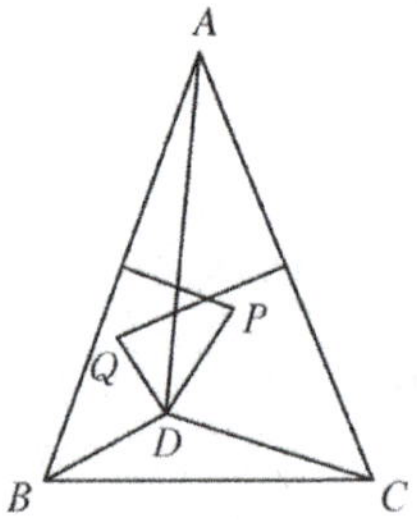

Exercise 13

13. As shown in the following figure, $AB = AC > BC$ in a triangle ABC. Let D be a point inside ΔABC such that $DA = DB + DC$. Suppose the perpendicular bisector of AB and the exterior angle bisector of $\angle ADB$ intersect at P, while the perpendicular bisector of AC and the exterior angle bisector of $\angle ADC$ intersect at Q. Prove that B, C, P, and Q are concyclic.

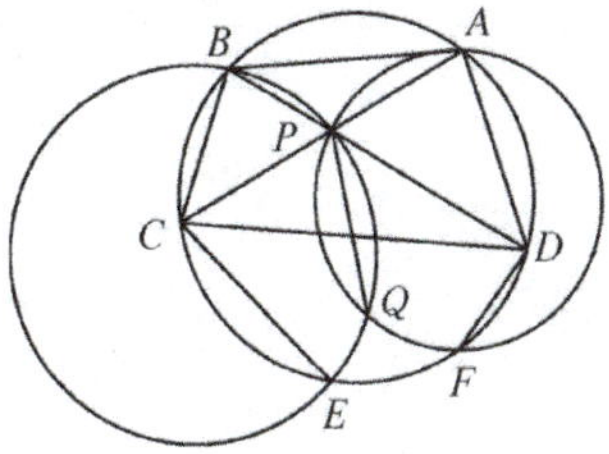

Exercise 14

14. Let $ABCD$ be a convex quadrilateral inscribed to $\odot O$, and its diagonals AC and BD intersect at P. A circle $\odot O_1$ passing through P and B and a circle $\odot O_2$ passing through P and A meet again at Q, and $\odot O_1$ and $\odot O_2$ intersect $\odot O$ again at E and F, respectively. Prove that the lines PQ, CE, and DF are either concurrent or pairwise parallel.

Chapter 31

The Extreme Principle

31.1 Key Points of Knowledge and Basic Methods

The extreme principle, in general, is a method of considering special objects, which are maximal, minimal, longest, shortest, etc., in order to crack a problem. The idea of the extreme principle has applications in various areas including geometry, number theory, combinatorics, graph theory, and so on. While the principle seems simple, it can be used to solve many problems regarding the existence of certain objects.

The method of extreme principle is based on the following facts:

1. Among finitely many real numbers, there is a maximum and a minimum.
2. Among infinitely many positive integers, there is a minimum.
3. Among infinitely many real numbers, there does not necessarily exist a maximum or minimum.

31.2 Illustrative Examples

Example 1. In a round-robin of ping-pong games, suppose there are $n(n \geq 3)$ players, and none of them win all the games. Prove that we can find three players A, B, and C such that A beats B, B beats C, and C beats A.

Proof. Let A be a player who wins the most games. Since A does not win all the games, there is some player C that beats A. Then we consider the players who lose the game to A. If all of them also lose the game to C, then we see that C wins more games than A, which is contradictory to the assumption. Hence, there is some player B who loses to A but beats C, and the players A, B, and C satisfy the desired condition.

Example 2. Let S_1, S_2, and S_3 be nonempty sets of integers such that for each permutation i, j, k of 1, 2, 3, if $x \in S_i$ and $y \in S_j$, then $x - y \in S_k$.

(1) Show that two of the sets S_1, S_2, and S_3 are identical.
(2) Is it possible that two of the three sets have no common numbers?

Solution. (1) By the condition, if $x \in S_i$ and $y \in S_j$, then

$$y - x \in S_k, (y - x) - y = -x \in S_i,$$

so each set has a nonnegative number.

If all the three sets consist of a single number 0, then the proposition holds trivially. Otherwise we assume that the minimal positive integer in S_1, S_2, and S_3 is a, and without loss of generality say $a \in S_1$. Then let b be the minimal nonnegative integer in S_2 and S_3, and assume that $b \in S_2$. Thus, $b - a \in S_3$.

If $b > 0$, then $0 \leq b - a < b$, which is contradictory to the assumption on b, so necessarily $b = 0$.

Now, choose an arbitrary $x \in S_1$. Since $0 \in S_2$, we have $x - 0 = x \in S_3$, so $S_1 \subseteq S_3$. Similarly, $S_3 \subseteq S_1$, and we get $S_1 = S_3$.

(2) It is possible. For example, let $S_1 = S_2 = \{\text{odd numbers}\}$ and $S_3 = \{\text{even numbers}\}$. Then the condition is satisfied, and S_2 has no common number with S_3.

Example 3. We call a set of points in a plane a "good" set, if no three points in the set are the vertices of an equilateral triangle. Let S be a set consisting of n points of the plane. Show that there is a subset T of S such that $|T| \geq \sqrt{n}$ and T is a "good" set.

Analysis. We choose a "good" subset of S with the greatest number of points. By the maximality assumption we can give an estimation of its number of points.

Proof. Let T be one with the greatest number of points among all the "good" subsets of S. It suffices to show that $|T| \geq \sqrt{n}$.

If $T = S$, then the proposition holds trivially.

If $T \neq S$, then for any point P in $S \backslash T$, by the maximality of T, there exist two points A and B in T such that ΔPAB is an equilateral triangle.

On the other hand, for any two points A and B in T, there are at most two points that form an equilateral triangle with them, so it follows that

$$|S \backslash T| \leq 2C_{|T|}^2.$$

Hence, $n - |T| \leq |T|^2 - |T|$, and $|T| \geq \sqrt{n}$.

Example 4. Suppose there are eight points in the space such that no four of them lie in the same plane. There are 17 line segments between some pairs of these points. Prove that these line segments form at least one triangle.

Proof. Without loss of generality assume that A is a point with the greatest number of line segments containing it, and let n denote this number. If the 17 line segments form no triangles, then no two of the n points connected to A have a line segment between them. For the other $7 - n$ points, each point is contained in no more than n line segments, so the total number of the line segments is at most

$$n + (7 - n)n = n(8 - n) \leq \left(\frac{n + 8 - n}{2}\right)^2 = 16.$$

However, since we have 17 line segments, there is a contradiction. Therefore, the line segments form at least one triangle.

Remark. The result in this problem can be strengthened to "at least four triangles". This is a harder problem, and the interested readers may have a try.

Example 5. Let $n \geq 3$ be a positive integer, and $a_1, a_2, \ldots, a_n$ are n distinct real numbers whose sum is positive. We call a permutation $b_1, b_2, \ldots, b_n$ of these numbers a "good" permutation, if $\sum_{i=1}^{k} b_i > 0$ for all $1 \leq k \leq n$. Find the minimum value of the number of "good" permutations.

Analysis. The answer is not hard to guess. We choose $a_1, a_2, \ldots, a_n$, such that one of them is a large positive number and the others are all negative with small absolute values; or one of them is a negative number with a large absolute value and the others are all small positive numbers. Then the answer is $(n - 1)!$. From this answer we may try to prove that among n permutations there has to be one that satisfies the condition, and it is natural to consider the pattern of circular permutations.

Proof. If $a_1 > 0$, while a_2, a_3, $\ldots$, a_n are all negative, then a_1 has to stay in the first position, so the total numbers of "good" permutations is $(n-1)!$.

Next, we show that there are at least $(n-1)!$ "good" permutations.

For each permutation $b_1, b_2, \ldots, b_n$ of $a_1, a_2, \ldots, a_n$, it suffices to show that at least one permutation among the following permutations is "good":

$$b_1, b_2, \ldots, b_n; b_2, b_3, \ldots, b_1; \ldots; b_n, b_1, \ldots, b_{n-1}.$$

In fact, for $1 \leq i \leq n$, let $S_i = b_1 + b_2 + \cdots + b_i$. Suppose S_k is the minimum among $S_1, S_2, \ldots, S_n$, where we take the largest index if the minimum occurs multiple times. Then for $i + 1 \leq m \leq n$,

$$b_{i+1} + \cdots + b_m = S_n - S_i > 0;$$

for $1 \leq m \leq i$,

$$b_{i+1} + \cdots + b_n + b_1 + \cdots + b_m = S_n - S_i + S_m \geq S_n > 0.$$

Hence, b_{i+1}, b_{i+2}, $\ldots$, b_i is a "good" permutation.

Therefore, the minimum value of the number of "good" permutations is $(n-1)!$.

Example 6. There are $n(n \geq 5)$ points in a plane and each of them is colored either red or blue. Suppose no three points with the same color are collinear. Prove that there exists a triangle such that

(1) its vertices have the same color;
(2) at least one side of the triangle contains no points with the opposite color.

Proof. Since $n \geq 5$ and there are two colors, by the pigeonhole principle at least three of them have the same color, so the set of triangles whose vertices have the same color is not empty.

We choose a triangle with vertices of the same color that has the least area (it exists since there are finitely many triangles). If each side of this triangle contains points of the opposite colors, then we can find a triangle with vertices of the same color that has a less area, which is contradictory to the assumption. Therefore, there exists a triangle with the desired property.

Example 7. For n points in a plane that do not lie on the same line, show that there exists a line that passes through exactly two of these points.

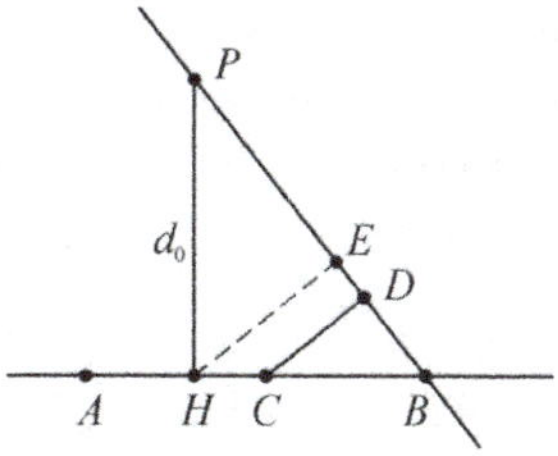

Figure 31.1

Proof. Since these points do not lie on the same line, each line determined by two of these points does not contain all the points (so that we can find a chosen point outside this line). For each line determined by two of the chosen points, we find the distances from the outside-points to this line. Then there are finitely many such distances (although the actual number may be large), so there is the minimal distance, which we denote as d_0.

Now, suppose the distance from point P to the line determined by A and B is d_0. We prove the result by contradiction.

Let H be the projection of P onto AB. Obviously $PH = d_0$. If the proposition does not hold, then the line AB contains another chosen point, namely C. At least two of A, B, and C lie on the same side of H on the line AB, and without loss of generality assume that they are B and C, and B is farther from H than C. Then we consider the distance from C to the line PB, as in Fiugire 31.1. It follows that

$$d_0 > HE \geq CD.$$

This is contradictory to the minimality assumption of d_0. Therefore, the line AB cannot contain the third chosen point, so it is the line that we desire.

Remark. This problem was raised by the British mathematician James J. Sylvester. While it seems simple, he failed to solve it during his life. Later many mathematicians tried to prove the proposition, but no one succeeded until 50 years later. As we appreciate this clever proof, it is important to see that considering the extreme case is the key to the solution.

Example 8. Suppose each cell of a $2 \times n$ table contains a positive number such that the two numbers in each column sum to 1. Prove that it is possible to erase one number from each column such that the sum of the remaining numbers in each row does not exceed $\frac{n+1}{4}$.

Proof. Suppose the numbers in the first row are $a_1, a_2, \ldots, a_n$, and by switching the order of columns we may assume that $a_1 \leq a_2 \leq \cdots \leq a_n$. Then the numbers in the second row are $b_1 = 1 - a_1$, $b_2 = 1 - a_2, \ldots$, $b_n = 1 - a_n$, and apparently $b_1 \geq b_2 \geq \cdots \geq b_n$. If

$$a_1 + a_2 + \cdots + a_n \leq \frac{n+1}{4},$$

then we can simply erase all the numbers in the second row, and the desired condition is satisfied. Otherwise, we can find the minimal k such that

$$a_1 + a_2 + \cdots a_k > \frac{n+1}{4}.$$

Now, we erase $a_k, a_{k+1}, \ldots, a_n$ in the first row and erase $b_1, b_2, \ldots, b_{k-1}$ in the second row. By the choice of k, it follows immediately that

$$a_1 + a_2 + \cdots + a_{k-1} \leq \frac{n+1}{4},$$

so it suffices to show that $b_k + b_{k+1} + \cdots + b_n \leq \frac{n+1}{4}$.

Since

$$a_k \geq \frac{a_1 + a_2 + \cdots + a_k}{k} > \frac{n+1}{4k},$$

we have

$$\begin{aligned}
& b_k + b_{k+1} + \cdots + b_n \\
& \leq (n+1-k)b_k = (n+1-k)(1-a_k) \\
& < (n+1-k)\left(1 - \frac{n+1}{4k}\right) \\
& = \frac{5}{4}(n+1) - \left[\frac{(n+1)^2 + (2k)^2}{4k}\right] \\
& \leq \frac{5}{4}(n+1) - \frac{2(n+1)(2k)}{4k} = \frac{n+1}{4}.
\end{aligned}$$

Therefore, the proposition is proven.

31.3 Exercises

Group A

1. Suppose there are 100 points in a plane such that the distance between any two of them is at most 1 and any three of them form an obtuse triangle. Show that these points can be covered by a circle with radius $\frac{1}{2}$.
2. Fifteen boys are playing with balls, and initially everyone has one ball in his hands. Suppose the distances between any two of them are distinct, and everyone throws his ball to the one closest to him.

   (1) Show that at least one boy has no ball in his hands.
   (2) Show that no one has more than five balls in his hands.

3. Prove that the equation $x^3 + 2y^3 = 4z^3$ has no positive real solutions (x, y, z).
4. Let n be a positive integer greater than 2 and $a_1 < a_2 < \cdots < a_k$ are all the positive integers less than n and relatively prime to n. Show that at least one of these k numbers is a prime number.
5. Let S be a set of people. Suppose for any two of these people, if they have the same number of friends in S, then they do not have a common friend in S. Show that there is a person in S who has exactly one friend. (We assume that at least one pair of people in S are friends.)

Group B

6. Suppose 23 people are having a soccer game with 11 people for each teamand the remaining one as the referee. For fairness, we require the total weights of the players from each team to be equal (and assume that the weight of every player is a positive integer). Suppose they can play the game no matter who serves as the referee. Prove that the 23 people all have the same weight.
7. There are $n (n \geq 7)$ circles in a plane such that we cannot find three of them, every two of which intersect (here two tangent circles are considered to be intersecting). Show that there is one circle that intersects at most five other circles.
8. Some people attend a party and some pairs of them are acquaintances. Suppose if two people have the same number of acquaintances among the attendees, then they do not have a common acquaintance. Suppose there is one attendee who has at least 2008 acquaintances. Show that there is someone who has exactly 2008 acquaintances.

9. For an arbitrary m by n matrix of real numbers,

$$\begin{pmatrix} a_{11} & a_{12} & \cdots & a_{1n} \\ a_{21} & a_{22} & \cdots & a_{2n} \\ \cdots & \cdots & \cdots & \cdots \\ a_{m1} & a_{m2} & \cdots & a_{mn} \end{pmatrix},$$

an operation means changing the signs of all the numbers in one column or one row, and leaving other numbers unchanged. Show that within finitely many operations, we can obtain a matrix such that the sum of the numbers in every column or row is nonnegative.

10. Find all integers k such that there exist positive integers a and b satisfying $\frac{b+1}{a} + \frac{a+1}{b} = k$.

11. There are $n(n > 3)$ players who are having a round-robin of ping-pong games. Suppose after some games, no two players have played against the same set of opponents. Prove that we can choose one player such that among the remaining players, every two of them still have played against different sets of opponents (in this case the chosen player is not counted).

12. Let X be a finite set. A mapping f assigns a real number $f(E)$ to each even subset E of X (i.e., a subset with an even number of elements). Suppose it satisfies the following conditions:

(1) there exists an even subset D such that $f(D) > 1990$;
(2) for any two disjoint even subsets A and B of X,

$$f(A \cup B) = f(A) + f(B) - 1990.$$

Prove that there exist subsets P and Q of X such that

(1) $P \cap Q = \emptyset$ and $P \cup Q = X$;
(2) for any nonempty even subset S of P, we have $f(S) > 1990$;
(3) for any even subset T of Q, the inequality $f(T) \leq 1990$ is satisfied.

Solutions

Solution 1

Concepts and Operations of Sets

1. 983. Since A and B are subsets of U, it follows that

$$0 \le a \le 31, \quad 1014 \le b \le 2012.$$

Hence, $A \cap B = \{x|b - 1014 \le x \le a + 1981\}$, namely $A \cap B = \{x|a \le x \le b$.

Therefore, the length of $A \cap B$ is minimal if and only if $a = 0$ and $b = 2012$ or $a = 31$ and $b = 2014$, and the minimum value is $1981 - 998 = 1014 - 31 = 983$.

2. $\{-3, 0, 2, 6\}$. Apparently, every element of A appears three times in all the three-element subsets of A. Hence,

$$3(a_1 + a_2 + a_3 + a_4) = (-1) + 3 + 5 + 8 = 15.$$

This means that $a_1 + a_2 + a_3 + a_4 = 5$, so the four elements of A are $5-(-1) = 6$, $5-3 = 2$, $5-5 = 0$, and $5-8 = -3$, thus $A = \{-3, 0, 2, 6\}$.

3. By discussing the signs of x, y, and z, we find that the desired set is $\{-3, -1, 1, 5\}$.

4. Since $A \cap B = \{2, 5\}$, we have $5 \in A$, so $a^3 - 2a^2 - a + 7 = 5$, that is $(a-2)(a^2-1) = 0$, from which $a = 2$ or $a = \pm 1$.

If $a = 2$, then $5a - 5 = 5$ and $-\frac{1}{2}a^2 + \frac{3}{2}a + 4 = 5$, and the set B has duplicating elements, which is impossible.

If $a = 1$, then $A = \{2, 4, 5\}$ and $B = \{1, 0, 5, 12\}$. In this case, $A \cap B = \{5\}$, which is contradictory to the condition.

If $a=-1$, then $A=\{2,4,5\}$ and $B=\{1,-10,2,4\}$. In this case, $A\cap B=\{2,4\}$, which is also contradictory to the condition.

Therefore, such a real number a does not exist.

5. Since X is a set with at most two elements and $X\cap A=\emptyset$, we have $1,7\notin X$. Also $X\subseteq\{4,10\}$ since $X\cap B=X$. If $X=\emptyset$, then $p^2-4q<0$; if $X=\{4\}$, then $x^2+px+q=(x-4)^2$, so $p=-8$ and $q=16$; if $X=\{10\}$, then $x^2+px+q=(x-10)^2$, so $p=-20$ and $q=100$; if $X=\{4,10\}$, then by Vieta's theorem $p=-14$ and $q=40$.

   Therefore, $(p,q)=(-8,16),(-20,100),(-14,40)$, or any pair of real numbers such that $p^2<4q$.
6. With the graph of the function, we see that the value range of a is $(-\infty,-1]\cup[1,+\infty)$.
7. Since $y=b^2-6b+10=(b-3)^2+1$, when $b=4,5,\ldots$, we see that y takes the same values as the function in A when $a=1,2,\ldots$. However, when $b=3$, we have $y=1\in B$, while $1\notin A$, so the relationship between A and B is $A\subsetneq B$.
8. Note that in the set N, the numbers x and y are both nonzero, so by $M=N$ it follows that $\lg(xy)=0$, namely $xy=1$, so $1\in N$. If $y=1$, then $x=1$, thus $N=\{0,1,1\}$, which is impossible. Similarly, if $x=1$ then $y=1$, and we also get a contradiction. If $x=-1$, then $y=-1$, which is valid. Therefore,

$$\left(x+\frac{1}{y}\right)+\left(x^2+\frac{1}{y^2}\right)+\cdots+\left(x^{2006}+\frac{1}{y^{2006}}\right)=0.$$

9. By $1\le a\le b\le 2$, we have $\frac{3}{a}+b\le\frac{3}{1}+2=5$. When $a=1$ and $b=2$, the maximum value $M=5$. On the other hand, since

$$\frac{3}{a}+b\ge\frac{3}{a}+a\ge 2\sqrt{\frac{3}{a}\cdot a}=2\sqrt{3},$$

   the minimum value $m=2\sqrt{3}$, attained when $a=b=\sqrt{3}$. Therefore, $M-m=5-2\sqrt{3}$.
10. (1) Since $s,t\in A$, we can assume that $s=m_1^2+n_1^2$ and $t=m_2^2+n_2^2$, where m_1,n_1,m_2 and n_2 are integers. Hence,

$$\begin{aligned}st&=\left(m_1^2+n_1^2\right)\left(m_2^2+n_2^2\right)\\&=m_1^2m_2^2+n_1^2n_2^2+m_1^2n_2^2+n_1^2m_2^2\\&=(m_1m_2+n_1n_2)^2+(m_1n_2-n_1m_2)^2,\end{aligned}$$

   which implies that $st\in A$.

(2) Since $s, t \in A$, by (1) we have $st \in A$, so we may assume that $st = m^2 + n^2$, where m and n are integers. For $t \neq 0$,

$$\frac{s}{t} = \frac{st}{t^2} = \frac{m^2}{t^2} + \frac{n^2}{t^2} = \left(\frac{m}{t}\right)^2 + \left(\frac{n}{t}\right)^2,$$

where $\frac{m}{t}$ and $\frac{n}{t}$ are rational numbers, and the assertion follows.

11. Note that for any integer x, if $x \equiv 0, \pm 1, \pm 2, \pm 3 \pmod 7$, then $x^2 \equiv 0, 1, 4, 2 \pmod 7$. This shows that if $x, y \in \mathbf{Z}$ and $x^2 + y^2 \equiv 0 \pmod 7$, then $x \equiv y \equiv 0 \pmod 7$. From this property, the set S may contain all the integers in $1, 2, \ldots, 50$ that are not divisible by 7 and one integer that is divisible by 7. Therefore, the maximum value of $|S|$ is 44.

12. First, if $x, y \in M (x < y)$, then $x + 1, x + 2, \ldots, y - 1$ all belong to M. In fact, if $y - x \geq 2$, then $x + 1 \leq \frac{x+y}{2} < \sqrt{\frac{x^2+y^2}{2}} < y$, so

$$x + 1 \leq \left[\sqrt{\frac{x^2 + y^2}{2}}\right] \leq y - 1.$$

This means that the interval $[x, y]$ contains the integer $\left[\sqrt{\frac{x^2+y^2}{2}}\right]$. Therefore, any two adjacent numbers have their difference at most 2. Hence, $2, 3, \ldots, 2005 \in M$. If $p \geq 2008$ and $p \in M$, then $2007 \in M$, which is a contradiction. Therefore, $M = \{1, 2, \ldots, 2006\}$ and the number of nonempty subsets of M is $2^{2006} - 1$.

13. It follows that if $x \in S_i$ and $y \in S_j$, then $y - x \in S_k$, so $(y - x) - y = -x \in S_i$. This means that S_i has nonnegative numbers, hence each of S_1, S_2 and S_3 contains nonnegative numbers. If one of the sets contains 0, say $0 \in S_1$, then $x - 0 \in S_3$ for each $x \in S_2$, so $S_2 \subseteq S_3$. For each $y \in S_3$ we also have $y - 0 \in S_2$, thus $S_3 \subseteq S_2$, and necessarily $S_2 = S_3$. Therefore, it suffices to show that one of the sets contains the number 0. In fact, if none of them contains 0, then choose the smallest positive integer a in $S_1 \cup S_2 \cup S_3$ and assume that $a \in S_1$. Also choose the smallest positive integer b in $S_2 \cup S_3$ and assume that $b \in S_2$. If $a = b$, then $b - a = 0 \in S_3$, a contradiction. If $b > a$, then $0 < b - a < b$ and $b - a \in S_3$, contradictory to the assumption on b. Therefore, one of S_1, S_2, and S_3 contains 0, and the assertion follows.

14. Note that $(A \cup B) \cap C = (A \cap C) \cup (B \cap C)$, while $A \cap C$ and $B \cap C$ are the solution sets to the systems of equations $\begin{cases} ax + y = 1 \\ x^2 + y^2 = 1 \end{cases}$ and

$\begin{cases} x+ay=1 \\ x^2+y^2=1 \end{cases}$, respectively. Solving the equations, we get

$$A \cap C = \left\{(0,1), \left(\frac{2a}{1+a^2}, \frac{1-a^2}{1+a^2}\right)\right\},$$

$$B \cap C = \left\{(1,0), \left(\frac{1-a^2}{1+a^2}, \frac{2a}{1+a^2}\right)\right\}.$$

Discussing different cases, we conclude that when $a = 0$ and 1, the set $(A \cup B) \cap C$ contains exactly two elements, and when $a = -1 \pm \sqrt{2}$, the set contains exactly three elements.

15. Since $A \cap B = \{a_1, a_4\}$, it follows that a_1 and a_4 are both perfect squares. Since $a_1 + a_4 = 10$, the only possibility is $a_1 = 1$ and $a_4 = 9$. Then by $a_4 < a_5$ we have $a_5 > 9$. On the other hand, the sum of numbers in $A \cup B$ is 224, thus

$$a_5^2 \leq 224 - 9^2 - (1+9) - 2^2 - 3^2 - (2+3),$$

which means that $a_5 < 11$, so $a_5 = 10$. In this case, $a_2^2 + a_3^2 + a_2 + a_3 = 32$, and one of a_2 and a_3 is 3. Therefore, $a_2 = 3$ and $a_3 = 4$, and $A = \{1, 3, 4, 9, 10\}$.

16. Note that the set $\{1, 2, 3; 17 \sim 80; 401 \sim 2000\}$ has no two numbers such that one is equal to five times the other. Hence, the maximum value of $|A|$ is at least 1667. On the other hand, suppose A is such a subset of $\{1, 2, \ldots, 2000\}$ and $|A| \geq 1668$. Note that the set $\{1, 2, \ldots, 2000\}$ has 1331 integers excluding the following pairs: $(1,5)\,(2,10)\,,(3,15)\,,(k,5k)$ and $(t, 5t)$, where $4 \leq k \leq 16$ and $81 \leq t \leq 400$. This implies that A contains at least 337 numbers in the above pairs, so two of them are in the same pair, which is a contradiction.

17. We claim that for each element a in A_1, this a belongs to at most 29 sets apart from A_1. Suppose not. Then assume that $a \in A_i$ for $i = 1, 2, \ldots, 31$. Since $A_1 \cap A_2 \cap \cdots \cap A_n = \emptyset$, there exists $j \geq 32$ such that $a \notin A_j$. Then using property (2) we see that any two of $A_1, \ldots, A_{31}$ have no common elements other than a. Also by (2), we see that $A_i \cap A_j$ has one element for $i = 1, 2, \ldots, 31$, which implies that A_j has at least 31 elements, a contradiction.

Therefore, each element of A_1 belongs to at most 29 other sets, hence $n \leq 29 \times 30 + 1 = 871$. Next, we construct 871 sets satisfying the conditions.

Let $A = \{a_1, a_2, \ldots, a_{29}\}$ and $B_i = \{a_0, a_{i,1}, a_{i,2}, \ldots, a_{i,29}\}$ with $1 \leq i \leq 29$;

$$A_{i,j} = \{a_i\} \cup \left\{a_{k,j+(k-1)(i-1)} | k = 1, 2, \ldots, 29\right\} \text{ with } 1 \leq i, j \leq 29.$$

Here $a_{k,s} = a_{k,s+29}$ while others are different elements if they have different indices. Thus, we have constructed $29^2 + 29 + 1 = 871$ sets. It is easy to verify that $A \cap B$, $A \cap A_{i,j}$, $B_i \cap B_j$, $B_r \cap A_{i,j}$, $A_{i,j} \cap A_{s,j}$ are all one-element sets, and for $i \neq s$ and $j \neq t$,

$$A_{i,j} \cap A_{s,t} = \left\{a_{k,j+(k-1)(i-1)}\right\}.$$

where k is the unique solution to the congruence equation

$$(x-1)(i-s) \equiv t - j (\bmod 29).$$

18. Note that $2000 = 11 \times 181 + 9$. We choose $A = \{x | x = 11t+1, 11t+4, 11t+6, 11t+7, 11t+9, 0 \leq t \leq 181, t \in \mathbf{Z}\}$. Then the difference between any two numbers of A is not 4 or 7. Therefore, the maximum value of $|A|$ is at least $5 \times 182 = 910$. On the other hand, let A be a set satisfying the condition with $|A| > 910$, then there exists some t such that the set $\{x | x = 11t + r, 1 \leq r \leq 11\}$ contains at least six numbers in A. We show that this is impossible. By subtracting $11t$ from each of the six numbers, we see that six numbers are in A among $1, 2, \ldots, 11$. In fact, we can arrange the eleven numbers on a circle as $1, 5, 9, 2, 6, 10, 3, 7, 11, 4, 8, 1$. Any two adjacent numbers differ by 4 or 7, and it is impossible to choose six numbers such that no two of them are adjacent. Therefore, the maximum value of $|A|$ is 910.
19. We discuss different cases as follows, based on the number of elements in T.

If $|T| = 2$, then there is only one such T, that is $\{\emptyset, A\}$.

If $|T| = 3$, then the collection Tcan be $\{\emptyset, A, \{a_1\}\}$, $\{\emptyset, A, \{a_2\}\}$, $\{\emptyset, A, \{a_3\}\}$, $\{\emptyset, A, \{a_1, a_2\}\}$, $\{\emptyset, A, \{a_2, a_3\}\}$, $\{\emptyset, A, \{a_1, a_3\}\}$, with six possibilities in all.

If $|T| = 4$, then since the intersection and union of any two members of T also belong to T, the two members apart from $\emptyset$ and A have two patterns: one is a proper subset of the other, like $\{\emptyset, A, \{a_1\}, \{a_1, a_3\}\}$; or the two sets are disjoint and their union is A, like $\{\emptyset, A, \{a_1\}, \{a_2, a_3\}\}$. There are $2 \times 3 + 3 = 9$ such collections.

If $|T| = 5$, then the three members apart from $\emptyset$ and A have the following two patterns: one of them is a proper subset of the other two, like $\{\emptyset, A, \{a_1\}, \{a_1, a_2\}, \{a_1, a_3\}\}$, and there are three such collections;

or one of them contains the other two, like $\{\emptyset, A, \{a_1\}, \{a_2\}, \{a_1, a_2\}\}$, and there are three such collections. Hence, there are six collections in this case.

If $|T| = 6$, then T has the form $\{\emptyset, A, \{a_1\}, \{a_2\}, \{a_1, a_2\}, \{a_1, a_3\}\}$, and there are six such collections.

If $|T| = 7$, then such T does not exist.

If $|T| = 8$, then T is the collection of all subsets of A, and there is only one possibility.

Overall, there are 29 such collections.

20. (1) We show by induction that there exists such a sequence of subsets $A_1, A_2, \ldots, A_{2^n}$, with $A_1 = \{1\}$ and $A_{2^n} = \emptyset$.

When $n = 2$, the sequence $\{1\}, \{1, 2\}, \{2\}, \emptyset$ satisfies the conditions.

Suppose such a sequence exists when $n = k$, written as $B_1, B_2, \ldots, B_{2^k}$. For $n = k + 1$, we construct a sequence as follows: $A_1 = B_1 = \{1\}; A_i = B_{i-1} \cup \{k + 1\}$, $i = 2, 3, \ldots, 2^k + 1; A_j = B_{j-2^k}, j = 2^k + 2, 2^k + 3, \ldots, 2^{k+1}$.

It is simple to verify that the sequence $A_1, A_2, \ldots, A_{2^{k+1}}$ satisfies the conditions. Therefore, such a sequence exists for each integer $n \geq 2$.

(2) Without loss of generality assume that $A_1 = \{1\}$. Since the number of integers in two adjacent sets differ by 1, it follows that one of them is odd and the other is even. Hence, $\sum_{i=0}^{2^n} (-1)^i S(A_i) = \sum_{A \in P} S(A) - \sum_{A \in Q} S(A)$, where P is the collection of all subsets of $\{1, 2, \ldots, n\}$ whose number of integers is even, and Q is the collection of all subsets of $\{1, 2, \ldots, n\}$ whose number of integers is odd.

For each $x \in \{1, 2, \ldots, n\}$, it appears exactly $\binom{n-1}{k-1}$ times in all the k-integer subsets, so its contribution in the expression $\sum_{A \in P} S(A) - \sum_{A \in Q} S(A)$ is

$$\begin{aligned} & \mathrm{C}_{n-1}^0 + \mathrm{C}_{n-1}^1 - \mathrm{C}_{n-1}^2 + \cdots + (-1)^n \mathrm{C}_{n-1}^{n-1} \\ & = -(1-1)^{n-1} = 0. \end{aligned}$$

Therefore, $\sum_{i=0}^{2^n} (-1)^i S(A_i) = 0$.

Solution 2

Number of Elements in a Finite Set

1. (1) $f : a_1 \mapsto b_1$, $a_2 \mapsto b_2$, $a_3 \mapsto b_3$ is an injection from A to B. The number of such maps is $4 \times 3 \times 2 = 24$.
   (2) The map $f : a_1 \mapsto b_1, a_2 \mapsto b_1, a_3 \mapsto b_1$ satisfies the condition. The number of such maps is $4^3 - 4 \times 3 \times 2 = 40$.
   (3) Since $|A| = 3$ and $|B| = 4$, there exist no surjections from A to B.
2. $3^3 = 27$.
3. If a, b and c are both harmonic and arithmetic, then $\begin{cases} \frac{1}{a} + \frac{1}{b} = \frac{2}{c} \\ a + c = 2b \end{cases}$, so $a = -2b$ and $c = 4b$. Hence, a good subset is a set of the form $\{-2b, b, 4b\}$ $(b \neq 0)$. Since it is also a subset of M, we have $-2013 \leq 4b \leq 2013$, hence $-503 \leq b \leq 503$ and $b \in \mathbf{Z}, b \neq 0$. There are 1006 different values of b, so the number of good subsets is 1006.
4. The set A consists of the four sides of the square whose vertices are $(a, 0), (0, a), (-a, 0)$, and $(0, -a)$. The set B consists of the four lines $x = \pm 1$ and $y = \pm 1$. Hence, $1 < a < 2$ or $a > 2$.

   (1) If $a > 2$, then the side length of the regular octagon can only be 2, so $\sqrt{2}a - 2\sqrt{2} = 2$, from which $a = 2 + \sqrt{2}$.
   (2) If $1 < a < 2$, then $a = \sqrt{2}$.

   In summary, $a = 2 + \sqrt{2}$ or $\sqrt{2}$.
5. From the assumption for each $k = 6, 7, \ldots, 105$, at least one of the numbers k and $19k$ does not belong to A. Hence, $|A| \leq 1995 - (105 - 6 + 1) = 1895$.

   On the other hand, let $B = \{1, 2, 3, 4, 5\}$ and $C = \{106, 107, \ldots, 1995\}$, and let $A = B \cup C$. Then $|A| = 1895$ and no element of A is 19 times another element of A. In fact, if $k \in B$, then $19 \leq 19k \leq 105$, so $19k \notin A$, while if $k \in C$, then $19k \geq 106 \times 19 > 1995$, so $19k \notin A$. Therefore, the maximum value of $|A|$ is 1895.

6. Let $A_1 = \{3k+1 | 0 \leq k \leq 33\}$ and $A_2 = \{3k+2 | 0 \leq k \leq 32\}$, and let $A = A_1 \cup A_2 \cup \{9, 18, 36, 45, 63, 72, 90, 99, 81\}$. Then $|A| = 76$ and no element of A is equal to 3 times another element of A. On the other hand, we consider the following 24 pairs: $(k, 3k)$, $k = 1, 2, 12, 13, \ldots, 33$. These pairs contain 48 different numbers, and 77 numbers from 1 through 100 must contain two numbers in the same pair, so one is three times the other. Therefore, the maximum value of $|A|$ is 76.
7. The set $\{1, 2, \ldots, 7\}$ has 127 nonempty subsets and each element appears in 64 different subsets. Since each of $1, 2, \ldots, 6$ appears 32 times in an odd position and 32 times in an even position, these numbers contribute 0 to the overall sum. The number 7 also appears 64 times, but always with a positive sign, so the overall sum of the alternating sums is $7 \times 64 = 448$.
8. If $S \subseteq \{1, 2, \ldots, 9\}$ with $|S| \geq 6$, then since the sum of any two numbers in T lies between 3 and 17, there can be at most 15 different sums. Also S has exactly $\binom{6}{2} = 15$ two-number subsets, so 3 and 17 must appear in the sums, and consequently $1, 2, 8, 9 \in S$. However, the equality $1+9 = 2+8$ provides a contradiction to the assumption. On the other hand, the set $S = \{1, 2, 3, 5, 8\}$ satisfies the condition. Therefore, the maximal number of integers in S is 5.
9. $|A \cap B| = 167$.
10. The set M has 2^{100} subsets, in which every integer of M appears 2^{99} times. Hence, the total sum of the numbers of all the subsets of M is

$$2^{99}(1+2+\cdots+100) = 5050 \cdot 2^{99}.$$

11. By the assumption, for $k = 9, 10, \ldots, 133$, at least one of the two numbers k and $15k$ does not belong to A. Hence, $|A| \leq 1995 - (133 - 9 + 1) = 1870$. On the other hand, choose

$$A = \{1, 2, \ldots, 8\} \cup \{134, 135, \ldots, 1995\}.$$

Then A satisfies the condition and $|A| = 1870$. Therefore, the maximum value of $|A|$ is 1870.

12. Let I be the set of all the teachers, and A, B, and C be the set of the teachers who can teach Literature, Mathematics, and English, respectively. Then $|I| = 120$, $|A| = 40$, $|B| = 50$, and $|C| = 45$. In addition, $|A \cap B| = 10$, $|B \cap C| = 15$, $|A \cap C| = 8$, and $|A \cap B \cap C| = 4$.

Hence, by the inclusion-exclusion principle, $|A \cup B \cup C| = |A| + |B| + |C| - |A \cap B| - |B \cap C| - |A \cap C| + |A \cap B \cap C| = 106$, hence $|I \backslash (A \cup B \cup C)| = 120 - 106 = 14$.

13. We divide the integers of E into the following 100 subsets: $E_1 = \{1, 200\}, E_2 = \{2, 199\}, \ldots, E_{100} = \{100, 101\}$. Then by the assumption, the two integers in the same E_i cannot both appear in G, so each E_i contains exactly one number in G. Since $2 + 4 + \cdots + 200 = 10100 \neq 10080$, the integers of G cannot be all even, and G can be obtained by replacing some numbers $a_1, a_2, \ldots, a_k$ of $\{2, 4, \ldots, 200\}$ by $201 - a_1, 201 - a_2, \ldots, 201 - a_k$, such that the sums of these numbers differ by $10100 - 10080 = 20$. Hence, $a_1^2 + a_2^2 + \cdots + a_{100}^2 = 2^2 + 4^2 + \cdots + 200^2 - (a_1^2 + a_2^2 + \cdots + a_k^2) + (201 - a_1)^2 + (201 - a_2)^2 + \cdots + (201 - a_k)^2 = 4(1^2 + 2^2 + \cdots + 100^2) - 201 \times 20$, which is a constant. Also since $a_1^2 + a_2^2 + \cdots + a_{100}^2 \equiv 0 \pmod 4$, while the square of an odd number is congruent to 1 modulo 4 and the square of an even number is congruent to 0, the number of odd integers in $a_1 a_2, \ldots, a_{100}$ is divisible by 4.

14. We first estimate the lower bound of $|A|$. Since the difference of any two of 1, 3, 6, and 8 is a prime number, it follows that $f(1), f(3), f(6)$, and $f(8)$ are pairwise distinct, so $|A| \geq 4$.

    Next, we construct an example. Let $A = \{0, 1, 2, 3\}$ and $f(x)$ be the remainder of x divided by 4.

    For any positive integers x and y, if $f(x) = f(y)$, then $|x - y|$ is divisible by 4, so it cannot be a prime. Therefore, the minimum value of $|A|$ is 4.

15. First, the set A has $\binom{10}{1}$, $\binom{10}{2}$, and $\binom{10}{3}$ subsets with exactly 1, 2, and 3 integers, respectively, so there are $\binom{10}{1} + \binom{10}{2} + \binom{10}{3} = 175$ such subsets, and for every two of them, their intersection has at most two numbers.

    On the other hand, if $k > 175$, then among these subsets at least $176 - \binom{10}{1} - \binom{10}{2} = 121$ have three or more numbers. We consider the three-integer subsets of all the B_i's. Since there are $\binom{10}{3} = 120$ different ways to choose three numbers from A, by the pigeonhole principle some three-number subset is contained in multiple B_i's, leading to a contradiction. Therefore, the maximum value of k is 175.

16. Let $A_2 = \{a \mid a \in S, 2 \mid a\}, A_3 = \{a \mid a \in S, 3 \mid a\}$, and $A_5 = \{a \mid a \in S, 5 \mid a\}$. Then $|A_2| = 50$, $|A_3| = 33$, $|A_5| = 20$, $|A_2 \cap A_3| = 16$, $|A_2 \cap A_5| = 10$, $|A_3 \cap A_5| = 6$, and $|A_2 \cap A_3 \cap A_5| = 3$.

Therefore, the number of integers in S that are divisible by at least one of 2, 3, and 5 is

$$|A_2 \cup A_3 \cup A_5| = 50 + 33 + 20 - 16 - 10 - 6 + 3 = 74.$$

If we choose four integers from the set $A_2 \cup A_3 \cup A_5$, then by the pigeonhole principle at least two of them belong to the same A_i ($i = 2, 3$, and 5), and they are not coprime. Hence, $n \geq 75$. Next, we show that $n = 75$ is valid. We construct four sets: $B_1 = \{1$ and the first 25 primes$\}$, $B_2 = \{2^2, 3^2, 5^2, 7^2\}, B_3 = \{2^3, 3^3, 5 \times 19, 7 \times 13\}$, and $B_4 = \{2^4, 3^4, 5 \times 17, 7 \times 11\}$. Then the intersection of any two of them is empty and any two numbers in the same set are coprime. Hence, $|B_1 \cup B_2 \cup B_3 \cup B_4| = 38$. Suppose $X \subseteq S$ with $|X| \geq 75$. Then X has at least $75 - (100 - 38) = 13$ numbers in the set $B_1 \cup B_2 \cup B_3 \cup B_4$, and by the pigeonhole principle four of them lie in the same B_i, which are pairwise coprime.

Therefore, the minimum value of n is 75.

17. Let $a \in A_1 \cup A_2 \cup \cdots \cup A_n$. We claim that a belongs to at most four of these sets. In fact, if every A_i contains a, then by condition (1) the other elements in these sets are all distinct, so $|A_1 \cup A_2 \cup \cdots \cup A_n| \geq 3n + 1 > n$, a contradiction. Hence, at least one A_i does not contain a. Without loss of generality assume that $a \notin A_1$, and suppose a belongs to at least five sets. Then A_1 has a common element with each of these sets, which are necessarily distinct since these sets have no common elements other than a. This implies that $|A_1| \geq 5$, a contradiction. Therefore, every element belongs to at most four sets, and since $|A_1| + |A_2| + \cdots + |A_n| = 4n$, it follows that every element belongs to exactly four sets. Assume that the element b belongs to the sets A_1, A_2, A_3, and A_4. Then $|A_1 \cup A_2 \cup A_3 \cup A_4| = 3 \times 4 + 1 = 13$. If $n > 13$, then there exists some element $c \notin A_1 \cup A_2 \cup A_3 \cup A_4$. Suppose A_5 contains c, which is different from A_1, A_2, A_3, or A_4. Then $b \notin A_5$ and A_5 has a common element with each of A_1, A_2, A_3 and A_4, which are different to each other and neither b nor c, so $|A_5| \geq 5$, a contradiction. Hence, $n \leq 13$.

On the other hand, consider the following 13 sets: $\{0, 1, 2, 3\}$, $\{0, 4, 5, 6\}$, $\{0, 7, 8, 9\}$, $\{0, 10, 11, 12\}$, $\{10, 1, 4, 7\}$, $\{10, 2, 5, 8\}$, $\{10, 3, 6, 9\}$, $\{11, 1, 5, 9\}$, $\{11, 2, 6, 7\}$, $\{11, 3, 4, 8\}$, $\{12, 1, 6, 8\}$, $\{12, 2, 4, 9\}$, $\{12, 3, 5, 7\}$. We can verify that they satisfy the conditions. Therefore, the maximum value of n is 13.

Solution 3

Quadratic Functions

1. (1) $y = 2x^2 - 2x - \frac{3}{2}$. Hint: We can assume that $y = 2(x+p)^2 + q$ after the translation, so $y = 2x^2 + 4px + 2p^2 + q$. By the assumption, if x_1 and x_2 are the roots of the equation $2x^2 + 4px + 2p^2 + q = 0$, then $|x_1 - x_2| = 2$, and we can plug the coordinates of the vertex into $y = -4x$, followed by solving for p and q.

   (2) 13. Hint: The problem is equivalent to finding all the integer values of the function $y = \frac{1}{3}x(3x - 8)$ for $x \in [0, 5]$.

   (3) $[5, 10]$. Hint: Let $f(-2) = uf(-1) + vf(-1)$, where $u, v \in \mathbf{R}$. Then $f(-1) = a - b$, $f(1) = a + b$, and $f(-2) = 4a - 2b$. Hence, $f(-2) = 3f(-1) + f(1)$. Therefore, $f(-2) \in [5, 10]$.

   (4) 0. Hint: Having the maximum value implies that the parabola opens downwards, so $a < 0$ and the vertex is $(3, 10)$. Since the distance between its two intersection points with the x-axis is 4, it follows that the intersection points are $(1, 0)$ and $(5, 0)$. Hence, $f(1) = 0$.

   (5) 7. Hint: $x_1 + x_2 = 2 \cdot \left(-\frac{b}{2a}\right) = -\frac{b}{a}$, so $f(x_1 + x_2) = a \cdot \frac{b^2}{a^2} - \frac{b^2}{a} + 7 = 7$.

2. Since $f(0) = 2$ and $f(-1) = -1$, we get $c = 2$ and $b = -(a+3)$. Hence,

$$f(x) = ax^2 - (a + 3)x + 2.$$

   Also since $|x_1 - x_2| = 2\sqrt{2}$, we can solve that $a = 1$ or $a = -\frac{9}{7}$. Therefore, the desired function is $y = x^2 - 4x + 2$ or $y = -\frac{9}{7}x^2 - \frac{12}{7}x + 2$.

3. By the conditions and the symmetry of quadratic functions, $\frac{a+b}{2} = -\frac{a}{2}$, that is $2a + b = 0$. Hence, $f(2) = 4 + 2a + b = 4$.

4. (1) The graph of $y = ax^2 + bx$ intersects the x-axis at $(0, 0)$ and the other intersection point $(x_0, 0)$ satisfies that $15 < x_0 < 16$, from which $b < 0$. Since $0 + x_0 = -\frac{b}{a}$, we get $15 < -\frac{b}{a} < 16$, which implies that $a \nmid b$.
   (2) The axis of symmetry of the graph of $y = ax^2 + bx$ is $x = \frac{x_0}{2}$, where $7.5 < \frac{x_0}{2} < 8$. When $n \geq 8 (n \in \mathbf{Z})$, the numbers $f(n)$ are increasing, and when $n \leq 7$ $(n \in \mathbf{Z})$, the numbers $f(n)$ are decreasing. Since $f(8) < f(7)$, we conclude that $f(n)$ is the minimal value when $n = 8$.
5. $y = 2x^2 + 1 - (x - 4)p$. When $x = 4$, we have $y = 33$, so the parabola always passes through the point $(4, 33)$.
6. $y = f(x) = (x - 1)^2 + 2$. In order to attain the minimum value 2, we need $x = 1$, from which $a \geq 1$. Also if $f(x) \leq 3$, then $0 \leq x \leq 2$, and since $f(0) = 3$, so necessarily $a \leq 2$. Therefore, the value range of a is $[1, 2]$.
7. $y = (x - 2)^2 + 2$. The axis of symmetry is $x = 2$. If $a < b < 2$, then the maximum and minimum values of the function on the interval $[a, b]$ are $f(a)$ and $f(b)$, respectively. Hence, $\begin{cases} a^2 - 4a + 6 = b \\ b^2 - 4b + 6 = a \end{cases}$. This system of equations has no real solutions. If $a \leq 2 \leq b$, then the minimum value of the function is $f(2)$ while the maximum value is either $f(a)$ or $f(b)$. Hence, $\begin{cases} a = 2 \\ b = f(a) \end{cases}$ or $\begin{cases} a = 2 \\ b = f(b) \end{cases}$. For the first case, $b = f(a) = f(2) = 2 = a$, a contradiction. For the second case, $a = 2$ and $b = 3$. If $2 < a < b$, then by a similar argument, $f(a) = a$ and $f(b) = b$, which has no solution. Therefore, $a = 2$ and $b = 3$.
8. We discuss the following three cases based on the interval $[a, b]$.
   (1) If $a < b \leq 0$, then the function $f(x)$ is increasing on $[a, b]$, so $f(a) = 2a$ and $f(b) = 2b$. Hence

$$\begin{cases} -\dfrac{1}{2}a^2 + \dfrac{13}{2} = 2a, \\[2ex] -\dfrac{1}{2}b^2 + \dfrac{13}{2} = 2b, \end{cases}$$

   which means that a and b are distinct roots of the equation $-\frac{1}{2}x^2 - 2x + \frac{13}{2} = 0$. But the two roots of this equation have opposite signs, which gives a contradiction.
   (2) If $a < 0 < b$, then $b = \frac{13}{4}$. Hence, $f(b) = -\frac{1}{2}\left(\frac{13}{4}\right)^2 + \frac{13}{2} = \frac{39}{12} > 0$. Also $a < 0$ and $f(b) \neq 2a$, so $f(a) = 2a$, in other words $-\frac{1}{2}a^2 + \frac{13}{2} = 2a$. Solving the equation, we get $a = -2 - \sqrt{17}$. In this case, $[a, b] = \left[-2 - \sqrt{17}, \frac{13}{4}\right]$.

(3) If $0 \le a < b$, then $f(x)$ is decreasing on $[a, b]$, so $f(a) = 2b$ and $f(b) = 2a$, or equivalently

$$\begin{cases} -\dfrac{1}{2}a^2 + \dfrac{13}{2} = 2b, \\ -\dfrac{1}{2}b^2 + \dfrac{13}{2} = 2a. \end{cases}$$

Solving the equations, we get $a = 1$ and $b = 3$, thus $[a, b] = [1, 3]$.

Therefore, the interval $[a, b]$ is $[1, 3]$ or $\left[-2 - \sqrt{17}, \frac{13}{4}\right]$.

9. Since $y = \begin{cases} (x-2)^2, & x \ge 0 \\ (x+2)^2, & x < 0 \end{cases}$, we have $b = 0$ and $c^2 - 4c + 4 = 4c$, which implies that $c = 4 + 2\sqrt{3}$. Hence, $b + c = 4 + 2\sqrt{3}$.

10. Let $f(x) = 7x^2 - (p + 13)x + p^2 - p + 2$. Then

$$\begin{cases} f(0) = p^2 - p - 2 > 0, \\ f(1) = p^2 - 2p - 8 < 0, \\ f(2) = p^2 - 3p > 0. \end{cases}$$

Solving the system of inequalities, we get $p \in (-2, -1) \cup (3, 4)$.

11. Let $f(x) = ax^2 + bx + c(a \ne 0)$. Then since $1 \le f(1) \le \frac{1+1}{2} = 1$, necessarily $f(1) = 1$. From $f(-1) = 0$ and $f(1) = 1$, we get $b = a + c = \frac{1}{2}$. Also since $ax^2 + bx + c \ge x$ for all real x, it follows that $a > 0$ and $\Delta = (b - 1)^2 - 4ac \le 0$, so $ac \ge \frac{1}{16}$. Therefore, $a = c = \frac{1}{4}$ and $b = \frac{1}{2}$, from which $f(x) = \frac{1}{4}x^2 + \frac{1}{2}x + \frac{1}{4}$.

12. (1) Let $A\left(\frac{y_1^2}{4}, y_1\right)$ and $B\left(\frac{y_2^2}{4}, y_2\right)$ $(y_1 > 0 > y_2)$. Suppose the slope of l_1 (it exists obviously) is k_1. Then the equation of l_1 is $y - y_1 = k_1\left(x - \frac{y_1^2}{4}\right)$. By $\begin{cases} y - y_1 = k_1\left(x - \frac{y_1^2}{4}\right) \\ y^2 = 4x \end{cases}$, we get $k_1y^2 - 4y + 4y_1 - k_1y_1^2 = 0$.

Since l_1 is tangent to the parabola, $\Delta = 16 - 4k_1(4y_1 - k_1y_1^2) = 0$. Hence, $k_1 = \frac{2}{y_1}$, so the equation of l_1 is $y = \frac{2}{y_1}x + \frac{1}{2}y_1$. Similarly, the equation of l_2 is $y = \frac{2}{y_2}x + \frac{1}{2}y_2$.

Aligning the equations for l_1 and l_2 we can solve for the coordinates of P, obtaining $P\left(\frac{y_1y_2}{4}, \frac{y_1+y_2}{2}\right)$. Since

$$k_{AB} = \frac{y_1 - y_2}{\frac{y_1^2}{4} - \frac{y_2^2}{4}} = \frac{4}{y_1 + y_2},$$

the equation of the line AB is $y - y_1 = \frac{4}{y_1+y_2}\left(x - \frac{y_1^2}{4}\right)$, and since the line AB passes through the focus $F(1, 0)$, so $-y_1 = \frac{4}{y_1+y_2}\left(1 - \frac{y_1^2}{4}\right)$, in other words $y_1y_2 = -4$.

Therefore, $x_P = \frac{y_1 y_2}{4} = -1$, and the point P lies on the line $x = -1$.

Alternative solution. Denote $A(x_1, y_1)$ and $B(x_2, y_2)$. Then the equation of l_1 is $y_1 y = 2(x+x_1)$ and the equation of l_2 is $y_2 y = 2(x+x_2)$.

Let $P(x_0, y_0)$. Then $y_0 y_1 = 2(x_0+x_1)$ and $y_0 y_2 = 2(x_0+x_2)$. Hence, the coordinates of A and B satisfy the equation $yy_0 = 2(x_0+x)$, which is exactly the equation of the line AB. Since AB passes through $F(1,0)$, we have $0 = 2(x_0+1)$, so $x_0 = -1$ and P lies on the fixed line $x = -1$.

(2) From (1) we see that the coordinates of C and D are $C\left(4, \frac{8}{y_1} + \frac{1}{2}y_1\right)$ and $D\left(4, \frac{8}{y_2} + \frac{1}{2}y_2\right)$. Hence,

$$|CD| = \left|\left(\frac{8}{y_1} + \frac{1}{2}y_1\right) - \left(\frac{8}{y_2} + \frac{1}{2}y_2\right)\right| = \left|\frac{(y_1 y_2 - 16)(y_1 - y_2)}{2y_1 y_2}\right|.$$

Therefore,

$$S_{\Delta PCD} = \frac{1}{2}\left|4 - \frac{y_1 y_2}{4}\right| \cdot \left|\frac{(y_1 y_2 - 16)(y_1 - y_2)}{2y_1 y_2}\right|.$$

Let $y_1 y_2 = -t^2 (t > 0)$ and $|y_1 - y_2| = m$. Then by

$$(y_1 + y_2)^2 = (y_1 - y_2)^2 + 4y_1 y_2 = m^2 - 4t^2 \geq 0,$$

we have $m \geq 2t$, where the equality holds if and only if $y_1 + y_2 = 0$. Hence,

$$S_{\Delta PCD} = \frac{1}{2}\left|4 + \frac{t^2}{4}\right| \cdot \left|\frac{(-t^2 - 16)m}{-2t^2}\right| = \frac{m \cdot (t^2+16)^2}{16t^2}$$
$$\geq \frac{2t \cdot (t^2+16)^2}{16t^2} = \frac{(t^2+16)^2}{8t}.$$

Let $f(t) = \frac{(t^2+16)^2}{8t}$. Then

$$f'(t) = \frac{2(t^2+16) \cdot 2t \cdot t - (t^2+16)^2}{8t^2} = \frac{(3t^2-16)(t^2+16)}{8t^2}.$$

When $0 < t < \frac{4\sqrt{3}}{3}$ we have $f'(t) < 0$, and $f'(t) > 0$ when $t > \frac{4\sqrt{3}}{3}$. Hence, the function $f(t)$ is decreasing on $\left(0, \frac{4\sqrt{3}}{3}\right]$ and increasing on $\left[\frac{4\sqrt{3}}{3}, +\infty\right)$. Consequently, when $t = \frac{4\sqrt{3}}{3}$, the function $f(t)$ attains its minimum value $\frac{128\sqrt{3}}{9}$.

To summarize, when $y_1 + y_2 = 0$ and $y_1y_2 = -\frac{16}{3}$, or equivalently $y_1 = \frac{4}{\sqrt{3}}$ and $y_2 = -\frac{4}{\sqrt{3}}$, the area of ΔPCD takes the minimum value $\frac{128\sqrt{3}}{9}$.

13. (1) Let $g(x) = f(x) - x$. Then the graph of $y = g(x)$ is symmetric with respect to the line $x = \frac{x_1+x_2}{2}$ and $g(x)$ is decreasing on $(0, x_1)$. Hence, when $x \in (0, x_1)$, we have $g(x) > g(x_1) = 0$, namely $f(x) > x$. Also let $h(x) = f(x) - x_1$. Then $h(x_1) = f(x_1) - x_1 = g(x_1) = 0$, and

$$h(0) = f(0) - x_1 = c - x_1 = ax_1x_2 - x_1 = x_1(ax_2 - 1) < 0.$$

Therefore, $h(x) < 0$ for all $x \in (0, x_1)$, namely $f(x) < x_1$.
(2) Let α be the other root of $h(x)$. Then $\alpha < 0$ by $h(0) < 0$. Hence, $x_0 = \frac{\alpha+x_1}{2} < \frac{x_1}{2}$.

14. Let $f(x) = x^2 - ax + b$. Then $\begin{cases} f(-1) = 1 + a + b \geq 0 \\ f(1) = 1 - a + b \leq 0 \\ f(2) = 4 - 2a + b \geq 0 \end{cases}$, so

$$-1 \leq a - 2b \leq 5.$$

15. Let x_1 and $x_2 (x_1 \neq x_2)$ be the roots of $ax^2 + bx + c = 0$, and

$$f(x) = ax^2 + bx + c + k\left(x + \frac{b}{2a}\right).$$

Then

$$f(x_1) = ax_1^2 + bx_1 + c + k\left(x_1 + \frac{b}{2a}\right) = k\left(x_1 + \frac{b}{2a}\right),$$

$$f(x_2) = ax_2^2 + bx_2 + c + \left(x_2 + \frac{b}{2a}\right) = k\left(x_2 + \frac{b}{2a}\right).$$

Hence, $f(x_1)f(x_2) = k^2\left[x_1x_2 + \frac{b}{2a}(x_1 + x_2) + \frac{b^2}{4a^2}\right] = \frac{k^2(4ac-b^2)}{4a^2} < 0$. This means that the equation $f(x) = 0$ has at least one root between x_1 and x_2.

16. (1) Plugging the coordinates of the three points into $y = ax^2 + bx + c$, we find that $a = 1$, $b = 2$, and $c = -1$, so the parabola has the formula $y = x^2 + 2x - 1$. If a point (x, y) on this parabola is equidistant to the two coordinate axes, then $|y| = |x|$, namely $y = \pm x$. Solving the equation gives $x_{1,2} = \frac{-3\pm\sqrt{13}}{2}$ and $x_{3,4} = \frac{-1\pm\sqrt{5}}{2}$. Therefore, there are

four points satisfying the conditions, namely

$$A\left(\frac{-3-\sqrt{13}}{2},\frac{3+\sqrt{13}}{2}\right),\quad B\left(\frac{-3+\sqrt{13}}{2},\frac{3-\sqrt{13}}{2}\right),$$

$$C\left(\frac{-1-\sqrt{5}}{2},\frac{-1-\sqrt{5}}{2}\right),\quad D\left(\frac{-1+\sqrt{5}}{2},\frac{-1+\sqrt{5}}{2}\right).$$

(2) Since A, B, C, and D have equal distances to the two axes, the line AB bisects the second and fourth quadrants, and the line CD bisects the first and third quadrants. Hence,

$$|AB| = \sqrt{2}|x_2 - x_1| = \sqrt{26}, |CD| = \sqrt{2}|x_4 - x_3| = \sqrt{10}.$$

Since $AB \perp CD$, we have $S_{ACBD} = \frac{1}{2}AB \cdot CD = \sqrt{65}$.

17. $y = -x^2 + 2ax + b$ has the vertex $(a, a^2 + b)$, so $ma - a^2 - b - 2m - 1 = 0$. Hence, $b = ma - a^2 - 2m + 1$. Also since the equation $-x^2 + 2ax + b = x^2$ has real solutions,

$$\Delta = (-2a)^2 - 8(-ma + a^2 + 2m - 1) \geq 0.$$

Thus, $a^2 - 2ma + 4m - 2 \leq 0$ and $\Delta' = 4m^2 - 4(4m - 2) \geq 0$. Solving the last inequality, we get that $m \leq 2 - \sqrt{2}$ or $m \geq 2 + \sqrt{2}$.

18. The condition is equivalent to saying that for any real numbers x and y,

$$(ax^2y^2 + b) + [a(x+y)^2 + b] \geq (ax^2 + b)(ay^2 + b). \text{ ①}$$

We first find necessary conditions for a and b.

Let $y = 0$ in ①. Then $b + (ax^2 + b) \geq (ax^2 + b)b$, so for all real numbers x we have $(1 - b)ax^2 + b(2 - b) \geq 0$. Since $a > 0$, we know that ax^2 can be arbitrarily large, so necessarily $1 - b \geq 0$, that is $0 < b \leq 1$.

Then let $y = -x$ in ①, and we have $(ax^4 + b) + b \geq (ax^2 + b)^2$, so

$$(a - a^2)x^4 - 2abx^2 + (2b - b^2) \geq 0 \text{ ②}$$

for all real x. We denote the left side of ② as $g(x)$. Apparently $a - a^2 \neq 0$ (or otherwise $a = 1$ since $a > 0$, but $g(x) = -2bx^2 + (2b - b^2)$ with $b > 0$, so $g(x)$ can take negative values, a contradiction). Thus,

$$\begin{aligned} g(x) &= (a - a^2)\left(x^2 - \frac{ab}{a - a^2}\right)^2 - \frac{(ab)^2}{a - a^2} + (2b - b^2) \\ &= (a - a^2)\left(x^2 - \frac{b}{1 - a}\right)^2 + \frac{b}{1 - a}(2 - 2a - b) \geq 0 \end{aligned}$$

for all real x. It follows that $a - a^2 \geq 0$, namely $0 < a < 1$.

Further, since $\frac{b}{1-a} > 0$ and $g\left(\sqrt{\frac{b}{1-a}}\right) = \frac{b}{1-a}(2-2a-b) \geq 0$, we get that $2a+b \leq 2$. Therefore, the numbers a and b need to satisfy:

$$0 < b \leq 1, \quad 0 < a < 1, \quad 2a+b \leq 2. \text{ ③}$$

Next, we show that if a and b satisfy the conditions in ③, then ① always holds, in other words,

$$h(x,y) = (a-a^2)x^2y^2 + a(1-b)(x^2+y^2) + 2axy + (2b-b^2) \geq 0$$

for all x and y.

In fact, if ③ is valid, then $a(1-b) \geq 0$, $a-a^2 > 0$, and $\frac{b}{1-a}(2-2a-b) \geq 0$. Combining them with $x^2+y^2 \geq -2xy$, we have

$$\begin{aligned} h(x,y) &\geq (a-a^2)x^2y^2 + a(1-b)(-2xy) + 2axy + (2b-b^2) \\ &= (a-a^2)x^2y^2 + 2abxy + (2b-b^2) \\ &= (a-a^2)\left(xy + \frac{b}{1-a}\right)^2 + \frac{b}{1-a}(2-2a-b) \geq 0. \end{aligned}$$

Therefore, all such pairs (a,b) are

$$\{(a,b) | 0 < b \leq 1, \ 0 < a < 1, \ 2a+b \leq 2\}.$$

Solution 4

Graphs and Properties of Functions

1. (1) $x - \frac{1}{2}$. Hint: By $f(x-1) = f(2-x)$ we have $f(x) = f(1-x)$, so $x = \frac{1}{2}$.

(2) $f(x) = \frac{1}{x^2-1}(x \neq \pm 1)$. Hint: Let $h(x) = f(x) + g(x) = \frac{1}{x-1}(x \neq \pm 1)$. Then

$$\left.\begin{aligned} h(x) &= \frac{1}{x-1} = f(x) + g(x) \\ h(-x) &= -\frac{1}{x+1} = f(x) - g(x) \end{aligned}\right\} \Rightarrow 2f(x) = \frac{1}{x-1} - \frac{1}{x+1}.$$

Hence, $f(x) = \frac{1}{2}\left(\frac{1}{x-1} - \frac{1}{x+1}\right) = \frac{1}{x^2-1}(x \neq \pm 1)$.

(3) $y = -\varphi(-x)$. Hint: The graphs of the third function and the first function are centrally symmetric with respect to the origin.

(4) $[-\sqrt{2}, \sqrt{2}]$. Hint: It is equivalent to solving the inequality $|x^2 - 1| \leq 1$.

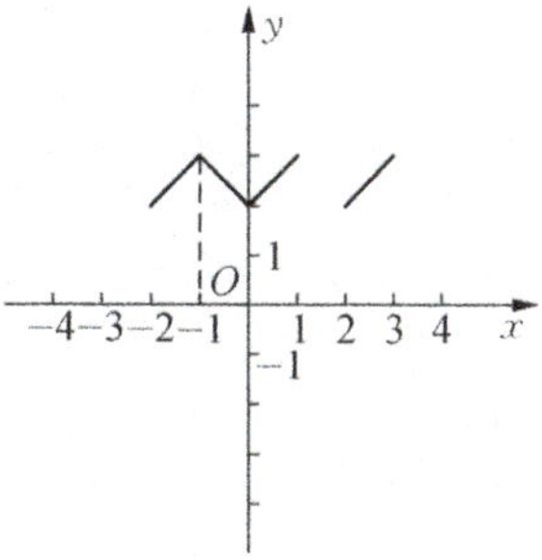

Exercise 1

2. 2015. Hint: Since $f(-10)+(-10)^3 = f(10)+10^3$, so $f(-10) = f(10)+2000 = 2015$.
3. From $f(1-x) = \frac{9^{1-x}}{9^{1-x}+3} = \frac{3}{9^x+3}$, we have $f(x) + f(1-x) = 1$, and hence $f\left(\frac{1}{2000}\right) + f\left(\frac{2}{2000}\right) + \cdots + f\left(\frac{1999}{2000}\right) = 999 + f\left(\frac{1000}{2000}\right) = 999\frac{1}{2}$.
4. The given condition says that $|x|^a - a^x = f(x) = f(-x) = |-x|^a - a^{-x}$. In other words, $a^x = a^{-x}$ for all real x, so $a = 1$. Thus the equation $f(x) = a$ becomes $|x| - 1 = 1$, whose solutions are $x = \pm 2$.
5. First, $f(1-m)$ and $f(1-m^2)$ are both defined, so $\begin{cases} -1 < 1-m < 1, \\ -1 < 1-m^2 < 1. \end{cases}$ Hence, $0 < m < \sqrt{2}$, and since $f(x)$ is odd, $f(1-m) = -f(m-1)$, from which $f(1-m^2) < -f(1-m) = f(m-1)$. By monotonicity $1-m^2 > m-1$, thus $-2 < m < 1$. Therefore, $0 < m < 1$.
6. $f(x) = (x^2+2x+7) + \frac{64}{x^2+2x+7} - 1$. Let $u = x^2+2x+7 = (x+1)^2+6$. Then $6 \le u \le 10$. The function $f(u) = u + \frac{64}{u} - 1$ is decreasing on $[6, 8]$ and increasing on $[8, 10]$. Since $f(6) > f(10)$, we have $f(8) \le f(u) \le f(6)$, or equivalently

$$15 \le f(u) \le 15\frac{2}{3}.$$

7. Since $f(x) - 2f\left(\frac{1}{x}\right) = x$ and $f\left(\frac{1}{x}\right) - 2f(x) = \frac{1}{x}$, by cancelling $f\left(\frac{1}{x}\right)$ we get $f(x) = -\frac{1}{3}\left(x + \frac{2}{x}\right)$, so $|f(x)| = \frac{1}{3}|x + \frac{2}{x}| \ge \frac{2\sqrt{2}}{3}$, and the range of $f(x)$ is $\left(-\infty, -\frac{2\sqrt{2}}{3}\right] \cup \left[\frac{2\sqrt{2}}{3}, +\infty\right)$.
8. Since $f(x)$ is defined on $(0, +\infty)$, by

$$\begin{cases} 2a^2 + a + 1 = 2\left(a + \dfrac{1}{4}\right)^2 + \dfrac{7}{8} > 0 \\ 3a^2 - 4a + 1 > 0, \end{cases},$$

we get $a > 1$ or $a < \frac{1}{3}$ ①. Since $f(x)$ is decreasing on $(0, +\infty)$,

$$2a^2 + a + 1 > 3a^2 - 4a + 1.$$

which gives that $0 < a < 5$. Combining it with ①, we conclude that $0 < a < \frac{1}{3}$ or $1 < a < 5$.
9. If $3 + x_0$ is a root of $f(x) = 0$, then $f(3 - x_0) = f(3 + x_0) = 0$, so $3 - x_0$ is also a root. Therefore, the six roots are $3 \pm x_1$, $3 \pm x_2$, and $3 \pm x_3$, which sum to 18.

10. By $f(x)+g(x)=\frac{1}{x^2-x+1}$, we have $f(x)-g(x)=\frac{1}{x^2+x+1}$, so

$$f(x)=\frac{1}{2}\left(\frac{1}{x^2-x+1}+\frac{1}{x^2+x+1}\right)=\frac{x^2+1}{(x^2-x+1)(x^2+x+1)}$$

$$g(x)=\frac{1}{2}\left(\frac{1}{x^2-x+1}-\frac{1}{x^2+x+1}\right)=\frac{x}{(x^2-x+1)(x^2+x+1)}$$

$$\frac{f(x)}{g(x)}=\frac{x^2+1}{x}=x+\frac{1}{x}.$$

When $x>0$, we have $\frac{f(x)}{g(x)}\geq 2$, and $\frac{f(x)}{g(x)}\leq -2$ when $x<0$. Therefore, the value range of $\frac{f(x)}{g(x)}$ is $(-\infty,-2]\cup[2,+\infty)$.

11. The inequality $f(1-kx-x^2)<f(2-k)$ holds for all $x\in[0,1]$ if and only if $1-kx-x^2<2-k$, namely $x^2+kx-k+1>0$ for all $x\in[0,1]$. Let $g(x)=x^2+kx-k+1$ for $x\in[0,1]$. Then $g(x)$ has the minimum value

$$g_{\min}(x)=\begin{cases} g(0)=1-k, & k>0,\\ g\left(-\dfrac{k}{2}\right)=-\dfrac{1}{4}k^2-k+1, & -2\leq k\leq 0,\\ g(1)=1, & k<-2. \end{cases}$$

Therefore, $k<1$.

12. (1) $f(x)=\sin^2 x+a\sin x+a-\frac{3}{a}$. Let $t=\sin x(-1\leq t\leq 1)$. Then $g(t)=t^2+at+a-\frac{3}{a}$. Note that $f(x)\leq 0$ for all real x if and only if

$$\begin{cases} g(-1)=1-\dfrac{3}{a}\leq 0,\\ g(1)=2+2a-\dfrac{3}{a}\leq 0. \end{cases}$$

Solving the inequalities, we get that $a\in(0,1]$.

(2) Since $a\geq 2$, we have $\frac{a}{2}\leq -1$. Hence, $g(t)_{\min}=g(-1)=1-\frac{3}{a}$. This implies that $f(x)_{\min}=1-\frac{3}{a}$. Therefore, $f(x)\leq 0$ for some real x if and only if $1-\frac{3}{a}\leq 0$, that is $0<a\leq 3$. The value range of a is thus $[2,3]$.

13. (1) Since $f(x+2)=\frac{1+f(x)}{1-f(x)}$, we have $f(x+4)=f[2+(x+2)]=\frac{1+f(x+2)}{1-f(x+2)}=-\frac{1}{f(x)}$, from which $f(x+8)=f(x+4+4)=-\frac{1}{f(x+4)}=f(x)$. Therefore, $f(x)$ is periodic with period 8.
(2) $f(1997)=f(8\times 249+5)=f(5)=f(1+4)=-\frac{1}{f(1)}=\sqrt{3}-2$ and $f(2001)=f(8\times 250+1)=f(1)=2+\sqrt{3}$.
14. Since $f(x+4)-f(x)\leq 2(x+1)$ for all real x,

$$\begin{aligned}f(x+12)-f(x)&=[f(x+12)-f(x+8)]+[f(x+8)-f(x+4)]\\&\quad+[f(x+4)-f(x)]\\&\leq 2[(x+8)+1]+2[(x+4)+1]+2(x+1)\\&=6x+30=6(x+5).\end{aligned}$$

Also $f(x+12)-f(x)\geq 6(x+5)$, so $f(x+12)-f(x)=6(x+5)$. Hence,

$$\begin{aligned}f(2016)&=[f(2016)-f(2004)]+[f(2004)-f(1992)]\\&\quad+\cdots+[f(12)-f(0)]+f(0)\\&=6\times 2009+6\times 1997+\cdots+6\times 5+1008\\&=6\times\frac{(2009+5)\times 168}{2}+1008=1008\times 1008.\end{aligned}$$

Therefore, $\frac{f(2016)}{2016}=\frac{1008}{2}=504$.
15. Since $3x^2=(x+1)^3-(x^3+3x+1)$, we have $f(x)=g(x)-h(x)$, where $g(x)=(x+1)^3$ and $h(x)=x^3+3x+1$ are both increasing polynomial functions.
16. If the equation $\sqrt{x^2-12}=y+3-2x$ for x has solutions in $(-\infty,-2\sqrt{3}]\cup[2\sqrt{3},+\infty)$, then $\begin{cases}3x^2-4(y+3)x+(y^2+6y+21)=0\\ x\leq\frac{1}{2}(y+3)\end{cases}$. Solving for y, we get $y\geq 4\sqrt{3}-3$ or $y\leq -9$. Therefore, the range of the function is $(-\infty,-9]\cup[4\sqrt{3}-3,+\infty)$.
17. (1) By the formula of the function,

$$\begin{aligned}f_1\left(\frac{2}{15}\right)&=\frac{2}{15}+\frac{1}{2}=\frac{19}{30}\\f_2\left(\frac{2}{15}\right)&=f\left(\frac{19}{30}\right)=2\left(1-\frac{19}{30}\right)=\frac{11}{15},\\f_3\left(\frac{2}{15}\right)&=f\left(\frac{11}{15}\right)=2\left(1-\frac{11}{15}\right)=\frac{8}{15},\end{aligned}$$

$$f_4\left(\frac{2}{15}\right) = f\left(\frac{8}{15}\right) = 2\left(1 - \frac{8}{15}\right) = \frac{14}{15},$$

$$f_5\left(\frac{2}{15}\right) = f\left(\frac{14}{15}\right) = 2\left(1 - \frac{14}{15}\right) = \frac{2}{15}.$$

Hence, the sequence $f_n\left(\frac{2}{15}\right)$ is periodic with period 5, in other words, $f_{5k+r}\left(\frac{2}{15}\right) = f_r\left(\frac{2}{15}\right)$. Therefore, $f_{2004}\left(\frac{2}{15}\right) = f_{5\times 400+4}\left(\frac{2}{15}\right) = f_4\left(\frac{2}{15}\right) = \frac{14}{15}$.

(2) Let $A = \{\frac{2}{15}, \frac{19}{30}, \frac{11}{15}, \frac{8}{15}, \frac{14}{15}\}$. Then by (1), $f_5(a) = a$ for $a \in A$, so $f_{15}(a) = a$, and hence $A \subseteq B$.

We draw the graph of $f(x)$, as shown in the figure. By $x = 2(1 - x)\left(\frac{1}{2} < x \le 1\right)$ we get $x = \frac{2}{3}$. Thus, $f(\frac{2}{3}) = \frac{2}{3}$ so $f_{15}\left(\frac{2}{3}\right) = \frac{2}{3}$, thus $\frac{2}{3} \in B$. Let $C = \{0, 1, \frac{1}{2}\}$. Then by $f(0) = \frac{1}{2}, f\left(\frac{1}{2}\right) = 1$, and $f(1) = 0$, we have $f_3(c) = c$ for $c \in C$, and consequently $f_{15}(c) = c$. This implies that $C \subseteq B$.

In summary, $\left\{\frac{2}{15}, \frac{19}{30}, \frac{11}{15}, \frac{8}{15}, \frac{14}{15}, \frac{2}{3}, 0, \frac{1}{2}, 1\right\} \subseteq B$, and the set B contains at least nine numbers.

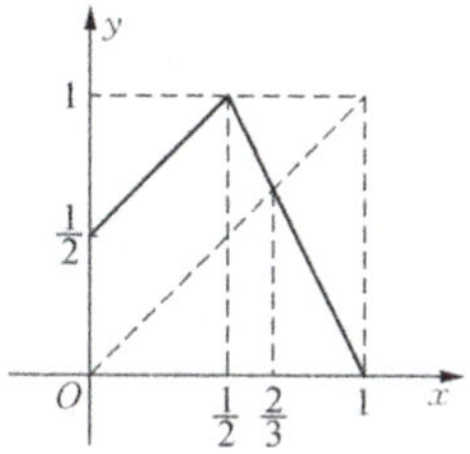

Exercise 17

18. If a, b and c are nonzero real numbers, then Ann has a winning strategy. Ann first chooses $b = 1$, and no matter which of a and c Bob chooses, Ann can always choose the other to make $1 - 4ac < 0$, so the equation $f(x) = 0$ has no real solutions.

However, if a, b and c can be any real numbers, then Bob has a winning strategy. If Ann first determines a or b, then Bob can let $c = 0$, so $x = 0$ is a root of the equation $f(x) = 0$. If Ann chooses $c \neq 0$, then Bob lets $a = -c$, thus $f\left(-\frac{\pi}{2}\right) \cdot f\left(\frac{\pi}{2}\right) = (a - b - a)(a + b - a) = -b^2 \le 0$, which means that $f(x) = 0$ has a real root in the interval $\left[-\frac{\pi}{2}, \frac{\pi}{2}\right]$.

Solution 5

Power Functions, Exponential Functions, and Logarithmic Functions

1. (1) 10. Hint: The original expression equals $5^{\lg 2\cdot 10}\cdot\left(\frac{1}{2}\right)^{\lg\frac{1}{2}} = 5^{\lg 2+1}\cdot 2^{\lg 2} = 5^{\lg 2}\cdot 2^{\lg 2}\cdot 5 = 10$.

   (2) 2. Hint: Make use of the property $\log_a b^m = m\log_a b$, and we have

$$f(x_1^2) - f(x_2^2) = 2\log_a x_1 - 2\log_a x_2 = 2(f(x_1) - f(x_2)) = 2$$

   (3) $x = 2$. Hint: The equation is equivalent to $\sqrt{4x^2 - 5x + 2} = \sqrt{4x}$, which shows that $x = 2$ or $\frac{1}{4}$. When $x = \frac{1}{4}$, we have $4x = 1$, which is impossible. Hence, $x = 2$.

   (4) $2^{90} - 1$. Hint: First solve the inequality $-1 \leq \log_{\frac{1}{x}} 10 < -\frac{1}{2}$ to get $10 \leq x < 100$ and $x \in \mathbf{N}_+$. Then the set

$$\left\{x \middle| -1 \leq \log_{\frac{1}{x}} 10 < -\frac{1}{2},\ x \in \mathbf{N}_+\right\} = \{x|10 \leq x < 100,\ x \in \mathbf{N}_+\}$$

   Therefore, it has $2^{90} - 1$ proper subsets.

   (5) 3. Hint: Finding $f^{-1}(30)$ is equivalent to finding the solution to the equation $30 = 3^x + \log_3 x + 2$. The solution is $x = 3$, so $f^{-1}(30) = 3$.

   (6) $-3 \leq m < 0$. Hint: Let $t = 5^{-|x+1|}$. Then $t \in (0, 1]$. The original function becomes $f(t) = t^2 - 4t - m$, where $t \in (0, 1]$. The graph of $f(t)$ intersects the x-axis, so $\Delta \geq 0$, $f(0) > 0$ and $f(1) \leq 0$. Therefore, $m \in [-3, 0)$.

2. If $a = 0$, then $x > -\frac{1}{2}$, and $y = \lg(ax^2 + 2x + 1)$ can take all real values. If $a > 0$, then y takes all real values if and only if $\Delta_x = 4 - 4a \geq 0$ in

$ax^2 + 2x + 1$, or equivalently $a \leq 1$. If $a < 0$, then the function cannot take all real values. Therefore, the value range of a is $[0, 1]$.

3. The function $f(x) = x^2 + 2x + 5$ has the range $[4, +\infty)$. If the inequality holds for all real x, then the value range of a is $(0, 1)$.
4. Let $x = \log_7(2\sqrt{2}+1) + \log_2(\sqrt{2}-1)$. Then $x + a = 1$. Hence, $x = 1 - a$.
5. Since $f(2) = a - 2$ and $f(4) = 2a$, we have $a - 2 < 2a$, that is $a > -2$.
6. By $\frac{1}{\log_x 3} + \frac{1}{\log_y 3} \geq 4$ we get that $\log_3(xy) \geq 4$, namely $xy \geq 81$. Now,

$$u = 2^x + 2^y \geq 2 \cdot 2^{\frac{x+y}{2}} \geq 2 \cdot 2^{\sqrt{xy}} \geq 2 \cdot 2^9 = 1024.$$

When $x = y = 9$, the equality holds. Therefore, the minimum value of u is 1024.

7. Since $f(g(x)) = 2^{x^2}$ and $g(f(x)) = 2^{2x}$, the inequality to be solved is

$$2^{x^2} < 3 \cdot 2^{2x} = 2^{2x+\log_2 3},$$

Or equivalently $x^2 < 2x + \log_2 3$. Therefore, the solution set is $1 - \sqrt{\log_2 6} < x < 1 + \sqrt{\log_2 6}$.

8. Let $u = 3^{-x^2+x-1}$. Then the equation becomes $u^2 - 18u - m = 0$, which has a positive real root. Since $-x^2 + x - 1 = -(x - \frac{1}{2})^2 - \frac{3}{4} \leq -\frac{3}{4}$, we have $0 < u \leq 3^{-\frac{3}{4}} = \frac{\sqrt[4]{3}}{3}$. Then $m = u^2 - 18u = (u-9)^2 - 81$, and hence $-6\sqrt[4]{3} + \frac{\sqrt{3}}{9} \leq m < 0$.
9. The solution set to the inequality $2(\log_{\frac{1}{2}} x)^2 + 7(\log_{\frac{1}{2}} x) + 3 \leq 0$ is $[\sqrt{2}, 8]$, and the function $f(x) = (\log_2 \frac{x}{2})(\log_2 \frac{x}{4}) = (\log_2 x - \frac{3}{2})^2 - \frac{1}{4}$. By $\sqrt{2} \leq x \leq 8$, we have $\frac{1}{2} \leq \log_2 x \leq 3$. Therefore, when $\log_2 x = 3$, that is $x = 8$, the function $f(x)$ has the maximum value 2, and when $\log_2 x = \frac{3}{2}$, namely $x = 2\sqrt{2}$, the function $f(x)$ has the minimum value $-\frac{1}{4}$.
10. Let $g(x) = f(2^x)$. Then

$$f(x) = g(\log_2 x) = (\log_2 x)^2 - 2a \log_2 x + a^2 - 1$$

Hence, the function $f(x)$ has the range $[-1, 0]$ on the interval $[2^{a-1}, 2^{a^2-2a+2}]$ if and only if $g(x) = x^2 - 2ax + a^2 - 1$ has the range $[-1, 0]$ on the interval $[a-1, a^2-2a+2]$.

Since $g(a) = -1 \in [-1, 0]$, we have $a \in [a-1, a^2-2a+2]$, and the maximum value of $g(x)$ on $[a-1, a^2-2a+2]$ should be attained at one of the endpoints. Also since $g(a-1) = 0$ is exactly the maximum value of $g(x)$ on the interval, it follows that a lies in the right half of the interval. In other words,

$$\frac{(a-1)+(a^2-2a+2)}{2} \leq a \leq a^2-2a+2.$$

Solving the inequalities, we get $\frac{3-\sqrt{5}}{2} \leq a \leq 1$ or $2 \leq a \leq \frac{3+\sqrt{5}}{2}$. Therefore, the value range of a is $[\frac{3-\sqrt{5}}{2}, 1] \cup [2, \frac{3+\sqrt{5}}{2}]$.

11. The original equation is equivalent to $\begin{cases} x-ak>0 \\ x^2-a^2>0 \\ (x-ak)^2=x^2-a^2 \end{cases}$. Hence, $k<-1$ or $0<k<1$, which is the value range of k.

12. The original inequality can be simplified to

$$\frac{1-(-2)^n}{3}\log_a x > \frac{1-(-2)^n}{3}\log_a(x^2-a)$$

(1) If n is odd, then $\log_a x > \log_a(x^2-a)$. Hence, $\begin{cases} x>0 \\ x^2-a>0 \\ x>x^2-a \end{cases}$, so $\sqrt{a}<x<\frac{1+\sqrt{1+4a}}{2}$.

(2) If n is even, then $\log_a x < \log_a(x^2-a)$. Hence, $\begin{cases} x>0 \\ x^2-a>0 \\ x<x^2-a \end{cases}$, so $x>\frac{1+\sqrt{1+4a}}{2}$.

13. Apparently $a>0$ and $a\neq 1$. Let $u=x^2+ax+5$. Then the inequality becomes

$$\frac{\log_3(\sqrt{u}+1)}{-\log_3 a}\cdot\log_5(u+1)+\frac{1}{\log_3 a}\geq 0.$$

(1) If $0<a<1$, then the inequality becomes

$$\log_3(\sqrt{u}+1)\cdot\log_5(u+1)\geq 1. \text{ ①}$$

Since $\log_3(\sqrt{u}+1)$ and $\log_5(u+1)$ are both nonnegative increasing functions when $u\geq 0$, their product function $f(u)=\log_3(\sqrt{u}+1)\cdot\log_5(u+1)$ is also increasing. Since $f(4)=\log_3(2+1)\cdot\log_5(4+1)=1$, the inequality ① is equivalent to $u\geq 4$, namely $x^2+ax+5\geq 4$. This inequality has infinitely many solutions.

(2) If $a>1$, then the inequality becomes

$$\log_3(\sqrt{u}+1)\cdot\log_5(u+1)\leq 1. \text{ ②}$$

By $f(4)=1$ we see that ② is equivalent to $0\leq u\leq 4$, that is $0\leq x^2+ax+5\leq 4$. Hence, there is only one solution x if and only if the equation $x^2+ax+5=4$ has a unique real solution. Equivalently, $\Delta=a^2-4=0$, from which $a=2$. In this case, the inequalities $0\leq x^2+2x+5\leq 4$ have only one solution $x=-1$.

Therefore, the original inequality has exactly one solution if and only if $a=2$.

14. Let $u = \max\{\frac{x}{y} + z, yz + \frac{1}{x}, \frac{1}{xz} + y\}$. Then

$$u \geq \frac{x}{y} + z, \quad u \geq \frac{1}{xz} + y(x, y, z > 0).$$

Hence, $u^2 \geq (\frac{x}{y} + z)(\frac{1}{xz} + y) = (x + \frac{1}{x}) + (yz + \frac{1}{yz}) \geq 2+2 = 4$, so $u \geq 2$ and $M \geq \lg 2$. On the other hand, $M = \lg 2$ when $x = y = z = 1$, thus the minimum value of $M$is $\lg 2$.

15. We construct the function

$$f(x) = \frac{x(a^x - 1)}{(a^x + 1)\log_a(\sqrt{x^2+1} - x)} - \ln(\sqrt{x^2+1} + x), \quad x \neq 0.$$

Then

$$\begin{aligned} f(-x) &= \frac{-x(a^{-x} - 1)}{(a^{-x} + 1)\log_a(\sqrt{x^2+1} + x)} - \ln(\sqrt{x^2+1} - x) \\ &= \frac{-x(1 - a^x)}{(1 + a^x)\log_a(\sqrt{x^2+1} - x)} - \ln(\sqrt{x^2+1} - x) \\ &= -\frac{x(a^x - 1)}{(a^x + 1)\log_a(\sqrt{x^2+1} - x)} + \ln(\sqrt{x^2+1} + x) \\ &= -f(x) \end{aligned}$$

Hence, $f(x)$ is an odd function on its domain.
Since $0 < a < 1$, we have $a^x - 1 < 0$ when $x > 0$, and

$$\log_a(\sqrt{x^2+1} - x) = \log_a \frac{1}{\sqrt{x^2+1} + x} > 0,$$

so $f(x) < 0$. Therefore, $f(x) = -f(-x) > 0$ when $x < 0$. In other words,

$$\frac{x(a^x - 1)}{(a^x + 1)\log_a(\sqrt{x^2+1} - x)} > \ln\left(\sqrt{x^2+1} + x\right).$$

Solution 6

Functions with Absolute Values

1. The zeros of $x+1=0$ and $x-2=0$ are $x=-1$ and $x=2$, respectively. When $x<-1$, we have $y=-2x+1$; when $-1\leq x<2$, we have $y=3$; when $x\geq 2$, we have $y=2x-1$. Therefore,

$$y=\begin{cases}-2x+1, & x<-1,\\ 3, & -1\leq x<2,\\ 2x-1, & x\geq 2,\end{cases}$$

and the graph is shown in the figure.

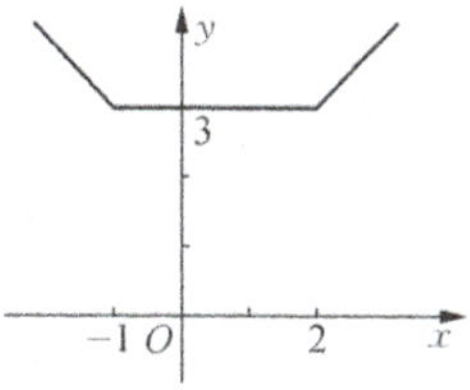

Exercise 1

2. First, draw the graph of $f(x)=x$ for $x\in[2,3]$ as the line segment AB in the following figure. The function $y=f(x)$ is periodic with period 2, so we can plot the graph of $f(x)$ for $x\in[0,1]\cup[-2,-1]$, which is the union of the line segments CD and EF. Since $f(x)$ is an even function, we can determine its graph on $[-1,0]$ by taking the symmetry with respect to the y-axis, which is the line segment FC. Hence, $f(x)=\begin{cases}x+4, & x\in[-2,-1]\\ -x+2, & x\in(-1,0]\end{cases}$, which is the same as $f(x)=3-|x+1|$. Therefore, the answer is C.

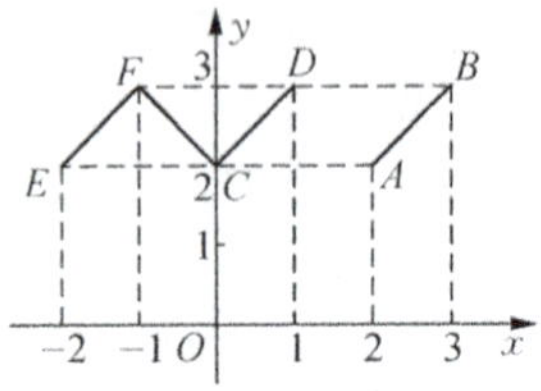

Exercise 2

3. If $|x| \geq 1$, then the equation is $x^2 - 1 = (4 - 2\sqrt{3})(x + 2)$, that is,

$$x^2 - (4 - 2\sqrt{3})x - 9 + 4\sqrt{3} = 0,$$

and its roots are $x_1 = \sqrt{3}$ and $x_2 = 4 - 3\sqrt{3}$, both of which satisfy $|x| \geq 1$.
If $|x| < 1$, then the equation is $1 - x^2 = (4 - 2\sqrt{3})(x + 2)$, namely

$$x^2 + (4 - 2\sqrt{3})x + 7 - 4\sqrt{3} = 0,$$

which has only one root $x_3 = \sqrt{3} - 2$ that satisfies $|x_3| < 1$.
Therefore, the equation has three roots, so the answer is C.

4. $y = x^2 - 2|x| - 1 = \begin{cases} x^2 - 2x - 1, & x \geq 0 \\ x^2 + 2x - 1, & x < 0 \end{cases}$, whose graph is shown in the following figure.

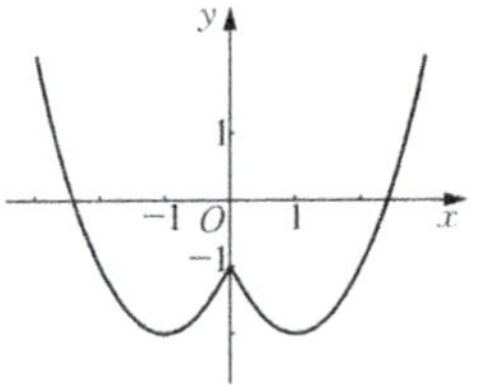

Exercise 4

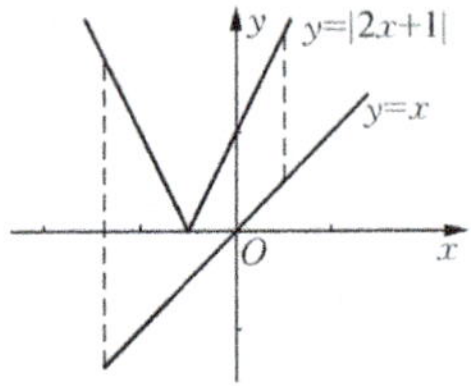

Exercise 5

5. We plot the graphs of $y = x$ and $y = |2x + 1|$, as shown in the figure. The geometric interpretation of solving the equation is that the vertical distance between the corresponding points on the graphs of the two functions at the solutions is 3. The graph of the second function is V-shaped, whose vertex is $(-\frac{1}{2}, 0)$, with slopes ± 2. When $x = -\frac{1}{2}$, the vertical distance between the two functions is $\frac{1}{2}$. When x moves to either the left or the right from $-\frac{1}{2}$, the vertical distance between the points on the graphs of the two functions increases from $\frac{1}{2}$ to infinity. Hence, there are exactly two values of x that satisfy the equation.

6. (1) $F(x) = \begin{cases} 2x, & 0 \le x \le 2 - \sqrt{3}, \\ (x-1)^2, & 2 - \sqrt{3} < x \le 1. \end{cases}$

   (2) $4 - 2\sqrt{3}$.

   (3) $x = \frac{1}{6}$ or $x = 1 - \frac{\sqrt{3}}{3}$.

7. (1) We remove the absolute value signs by dividing the domain into subintervals:

$$y = \begin{cases} 1 - \dfrac{2}{x+1}, & x < -1, \\ -1 + \dfrac{2}{x+1}, & -1 < x \le 0, \\ 1, & x > 0, \end{cases}$$

and the graph is shown in the figure. This is an example to show that not all functions with absolute values can be plotted by taking symmetries.

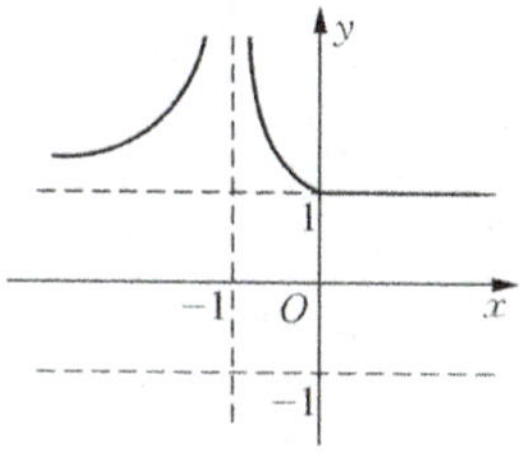

Exercise 7(1)

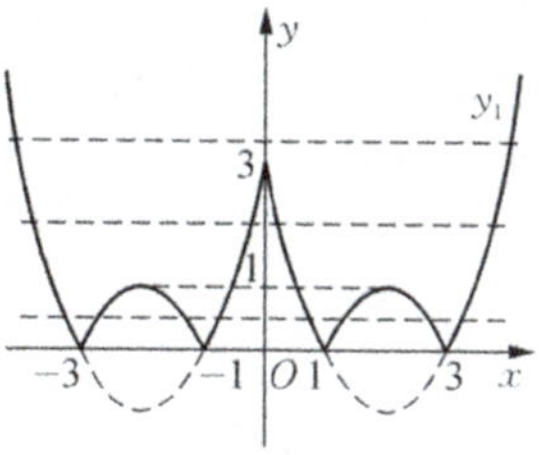

Exercise 7(2)

   (2) It is rather complicated to discuss the problem algebraically, so we may consider combining the numbers with the graph. We view the solutions to the equation as the x-coordinates of the intersections points of $y_1 = |x^2 - 4|x| + 3|$ and $y_2 = a$, and the number of solutions to the equation is the number of the intersection points they have. From the graph, we can see that when $a < 0$, the equation has no solution; when $a = 0$ or $1 < a < 3$, the equation has four solutions; when $0 < a < 1$, the equation has eight solutions; when $a = 1$, the equation has six solutions; when $a = 3$, the equation has three solutions; when $a > 3$, the equation has two solutions.

8. Let $f(x,y) = x + y + \frac{|x-1|}{y} + \frac{|y-1|}{x}$.
If $x \geq 1$ and $y \geq 1$, then $f(x,y) \geq x + y \geq 2$.
If $0 < x \leq 1$ and $0 < y \leq 1$, then

$$f(x,y) = x + y + \frac{1-x}{y} + \frac{1-y}{x} \geq x + y + 1 - x + 1 - y = 2.$$

Otherwise, we assume that $0 < x < 1 < y$, so

$$\begin{aligned} f(x,y) &= x + y + \frac{1-x}{y} + \frac{y-1}{x} \\ &= y + \frac{1}{y} + \frac{xy-x}{y} + \frac{y-1}{x} \\ &= y + \frac{1}{y} + (y-1)\left(\frac{x}{y} + \frac{1}{x}\right) \\ &> 2\sqrt{y \cdot \frac{1}{y}} + 0 = 2. \end{aligned}$$

Therefore, $f(x,y) \geq 2$ for all $x > 0$ and $y > 0$, and since $f(1,1) = 2$, the minimum value of the function is 2.

9. Let $t = x - a$. Then $f(x) = \frac{|t|}{t(t+a)+1}$. Note that the discriminant of $t^2 + at + 1$ is $\Delta = a^2 - 4 < 0(|a| < 2)$, so $t^2 + at + 1 > 0$. Hence, $f(x) \geq 0$, where the equality holds when $x = a$, so the minimum value of $f(x)$ is 0. Also

$$f(x) = \frac{1}{|t| + \frac{1}{|t|} + a \cdot \frac{t}{|t|}} \leq \frac{1}{2 + a \cdot \frac{t}{|t|}}.$$

When $x > a$, the maximum value of $f(x)$ is $\frac{1}{2+a}$, where the equality holds when $x = a + 1$, and when $x < a$, the maximum value of $f(x)$ is $\frac{1}{2-a}$, where the equality holds when $x = a - 1$. Therefore, if $a \in (-2, 0]$, then the function $f(x)$ has the maximum value $\frac{1}{2+a}$, and if $a \in [0, 2)$, then the function $f(x)$ has the maximum value $\frac{1}{2-a}$.

10. In order to plot the graph of $y = ||x^2 - 6x - 16| - 10|$, we take two steps. First, we draw the graph of $y = |x^2 - 6x - 16|$, or equivalently $y = |(x-3)^2 - 25|$ (the solid curve in the first figure below). Then we translate the graph downwards by 10 units, and reflect the part below the x-axis upwards. This gives us the graph of $y = ||x^2 - 6x - 16| - 10|$ (as shown in the second figure). From the graph, we see that only when

$a = 10$, the graph of the function and the line $y = a$ have six intersection points. Therefore, $a = 10$.

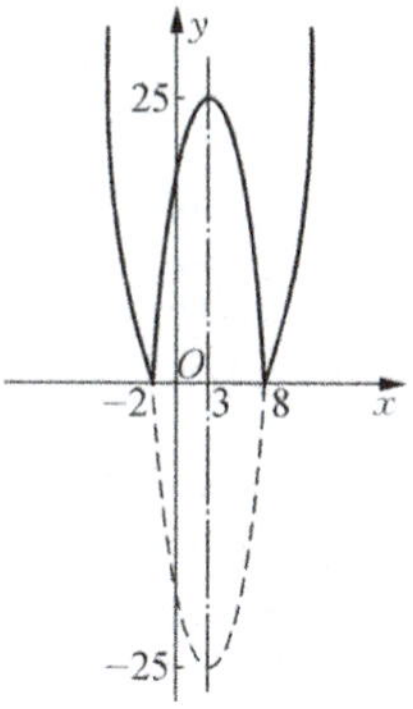

Exercise 10(1)

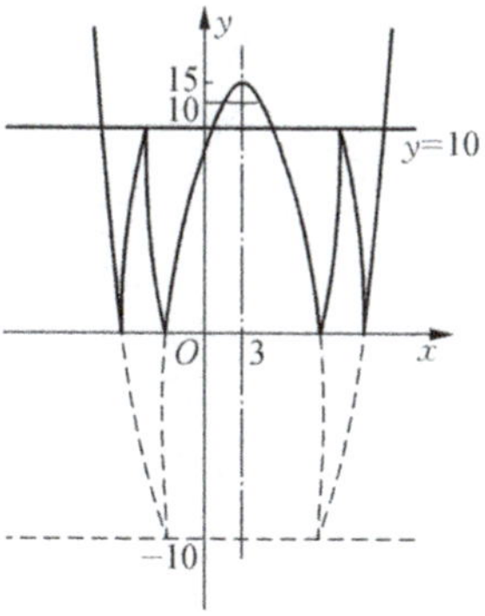

Exercise 10(2)

11. By the definition of the number M, we have that $M \geq |f(-1)| = |1 - p + q|, M \geq |f(1)| = |1 + p + q|$, and $M \geq |f(0)| = |q|$. Hence,

$$4M \geq |1 - p + q| + |1 + p + q| + 2|q|$$
$$\geq |(1 - p + q) + (1 + p + q) - 2q| = 2.$$

so $M \geq \frac{1}{2}$. If we take $p = 0$ and $q = -\frac{1}{2}$, then $f(x) = x^2 - \frac{1}{2}$. The graph of this function is shown in the figure below, and in this case $M = \frac{1}{2}$. Therefore, the minimum value of $M$is $\frac{1}{2}$.

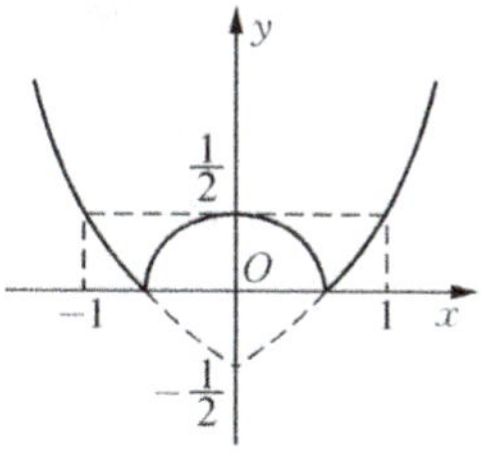

Exercise 11

Solution 7

Maximum and Minimum Values of Functions

1. (1) 211. Hint: It follows that $x-a \geq 0$, $x+19 \geq 0$, and $x-a-96 \leq 0$. Hence, $y = x-a+x+19+96+a-x = x+19+96$ and $y_{\max} = 211$.

   (2) 10. Hint: Using the property of quadratic functions we see that the axis of symmetry $x = 2$ lies in the interval $[0, 3]$.

   (3) $\frac{1}{2}$. Hint: Rewrite $y(x^2+3) = x+1$ as

   $$yx^2 - x + 3y - 1 = 0,$$

   where

   $$\Delta = 1 - 4y(3y-1) \geq 0.$$

   Hence, $-\frac{1}{6} \leq y \leq \frac{1}{2}$, so $y_{\max} = \frac{1}{2}$.

   (4) 16. Hint: $|x+1| \leq 6$, so $x \in [-7, 5]$. Hence,

   $$y = \begin{cases} x^2 - 2x + 1, & x \in [0, 5], \\ -x^2 - 2x + 1, & x \in [-7, 0). \end{cases}$$

   Therefore, $y_{\max} = 16$.

   (5) $\frac{41}{8}$. Hint: $y = -2(x+1) + 5\sqrt{x+1} + 2 = -2(\sqrt{x+1})^2 + 5\sqrt{x+1} + 2$. Since $\sqrt{x+1} \in [1, \sqrt{2}]$, we have $y_{\max} = y|_{\sqrt{x+1}=\frac{5}{4}} = \frac{41}{8}$.

(6) $\frac{4}{9}$. Hint: $2y^2 = 2x - 3x^2$, so

$$\begin{cases} x \geq 0 \\ 2x - 3x^2 \geq 0 \end{cases} \Rightarrow 0 \leq x \leq \frac{2}{3},$$

$$x^2 + y^2 = x^2 + \frac{1}{2}(2x - 3x^2) = -\frac{1}{2}x^2 + x = -\frac{1}{2}(x-1)^2 + \frac{1}{2}.$$

Since $0 \leq x \leq \frac{2}{3}$, we have $(x^2+y^2)_{\max} = \frac{4}{9}$.

2. $y = (x-1)^2 + (\sqrt{x} - \frac{1}{\sqrt{x}})^2 + 1 \geq 1$, and the equality is satisfied when $x = 1$. Hence, $y_{\min} = 1$.
3. 2. Hint: Since

$$f(x) = \frac{x^2 + 2\sqrt{2013}x + 2013 + \sin 2013x}{x^2 + 2013} = 1 + \frac{2\sqrt{2013}x + \sin 2013x}{x^2 + 2013}$$

and $g(x) = \frac{2\sqrt{2013}x + \sin 2013x}{x^2+2013}$ is an odd function, by the property of an odd function, $g(x)_{\max} + g(x)_{\min} = 0$. Hence, $M + m = 2$.

4. $f(x,y) = \left(\frac{x^2}{y^2} - 1\right)^2 + \left(\frac{y^2}{x^2} - 1\right)^2 + \left(\frac{x}{y} - \frac{y}{x}\right)^2 + \left(\sqrt{\frac{x}{y}} - \sqrt{\frac{x}{y}}\right)^2 + 2 \geq 2$, and the equality holds when $x = y$, so the minimum value of $f(x,y)$ is 2.
5. $-\frac{1}{3}$. Hint: Let $u = x + y$ and $v = x - y$. Then $x = \frac{u+v}{2}$ and $y = \frac{u-v}{2}$. Hence,

$$\begin{aligned} & x^2 + xy + y^2 - x - y \\ &= \left(\frac{u+v}{2}\right)^2 + \frac{u+v}{2} \cdot \frac{u-v}{2} + \left(\frac{u-v}{2}\right)^2 - u \\ &= \frac{3u^2 - 4u + v^2}{4} \\ &= \frac{3(u - \frac{2}{3})^2}{4} + \frac{v^2}{4} - \frac{1}{3} \geq -\frac{1}{3}. \end{aligned}$$

6. Note that

$$\begin{aligned} y &= (x+1)(x+4)(x+2)(x+3) + 5 \\ &= (x^2+5x+4)(x^2+5x+6) + 5 \\ &= (x^2+5x+5)^2 + 4. \end{aligned}$$

Let $t = x^2+5x+5$ for $x \in [-6, 6]$. Then $-\frac{5}{4} \leq t \leq 71$. Hence, $y = t^2+4$ for $-\frac{5}{4} \leq t \leq 71$, so $y_{\min} = 4$ and $y_{\max} = 5045$.

7. $y = \frac{5}{(x^2+1)^2} - \frac{1}{x^2+1} + 1$. Let $t = \frac{1}{x^2+1}$. Then

$$y = 5t^2 - t + 1 = 5\left(t - \frac{1}{10}\right)^2 + \frac{19}{20}, \quad 0 < t \le 1.$$

Therefore, when $x = \pm 3$, the function y attains its minimum value $\frac{19}{20}$, and when $x = 0$, the function y attains its maximum value 5.

8. Plugging $y = u - x$ into the equation, we get $4x^2 - 4ux + u^2 - \sqrt{2}u + 6 = 0$. Solving the inequality

$$\Delta = 16u^2 - 16(u^2 - \sqrt{2}u + 6) \ge 0.$$

we have $u \ge 3\sqrt{2}$. Since $u = 3\sqrt{2}$ when $x = y = \frac{3\sqrt{2}}{2}$, the minimum value of u is $3\sqrt{2}$.

9. Since $\Delta = (k-2)^2 - 4(k^2 + 3k + 5) \ge 0$, we have $k \in \left[-4, -\frac{4}{3}\right]$. By Vieta's theorem, $x_1^2 + x_2^2 = (x_1 + x_2)^2 - 2x_1x_2 = -k^2 - 10k - 6 = -(k+5)^2 + 19$. Hence, the minimum value of $x_1^2 + x_2^2$ is $\frac{50}{9}$ and the maximum value is 18.

10. $m = a^2 - 3a$. By $0 \le a^2 - 4a - 2 \le 10$, we get $-2 \le a \le 2 - \sqrt{6}$ or $2 + \sqrt{6} \le a \le 6$. Hence, m has the maximum value 18 when $a = 6$.

11. Since

$$\begin{aligned} f(x) &= \frac{x}{\sqrt{x^4 + x^2 + 1} + \sqrt{x^4 + 1}} \\ &= \frac{x}{|x|\left(\sqrt{\left(x - \frac{1}{x}\right)^2 + 3} + \sqrt{\left(x - \frac{1}{x}\right)^2 + 2}\right)}, \end{aligned}$$

we see that $f(x)$ is maximal when $x = \frac{1}{x} > 0$, or equivalently $x = 1$, and the maximum value is $\sqrt{3} - \sqrt{2}$.

12. The given function $u = \left(x + \frac{y-1}{2}\right)^2 + \frac{3}{4}(y-1)^2 + 2 \ge 2$. Also $u = 2$ when $x = 0$ and $y = 1$. Hence, $u_{\min} = 2$.

13. The discriminant of $yx^2 - ax + (y - b) = 0$ is $\Delta = (-a)^2 - 4y(y-b) \ge 0$, so $y^2 - by - \frac{a^2}{4} \le 0$. Also $(y+1)(y-4) \le 0$, namely $y^2 - 3y - 4 \le 0$. Hence, $b = 3$ and $\frac{a^2}{4} = 4$, in other words, $a = \pm 4$ and $b = 3$.

14. Since $f(x) \ge 0$ for all real x, we have $\Delta = b^2 - 4ac \le 0$. Also since $a > 0$, we get $c \ge \frac{b^2}{4a}$. Hence,

$$\begin{aligned} \frac{f(1)}{f(0) - f(-1)} &= \frac{a + b + c}{c - (a - b + c)} = \frac{a + b + c}{b - a} \\ &\ge \frac{a + b + \frac{b^2}{4a}}{b - a} = \frac{1 + \frac{b}{a} + \frac{b^2}{4a^2}}{\frac{b}{a} - 1} \end{aligned}$$

Let $t = \frac{b}{a}$. Then $\frac{f(1)}{f(0)-f(-1)} = \frac{1+t+\frac{1}{4}t^2}{t-1}$ $(t > 2)$. Further, letting $s = t - 1$ gives that

$$\frac{f(1)}{f(0)-f(-1)} = \frac{\frac{1}{4}s^2+\frac{3}{2}s+\frac{9}{4}}{s} = \frac{s}{4}+\frac{9}{4s}+\frac{3}{2}\ (s>1)$$

$$\geq 2\sqrt{\frac{s}{4}\cdot\frac{9}{4s}}+\frac{3}{2} = 2\cdot\frac{3}{4}+\frac{3}{2} = 3.$$

The equality holds when $\frac{s}{4} = \frac{9}{4s}$ $(s > 1)$, or equivalently $s = 3$.

15. The domain of the function is $(-\infty, 0] \cup [2, +\infty)$. Since $2x^2 - 3x + 4$ and $x^2 - 2x$ are decreasing on $(-\infty, 0]$ and increasing on $[2, +\infty)$, the function

$$f(x) = \sqrt{2x^2-3x+4}+\sqrt{x^2-2x}$$

is also decreasing on $(-\infty, 0]$ and increasing on $[2, +\infty)$. Therefore,

$$f_{\min}(x) = \min\{f(0), f(2)\} = \min\{2, \sqrt{6}\} = 2.$$

16. The given function $f(x) = -9\left(x+\frac{a}{3}\right)^2 + 2a$ for $x \in \left[-\frac{1}{3}, \frac{1}{3}\right]$.

(1) If $-\frac{a}{3} < -\frac{1}{3}$, then $f\left(-\frac{1}{3}\right) = -3$, so $a = 2+\sqrt{6}$.
(2) If $-\frac{1}{3} \leq -\frac{a}{3} \leq \frac{1}{3}$, then $f\left(-\frac{a}{3}\right) = -3$, which is impossible.
(3) If $-\frac{a}{3} > \frac{1}{3}$, then $f\left(\frac{1}{3}\right) = -3$, so $a = -\sqrt{2}$.

Therefore,$a = -\sqrt{2}$ or $2+\sqrt{6}$.

17. Let $\left[\frac{1}{x}+\frac{1}{2}\right] = n$ and $\left\{\frac{1}{x}+\frac{1}{2}\right\} = \alpha$. Then

$$f(x) = \left|\frac{1}{x}-n\right| = \left|\alpha-\frac{1}{2}\right|.$$

Since $0 \leq \alpha < 1$, we have $f(x) \leq \frac{1}{2}$. Also $f(x) = \frac{1}{2}$ when $x = \frac{2}{2k-1}$ $(k \in \mathbf{Z})$, so $f_{\max}(x) = \frac{1}{2}$.

18. $f(x) = \sqrt{(x-3)^2+(x^2-2)^2} - \sqrt{(x-0)^2+(x^2-1)^2}$. In the coordinate plane, denote $P(x, x^2)$, $A(3, 2)$ and $B(0, 1)$. Then

$$f(x) = |PA| - |PB| \leq |AB| = \sqrt{3^2+1^2} = \sqrt{10}.$$

Also $f(x) = \sqrt{10}$ when $x = \frac{-\sqrt{37}+1}{6}$. Therefore, $f_{\max}(x) = \sqrt{10}$.

19. $f(x) = a\left(x+\frac{4}{a}\right)^2 + 3 - \frac{16}{a}$.

(1) If $3 - \frac{16}{a} > 5$, or equivalently $-8 < a < 0$, then $l(a)$ is the smaller root of the equation $ax^2 + 8x + 3 = 5$, so $l(a) = \frac{-8+\sqrt{64+8a}}{2a}$.

(2) If $3-\frac{16}{a}\le 5$, or equivalently $a\le -8$, then $l(a)$ is the larger root of the equation $ax^2+8x+3=-5$, so $l(a)=\frac{-8-\sqrt{64-32a}}{2a}$.

Hence, $l(a)=\begin{cases}\frac{-8-\sqrt{64-32a}}{2a}, a\le -8\\ \frac{-8+\sqrt{64+8a}}{2a}, \quad -8<a<0\end{cases}$.

When $a\le -8$,

$$l(a)=\frac{-8-\sqrt{64-32a}}{2a}=\frac{4}{\sqrt{4-2a}-2}\le\frac{4}{\sqrt{20}-2}=\frac{\sqrt{5}+1}{2}.$$

When $-8<a<0$,

$$l(a)=\frac{2}{\sqrt{16+2a}+4}<\frac{2}{4}<\frac{\sqrt{5}+1}{2}.$$

Therefore, $l(a)$ has the maximum value $\frac{\sqrt{5}+1}{2}$ when $a=-8$.

20. Since

$$\begin{aligned}f(x)&=\frac{1-\cos 2\omega x}{2}+\frac{\sqrt{3}}{2}\sin 2\omega x\\&=\frac{\sqrt{3}}{2}\sin 2\omega x-\frac{1}{2}\cos 2\omega x+\frac{1}{2}\\&=\sin\left(2\omega x-\frac{\pi}{6}\right)+\frac{1}{2},\end{aligned}$$

we have $T=\frac{2\pi}{2\omega}=\frac{\pi}{2}$, so $\omega=2$. Hence, $f(x)=\sin(4x-\frac{\pi}{6})+\frac{1}{2}$.

When $\frac{\pi}{8}\le x\le\frac{\pi}{4}$, we have $\frac{\pi}{3}\le 4x-\frac{\pi}{6}\le\frac{5\pi}{6}$, so $\frac{1}{2}\le\sin\left(4x-\frac{\pi}{6}\right)\le 1$, or equivalently

$$1\le\sin\left(4x-\frac{\pi}{6}\right)+\frac{1}{2}\le\frac{3}{2}.$$

Therefore, $f(x)$ has the maximum value $\frac{3}{2}$ when $x=\frac{\pi}{6}$ and the minimum value 1 when $x=\frac{\pi}{4}$.

21. (1) Without loss of generality we assume that $\alpha\le x_1<x_2\le\beta$. By the condition, $\alpha+\beta=\frac{t}{2}$ and $\alpha\beta=1$. Since $(x_1-\alpha)(x_2-\beta)\le 0$, we see that $x_1x_2-(\alpha x_2+\beta x_1)+\alpha\beta\le 0$, or equivalently $4x_1x_2-4(\alpha x_2+\beta x_1)-4\le 0$. Hence,

$$\begin{aligned}4x_1x_2-t(x_1+x_2)-4&\le 4(\alpha x_2+\beta x_1)-t(x_1+x_2)\\&=4(\alpha x_2+\beta x_1)-2(\alpha+\beta)(x_1+x_2)\\&=2(\alpha x_2+\beta x_1)-2(\alpha x_1+\beta x_2)\\&=2(x_2-x_1)(\alpha-\beta)<0.\end{aligned}$$

(2) We have $\alpha = \frac{t-\sqrt{t^2+16}}{4}, \beta = \frac{t+\sqrt{t^2+16}}{4}$, $f(\alpha) = \frac{-8}{\sqrt{t^2+16}-t}$, and $f(\beta) = \frac{8}{\sqrt{t^2+16}+t}$. Let $x_1 x_2 \in [\alpha, \beta]$ with $x_1 < x_2$. Then

$$f(x_2) - f(x_1) = -\frac{4x_1x_2 - t(x_1 + x_2) - 4}{(x_1^2+1)(x_2^2+1)}(x_2 - x_1) > 0,$$

so $f(x)$ is an increasing function on $[\alpha, \beta]$. Therefore, $g(t) = f(\beta) - f(\alpha) = \sqrt{t^2+16} \geq 4$, and the minimum value of $g(t)$ is 4.

Solution 8

Properties of Inequalities

1. $\sin 1 < \log_3 \sqrt{7}$. Hint: Apparently $\sin 1 < \sin \frac{\pi}{3} = \frac{\sqrt{3}}{2} < \frac{7}{8}$ and $3^7 < 7^4$, so $\sin 1 < \frac{7}{8} < \log_3 \sqrt{7}$.
2. $a < b$. Hint: This follows from $a = \frac{1}{\sqrt{m+1}+\sqrt{m}} < \frac{1}{\sqrt{m}+\sqrt{m-1}} = b$.
3. $f(x) < f(x^2) < f^2(x)$. Hint: When $0 < x < 1$, we have $f(x) = \frac{x}{\lg x} < 0$, $f^2(x) > 0$, and $\frac{x}{\lg x} - \frac{x^2}{\lg^2 x} = \frac{2x-x^2}{2\lg x} = \frac{(2-x)x}{\lg x^2} < 0$. Hence, $f(x) < f(x^2) < f^2(x)$.
4. $a > b > 1$. Hint: Since $a^n = a + 1 > 1$, we have $a > 1$. Also $b^{2n} = b + 3a > 1$, so $b > 1$. Since $a^{2n} - b^{2n} = (a+1)^2 - (b+3a) = a^2 - a - b + 1$ and

$$a^{2n} - b^{2n} = (a+b)(a^{2n-1} + a^{2n-2} \cdot b + \cdots + b^{2n-1}),$$

it follows that $\frac{a^2-a-b+1}{a-b} = a^{2n-1} + \cdots + b^{2n-1} > 1$, namely $\frac{(a-1)^2}{a-b} > 0$. Therefore, $a > b$.
5. It is easy to see that $A - B = \frac{(a^p-b^p)(a^q-b^q)}{4}$. If $a = b$, then $A = B$; if $a \neq b$, then since $p, q > 0$, we have $(a^p - b^p)(a^q - b^q) > 0$, so $A > B$.
6. (1) Taking a difference, we get

$$2\left(a^2b^2 - \frac{1}{2}\right)^2 + 2\left(ab - \frac{1}{2}\right)^2 + (a - b - c)^2 > 0,$$

so $2a^4b^4 + a^2 + b^2 + c^2 + 1 > 4ab - 2bc + 2ac$.

(2) Since $\frac{a^a b^b c^c}{(abc)^{\frac{a+b+c}{3}}} = \left(\frac{a}{b}\right)^{\frac{a-b}{3}} \left(\frac{b}{c}\right)^{\frac{b-c}{3}} \left(\frac{c}{a}\right)^{\frac{c-a}{3}} \geq 1$, then $a^a b^b c^c = (abc)^{\frac{a+b+c}{3}}$ when $a = b = c$, and $a^a b^b c^c > (abc)^{\frac{a+b+c}{3}}$ otherwise.

7. Since $a > b > c$,

$$
\begin{aligned}
& a^2(b-c) + c^2(a-b) - b^2(a-c) \\
&= a^2(b-a+a-c) + c^2(a-b) - b^2(a-c) \\
&= (a^2-b^2)(a-c) + (a-b)(c^2-a^2) \\
&= (a-b)(a+b)(a-c) + (a-b)(c+a)(c-a) \\
&= (a-b)(a-c)(b-c) > 0.
\end{aligned}
$$

Therefore, the inequality is valid.

8. By the definition of S, $\frac{a_1^2}{a_1-1} > S \Leftrightarrow a_1^2 > (a_1+a_2+a_3)(a_1-1) \Leftrightarrow \frac{1}{a_2+a_3} > \frac{a_1}{a_1+a_2+a_3}$.
Summing the three similar inequalities, we get the desired result.

9. Since $|x-y| \leq 2$, we have $x^2 - 2xy + y^2 \leq 4$, so $x^2 + 2xy + y^2 \leq 4 + 4xy$, or equivalently $2\sqrt{xy+1} \geq x+y$.
Similarly, we can get $2\sqrt{yz+1} \geq y+z$ and $2\sqrt{xz+1} \geq x+z$. Summing the three inequalities gives

$$\sqrt{xy+1} + \sqrt{yz+1} + \sqrt{xz+1} \geq x+y+z.$$

It remains to show that the equality above cannot be satisfied. We can assume that $x \geq y \geq z$. Then the equality holds if and only if $\begin{cases} x-y=2 \\ y-z=2 \\ x-z=2 \end{cases}$, but this is impossible. Therefore, the desired inequality is true.

10. Note that $LHS - RHS$

$$
\begin{aligned}
&= x^2+y^2+z^2 - 2\sqrt{\frac{ab}{(b+c)(c+a)}}xy - 2\sqrt{\frac{bc}{(a+b)(a+c)}}yz \\
&\quad -2\sqrt{\frac{ac}{(a+b)(a+c)}}zx \\
&= \frac{b+c}{b+c}x^2 + \frac{c+a}{c+a}y^2 + \frac{a+b}{a+b}z^2 - 2\sqrt{\frac{ab}{(b+c)(c+a)}}xy \\
&\quad -2\sqrt{\frac{bc}{(a+b)(a+c)}}yz - 2\sqrt{\frac{ac}{(a+b)(a+c)}}zx
\end{aligned}
$$

$$= \left(\sqrt{\frac{b}{b+c}}x - \sqrt{\frac{a}{c+a}}y\right)^2 + \left(\sqrt{\frac{c}{c+a}}y - \sqrt{\frac{b}{a+b}}z\right)^2$$
$$+ \left(\sqrt{\frac{a}{a+b}}z - \sqrt{\frac{c}{b+c}}x\right)^2 \geq 0.$$

Therefore, the inequality is valid.

11. Since

$$\begin{aligned}
&2a^2b^2 + 2b^2c^2 + 2a^2c^2 - a^4 - b^4 - c^4 \\
&= c^2(2b^2 + 2a^2) - (a^2 - b^2)^2 - c^4 \\
&= c^2(a^2 + b^2 + 2ab - 2ab + a^2 + b^2) - (a+b)^2(a-b)^2 - c^4 \\
&= c^2(a+b)^2 + c^2(a-b)^2 - (a+b)^2(a-b)^2 - c^4 \\
&= [(a+b)^2 - c^2] \cdot [c^2 - (a-b)^2] \\
&= (a+b+c)(a+b-c)(a+c-b)(c+b-a) > 0,
\end{aligned}$$

the desired inequality follows.

12. The original inequality is equivalent to $a^2b+a^2c+b^2c+b^2a+c^2a+c^2b-6abc \geq 0$, which is equivalent to $a(b-c)^2 + b(a-c)^2 + c(a-b)^2 \geq 0$.

13. Since $\frac{a_1^2-a_2^2}{a_1+a_2} + \frac{a_2^2-a_3^2}{a_2+a_3} + \cdots + \frac{a_{n-1}^2-a_n^2}{a_{n-1}+a_n} + \frac{a_n^2-a_1^2}{a_n+a_1} = 0$,

$$\frac{a_1^2}{a_1+a_2} + \frac{a_2^2}{a_2+a_3} + \cdots + \frac{a_n^2}{a_n+a_1} = \frac{a_2^2}{a_1+a_2} + \frac{a_3^2}{a_2+a_3} + \cdots + \frac{a_1^2}{a_n+a_1}$$

Also $\frac{a_1^2+a_2^2}{a_1+a_2} \geq \frac{1}{2}(a_1+a_2)$, $\frac{a_2^2+a_3^2}{a_2+a_3} \geq \frac{1}{2}(a_2+a_3), \ldots, \frac{a_n^2+a_1^2}{a_n+a_1} \geq \frac{1}{2}(a_n+a_1)$, so $\frac{a_1^2}{a_1+a_2} + \frac{a_2^2}{a_2+a_3} + \cdots + \frac{a_n^2}{a_n+a_1} \geq \frac{1}{2}(a_1 + a_2 \cdots + a_n) = \frac{1}{2}$.

14. Since $af - be = 1$,

$$\begin{aligned}
d = d(af - be) &= adf - bcf + bcf - bed \\
&= f(ad - bc) + b(cf - ed).
\end{aligned}$$

Since $\frac{a}{b} > \frac{c}{d} > \frac{e}{f}$, and $a, b, c, d, e,$ and f are all positive integers, it follows that

$$ad - bc \geq 1, \quad cf - ed \geq 1.$$

Therefore, $d \geq f + b$.

Remark. Since $a, b, c,$ and d are positive integers, $ad - bc$ is also an integer, and since it is positive, it is at least 1. Such a method is commonly used in inequalities regarding integers.

15. By $xy + yz + zx = -1$,

$$\begin{aligned} &x^2 + 5y^2 + 8z^2 - 4 \\ &\quad = x^2 + 5y^2 + 8z^2 + 4xy + 4yz + 4zx \\ &\quad = (x^2 + 4y^2 + 4z^2 + 4xy + 8yz + 4zx) + (y^2 + 4z^2 - 4yz) \\ &\quad = (x + 2y + 2z)^2 + (y - 2z)^2 \geq 0. \end{aligned}$$

Hence, $x^2 + 5y^2 + 8z^2 \geq 4$, where the equality holds if and only if

$$\begin{cases} x + 2y + 2z = 0, \\ y - 2z = 0, \\ xy + yz + zx = -1. \end{cases}$$

Solving the equations, we have $(x, y, z) = \left(-\frac{3}{2}, \frac{1}{2}, \frac{1}{4}\right)$ or $\left(\frac{3}{2}, -\frac{1}{2}, -\frac{1}{4}\right)$.

Solution 9

Basic Inequalities

1. $x = \frac{\sqrt{6}}{3}$ and $y = -\frac{\sqrt{6}}{2}$, or $x = -\frac{\sqrt{6}}{3}$ and $y = \frac{\sqrt{6}}{2}$. Hint:

$$\frac{4}{4-x^2} + \frac{9}{9-y^2} \geq 2\sqrt{\frac{4}{4-x^2} \cdot \frac{9}{9-y^2}} = \frac{12}{\sqrt{37-(9x^2+4y^2)}}$$
$$\geq \frac{12}{\sqrt{37-2\sqrt{36(xy)^2}}} = \frac{12}{5}.$$

where the equality holds if and only if $\frac{4}{4-x^2} = \frac{9}{9-y^2}$ and $xy = -1$ with $x, y \in (-2, 2)$. Solving the equations gives the answer.

2. $\frac{m^2}{n^2} \neq \frac{bc}{ad}$. Hint:

$$Q^2 = ab + \frac{nbc}{m} + \frac{mad}{n} + cd \geq ab + 2\sqrt{abcd} + cd = P^2,$$

where the equality holds if and only if $\frac{nbc}{m} = \frac{mad}{n}$.

3. $\sqrt{10}$. Hint: By Cauchy's inequality,

$$(2\sqrt{x-3} + \sqrt{5-x})^2 \leq (2^2+1^2)(x-3+5-x) = 10,$$

where the equality holds if and only if $\frac{\sqrt{x-3}}{2} = \frac{\sqrt{5-x}}{1}$, or equivalently $x = \frac{23}{5}$.
Therefore, the maximum value of $f(x) = 2\sqrt{x-3} + \sqrt{5-x}$ is $\sqrt{10}$.

4. $\frac{5}{6}$. Hint: By Cauchy's inequality,

$$25 \times 36 = (a^2+b^2+c^2)(x^2+y^2+z^2) \geq (ax+by+cz)^2 = 30^2,$$

and the equality holds. Hence, by the equality condition of Cauchy's inequality, $\frac{a}{x} = \frac{b}{y} = \frac{c}{z} = k$. Now, $k^2 = \frac{a^2+b^2+c^2}{x^2+y^2+z^2} = \frac{25}{36}$, so $k = \frac{5}{6}$ (here

$k = -\frac{5}{6}$ is not valid since in this case $ax + by + cz = -30$). Therefore, $k = \frac{a+b+c}{x+y+z} = \frac{5}{6}$.

5. It follows that $A + B + C = \pi$. By Cauchy's inequality,

$$(A + B + C)\left(\frac{1}{A} + \frac{1}{B} + \frac{1}{C}\right) \geq 3^2,$$

so $\frac{1}{A} + \frac{1}{B} + \frac{1}{C} \geq \frac{9}{\pi}$. Also since $(1^2 + 1^2 + 1^2)(A^2 + B^2 + C^2) \geq (A+B+C)^2$, we get $A^2 + B^2 + C^2 \geq \frac{\pi^2}{3}$.

6. Adding and subtracting the two equations, we get $\begin{cases} x^2 + y^2 + z^2 = (\frac{3}{2})^2 \\ -8x + 6y - 24z = 39 \end{cases}$. By Cauchy's inequality,

$$39 = -8x + 6y - 24z \leq \sqrt{(-8)^2 + 6^2 + (-24)^2} \cdot \sqrt{x^2 + y^2 + z^2}$$
$$= \sqrt{676} \times \sqrt{\frac{9}{4}} = 39.$$

Hence, the equality condition for Cauchy's inequality holds, and

$$\frac{x}{-8} = \frac{y}{6} = \frac{z}{-24} = \frac{-8x + 6y - 24z}{64 + 36 + 576} = \frac{39}{676} = \frac{3}{52}.$$

Therefore, the real solution to the system of equations is $(x, y, z) = \left(-\frac{6}{13}, \frac{9}{26}, -\frac{18}{13}\right)$.

7. $8 = 1 + (a+b+c) + (ab+bc+ca) + abc \geq 1 + 3\sqrt[3]{abc} + 3(\sqrt[3]{abc})^2 + abc = (1 + \sqrt[3]{abc})^3$. Hence, $abc \leq 1$.

8. By the two-dimensional triangle inequality,

$$\sqrt{x^2 + xy + y^2} + \sqrt{x^2 + xz + z^2}$$
$$= \sqrt{\left(x + \frac{y}{2}\right)^2 + \left(\frac{\sqrt{3}}{2}y\right)^2} + \sqrt{\left(x + \frac{z}{2}\right)^2 + \left(-\frac{\sqrt{3}}{2}z\right)^2}$$
$$\geq \sqrt{\left(x + \frac{y}{2} - x - \frac{z}{2}\right)^2 + \left(\frac{\sqrt{3}}{2}y + \frac{\sqrt{3}}{2}z\right)^2}$$
$$= \sqrt{\frac{1}{4}(y^2 - 2yz + z^2) + \frac{3}{4}(y^2 + 2yz + z^2)}$$
$$= \sqrt{y^2 + yz + z^2}.$$

9. Since $abc = 1$, we have $\frac{1}{a}+\frac{1}{b}+\frac{1}{c} = bc+ac+ab$. By Cauchy's inequality,

$$\left[\frac{1}{a^3(b+c)}+\frac{1}{b^3(a+c)}+\frac{1}{c^3(a+b)}\right]\cdot[a(b+c)+b(a+c)+c(a+b)]$$
$$\geq\left(\frac{1}{a}+\frac{1}{b}+\frac{1}{c}\right)^2.$$

Hence,

$$\left[\frac{1}{a^3(b+c)}+\frac{1}{b^3(a+c)}+\frac{1}{c^3(a+b)}\right]$$
$$\geq\frac{1}{2}\left(\frac{1}{a}+\frac{1}{b}+\frac{1}{c}\right)\geq\frac{3}{2}\cdot\sqrt[3]{\frac{1}{abc}}=\frac{3}{2}.$$

10. If $x = y = z = 0$, then the inequality holds obviously. Otherwise, there is a positive number and a negative number in x, y, and z. Without loss of generality, assume that $xy < 0$. Then

$$\begin{aligned}6(x^3+y^3+z^3)^2 &= 6[(x^3+y^3)-(x+y)^3]^2 = 54x^2y^2z^2\\ &= 54|xy|\cdot|xy|\cdot z^2\\ &\leq (2z^2+2|xy|)^3.\end{aligned}$$

On the other hand, $x^2+y^2 = (x+y)^2-2xy = z^2+2|xy|$, so

$$(2z^2+2|xy|)^3 = (x^2+y^2+z^2)^3,$$

and the desired inequality follows.

11. Let $f(x) = ax^2+bx+c$. Then

$$\begin{aligned}|f(0)| &= |c| \leq 1,\\ |f(1)| &= |a+b+c| \leq 1,\\ |f(-1)| &= |a-b+c| \leq 1.\end{aligned}$$

Since $cx^2+bx+a = c(x^2-1)+(a+b+c)\frac{1+x}{2}+(a-b+c)\frac{1-x}{2}$, for all $x\in[-1,1]$,

$$\begin{aligned}|cx^2+bx+a| &\leq |c|\cdot|x^2-1|+|a+b+c|\cdot\left|\frac{1+x}{2}\right|\\ &\quad+|a-b+c|\cdot\left|\frac{1-x}{2}\right|\\ &\leq |x^2-1|+\left|\frac{1+x}{2}\right|+\left|\frac{1-x}{2}\right|\end{aligned}$$

$$= 1 - x^2 + \frac{1+x}{2} + \frac{1-x}{2}$$
$$= 2 - x^2 \leq 2.$$

12. For $1 \leq k \leq n$, since $1+a_1 \geq 2\sqrt{a_1}$, $1+a_2 \geq 2\sqrt{a_2}, \ldots, 1+a_k \geq 2\sqrt{a_k}$, we have $(1+a_1)(1+a_2)\cdots(1+a_k) \geq 2^k\sqrt{a_1a_2\cdots a_k} \geq 2^k$. Hence,

$$\sum_{k=1}^{n} \frac{k}{(1+a_1)\cdots(1+a_k)} \leq \sum_{k=1}^{n} \frac{k}{2^k}.$$

Let $S = \sum_{k=1}^{n} \frac{k}{2^k}$. Then $2S = \sum_{k=1}^{n} \frac{k}{2^{k-1}} = \sum_{k=0}^{n-1} \frac{k+1}{2^k}$, so

$$S = 2S - S = 1 - \frac{n}{2^n} + \sum_{k=1}^{n-1} \frac{1}{2^k} = 2 - \frac{n}{2^n} - \frac{1}{2^{n-1}} < 2.$$

Therefore,

$$\sum_{k=1}^{n} \frac{k}{(1+a_1)\cdots(1+a_k)} \leq S < 2.$$

Solution 10

Solutions of Inequalities

1. The empty set. Hint: $4x^2 - 4x - 15 < 0$ implies that $-\frac{3}{2} < x < \frac{5}{2}$. Also from $\sqrt{x-1} < x - 3$ we get $x > 5$. Hence, there is no solution.
2. $[2, 4)$. Hint: It is necessary that $\log_2 x - 1 \geq 0$. Let $t = \sqrt{\log_2 x - 1}$ (≥ 0). Then the inequality becomes $t - \frac{3}{2}t^2 + \frac{1}{2} > 0$, so $0 \leq t < 1$. Hence,

$$0 \leq \log_2 x - 1 < 1,$$

or equivalently $2 \leq x < 4$.
3. $-4 \leq a \leq -1$. Hint: We easily get $A = (1, 3)$. Let

$$f(x) = 2^{1-x} + a, \quad g(x) = x^2 - 2(a+7)x + 5.$$

If $A \subseteq B$, then for $1 < x < 3$, the graphs of $f(x)$ and $g(x)$ both lie below the x-axis. Since $f(x)$ is a decreasing function and $g(x)$ is a quadratic function, we see that $A \subseteq B$ if and only if $f(1) \leq 0, g(1) \leq 0$, and $g(3) \leq 0$ are satisfied simultaneously. Solving the inequalities, we get $-4 \leq a \leq -1$.
4. $\{x|x > 1\}$. Hint: The original inequality is equivalent to $(\frac{1}{3})^x + (\frac{2}{3})^x < 1$. Let

$$f(x) = \left(\frac{1}{3}\right)^x + \left(\frac{2}{3}\right)^x.$$

Then $f(x)$ is decreasing on $\mathbf{R}$. Note that $f(1) = 1$, and so the inequality $f(x) < f(1)$ has the solution set $\{x|x > 1\}$.
5. $-\frac{3}{2}$. Hint: By $x^2 + (b+1)x \leq 0$ and $b < -1$, we have $0 \leq x \leq -(b+1)$. Also $f(x) = \left(x + \frac{b}{2}\right)^2 - \frac{b^2}{4}$. If $-2 < b < -1$, then $-(b+1) < -\frac{b}{2}$ and

$f(x)_{\min} = f(-b-1) = -\frac{1}{2}$, so $b = -\frac{3}{2}$. If $b \le -2$, then $-\frac{b}{2} \le (b+1)$ and $f(x)_{\min} = f\left(-\frac{b}{2}\right) = -\frac{1}{2}$, so $b = \pm\sqrt{2} \notin (-\infty, -2)$, which is impossible.

6. (1) $\left(-\infty, -\frac{1}{2}\right] \cup \left[-\frac{1}{3}, 0\right] \cup \{1\} \cup [2, +\infty)$.

   (2) The original inequality is equivalent to $\frac{2x^2-3x+1}{3x^2-7x+2} \ge 0$, which holds if and only if $\left(x - \frac{1}{3}\right)\left(x - \frac{1}{2}\right)(x-1)(x-2) \ge 0$, and $x \ne -\frac{1}{3}$ or 2. Hence, $x \in \left(-\infty, \frac{1}{3}\right) \cup \left[\frac{1}{2}, 1\right] \cup (2, +\infty)$.

7. (1) The original inequality is equivalent to $\sqrt{x^2+x-2} > 2x^2 + 2x - 10$. Let $\sqrt{x^2+x-2} = t \ge 0$. Then the inequality becomes $t > 2(t^2+2) - 10$. Solving this inequality, we get $0 \le t < 2$, so $0 \le \sqrt{x^2+x-2} < 2$. Hence, $-3 < x \le -2$ or $1 \le x < 2$.

   (2) It is necessary that $1 - \sqrt{1+2x} \ne 0$, or equivalently $x \ge -\frac{1}{2}$ and $x \ne 0$. When $x \ne 0$, the original inequality becomes $2 + 2x + 2\sqrt{1+2x} < 2x + 9$, whose solution set is $-\frac{1}{2} \le x < \frac{45}{8}$. Therefore, the solution set to the original inequality is $\left[-\frac{1}{2}, 0\right) \cup \left(0, \frac{45}{8}\right)$.

   (3) The original inequality is equivalent to $\sqrt{3-x} > \frac{1}{2} + \sqrt{x+1}$

$$\Leftrightarrow \begin{cases} x+1 \ge 0 \\ 3 - x > \left(\dfrac{1}{2} + \sqrt{x+1}\right)^2 \end{cases} \Leftrightarrow \begin{cases} x \ge -1 \\ \sqrt{x+1} < \dfrac{7}{4} - 2x \end{cases}$$

$$\Leftrightarrow \begin{cases} x \ge -1 \\ \dfrac{7}{4} - 2x > 0 \\ x + 1 < \left(\dfrac{7}{4} - 2x\right)^2 \end{cases} \Leftrightarrow -1 \le x < \frac{8 - \sqrt{31}}{8}.$$

8. (1) From the given inequality, it is necessary that $\log_3 x > 0$ and $\log_3(9\sqrt[3]{x}) > 0$, from which $x > 1$. Then the original inequality reduces to $\frac{(\log_3 x+2)(\log_3 x-3)}{\log_3 x+6} \ge 0$, so $-6 < \log_3 x \le -2$ or $\log_3 x \ge 3$. Hence, the solution set is $x \ge 27$.

   (2) The original inequality is equivalent to $\log_2(2^x - 1) \cdot [1 + \log_2(2^x - 1)] < 2 \Leftrightarrow -2 < \log_2(2^x-1) < 1 \Leftrightarrow \log_2 \frac{5}{4} < x < \log_2 3$.

9. (1) $(0, 1) \cup (1, 2^{\frac{2}{7}}) \cup (4, +\infty)$.

   (2) $|x+3|$ and $|x-3|$ have the zeros $x = -3$ and $x = 3$, respectively. Discussing the intervals $(-\infty, -3]$, $(-3, 3]$ and $(3, +\infty)$, we may remove the absolute value signs and get $x < -\frac{3}{2}$ or $x > \frac{3}{2}$. It is also possible to remove the absolute value signs by taking squares on both sides.

10. By $a < |a|$ we have $a < 0$, so the inequality becomes

$$1 > \frac{9a^3 - 11a}{a} > \frac{|a|}{a} = -1,$$

namely $-1 < 9a^2 - 11 < 1$. Hence, $a^2 \in \left(\frac{10}{9}, \frac{4}{3}\right)$. Since $a < 0$, we have

$$a \in \left(-\frac{2\sqrt{3}}{3}, -\frac{\sqrt{10}}{3}\right).$$

11. (1) The original inequality is equivalent to $a^{x^4-2x^2} > a^{-a^2}$. If $0 < a < 1$, then $x^4 - 2x^2 + a^2 < 0 \Leftrightarrow 1 - \sqrt{1-a^2} < x^2 < 1 + \sqrt{1-a^2}$, so $-\sqrt{1+\sqrt{1-a^2}} < x < -\sqrt{1-\sqrt{1-a^2}}$ or $\sqrt{1-\sqrt{1-a^2}} < x < \sqrt{1+\sqrt{1-a^2}}$. If $a > 1$, then $x^4 - 2x^2 + a^2 > 0$ for all $x \in \mathbf{R}$, so $\Delta = 4 - 4a^2 < 0$.

(2) The original inequality is equivalent to $\frac{a(x+1)}{x+2} \geq 1 \Leftrightarrow \frac{(a-1)x+a-2}{x+2} \geq 0 \Leftrightarrow (x+2)[(a-1)x + a - 2] \geq 0 (x \neq -2)$.

If $a > 1$, then $\frac{2-a}{a-1} > -2$, so $x < -2$ or $x \geq \frac{2-a}{a-1}$.

If $a = 1$, then the solution set is $x < -2$.

If $a = 0$, then the inequality has no solution.

If $0 < a < 1$, then $\frac{2-a}{a-1} < -2$, and the solution set is $\frac{2-a}{a-1} \leq x < -2$.

If $a < 0$, then $\frac{2-a}{a-1} > -2$, and the solution set is $-2 < x \leq \frac{2-a}{a-1}$.

12. (1) Since $a > 1$, the original inequality is equivalent to

$$\begin{cases} x - 1 > 0, \\ 1 + a(x-2) > 0, \\ (x-1)^2 > 1 + ax - 2a \end{cases} \Leftrightarrow \begin{cases} x > 2 - \dfrac{1}{a}, \\ \\ (x-2)(x-a) > 0. \end{cases}$$

If $1 < a < 2$, then

$$a - \left(2 - \frac{1}{a}\right) = a + \frac{1}{a} - 2 = \frac{(a-1)^2}{a} > 0,$$

$$2 - \left(2 - \frac{1}{a}\right) = \frac{1}{a} > 0.$$

Hence, the solution set to the inequality is $\left(2 - \frac{1}{a}, a\right) \cup (2, +\infty)$.

If $a \geq 2$, then $2 > 2 - \frac{1}{a}$, so the solution set is $(2 - \frac{1}{a}, 2) \cup (a, +\infty)$.

(2) By $\begin{cases} ax > 0 \text{ and } ax \neq 1 \\ x > 0 \text{ and } x \neq 1 \end{cases}$, we have $a > 0$. If $a = 1$, then the inequality holds constantly, and the solution set is $x > 0$ and $x \neq 1$.

If $a \neq 1$, then the original inequality reduces to $\frac{\log_a x}{1+\log_a x} + \frac{2(1+\log_a x)}{\log_a x} > 0$, so $\log_a x < -1$ or $\log_a x > 0$. If $a > 1$, then the solution set is $\left(0, \frac{1}{a}\right) \cup (1, +\infty)$. If $0 < a < 1$, then the solution set is $(0, 1) \cup \left(\frac{1}{a}, +\infty\right)$.

13. Apparently $a > 0$ and $a \neq 1$. The two numbers in the solution set are necessarily nonzero and sum to 0. Let $u = x^2 - a|x| + 5$. Then the original inequality becomes

$$\frac{\log_3(\sqrt{u}+1)}{-\log_3 a} \cdot \log_5(u+1) + \frac{1}{\log_3 a} \geq 0. \text{ ①}$$

(1) If $0 < a < 1$, then since $u > 0$, the inequality reduces to $\log_3(\sqrt{u}+1) \cdot \log_5(u+1) \geq 1$. Since $f(u) = \log_3(\sqrt{u}+1) \cdot \log_5(u+1)$ is an increasing function, the solution set to ① is $u \geq 4$, so $x^2 - a|x| + 5 \geq 4$. This inequality has infinitely many solutions, thus this case is impossible.

(2) If $a > 1$, then the inequality reduces to $\log_3(\sqrt{u}+1) \cdot \log_5(u+1) \leq 1$. The solution set to ① is $0 \leq u \leq 4$. Hence, $0 \leq x^2 - a|x| + 5 \leq 4$ has exactly two solutions, which sum to 0. This implies that the equation $t^2 - at + 5 = 4$ has two equal positive roots, thus $\Delta = a^2 - 4 = 0$. Since $a > 0$, we get $a = 2$. In this case, the inequality $0 \leq x^2 - a|x| + 5 \leq 4$ has exactly two solutions $x = \pm 1$.

Therefore, $a = 2$.

14. For $x = 0$, the inequality holds obviously. For $x \neq 0$, let $f(x) = x^2 + \frac{1}{x}$.

If $x \in [-1, 0)$, then the inequality reduces to $f(x) \leq a$. Note that $f(x)$ is decreasing on $[-1, 0)$, so its maximum value is $f(-1) = 0$, and we get $0 \leq a$.

If $x \in (0, 1]$, then the inequality reduces to $f(x) \geq a$. Since

$$x^2 + \frac{1}{x} = x^2 + \frac{1}{2x} + \frac{1}{2x} \geq 3\sqrt[3]{x^2 \cdot \frac{1}{2x} \cdot \frac{1}{2x}} = \frac{3}{\sqrt[3]{4}},$$

we see that the minimum value of $f(x)$ on $(0, 1]$ is $\frac{3}{\sqrt[3]{4}} \geq a$.

Therefore, the value range of a is $\left[0, \frac{3\sqrt[3]{2}}{2}\right]$.

Solution 11

Synthetical Problems of Inequalities

1. $\sqrt{6}$. Hint: Let $y = \sqrt{x-3} + \sqrt{6-x}$ with $3 \le x \le 6$. Then
$$y^2 = (x-3) + (6-x) + 2\sqrt{(x-3)(6-x)} \le 2[(x-3)+(6-x)] = 6.$$
Hence, $0 < y \le \sqrt{6}$, and the maximum value of k is $\sqrt{6}$.
2. $-\frac{1}{2} < a < \frac{3}{2}$. Hint: $a - a^2 > -\frac{(1+2^x)}{4^x} = -\left(\frac{1}{4^x} + \frac{1}{x^2}\right)$. By the condition,
$$a - a^2 > \max_{x \le 1}\left[-\left(\frac{1}{4^x} + \frac{1}{2^x}\right)\right] = -\frac{3}{4},$$
so $4a^2 - 4a - 3 < 0$. Solving the inequality, we get $-\frac{1}{2} < a < \frac{3}{2}$.
3. 3. Hint: $3^{\cos^2 x} + 3^{\sin^2 x} \ge 2\sqrt{3^{\cos^2 x + \sin^2 x}} = 2\sqrt{3} > 3$. Since $y = 3^t + 3^{1-t}$ is decreasing on $\left(0, \frac{1}{2}\right)$ and increasing on $\left(\frac{1}{2}, 1\right)$, we have $y < \max\left\{3^0 + \frac{3}{3^0}, 3^1 + \frac{3}{3^1}\right\} = 4$. Hence, $3 < 3^{\cos^2 x} + 3^{\sin^2 x} < 4$.
4. $7\sqrt{3}$. Hint: The function $y = 4S + \frac{9}{S}$ is decreasing on $\left(0, \frac{3}{2}\right)$. Since the equilateral triangle has the greatest area among all triangles inscribed to the unit circle, which is equal to $\frac{3\sqrt{3}}{4}$, we have $0 < S < \frac{3\sqrt{3}}{4} < \frac{3}{2}$. Hence,
$$y_{\min} = 4 \times \frac{3\sqrt{3}}{4} + \frac{9}{\frac{3\sqrt{3}}{4}} = 7\sqrt{3}.$$
5. $(1, 4)$. Hint: $f(x) = 1 + 2\sin x$. For $\frac{\pi}{6} \le x \le \frac{2\pi}{3}$, the inequality $|f(x) - m| < 2$ always holds, so $f(x) - 2 < m < f(x) + 2$. Hence, $[f(x) - 2]_{\max} < m < [f(x)+2]_{\min}$. Since $f(x)_{\max} = 3$ and $f(x)_{\min} = 2$, we get $1 < m < 4$.

6. $|ax+b|<2 \Leftrightarrow -2<ax+b<2 \Leftrightarrow -2-b<ax<2-b$. If $a>0$, then $\frac{-2-b}{a}<x<\frac{2-b}{a}$. From $\frac{-2-b}{a}=2$ and $\frac{2-b}{a}=6$, we get $a=1$ and $b=-4$.

   If $a<0$, then $\frac{2-b}{a}<x<\frac{-2-b}{a}$. By $\frac{2-b}{a}=2$ and $\frac{-2-b}{a}=6$ we see that $a=-1$ and $b=4$.
7. Let

$$M=\{(x,y)|x+y\geq x^2+y^2>1\},$$
$$N=\{(x,y)|0<x+y<x^2+y^2<1\}.$$

   Then $M\cup N$ is the solution set to the inequality.

   If $(x,y)\in N$, then $y<1$. If $(x,y)\in M$, then by $\left(x-\frac{1}{2}\right)^2+\left(y-\frac{1}{2}\right)^2\leq\frac{1}{2}$ we have $y\leq\frac{1}{2}+\frac{\sqrt{2}}{2}$. On the other hand, apparently $\left(\frac{1}{2},\frac{1}{2}+\frac{\sqrt{2}}{2}\right)\in M$. Therefore, the maximum value of y is $\frac{1}{2}+\frac{\sqrt{2}}{2}$.
8. Since $S\leq x_1+(-x_2+x_3)+(-x_4+x_5)+\cdots+(-x_{98}+x_{99})\leq x_1+x_2+x_4+\cdots+x_{98}\leq 1+2+2^3+\cdots+2^{97}$, and the equality can be reached by taking $x_1=1$, $x_2=2$, $x_3=2^2,\ldots,x_{98}=2^{97},x_{99}=2^{98},x_{100}=0$. They are the desired numbers for the maximum value.
9. If $x_1=1$, then $\frac{1}{x_1^3}=1$, so $\frac{1}{x_1^3}+\frac{1}{x_2^3}+\cdots+\frac{1}{x_m^3}>1$, a contradiction. If $x_1\geq 2$, then $x_i\geq i+1$ for $i=1,2,\ldots,m$. Note that

$$\frac{1}{x_i^3}\leq\frac{1}{(i+1)^3}<\frac{1}{(i+1)^2}=\frac{1}{i}-\frac{1}{i+1}.$$

   Hence,

$$\sum_{i=1}^{m}\frac{1}{x_i^3}<\left(1-\frac{1}{2}\right)+\left(\frac{1}{2}-\frac{1}{3}\right)+\cdots+\left(\frac{1}{m}-\frac{1}{m+1}\right)=1-\frac{1}{m+1}<1,$$

   which also gives a contradiction.
10. There are two solutions at most. Otherwise, suppose three different integers x_1, x_2, and x_3 satisfy $|f(x_i)|\leq 50$ with $i=1,2,$ and 3. Assume that $x_1>x_2>-\frac{b}{2a}$. Then since x_1 and x_2 are integers, $x_1+x_2+\frac{b}{a}\geq 1$. Hence,

$$|f(x_1)-f(x_2)|=|a(x_1-x_2)\left(x_1+x_2+\frac{b}{a}\right)|\geq a>100.$$

   This is contradictory to $|f(x_1)|\leq 50$ and $|f(x_2)|\leq 50$. Similarly, we can show that the case when two of x_1, x_2 and x_3 are less than or equal to $-\frac{b}{2a}$ is also impossible.

11. Let $t = \log_a x$. Then $x = a^t$, so $f(t) = \frac{a(a^{2t}-1)}{a^t(a^2-1)}$. Hence,

$$f(n) = \frac{a(a^{2n}-1)}{a^n(a^2-1)} = \frac{a}{a^n}[1 + a^2 + a^4 + \cdots + a^{2(n-1)}].$$

Since $a > 0$ and $a \neq 1$,

$$a^{2k} + a^{2(n-k-1)} > 2\sqrt{a^{2k} \cdot a^{2(n-k-1)}} = 2a^{n-1} \quad (k = 0, 1, 2, \ldots, n-1).$$

Adding the n inequalities, we get

$$2[1 + a^2 + a^4 + \cdots + a^{2(n-1)}] > 2na^{n-1}.$$

Therefore, $f(n) > \frac{a}{a^n} \cdot na^{n-1} = n$.

12. Suppose there exists such a pair of real numbers (p, q). Then $|f(x)| \leq 2$ for all $x \in [1, 5]$. We choose $x = 1, 3$, and 5 in succession. Then

$$|1 + p + q| \leq 2, \quad |9 + 3p + 1| \leq 2, \quad |25 + 5p + q| \leq 2.$$

Also $8 = f(5) - 2f(3) + f(1)$, from which

$$\begin{aligned} 8 &= |f(5) - 2f(3) + f(1)| \\ &\leq |f(5)| + 2|f(3)| + |f(1)| \\ &\leq 2 + 2 \times 2 + 2 = 8. \end{aligned}$$

Hence, $f(5) = f(1) = 2$ and $f(3) = -2$. Solving the equations, we get $p = -6$ and $q = 7$. In this case, $f(x) = x^2 - 6x + 7$, and $|f(x)| \leq 2$ for all $x \in [1, 5]$.

13. Apparently $a \neq 0$. Since replacing a, b, and c with $-a, -b$ and $-c$ respectively does not change the condition or the result, we may assume that $a > 0$.

   Note that $|a(b-c)| > |b^2 - ac| + |c^2 - ab|$ and

$$\begin{aligned} |b^2 - ac| + |c^2 - ab| &\geq |(b^2 - ac) - (c^2 - ab)| \\ &= |(b-c)(a+b+c)|, \end{aligned}$$

From which $|b-c| \cdot |a+b+c| < a \cdot |b-c|$. Hence, $b \neq c$ and $|a+b+c| < a$. In other words, $-a < a + b + c < a$, so $b + c < 0$ and $2a + b + c > 0$.

   Let $f(x) = ax^2 + bx + c$. Then $f(0) = c$ and $f(2) = 4a + 2b + c$.

   If $b < c$, then by

$$\begin{aligned} (b^2 - ac) + (c^2 - ab) &\leq |b^2 - ac| + |c^2 - ab| \\ &= |a(b-c)| = a(c-b). \end{aligned}$$

we get $b^2 + c^2 < 2ac$. Since $f(x) = 0$ has two real roots, necessarily $\Delta = b^2 - 4ac > 0$, hence $b^2 \geq 4ac > 2(b^2 + c^2)$, namely $b^2 + 2c^2 < 0$. This is impossible.

If $b > c$, then by $b + c < 0$ we have $c < 0$. In this case, $f(0) = c < 0$, and since $2a + b > 2a + b + c > 0$, we have $f(2) = 4a + 2b + c > 0$. By the property of continuous functions, the equation $f(x) = 0$ has at least one root in $(0, 2)$.

14. We have $f(x) = f\left(\frac{x}{2} + \frac{x}{2}\right) = f^2\left(\frac{x}{2}\right) \geq 0$. If there exists $x_0 \in \mathbf{R}$ such that $f(x_0) = 0$, then for all real x we have $f(x) = f(x_0 + x - x_0) = f(x_0)f(x - x_0) = 0$, which is contradictory to the condition. Hence, $f(x) > 0$ for all x.

Let $x = y = 0$ in $f(x + y) = f(x)f(y)$. Then $f(0)^2 = f(0)$. Since $f(0) > 0$ we have $f(0) = 1$.

Next, we show that $f(x)$ is increasing on $\mathbf{R}$.

If $x_2 > x_1$, then $x_2 - x_1 > 0$, so $f(x_2 - x_1) > 1$. Hence,

$$f(x_1)f(x_2 - x_1) > f(x_1)$$

and $f(x_2) = f(x_1 + x_2 - x_1) = f(x_1)f(x_2 - x_1) > f(x_1)$.

Since $f(x) > 0$ and $f(x)$ is increasing on $\mathbf{R}$,

$$f(x) \leq \frac{1}{f(x+1)} \Leftrightarrow f(x)f(x+1) \leq 1 \Leftrightarrow f(2x+1) \leq f(0) \Leftrightarrow 2x+1 \leq 0$$

Therefore, the solution set is $\left(-\infty, -\frac{1}{2}\right]$.

Solution 12

Concepts and Properties of Trigonometric Functions

1. It follows that $k > 0$. Further, from $0 < \beta < \frac{\pi}{2} - \alpha < \frac{\pi}{2}$,

$$\sin\frac{\alpha}{2} = k\cos\beta > k\cos\left(\frac{\pi}{2} - \alpha\right) = k\sin\alpha.$$

Hence, $1 > 2k\cos\frac{\alpha}{2} > 2k\cos\frac{\pi}{4} = \sqrt{2}k$ (since $0 < \alpha < \frac{\pi}{2}$ and $0 < \frac{\alpha}{2} < \frac{\pi}{4}$). This implies that $k < \frac{\sqrt{2}}{2}$. On the other hand, $k \to \frac{\sqrt{2}}{2}$ when $\alpha \to \frac{\pi}{2}$, and $k \to 0$ when $\alpha \to 0$. Therefore, the value range of k is $\left(0, \frac{\sqrt{2}}{2}\right)$.

2. By the formula of the auxiliary angle, $T = a\sin t + b\cos t = \sqrt{a^2+b^2}\sin(t+\varphi)$, where φ satisfies $\sin\varphi = \frac{b}{\sqrt{a^2+b^2}}$ and $\cos\varphi = \frac{a}{\sqrt{a^2+b^2}}$. Hence, the range of T is $[-\sqrt{a^2+b^2}, \sqrt{a^2+b^2}]$ and the maximal difference of temperatures is $2\sqrt{a^2+b^2} = 10$, so $a^2+b^2 = 25$. Therefore, $a + b \le \sqrt{2(a^2+b^2} = 5\sqrt{2}$, where the equality holds if and only if $a = b = \frac{5}{2}\sqrt{2}$.

3. Since $f(x) + f\left(\frac{1}{x}\right) = \frac{1}{x^2+1} + \frac{1}{\left(\frac{1}{x}\right)^2+1} = \frac{1}{x+1} + \frac{x^2}{x^2+1} = 1$,

$$\begin{aligned} &f(\tan 5^\circ) + f(\tan 10^\circ) + \cdots + f(\tan 85^\circ) \\ &\quad = (f(\tan 5^\circ) + f(\cot 5^\circ)) + (f(\tan 10^\circ) + f(\cot 10^\circ)) \\ &\qquad + \cdots + (f(\tan 40^\circ) + f(\cot 40^\circ)) + f(\tan 45^\circ) \\ &\quad = 8 + f(1) = 8 + \frac{1}{2} = \frac{17}{2}. \end{aligned}$$

4. By the boundedness of sine and cosine functions, $|\cos\alpha_1\cos\alpha_2\cdots\cos\alpha_n + \sin\alpha_1\sin\alpha_2\cdots\sin\alpha_n| \le |\cos\alpha_1\cos\alpha_2\cdots\cos\alpha_n| + |\sin\alpha_1$

$\sin \alpha_2 \cdots \sin \alpha_n| \leq |\cos \alpha_1| + |\sin \alpha_1|$, where $|\cos \alpha_1| + |\sin \alpha_1| = |\cos \alpha_1 + \sin \alpha_1|$ or $|\cos \alpha_1 - \sin \alpha_1|$, both of which have the maximum value $\sqrt{2}$. Therefore, the proposition follows.

5. In fact, it suffices to prove that $\cos x^2 = 1, \cos y^2 = 1$, and $\cos xy = -1$ cannot be satisfied simultaneously. If they hold simultaneously, then there exist integers m, n, and k such that $x^2 = 2m\pi$, $y^2 = 2n\pi$, and $xy = (2k+1)\pi$. Hence, $(2m\pi)(2n\pi) = (2k+1)^2\pi^2$, namely $4mn = (2k+1)^2$, which is impossible since one side of the equality is even and the other side is odd. Therefore, the proposition follows.
6. The problem is equivalent to finding the value range of a such that for all real x,

$$\sin^6 x + \cos^6 x + a \sin x \cos x \geq 0.$$

Note that

$$\begin{aligned}
&\sin^6 x + \cos^6 x + a \sin x \cos x \\
&\quad = (\sin^2 x + \cos^2 x)(\sin^4 x - \sin^2 x \cos^2 x + \cos^4 x) + a \sin x \cos x \\
&\quad = (\sin^2 x + \cos^2 x)^2 - 3 \sin^2 x \cos^2 x + a \sin x \cos x \\
&\quad = 1 + a \sin x \cos x - 3 \sin^2 x \cos^2 x.
\end{aligned}$$

If we let $t = \sin x \cos x$, then the problem reduces to finding the value range of a such that $f(t) = 1 + at - 3t^2 \geq 0$ for all $t \in \left[-\frac{1}{2}, \frac{1}{2}\right]$.

This is an extreme value problem of a quadratic function on a finite interval. Since the graph of $f(t)$ opens downwards, the value range of a is determined by $\begin{cases} f\left(\frac{1}{2}\right) \geq 0 \\ f\left(-\frac{1}{2}\right) \geq 0 \end{cases}$. Hence, $-\frac{1}{2} \leq a \leq \frac{1}{2}$.

7. (1) Since $f(x) = \sin(\cos x)$ is decreasing on $\left(0, \frac{\pi}{2}\right)$, while $g(x) = x$ is increasing, and $f(0) = \sin 1 > g(0)$ and $f\left(\frac{\pi}{2}\right) = 0 < g\left(\frac{\pi}{2}\right)$, we see that the graphs of $f(x)$ and $g(x)$ have exactly one intersection point over $\left(0, \frac{\pi}{2}\right)$. Hence, the equation $\sin(\cos x) = x$ has exactly one solution in $\left(0, \frac{\pi}{2}\right)$. By a similar method we can show that $\cos(\sin x) = x$ has a unique solution in $\left(0, \frac{\pi}{2}\right)$.

   (2) By the conclusion in (1), for $x \in (0, b)$ we have $\sin(\cos x) > x$ and for $x \in \left(b, \frac{\pi}{2}\right)$ we have $\sin(\cos x) < x$, and vice versa. Since $\sin(\cos a) = \sin a < a$, it follows that $a > b$. On the other hand, by a similar analysis, $\cos x > x$ if and only if $0 < x < a$. Now, $\cos(\sin c) = c > \sin c$, so $\sin c < a$. Hence, $\cos(\sin c) > \cos a = a$, namely $c > a$. Therefore, $c > a > b$.

8. It follows that

$$\frac{\sin(a-c)}{\sin(b-d)} = \frac{\sin a - \sin c}{\sin b - \sin d} = \frac{2\sin\frac{a-c}{2}\cos\frac{a+c}{2}}{2\sin\frac{b-d}{2}\cos\frac{b+d}{2}},$$

$$\frac{\sin(a-c)}{\sin(b-d)} = \frac{\sin a + \sin c}{\sin b + \sin d} = \frac{2\sin\frac{a+c}{2}\cos\frac{a-c}{2}}{2\sin\frac{b+d}{2}\cos\frac{b-d}{2}}.$$

We apply the double angle formula on the left side of both euqalities and cancel the duplicating terms, getting

$$\frac{\cos\frac{a-c}{2}}{\cos\frac{b-d}{2}} = \frac{\cos\frac{a+c}{2}}{\cos\frac{b+d}{2}}, \frac{\sin\frac{a-c}{2}}{\sin\frac{b-d}{2}} = \frac{\sin\frac{a+c}{2}}{\sin\frac{b+d}{2}}.$$

Now, we multiply the two equalities to obtain that

$$\frac{\sin(a-c)}{\sin(b-d)} = \frac{\sin(a+c)}{\sin(b+d)} = \frac{\sin(a+c)+\sin(a-c)}{\sin(b+d)+\sin(b-d)} = \frac{2\sin a\cos c}{2\sin b\cos d}.$$

Combining the above with $\frac{\sin a}{\sin b} = \frac{\sin(a-c)}{\sin(b-d)}$, we have $\frac{\cos c}{\cos d} = 1$, that is $\cos c = \cos d$. Since $c, d \in (0, \pi)$ and the cosine function is decreasing on this interval, it follows that $c = d$. By a similar analysis, $\cos a = \cos b$, hence $a = b$.

9. The key of this problem is to show the following statement: If x and y are positive real numbers such that $x + y < \pi$, then $\sin x < \sin y$ if and only if $x < y$.

In fact, if $0 < x < y < \frac{\pi}{2}$, then since the sine function is increasing on $\left[0, \frac{\pi}{2}\right]$, we get $\sin x < \sin y$.

If $y \geq \frac{\pi}{2}$, then $\frac{\pi}{2} \leq y < \pi - x < \pi$. Since the sine function is decreasing on $\left[\frac{\pi}{2}, \pi\right]$, we have $\sin x = \sin(\pi - x) < \sin y$. This shows that $\sin x < \sin y$ when $x < y$.

Conversely, assume that $\sin x < \sin y$. If $x < \frac{\pi}{2}$ and $y \geq \frac{\pi}{2}$, then apparently $x < y$. If $x < \frac{\pi}{2}$ and $y < \frac{\pi}{2}$, then by monotonicity $x < y$. If $x \geq \frac{\pi}{2}$, then by $x+y < \pi$ we have $\frac{\pi}{2} \leq x < \pi - y < \pi$, so by monotonicity $\sin x > \sin y$, a contradiction. Therefore, the above statement is valid.

Return to the original problem. Since $0 < \alpha < \beta$ and $\alpha + \beta < \pi$, we have $\sin\alpha < \sin\beta$. Combining it with the condition $\frac{\sin a}{\sin b} \leq \frac{\sin\alpha}{\sin\beta}$ we also have $\sin a < \sin b$, and since $a + b < \pi$ and $a, b > 0$ we conclude that $a < b$.

10. By $\sin\omega a + \sin\omega b = 2$ we have $\sin\omega a = \sin\omega b = 1$. Since $[\omega a, \omega b] \subseteq [\omega\pi, 2\omega\pi]$, the condition in the problem is equivalent to the following:

there exist integers k and $l(k < l)$ such that

$$\omega\pi \le 2k\pi + \frac{\pi}{2} < 2l\pi + \frac{\pi}{2} \le 2\omega\pi.$$

If $\omega \ge 4$, then the interval $[\omega\pi, 2\omega\pi]$ has its length at least 4π, so the integers k and l are guaranteed to exist.

If $0 < \omega < 4$, then since $[\omega\pi, 2\omega\pi] \subseteq (0, 8\pi)$, it suffices to consider the following cases:

(1) $\omega\pi \le \frac{\pi}{2} < \frac{5\pi}{2} \le 2\omega\pi$. In this case, $\omega \le \frac{1}{2}$ and $\omega \ge \frac{5}{4}$, which is impossible.
(2) $\omega\pi \le \frac{5\pi}{2} < \frac{9\pi}{2} \le 2\omega\pi$. In this case, $\frac{9}{4} \le \omega \le \frac{5}{2}$.
(3) $\omega\pi \le \frac{9\pi}{2} < \frac{13\pi}{2} \le 2\omega\pi$. In this case, $\frac{13}{4} \le \omega < 4$.

In summary, the value range of ω is $\left[\frac{9}{4}, \frac{5}{2}\right] \cup \left[\frac{13}{4}, +\infty\right]$.

11. Let $x = 0$ and $x = 1$ in succession. The given condition implies that $\sin\theta > 0$ and $\cos\theta > 0$. Note that

$$\begin{aligned} & x^2\cos\theta - x(1-x) + (1-x)^2\sin\theta \\ &\quad = (1+\cos\theta+\sin\theta)x^2 - (1+2\sin\theta)x + \sin\theta \\ &\quad = (1+\cos\theta+\sin\theta)\left(x - \frac{1+2\sin\theta}{2+2\cos\theta+2\sin\theta}\right)^2 \\ &\qquad + \sin\theta - \frac{(1+2\sin\theta)^2}{4(1+\cos\theta+\sin\theta)}, \end{aligned}$$

and since $\sin\theta > 0$ and $\cos\theta > 0$, we have $1+\cos\theta+\sin\theta > 0$ and

$$0 < \frac{1+2\sin\theta}{2+2\cos\theta+2\sin\theta} < 1.$$

Since the original inequality holds for all $x \in [0,1]$, it holds for $x = \frac{1+2\sin\theta}{2+2\cos\theta+2\sin\theta}$. Hence, $\sin\theta - \frac{(1+2\sin\theta)^2}{4(1+\cos\theta+\sin\theta)} > 0$. Conversely, if this inequality holds, then the original inequality holds for all $x \in [0,1]$. Therefore, the problem's condition is equivalent to

$$\begin{cases} \sin\theta > 0, \\ \cos\theta > 0, \\ \sin\theta - \dfrac{(1+2\sin\theta)^2}{4(1+\cos\theta+\sin\theta)} > 0. \end{cases}$$

By $\sin\theta - \frac{(1+2\sin\theta)^2}{4(1+\cos\theta+\sin\theta)} > 0$ we have $4\sin\theta\cos\theta > 1$, that is $\sin 2\theta > \frac{1}{2}$. Combining the above with $\sin\theta > 0$ and $\cos\theta > 0$, we finally get

$$2k\pi + \frac{\pi}{12} < \theta < 2k\pi + \frac{5}{12}\pi, \quad k \in \mathbf{Z}.$$

12. It follows from the condition that $B \subseteq C(r)$. Hence, for $0 < t < 2\pi$,

$$\sin^2 t + (2\sin t\cos t)^2 \le r^2.$$

Let $g(t) = \sin^2 t + 4\sin^2 t\cos^2 t$. Then on the interval $0 < t < 2\pi$ we have $(g(t))_{\max} \le r^2$. Note that

$$\begin{aligned} g(t) &= \sin^2 t(1 + 4\cos^2 t) = \sin^2 t(5 - \sin^2 t) \\ &= -4(\sin^2 t)^2 + 5\sin^2 t = -4\left(\sin^2 t - \frac{5}{8}\right)^2 + \frac{25}{16} \le \frac{25}{16}, \end{aligned}$$

where the equality holds if and only if $\sin^2 t = \frac{5}{8}$.

If $t = \arcsin\sqrt{\frac{5}{8}}$, then apparently $t \in (0, 2\pi)$ and $\sin^2 t = \frac{5}{8}$, so $(g(t))_{\max} = \frac{25}{16}$. Therefore, the minimum value of r is $\frac{5}{4}$.

13. Let $y = 3 + \sin x$. Then the function y has the range $[2, 4]$. Note that the function $g(y) = y + \frac{2}{y}$ is increasing on $[2, 4]$ (it needs to be proven, and we leave it for the readers to complete), so the range of $g(y)$ is $[g(2), g(4)]$, i.e., $\left[3, \frac{9}{2}\right]$.

Hence, the function

$$f(b) = \max|g(y) + b - 3| = \max\left\{|b|, |b + \frac{3}{2}|\right\}$$

has the minimum value $\frac{3}{4}$, attained at $b = -\frac{3}{4}$. It is also easy to see that $b = -\frac{3}{4}$ is the only point that makes $f(b)$ reach the minimum value.

14. We prove it by contradiction. If there exists $T > 0$ such that $\sin(x + T)^2 = \sin x^2$ for all $x \in \mathbf{R}$, then we let $x = 0$, $\sqrt{\pi}$, and $\sqrt{2\pi}$ in the identity above, and get $\sin T^2 = \sin(\sqrt{\pi} + T)^2 = \sin(\sqrt{2\pi} + T)^2 = 0$. Hence, there exist positive integers m, n, and k such that

$$T^2 = m\pi, (\sqrt{\pi} + T)^2 = n\pi, (\sqrt{2\pi} + T)^2 = k\pi.$$

This implies that $\pi + 2\sqrt{\pi}T = (n - m)\pi$ and $2\pi + 2\sqrt{2\pi}T = (k - m)\pi$, and consequently $\frac{(k-m-2)\pi}{(n-m-1)\pi} = \frac{2\sqrt{2\pi}T}{2\sqrt{\pi}T} = \sqrt{2}$, that is $\sqrt{2} = \frac{k-m-2}{n-m-1}$. However, this is impossible since $\sqrt{2}$ is irrational. Therefore, $y = \sin x^2$ is not a periodic function.

15. Note that $a\sin x + b\cos x = \sqrt{a^2+b^2}\sin(x+\varphi)$, where φ is determined by $\sin\varphi = \frac{b}{\sqrt{a^2+b^2}}$ and $\cos\varphi = \frac{a}{\sqrt{a^2+b^2}}$. Now, if $a^2+b^2 \geq \frac{\pi^2}{4}$, then $\sqrt{a^2+b^2} \geq \frac{\pi}{2}$, so $0 < \frac{\pi}{2\sqrt{a^2+b^2}} \leq 1$. In this case, there exists $\beta \in \mathbf{R}$ such that $\sin(\beta+\varphi) = \frac{\pi}{\sqrt{a^2+b^2}}$. For this β, we have $a\sin\beta + b\cos\beta = \frac{\pi}{2}$, so $\cos(a\sin\beta) = \sin(b\cos\beta)$, giving a contradiction. Therefore, $a^2 + b^2 < \frac{\pi^2}{4}$.

Solution 13

Deformation via Trigonometric Identities

1. (1) The original expression

$$= \frac{(\tan\alpha + \cot\alpha)\sin\alpha\cos\alpha}{\frac{1}{\sec^2\alpha} + \frac{1}{\csc^2\alpha}} = \frac{(\tan\alpha + \cot\alpha)\sin\alpha\cos\alpha}{\cos^2\alpha + \sin^2\alpha}$$

$$= (\tan\alpha + \cot\alpha)\sin\alpha\cos\alpha = \left(\frac{\sin\alpha}{\cos\alpha} + \frac{\cos\alpha}{\sin\alpha}\right)\sin\alpha\cos\alpha$$

$$= \sin^2\alpha + \cos^2\alpha = 1.$$

(2) The original expression

$$= \frac{\sin^2\alpha - \sin^2\beta + \cos^2\alpha - \cos^2\beta}{(\cos\alpha + \cos\beta)(\sin\alpha - \sin\beta)}$$

$$= \frac{1 - 1}{(\cos\alpha + \cos\beta)(\sin\alpha - \sin\beta)} = 0.$$

(3) The original expression equals $\cos A(\sin B\cos C - \cos B\sin C) + \cos B(\sin C\cos A - \cos C\sin A) + \cos C(\sin A\cos B - \cos A\sin B)$, and expanding the expression gives 0.

2. (1) The original expression

$$= \frac{1 - 4\sin 10^\circ \sin 70^\circ}{2\sin 10^\circ} = \frac{1 - 2(\cos 60^\circ - \cos 80^\circ)}{2\sin 10^\circ}$$

$$= \frac{1 - 2\left(\frac{1}{2} - \sin 10^\circ\right)}{2\sin 10^\circ} = \frac{2\sin 10^\circ}{2\sin 10^\circ} = 1.$$

(2) The original expression

$$\begin{aligned}
&= \sin 20^\circ \cos 50^\circ + \cos^2 50^\circ + \sin^2 20^\circ \\
&= \cos 50^\circ(\sin 20^\circ + \sin 40^\circ) + \sin^2 20^\circ \\
&= 2\cos 50^\circ \sin 30^\circ \cos 10^\circ + \frac{1-\cos 40^\circ}{2} \\
&= \cos 50^\circ \cos 10^\circ - \frac{\cos 40^\circ}{2} + \frac{1}{2} \\
&= \frac{\cos 60^\circ + \cos 40^\circ}{2} - \frac{\cos 40^\circ}{2} + \frac{1}{2} = \frac{3}{4}.
\end{aligned}$$

(3) The original expression

$$\begin{aligned}
&= \frac{\sin 7^\circ + \frac{1}{2}(\sin 23^\circ - \sin 7^\circ)}{\cos 7^\circ - \frac{1}{2}(\cos 7^\circ - \cos 23^\circ)} = \frac{\sin 7^\circ + \sin 23^\circ}{\cos 7^\circ + \cos 23^\circ} \\
&= \frac{2\sin 15^\circ \cos 8^\circ}{2\cos 15^\circ \cos 8^\circ} \\
&= \tan 15^\circ = \frac{1-\cos 30^\circ}{\sin 30^\circ} = 2-\sqrt{3}.
\end{aligned}$$

(4) Note that for $1 \le k \le 44$,

$$\begin{aligned}
(1+\tan k^\circ)(1+\tan(45^\circ - k^\circ)) &= \left(1+\tan k^\circ\right)\cdot\left(1+\frac{1-\tan k^\circ}{1+\tan k^\circ}\right) \\
&= (1+\tan k^\circ)\cdot\frac{2}{1+\tan k^\circ} = 2.
\end{aligned}$$

Hence, the original expression equals

$$\begin{aligned}
&(1+\tan 44^\circ)(1+\tan 43^\circ)\cdot\dots\cdot(1+\tan 1^\circ) \\
&= [(1+\tan 44^\circ)(1+\tan 1^\circ)]\cdot\dots\cdot \\
&\quad [(1+\tan 23^\circ)(1+\tan 22^\circ)] = 2^{22}.
\end{aligned}$$

3. Squaring both sides of the condition's equality, we have $1-2\sin\theta\cos\theta = \frac{1}{4}$, so $\sin\theta\cos\theta = \frac{3}{8}$. Further,

$$\begin{aligned}
\sin^3\theta - \cos^3\theta &= (\sin\theta - \cos\theta)(\sin^2\theta + \sin\theta\cos\theta + \cos^2\theta) \\
&= \frac{1}{2}\times\left(1+\frac{3}{8}\right) = \frac{11}{16}.
\end{aligned}$$

4. Note that

$$\frac{\sin 3\alpha}{\sin\alpha} - \frac{\cos 3\alpha}{\cos\alpha} = \frac{\sin 3\alpha\cos\alpha - \cos 3\alpha\sin\alpha}{\sin\alpha\cos\alpha} = \frac{\sin(3\alpha-\alpha)}{\frac{1}{2}\sin 2\alpha} = 2.$$

Hence, $\frac{\sin 3\alpha}{\sin\alpha} = \frac{\cos 3\alpha}{\cos\alpha} + 2 = \frac{7}{3}$.

5. The condition implies that $\sin(\alpha+\beta)=\sqrt{1-\cos^2(\alpha+\beta)}=\frac{3}{5}$. Since $0<2\alpha+\beta<\frac{3}{2}\pi$ and $\cos(2\alpha+\beta)=\frac{3}{5}>0$, we see that $2\alpha+\beta$ is an acute angle and

$$\sin(2\alpha+\beta)=\sqrt{1-\cos^2(2\alpha+\beta)}=\frac{4}{5}.$$

Consequently,

$$\begin{aligned}\cos\alpha &= \cos[(2\alpha+\beta)-(\alpha+\beta)]\\ &= \cos(2\alpha+\beta)\cos(\alpha+\beta)+\sin(2\alpha+\beta)\cdot\sin(\alpha+\beta)\\ &= \frac{3}{5}\times\frac{4}{5}+\frac{4}{5}\times\frac{3}{5}=\frac{24}{25}.\end{aligned}$$

6. By the condition, $\gamma-\alpha=\frac{2\pi}{3}$. Hence, $\tan(\beta-\alpha)=\sqrt{3}, \tan(\gamma-\beta)=\sqrt{3}$, and $\tan(\gamma-\alpha)=-\sqrt{3}$, so $\tan\beta-\tan\alpha=\sqrt{3}(1+\tan\alpha\tan\beta), \tan\gamma-\tan\beta=\sqrt{3}(1+\tan\beta\tan\gamma)$, and $\tan\alpha-\tan\gamma=\sqrt{3}(1+\tan\gamma\tan\alpha)$. Adding the three equalities, we get that

$$\tan\alpha\tan\beta+\tan\beta\tan\gamma+\tan\gamma\tan\alpha=-3.$$

7. By the condition,

$$\begin{aligned}&\sin\alpha\cos(\alpha+10^\circ)\cos(\alpha+20^\circ)\cos(\alpha+30^\circ)\\ &\quad=\cos\alpha\sin(\alpha+10^\circ)\sin(\alpha+20^\circ)\sin(\alpha+30^\circ).\end{aligned}$$

Multiplying both sides by 4 and using the product-to-sum formula, we obtain that

$$2\sin(2\alpha+20^\circ)\cos(2\alpha+40^\circ)=2\sin 20^\circ\cos 20^\circ$$

$$\sin(4\alpha+60^\circ)-\sin 20^\circ=\sin 40^\circ,$$

$$\sin(4\alpha+60^\circ)=\sin 40^\circ+\sin 20^\circ=2\sin 30^\circ\cos 10^\circ=\cos 10^\circ.$$

Therefore, $\sin(4\alpha+60^\circ)=\sin 80^\circ$, and since α is an acute angle, $\alpha=5^\circ$ or 10°.

8. (1) By the condition, $2\sin B=\sin A+\sin C$, or equivalently

$$2\sin(A+C)=\sin A+\sin C.$$

Hence, $4\sin\frac{A+C}{2}\cos\frac{A+C}{2}=2\sin\frac{A+C}{2}\cos\frac{A-C}{2}$, so $2\cos\frac{A+C}{2}=\cos\frac{A-C}{2}$.

Consequently, $2\left(\cos\frac{A}{2}\cos\frac{C}{2}-\sin\frac{A}{2}\sin\frac{C}{2}\right)=\cos\frac{A}{2}\cos\frac{C}{2}+\sin\frac{A}{2}\sin\frac{C}{2}$, and this implies that $\cot\frac{A}{2}\cot\frac{C}{2}=3$.

(2) By the condition, $\sin C = \frac{\sin A+\sin B}{\cos A+\cos B} = \frac{2\sin\frac{A+B}{2}\cos\frac{A-B}{2}}{2\cos\frac{A+B}{2}\cos\frac{A-B}{2}} = \frac{\cos\frac{C}{2}}{\sin\frac{C}{2}}$, from which $2\sin^2\frac{C}{2} = 1$. Equivalently, $1 - 2\sin^2\frac{C}{2} = 0$, namely $\cos C = 0$. Therefore, $C = \frac{\pi}{2}$ and ΔABC is a right triangle.

(3) By the condition, $\sin A+\sin B+\sin C = \sqrt{3}(\cos A+\cos B+\cos C)$, or equivalently

$$(\sin A - \sqrt{3}\cos A) + (\sin B - \sqrt{3}\cos B) + (\sin C - \sqrt{3}\cos C) = 0.$$

Hence, $2\left(\sin(A-\frac{\pi}{3}) + \sin(B-\frac{\pi}{3}) + \sin\left(C-\frac{\pi}{3}\right)\right) = 0$. Let $\frac{\pi}{3} - A = \alpha, \frac{\pi}{3} - B = \beta$, and $\frac{\pi}{3} - C = \gamma$. Then by $A+B+C = \pi$ we have $\alpha+\beta+\gamma = 0$. Combining it with the earlier equality, we have $\sin\alpha+\sin\beta = \sin(\alpha+\beta)$, so $2\sin\frac{\alpha+\beta}{2}\cos\frac{\alpha-\beta}{2} = 2\sin\frac{\alpha+\beta}{2}\cos\frac{\alpha+\beta}{2}$. Hence,

$$\sin\frac{\alpha+\beta}{2}\left(\cos\frac{\alpha-\beta}{2} - \cos\frac{\alpha+\beta}{2}\right) = 0,$$

$$\Rightarrow 2\sin\frac{\alpha+\beta}{2}\sin\frac{\alpha}{2}\sin\frac{\beta}{2} = 0.$$

In other words, $\sin\left(\frac{\pi}{6}-\frac{A}{2}\right)\sin\left(\frac{\pi}{6}-\frac{B}{2}\right)\sin\left(\frac{\pi}{6}-\frac{C}{2}\right) = 0$. Note that $-\frac{\pi}{3} < \frac{\pi}{6} - \frac{A}{2} < \frac{\pi}{6}, -\frac{\pi}{3} < \frac{\pi}{6} - \frac{B}{2} < \frac{\pi}{6}$, and $-\frac{\pi}{3} < \frac{\pi}{6} - \frac{C}{2} < \frac{\pi}{6}$. Combining the above with the property of the sine function, we conclude that at least one of $\frac{\pi}{6} - \frac{A}{2}, \frac{\pi}{6} - \frac{B}{2}$, and $\frac{\pi}{6} - \frac{C}{2}$ is 0, which is the conclusion that we wanted to prove.

9. Trigonometric identities for triangles are a special kind of trigonometric identities, and they all have an implicit condition: $A+B+C = \pi$.

(1) We use the double angle formula and the sum-to-product formula:

$$\begin{aligned}
&\sin A + \sin B + \sin C \\
&= 2\sin\frac{A+B}{2}\cos\frac{A-B}{2} + \sin(A+B) \\
&= 2\sin\frac{A+B}{2}\cos\frac{A-B}{2} + 2\sin\frac{A+B}{2}\cos\frac{A+B}{2} \\
&= 2\sin\frac{A+B}{2}\left(\cos\frac{A-B}{2} + \cos\frac{A+B}{2}\right) \\
&= 4\sin\frac{A+B}{2}\cos\frac{A}{2}\cos\frac{B}{2} \\
&= 4\cos\frac{A}{2}\cos\frac{B}{2}\cos\frac{C}{2}.
\end{aligned}$$

(2) The method is similar to (1):

$$\begin{aligned}
&\cos A + \cos B + \cos C \\
&= 2\cos\frac{A+B}{2}\cos\frac{A-B}{2} - \cos(A+B) \\
&= 2\cos\frac{A+B}{2}\cos\frac{A-B}{2} - 2\cos^2\frac{A+B}{2} + 1 \\
&= 1 + 2\cos\frac{A+B}{2}\left(\cos\frac{A-B}{2} - \cos\frac{A+B}{2}\right) \\
&= 1 + 4\cos\frac{A+B}{2}\sin\frac{A}{2}\sin\frac{B}{2} \\
&= 1 + 4\sin\frac{A}{2}\sin\frac{B}{2}\sin\frac{C}{2}.
\end{aligned}$$

(3) By $C = \pi - A - B$ we have $\tan C = -\tan(A+B)$, so

$$\tan C = -\frac{\tan A + \tan B}{1 - \tan A \tan B}.$$

Multiplying both sides by $1 - \tan A \tan B$, we get

$$\tan C - \tan A \tan B \tan C = -\tan A - \tan B,$$

$$\tan A + \tan B + \tan C = \tan A \tan B \tan C.$$

(4) $\sin^2 A + \sin^2 B + \sin^2 C = \frac{1-\cos 2A}{2} + \frac{1-\cos 2B}{2} + \sin^2 C$

$$\begin{aligned}
&= 1 - \frac{1}{2}(\cos 2A + \cos 2B) + 1 - \cos^2(A+B) \\
&= 2 - \cos(A+B)\cos(A-B) - \cos^2(A+B) \\
&= 2 - \cos(A+B)(\cos(A-B) + \cos(A+B)) \\
&= 2 - 2\cos(A+B)\cos A \cos B \\
&= 2 + 2\cos A \cos B \cos C.
\end{aligned}$$

(5) $\cos^2 A + \cos^2 B + \cos^2 C = 3 - (\sin^2 A + \sin^2 B + \sin^2 C)$, so using the conclusion of (4) we see that the equality holds.

(6) We have $\tan A + \tan B + \tan C = \tan A \tan B \tan C$ by (3). Dividing both sides by $\tan A \tan B \tan C$ gives the desired equality.

(7) Let $\alpha = \frac{\pi}{2} - \frac{A}{2}$, $\beta = \frac{\pi}{2} - \frac{B}{2}$, and $\gamma = \frac{\pi}{2} - \frac{C}{2}$. Then $\alpha + \beta + \gamma = \pi$. Substituting α, β, and γ for A, B, and C respectively in (6), we get $\cot\alpha\cot\beta + \cot\beta\cot\gamma + \cot\gamma\cot\alpha = 1$. Hence, $\tan\frac{A}{2}\tan\frac{B}{2} + \tan\frac{B}{2}\tan\frac{C}{2} + \tan\frac{C}{2}\tan\frac{A}{2} = 1$.

10. We have $\sin 3x = 3\sin x - 4\sin^3 x$, so

$$\begin{aligned}
&3^n \sin\frac{\alpha}{3^n} - 4\left(3^{n-1}\sin^3\frac{\alpha}{3^n} + 3^{n-2}\sin^3\frac{\alpha}{3^{n-1}} + \cdots + \sin^3\frac{\alpha}{3}\right)\\
&= 3^{n-1}\left(3\sin\frac{\alpha}{3^n} - 4\sin^3\frac{\alpha}{3^n}\right)\\
&\quad -4\left(3^{n-2}\sin^3\frac{\alpha}{3^{n-1}} + \cdots + \sin^3\frac{\alpha}{3}\right)\\
&= 3^{n-1}\sin\frac{\alpha}{3^{n-1}} - 4\left(3^{n-2}\sin^3\frac{\alpha}{3^{n-1}} + \cdots + \sin^3\frac{\alpha}{3}\right)\\
&= \cdots\\
&= 3\sin\frac{\alpha}{3} - 4\sin^3\frac{\alpha}{3} = \sin\alpha.
\end{aligned}$$

Hence, the proposition is true.

11. (1) Note that

$$\begin{aligned}
&\cot\alpha - 4\cot 4\alpha\\
&= \frac{1}{\tan\alpha} - \frac{2(1-\tan^2 2\alpha)}{\tan 2\alpha} = \frac{1}{\tan\alpha} - \frac{2}{\tan 2\alpha} + 2\tan 2\alpha\\
&= \frac{1}{\tan\alpha} - \frac{1-\tan^2\alpha}{\tan\alpha} + 2\tan 2\alpha = \tan\alpha + 2\tan 2\alpha.
\end{aligned}$$

Hence, the equality is satisfied.

(2) By (1),

$$\begin{aligned}
&\cot\alpha - (\tan\alpha + 2\tan 2\alpha + \cdots + 2^{n-1}\tan(2^{n-1}\alpha))\\
&= (\cot\alpha - \tan\alpha - 2\tan 2\alpha)\\
&\quad -(2^2\tan 2^2\alpha + \cdots + 2^{n-1}\tan(2^{n-1}\alpha))\\
&= 4\cot 2^2\alpha - (2^2\tan 2^2\alpha + \cdots + 2^{n-1}\tan(2^{n-1}\alpha))\\
&= 2^2[\cot\beta - (\tan\beta + 2\tan 2\beta + \cdots + 2^{n-3}\tan(2^{n-3}\beta))],
\end{aligned}$$

where $\beta = 2^2\alpha$. By induction, together with the fact that n is even, $\cot\alpha - (\tan\alpha + 2\tan 2\alpha + \cdots + 2^{n-1}\tan 2^{n-1}\alpha) = 2^n\cot 2^n\alpha$, so the equality holds.

12. Let AD be an altitude of ΔABC. Then

$$BC = BD + DC = c\cos B + b\cos C,$$

or equivalently $a = c\cos B + b\cos C$. Similarly, $b = a\cos C + c\cos A$ and $c = b\cos A + a\cos B$. Adding the three equalities above gives us the desired conclusion.

Remark. Here the equality $BC = BD + DC$ in the proof means the sum of the oriented line segments.

13. By the given condition, $(\cos x \cos A - \sin x \sin A)(\cos x \cos B - \sin x \sin B)\,(\cos x \cos C - \sin x \sin C) + \cos^3 x = 0$. Note that $\cos x \neq 0$, or otherwise $\sin A \sin B \sin C = 0$, a contradiction. Hence,

$$(\cos A - \tan x \sin A)(\cos B - \tan x \sin B)(\cos C - \tan x \sin C) + 1 = 0,$$

and also $(\tan x - \cot A)(\tan x - \cot B)(\tan x - \cot C) = \frac{1}{\sin A \sin B \sin C}$. Note that (by (6) in Exercise 9)

$$\cot A \cot B + \cot B \cot C + \cot C \cot A = 1.$$

If we let $k = \cot A + \cot B + \cot C$, then

$$\begin{aligned} k &= \cot A + \frac{\sin(B+C)}{\sin B \sin C} = \frac{\cos A \sin B \sin C + \sin^2 A}{\sin A \sin B \sin C} \\ &= \frac{1 + \cos A(\sin B \sin C - \cos A)}{\sin A \sin B \sin C} \\ &= \frac{1 + \cos A(\sin B \sin C + \cos(B+C))}{\sin A \sin B \sin C} \\ &= \frac{1 + \cos A \cos B \cos C}{\sin A \sin B \sin C}. \end{aligned}$$

Combining the discussion above, we get

$$\tan^3 x - k \tan^2 x + \tan x - k = 0,$$

in other words $(\tan x - k)(\tan^2 x + 1) = 0$, and

$$\tan x = k = \cot A + \cot B + \cot C.$$

Squaring both sides and adding 1 to each side, we get

$$\begin{aligned} 1 + \tan^2 x &= \cot^2 A + \cot^2 B + \cot^2 C \\ &\quad + 2(\cot A \cot B + \cot B \cot C + \cot C \cot A) + 1 \\ &= \cot^2 A + \cot^2 B + \cot^2 C + 3. \end{aligned}$$

Therefore, $\sec^2 x = \csc^2 A + \csc^2 B + \csc^2 C$.

14. Note that for any real number α,

$$
\begin{aligned}
&(4\sin^2\alpha - 3)\left(4\sin^2\left(\frac{\pi}{2}-\alpha\right)-3\right)\\
&\quad = (4\sin^2\alpha - 3)(4\cos^2\alpha - 3)\\
&\quad = 16\sin^2\alpha\cos^2\alpha - 12(\sin^2\alpha+\cos^2\alpha)+9\\
&\quad = 4\sin^2 2\alpha - 3.
\end{aligned}
$$

Consequently,

$$
\begin{aligned}
A_n &= \prod_{k=0}^{2^n-1}\left(4\sin^2\frac{k\pi}{2^n}-3\right)\\
&= \left(\prod_{k=0}^{2^{n-2}-1}\left(4\sin^2\frac{k\pi}{2^n}-3\right)\left(4\sin^2\left(\frac{\pi}{2}-\frac{k\pi}{2^n}\right)-3\right)\right)\\
&\quad \cdot\left(4\sin^2\frac{\pi}{4}-3\right)\\
&= -\prod_{k=0}^{2^{n-2}-1}\left(4\sin^2\frac{k\pi}{2^{n-1}}-3\right) = -\prod_{k=0}^{2^{n-2}}\left(4\sin^2\frac{k\pi}{2^{n-1}}-3\right).
\end{aligned}
$$

The last equality follows from $4\sin^2\frac{\pi}{2}-3=1$. This shows that $A_n = -A_{n-1}$. Combining it with $A_1=-3$, we conclude that

$$
A_n = \begin{cases} -3, & n \text{ is odd},\\ 3, & n \text{ is even}. \end{cases}
$$

Solution 14

Trigonometric Inequalities

1. If $\theta \in \left(0, \frac{\pi}{2}\right)$, then $0 < \sin\theta$ and $\cos\theta < 1$, so $f(x) = \log_{\sin\theta} x$ is decreasing on $(0, +\infty)$. Also $(\sin\theta + \cos\theta)^2 = 1 + 2\sin\theta\cos\theta = 1 + \sin 2\theta \geq 2\sin 2\theta$, or equivalently $\frac{\sin\theta+\cos\theta}{2} \geq \frac{\sin 2\theta}{\sin\theta+\cos\theta}$. Hence, $f\left(\frac{\sin\theta+\cos\theta}{2}\right) \leq f\left(\frac{\sin 2\theta}{\sin\theta+\cos\theta}\right)$.
2. It follows that $2 > \left(\frac{\cos\alpha}{\sin\beta}\right)^x + \left(\frac{\cos\beta}{\sin\alpha}\right)^x \geq 2\left(\frac{\cos\alpha\cos\beta}{\sin\alpha\sin\beta}\right)^{\frac{x}{2}}$. Hence, for all $x > 0$, we have $\left(\frac{\cos\alpha\cos\beta}{\sin\alpha\sin\beta}\right)^{\frac{x}{2}} < 1$. This requires that $\frac{\cos\alpha\cos\beta}{\sin\alpha\sin\beta} < 1$, that is $\cos\alpha\cos\beta < \sin\alpha\sin\beta$, so $\cos(\alpha+\beta) < 0$. Hence, $\alpha + \beta \in \left(\frac{\pi}{2}, \pi\right)$, from which the third angle (it is $\pi - (\alpha+\beta)$) of this triangle is also acute.
3. We have $\sin^2\alpha = \cos\alpha\cos\beta + \sin\alpha\sin\beta$. Hence, $\sin^2\alpha > \sin\alpha\sin\beta$, so $\sin\alpha > \sin\beta$. Thus, $\alpha > \beta$.
4. From $\cos(\sin x) - \sin(\cos x) = \sin\left(\frac{\pi}{2} - \sin x\right) - \sin(\cos x) = 2\cos\left[\frac{\pi}{4} - \frac{1}{2}(\sin x - \cos x)\right]\sin\left[\frac{\pi}{4} - \frac{1}{2}(\sin x + \cos x)\right]$ and the fact

$$|\sin x \pm \cos x| \leq \sqrt{2} < \frac{\pi}{2}.$$

We obtain $0 < \frac{\pi}{4} - \frac{\sqrt{2}}{2} \leq \frac{\pi}{4} - \frac{1}{2}(\sin x \pm \cos x) \leq \frac{\pi}{4} + \frac{\sqrt{2}}{2} < \frac{\pi}{2}$. Hence,

$$\cos\left(\frac{\pi}{4} + \frac{\sqrt{2}}{2}\right) \leq \cos\left[\frac{\pi}{4} - \frac{1}{2}(\sin x - \cos x)\right] \leq \cos\left(\frac{\pi}{4} - \frac{\sqrt{2}}{2}\right),$$

$$\sin\left(\frac{\pi}{4} - \frac{\sqrt{2}}{2}\right) \leq \sin\left[\frac{\pi}{4} - \frac{1}{2}(\sin x + \cos x)\right] \leq \sin\left(\frac{\pi}{4} + \frac{\sqrt{2}}{2}\right).$$

Therefore, the proposition follows.

5. (1) Note that at least one of A, B, and C is acute. Assume that C is an acute angle. Then

$$\begin{aligned}\cos A\cos B\cos C &= \frac{1}{2}\cos C(\cos(A+B)+\cos(A-B))\\ &\le \frac{1}{2}\cos C(1+\cos(A+B)) = \frac{1}{2}\cos C(1-\cos C)\\ &\le \frac{1}{2}\left(\frac{\cos C+1-\cos C}{2}\right)^2 = \frac{1}{8}.\end{aligned}$$

(2) By (2) in Exercise 9 of Chapter 13,

$$\cos A+\cos B+\cos C = 1+4\sin\frac{A}{2}\sin\frac{B}{2}\sin\frac{C}{2}.$$

Since $\sin\frac{A}{2}>0, \sin\frac{B}{2}>0$, and $\sin\frac{C}{2}>0$, we have $\cos A+\cos B+\cos C>1$.

Then by Example 5 in this chapter,

$$\cos A+\cos B+\cos C = 1+4\sin\frac{A}{2}\sin\frac{B}{2}\sin\frac{C}{2} \le 1+4\times\frac{1}{8} = \frac{3}{2}.$$

(3) Note that

$$\begin{aligned}&\sin A+\sin B+\sin C+\sin\frac{A+B+C}{3}\\ &= 2\sin\frac{A+B}{2}\cos\frac{A-B}{2}+2\sin\frac{A+B+4C}{6}\cos\frac{A+B-2C}{6}\\ &\le 2\sin\frac{A+B}{2}+2\sin\frac{A+B+4C}{6}\\ &= 4\sin\frac{A+B+C}{3}\cos\frac{A+B-2C}{6}\\ &\le 4\sin\frac{A+B+C}{3}.\end{aligned}$$

Hence, $\sin A+\sin B+\sin C \le 3\sin\frac{A+B+C}{3} = \frac{3\sqrt{3}}{2}$.

6. (1) Note that

$$\begin{aligned}&(1+\cos A+\cos B+\cos C)-(\sin A+\sin B+\sin C)\\ &= 2\cos^2\frac{A}{2}+2\cos\frac{B+C}{2}\cos\frac{B-C}{2}\\ &\quad -2\left(\sin\frac{A}{2}\cos\frac{A}{2}+\sin\frac{B+C}{2}\cos\frac{B-C}{2}\right)\end{aligned}$$

$$= 2\left[\cos\frac{A}{2}\left(\cos\frac{A}{2} - \sin\frac{A}{2}\right) + \cos\frac{B-C}{2}\right.$$
$$\left.\cdot\left(\cos\frac{B+C}{2} - \sin\frac{B+C}{2}\right)\right]$$
$$= 2\left[\cos\frac{A}{2}\left(\cos\frac{A}{2} - \sin\frac{A}{2}\right) + \cos\frac{B-C}{2}\left(\sin\frac{A}{2} - \cos\frac{A}{2}\right)\right]$$
$$= 2\left(\cos\frac{A}{2} - \sin\frac{A}{2}\right)\left(\cos\frac{A}{2} - \cos\frac{B-C}{2}\right)$$
$$= -4\left(\cos\frac{A}{2} - \sin\frac{A}{2}\right)\sin\frac{A+B-C}{4}\cdot\sin\frac{A+C-B}{4}.$$

Since A, B, and C are acute angles, $0 < \frac{A}{2} < \frac{\pi}{4}, 0 < \frac{A+B-C}{4} < \frac{\pi}{4}$ and $0 < \frac{A+C-B}{4} < \frac{\pi}{4}$. Combined with the equality above, the desired inequality follows.

(2) We have $\cot A\cot B + \cot B\cot C + \cot C\cot A = 1$. By the AM-GM inequality, $3\sqrt[3]{(\cot A\cot B\cot C)^2} \le 1$ (here we have used the fact that ΔABC is an acute triangle, so $\cot A\cot B$ and $\cot C$ are all positive), thus

$$\sqrt[3]{\tan A\tan B\tan C} \ge \sqrt{3}.$$

Hence,

$$\begin{aligned}\tan^n A + \tan^n B + \tan^n C &\ge 3\sqrt[3]{(\tan A\tan B\tan C)^n}\\ &\ge 3(\sqrt{3})^n > 3\left(1+\frac{1}{2}\right)^n\\ &\ge 3\left(1+\frac{n}{2}\right) = 3 + \frac{3n}{2}.\end{aligned}$$

7. It follows that

$$f(x) = 1 - \sqrt{a^2+b^2}\cos(x-\beta) - \sqrt{A^2+B^2}\cos 2(x-\gamma), \quad ①$$

where β and γ are constants determined by $\cos\beta = \frac{a}{\sqrt{a^2+b^2}}$ $\sin\beta = \frac{b}{\sqrt{a^2+b^2}}$, $\cos 2\gamma = \frac{A}{\sqrt{A^2+B^2}}$ and $\sin 2\gamma = \frac{B}{\sqrt{A^2+B^2}}$. We let $x = \beta + \frac{\pi}{4}$, $\beta - \frac{\pi}{4}$, γ, and $\gamma + \pi$ in in sucession, and get the following inequalities:

$$1 - \sqrt{a^2+b^2}\cos\frac{\pi}{4} - \sqrt{A^2+B^2}\cos 2\left(\beta - \gamma + \frac{\pi}{4}\right) \ge 0,$$
$$1 - \sqrt{a^2+b^2}\cos\frac{\pi}{4} - \sqrt{A^2+B^2}\cos 2\left(\gamma - \beta + \frac{\pi}{4}\right) \ge 0,$$

$$1 - \sqrt{a^2+b^2}\cos(\beta-\gamma) - \sqrt{A^2+B^2} \geq 0,$$
$$1 + \sqrt{a^2+b^2}\cos(\beta-\gamma) - \sqrt{A^2+B^2} \geq 0.$$

Adding the first two inequalities gives $\sqrt{2}\cdot\sqrt{a^2+b^2} \leq 2$, so $a^2+b^2 \leq 2$. Adding the last two inequalities involving A and B gives $2\sqrt{A^2+B^2} \leq 2$, namely $A^2+B^2 \leq 1$.

8. By symmetry we may assume that $A \geq B \geq C$, so

$$\frac{\sin C \sin(C-A)\sin(C-B)}{\sin(A+B)} \geq 0.$$

The sum of the other two expressions is

$$\frac{\sin A\sin(A-B)\sin(A-C)}{\sin(B+C)} + \frac{\sin B\sin(B-C)\sin(B-A)}{\sin(C+A)}$$
$$= \frac{\sin(A-B)[\sin A\sin(A-C)\sin(A+C) - \sin B\sin(B-C)\sin(B+C)]}{\sin(B+C)\sin(C+A)}. \quad ①$$

Since

$$\begin{aligned}\sin(x-y)\sin(x+y) &= \frac{1}{2}(\cos 2y - \cos 2x)\\ &= \frac{1}{2}[(1-2\sin^2 y) - (1-2\sin^2 x)]\\ &= \sin^2 x - \sin^2 y.\end{aligned}$$

the numerator of the right side of ① equals

$$\begin{aligned}&\sin(A-B)[\sin A(\sin^2 A - \sin^2 C) - \sin B(\sin^2 B - \sin^2 C)]\\ &= \sin(A-B)[(\sin^3 A - \sin^3 B) - \sin^2 C(\sin A - \sin B)]\\ &= \sin(A-B)(\sin A - \sin B)(\sin^2 A + \sin^2 B + \sin A\sin B - \sin^2 C).\end{aligned}$$

Combined with $A \geq B \geq C$, each term in the above expression is nonnegative, so the right side of ① is nonnegative, and hence the original inequality is true.

9. This inequality is a strengthened version of Example 5 in this chapter, while the approach is somewhat different. Note that $\cos\frac{A-B}{2} - \cos\frac{A+B}{2} = 2\sin\frac{A}{2}\sin\frac{B}{2}$ and

$$\cos\frac{A-B}{2} = 2\sin\frac{A}{2}\sin\frac{B}{2} + \sin\frac{C}{2} \geq 2\sqrt{2\sin\frac{A}{2}\sin\frac{B}{2}\sin\frac{C}{2}}.$$

Hence,

$$\cos^2\frac{A-B}{2}\geq 8\sin\frac{A}{2}\sin\frac{B}{2}\sin\frac{C}{2}.$$

Similarly,

$$\cos^2\frac{B-C}{2}\geq 8\sin\frac{A}{2}\sin\frac{B}{2}\sin\frac{C}{2},$$
$$\cos^2\frac{C-A}{2}\geq 8\sin\frac{A}{2}\sin\frac{B}{2}\sin\frac{C}{2}.$$

Multiplying the three inequalities gives the desired conclusion.

10. By the law of sines and trigonometric identities,

$$\begin{aligned}
&(a+b)\left(\frac{1}{a}+\frac{1}{b}+\frac{1}{c}\right)-\left(4+\frac{1}{\sin\frac{C}{2}}\right)\\
&=\frac{(a+b)^2}{ab}+\frac{a+b}{c}-4-\frac{1}{\sin\frac{C}{2}}\\
&=\frac{(\sin A+\sin B)^2}{\sin A\sin B}+\frac{\sin A+\sin B}{\sin C}-4-\frac{1}{\sin\frac{C}{2}}\\
&=\frac{4\sin^2\frac{A+B}{2}\cos^2\frac{A-B}{2}}{\frac{1}{2}[\cos(A-B)-\cos(A+B)]}+\frac{2\sin\frac{A+B}{2}\cos\frac{A-B}{2}}{2\sin\frac{C}{2}\cos\frac{C}{2}}-4-\frac{1}{\sin\frac{C}{2}}\\
&=\frac{4\sin^2\frac{A+B}{2}\cos^2\frac{A-B}{2}}{\cos^2\frac{A-B}{2}-\cos^2\frac{A+B}{2}}-4+\frac{\cos\frac{A-B}{2}}{\sin\frac{C}{2}}-\frac{1}{\sin\frac{C}{2}}\\
&=\frac{4\cos^2\frac{A+B}{2}-4\cos^2\frac{A+B}{2}\cos^2\frac{A-B}{2}}{\cos^2\frac{A-B}{2}-\cos^2\frac{A+B}{2}}-\frac{1}{\sin\frac{C}{2}}\left(1-\cos\frac{A-B}{2}\right)\\
&\geq\frac{4\cos^2\frac{A+B}{2}\left(1-\cos\frac{A-B}{2}\right)}{\cos\frac{A+B}{2}-\cos\frac{A+B}{2}}-\frac{1}{\sin\frac{C}{2}}\left(1-\cos\frac{A-B}{2}\right) \quad ①\\
&=\frac{1-\cos\frac{A-B}{2}}{\left(\cos\frac{A-B}{2}-\cos\frac{A+B}{2}\right)\sin\frac{C}{2}}\left[4\sin^3\frac{C}{2}-\cos\frac{A-B}{2}+\sin\frac{C}{2}\right]\\
&\geq\frac{1-\cos\frac{A-B}{2}}{\left(\cos\frac{A-B}{2}-\cos\frac{A+B}{2}\right)\sin\frac{C}{2}}\left[4\times\left(\frac{1}{2}\right)^3-\cos\frac{A-B}{2}+\frac{1}{2}\right]\\
&=\frac{\left(1-\cos\frac{A-B}{2}\right)^2}{\left(\cos\frac{A-B}{2}-\cos\frac{A+B}{2}\right)\sin\frac{C}{2}}\geq 0,
\end{aligned}$$

where the inequality ① holds because $\cos\frac{A-B}{2} - \cos\frac{A+B}{2} = 2\sin\frac{A}{2}\sin\frac{B}{2} > 0$ and $1 + \cos\frac{A-B}{2} \geq \cos\frac{A-B}{2} + \cos\frac{A+B}{2}$.

Therefore, the original inequality follows.

11. The given condition implies that

$$\begin{aligned}
&x^2 + y^2 + z^2 - 2yz\cos A - 2zx\cos B - 2xy\cos C \\
&\quad = x^2 - 2x(y\cos C + z\cos B) + y^2 + z^2 - 2yz\cos A \\
&\quad = (x - y\cos C - z\cos B)^2 + y^2 + z^2 \\
&\qquad -2yz\cos A - (y\cos C + z\cos B)^2 \\
&\quad = (x - y\cos C - z\cos B)^2 + y^2\sin^2 C \\
&\qquad -2yz(\cos A + \cos B\cos C) + z^2\sin^2 B \\
&\quad = (x - y\cos C - z\cos B)^2 + y^2\sin^2 C \\
&\qquad -2yz(\cos B\cos C - \cos(B + C)) + z^2\sin^2 B \\
&\quad = (x - y\cos C - z\cos B)^2 + y^2\sin^2 C - 2yz\sin B\sin C + z^2\sin^2 B \\
&\quad = (x - y\cos C - z\cos B)^2 + (y\sin C - z\sin B)^2 \geq 0.
\end{aligned}$$

Hence, the conclusion follows. This is called the embedding inequality, which is often used in geometric inequalities.

12. Let α be a real number that satisfies the condition. First, we show that $\cos\alpha \leq -\frac{1}{4}$. Suppose not. Then $-\frac{1}{4} < \cos\alpha < 0$, so

$$\cos 2\alpha = 2\cos^2\alpha - 1 < 2 \times \left(-\frac{1}{4}\right)^2 - 1 = -\frac{7}{8},$$

$$\cos 4\alpha = 2\cos^2 2\alpha - 1 > 2 \times \left(-\frac{7}{8}\right)^2 - 1 = \frac{17}{32} > 0,$$

which is a contradiction. By a similar argument we can show that for any nonnegative integer n, it is true that $\cos 2^n\alpha \leq -\frac{1}{4}$. Thus $|\cos 2^n\alpha - \frac{1}{2}| \geq \frac{3}{4}$. Hence,

$$\left|\cos 2\alpha + \frac{1}{2}\right| = 2\left|\cos\alpha - \frac{1}{2}\right| \cdot \left|\cos\alpha + \frac{1}{2}\right| \geq \frac{3}{2}\left|\cos\alpha + \frac{1}{2}\right|,$$

or equivalently $|\cos\alpha + \frac{1}{2}| \leq \frac{2}{3}|\cos 2\alpha + \frac{1}{2}|$.

Continuing this process gives that

$$\left|\cos\alpha + \frac{1}{2}\right| \le \frac{2}{3}\left|\cos 2\alpha + \frac{1}{2}\right| \le \cdots \le \left(\frac{2}{3}\right)^n \left|\cos 2^n\alpha + \frac{1}{2}\right| \le \frac{1}{2}\times\left(\frac{2}{3}\right)^n.$$

As n can be arbitrary, $|\cos\alpha + \frac{1}{2}| = 0$ (since $\frac{1}{2}\times\left(\frac{2}{3}\right)^n \to 0$ as $n\to\infty$). Hence, $\cos\alpha = -\frac{1}{2}$ and $\alpha = 2k\pi \pm \frac{2}{3}\pi$ with $k \in \mathbf{Z}$.

On the other hand, it is easy to verify that in this case, every term in the sequence $\cos\alpha \cos 2\alpha \cos 4\alpha, \ldots$ equals $-\frac{1}{2}$. Therefore, the set of all such α is $\{\alpha | \alpha = 2k\pi \pm \frac{2}{3}\pi, k \in \mathbf{Z}\}$.

13. Let $\alpha = \frac{\pi}{6}$ and $\beta = \frac{\pi}{3}$. Then $x_1 = \cot\alpha$ and $y_1 = \tan\beta$, and

$$x_2 = \cot\alpha + \csc\alpha = \frac{1+\cos\alpha}{\sin\alpha} = \frac{2\cos^2\frac{\alpha}{2}}{2\sin\frac{\alpha}{2}\cos\frac{\alpha}{2}} = \cot\frac{\alpha}{2},$$

$$y_2 = \frac{\tan\beta}{1+\sec\beta} = \frac{\sin\beta}{1+\cos\beta} = \frac{2\sin\frac{\beta}{2}\cos\frac{\beta}{2}}{2\cos^2\frac{\beta}{2}} = \tan\frac{\beta}{2}.$$

In general, if $x_n = \cot\frac{\alpha}{2^{n-1}}$ and $y_n = \tan\frac{\beta}{2^{n-1}}$, then similarly we can show that

$$x_{n+1} = \cot\frac{\alpha}{2^n}, \quad y_{n+1} = \tan\frac{\beta}{2^n}.$$

By induction, for each positive integer n,

$$x_n = \cot\frac{\alpha}{2^{n-1}}, y_n = \tan\frac{\beta}{2^{n-1}}.$$

Hence, $x_n y_n = \cot\frac{\pi}{3\times 2^n}\tan\frac{\pi}{3\times 2^{n-1}}$, from which $x_n y_n = \frac{2}{1-\tan^2\frac{\pi}{3\times 2^n}}$. Note that for $n > 1$, we have $0 < \tan^2\frac{\pi}{3\times 2^n} < \tan^2\frac{\pi}{6} = \frac{1}{3}$ and $2 < x_n y_n < 3$. The problem's conclusion is proved.

14. Since $x < 4\tan\frac{x}{4} = \sin x \sec\frac{x}{2}\sec^2\frac{x}{4}$, we have $x\csc x < \sec\frac{x}{2}\sec^2\frac{x}{4}$. Hence,

$$\frac{x\csc x + y\csc y}{2} < \frac{1}{2}\left(\sec\frac{x}{2}\sec^2\frac{x}{4} + \sec\frac{y}{2}\sec^2\frac{y}{4}\right) \quad ①$$

for all $0 < x, y < \frac{\pi}{2}$. Without loss of generality we assume that $x \le y$. Then

$$\text{RHS of } ① = \sec\frac{x}{2}\sec\frac{y}{2}\left(\frac{\cos\frac{y}{2}\sec^2\frac{x}{4} + \cos\frac{x}{2}\sec^2\frac{y}{4}}{2}\right)$$

$$= \sec\frac{x}{2}\sec\frac{y}{2}\left(\frac{\cos\frac{y}{2}}{1+\cos\frac{x}{2}} + \frac{\cos\frac{x}{2}}{1+\cos\frac{y}{2}}\right)$$

$$\begin{aligned}
&\le \sec\frac{x}{2}\sec\frac{y}{2}\left(\frac{\cos\frac{y}{2}}{1+\cos\frac{x}{2}}+\frac{1}{1+\cos\frac{y}{2}}\right)\\
&= \sec\frac{x}{2}\sec\frac{y}{2} = \frac{1}{\cos\frac{x}{2}\cos\frac{y}{2}}\\
&= \frac{1}{\cos\frac{x+y}{2}+\sin\frac{x}{2}\sin\frac{y}{2}}\\
&< \frac{1}{\cos\frac{x+y}{2}} = \sec\frac{x+y}{2}.
\end{aligned}$$

Therefore, the desired inequality holds.

15. Let $f(x) = |\sin x|\cdot|\sin 2x|^{\frac{1}{2}}$. Then

$$f(x) = \sqrt{2}|\sin x|^{\frac{3}{2}}\cdot|\cos x|^{\frac{1}{2}}.$$

Hence,

$$\begin{aligned}
f(x)^4 &= 2^2(\sin x)^6(\cos x)^2\\
&= \frac{2^2}{3}(\sin^2 x)(\sin^2 x)(\sin^2 x)(3\cos^2 x)\\
&\le \frac{2^2}{3}\cdot\left(\frac{3\sin^2 x+3\cos^2 x}{4}\right)^4 = \frac{2^2\times 3^3}{4^4}. \quad ①
\end{aligned}$$

Here the last inequality used the AM-GM inequality. Now, let $g(\theta) = \prod_{i=0}^{n}|\sin 2^i\theta|$. Then

$$g(\theta) \le \prod_{i=0}^{n-1}|\sin 2^i\theta|, g(\theta)^{\frac{1}{2}} \le \left(\prod_{i=0}^{n-1}|\sin 2^{i+1}\theta|\right)^{\frac{1}{2}}.$$

Hence,

$$g(\theta)^{\frac{3}{2}} \le \prod_{i=0}^{n-1}|\sin 2^i\theta|\cdot|\sin 2^{i+1}\theta|^{\frac{1}{2}} \le \prod_{i=0}^{n-1}\left(\frac{2^2\times 3^3}{4^4}\right)^{\frac{1}{4}} = \left(\frac{\sqrt{3}}{2}\right)^{\frac{3}{2}n}.$$

The last inequality used ①. Therefore, $g(\theta) \le \left(\frac{\sqrt{3}}{2}\right)^n$.

Solution 15

Extreme Value Problems of Trigonometric Functions

1. By $1 = \sin^2\alpha + \cos^2\alpha \geq 2\sin\alpha\cos\alpha$, we have $\sin\alpha\cos\alpha \leq \frac{1}{2}$. Since α is acute, $0 < \sin\alpha\cos\alpha$. Hence, $\tan\alpha + \cot\alpha + \sec\alpha + \csc\alpha \geq 2\sqrt{\tan\alpha\cot\alpha} + 2\sqrt{\frac{1}{\sin\alpha\cos\alpha}} \geq 2 + 2\sqrt{2}$, where the equality holds when $\alpha = \frac{\pi}{4}$. Therefore, the minimum value is $2 + 2\sqrt{2}$.
2. Since $\frac{1}{\cos^2\alpha} + \frac{1}{\sin^2\alpha\sin^2\beta\cos^2\beta} = \frac{1}{\cos^2\alpha} + \frac{4}{\sin^2\alpha\sin^2 2\beta} \geq \frac{1}{\cos^2\alpha} + \frac{4}{\sin^2\alpha} = 1 + \tan^2\alpha + 4(1+\cot^2\alpha) = 5 + \tan^2\alpha + 4\cot^2\alpha \geq 5 + 2\sqrt{\tan^2\alpha \cdot 4\cot^2\alpha} = 9$, and the equality holds when $\beta = \frac{\pi}{4}$ and $\tan\alpha = \sqrt{2}$, the desired minimum value is 9.
3. It follows that $0 < \sin\beta \leq 1$, so $\cos^2\alpha\sin\beta + \frac{1}{\sin\beta} \geq \frac{1}{\sin\beta} \geq 1$, where the equality holds when $\alpha = \beta = \frac{\pi}{2}$. Hence, the minimum value is 1.
4. Note that $(\cos x + \cos y)^2 + (\sin x + \sin y)^2 = 2 + \cos(x - y) \leq 4$, so $(\cos x + \cos y)^2 \leq 3$, from which

$$-\sqrt{3} \leq \cos x + \cos y \leq \sqrt{3}.$$

On the other hand, $\cos x + \cos y = \sqrt{3}$ when $x = y = \frac{\pi}{6}$, and $\cos x + \cos y = -\sqrt{3}$ when $x = y = \frac{5\pi}{6}$. Therefore, the maximum and minimum values of $\cos x + \cos y$ are $\sqrt{3}$ and $-\sqrt{3}$, respectively.
5. If $a = 0$ and $b = 1$, then the conditions are satisfied. Now, suppose a and b are real numbers that satisfy the condition. Then let $x = 0$ and $\frac{2\pi}{3}$, and we get $a + b \leq 1$ and $-\frac{1}{2}a + b \leq 1$. Cancelling a in these two inequalities we have $3b \leq 3$, so $b \leq 1$. Therefore, the maximum value of b is 1.

6. Note that $\cot\frac{A}{2} - 2\cot A = \frac{1}{\tan\frac{A}{2}} - \frac{1-\tan^2\frac{A}{2}}{\tan\frac{A}{2}} = \tan\frac{A}{2}$, so the condition reduces to $\tan\frac{A}{2} + \tan\frac{B}{2} + \tan\frac{C}{2} \geq m$. By (7) in Exercise 9 of Chapter 13,

$$\tan\frac{A}{2}\tan\frac{B}{2} + \tan\frac{B}{2}\tan\frac{C}{2} + \tan\frac{C}{2}\tan\frac{A}{2} = 1,$$

thus $\left(\tan\frac{A}{2} + \tan\frac{B}{2} + \tan\frac{C}{2}\right)^2 \geq 3\left(\tan\frac{A}{2}\tan\frac{B}{2} + \tan\frac{B}{2}\tan\frac{C}{2} + \tan\frac{C}{2}\tan\frac{A}{2}\right) = 3$, and combining it with $\tan\frac{A}{2}, \tan\frac{B}{2}, \tan\frac{C}{2} > 0$, we have

$$\tan\frac{A}{2} + \tan\frac{B}{2} + \tan\frac{C}{2} \geq \sqrt{3}.$$

Therefore, the minimum value of m is $\sqrt{3}$.

7. It follows from the condition that

$$\begin{aligned} 2\sin\frac{A+C}{2}\cos\frac{A-C}{2} &= 2\sin(A+C)\cos\frac{A+C}{2}\cos\frac{A-C}{2} \\ &= 4\sin\frac{A+C}{2}\cos^2\frac{A+C}{2}\cos\frac{A-C}{2}. \end{aligned}$$

Hence, $\cos^2\frac{A+C}{2} = \frac{1}{2}$ and $\frac{A+C}{2} = \frac{\pi}{4}$. This implies that $B = \frac{\pi}{2}$. Combining it with $S_{\Delta ABC} = 4$ we get $\frac{1}{2}ac = 4$, so $ac = 8$. Thus,

$$a + b + c = a + c + \sqrt{a^2 + c^2} \geq 2\sqrt{ac} + \sqrt{2ac} = 4 + 4\sqrt{2},$$

where the equality holds when $a = c = 2\sqrt{2}$ and $b = 4$. Therefore, the minimal perimeter of ΔABC is $4 + 4\sqrt{2}$.

8. We assume that $x = r\cos\theta$ and $y = r\sin\theta$ with $\theta \in [0, 2\pi)$. Then

$$4r^2 - 5r^2\sin\theta\cos\theta = 19,$$

or equivalently $r^2\left(4 - \frac{5}{2}\sin 2\theta\right) = 19$. Note that $4 - \frac{5}{2}\sin 2\theta \in \left[\frac{3}{2}, \frac{13}{2}\right]$, so $\frac{38}{13} \leq r^2 \leq \frac{38}{3}$, where the maximum and minimum values are attained at $\theta = \frac{\pi}{4}$ and $\theta = \frac{3\pi}{4}$, respectively.

9. Let $x = t + 3$. Then $f(x) = \sqrt{3(t+5)} + \sqrt{5-t}$, so $-5 \leq t \leq 5$. Now, assume that $t = 5\cos\alpha$ for $0 \leq \alpha \leq \pi$. Then

$$\begin{aligned} f(x) &= \sqrt{15(1+\cos\alpha)} + \sqrt{5(1-\cos\alpha)} \\ &= \sqrt{30}\cos\frac{\alpha}{2} + \sqrt{10}\sin\frac{\alpha}{2} = 2\sqrt{10}\sin\left(\frac{\alpha}{2} + \frac{\pi}{3}\right). \end{aligned}$$

Therefore, the maximum value of $f(x)$ is $2\sqrt{10}$, attained at $\alpha = \frac{\pi}{3}$, and the minimum value is $\sqrt{10}$, attained at $\alpha = \pi$.

10. Note that when $A_1 = A_2 = \cdots = A_n = \frac{\pi}{4}$, we have $\tan A_1 \tan A_2 \cdots \tan A_n = 1$ and $\sin A_1 \sin A_2 \cdots \sin A_n = 2^{-\frac{n}{2}}$. On the other hand,

$$\sec^2 A_1 \sec^2 A_2 \cdots \sec^2 A_n = (1+\tan^2 A_1)\cdots(1+\tan^2 A_n)$$
$$\geq (2\tan A_1)(2\tan A_2)\cdots(2\tan A_n) = 2^n.$$

Hence, $|\sec A_1 \sec A_2 \cdots \sec A_n| \geq 2^{\frac{n}{2}}$, so $|\cos A_1 \cos A_2 \cdots \cos A_n| \leq 2^{-\frac{n}{2}}$. By the condition, $\tan A_1 \tan A_2 \cdots \tan A_n = 1$ thus $\cos A_1 \cos A_2 \cdots \cos A_n = \sin A_1 \sin A_2 \cdots \sin A_n$, from which

$$\sin A_1 \sin A_2 \cdots \sin A_n \leq |\cos A_1 \cos A_2 \cdots \cos A_n| \leq 2^{-\frac{n}{2}}.$$

Therefore, the desired maximum value is $2^{-\frac{n}{2}}$.

11. Since $F(x) = |\cos 2x + \sin 2x + Ax + B| = |\sqrt{2}\sin\left(2x + \frac{\pi}{4}\right) + Ax + B|$, with $f(x) = \sqrt{2}\sin\left(2x+\frac{\pi}{4}\right)$ and $g(x) = Ax + B$, for $x \in \left[0, \frac{3\pi}{2}\right]$,

$$f_{\max} = f\left(\frac{\pi}{8}\right) = f\left(\frac{9\pi}{8}\right) = \sqrt{2}, f_{\min} = f\left(\frac{5\pi}{8}\right) = -\sqrt{2}.$$

Consequently, $F_{\max} = |f|_{\max} = \sqrt{2}$ when $A = B = 0$. Next, we show that for any real numbers A and B that are not both zero, $F_{\max} > \sqrt{2}$.

In fact, if A and B are not both zero, then $g\left(\frac{\pi}{8}\right)$ and $g\left(\frac{9\pi}{8}\right)$ are not both 0. If one of the two numbers is positive, say $g\left(\frac{\pi}{8}\right) > 0$, then $F\left(\frac{\pi}{8}\right) = f\left(\frac{\pi}{8}\right) + g\left(\frac{\pi}{8}\right) > \sqrt{2}$, which proves the assertion. If both of $g\left(\frac{\pi}{8}\right)$ and $g\left(\frac{9\pi}{8}\right)$ are nonpositive, then by monotonicity of linear functions, $g\left(\frac{5\pi}{8}\right) < 0$, so

$$F\left(\frac{5\pi}{8}\right) = \left|f\left(\frac{5\pi}{8}\right) + g\left(\frac{5\pi}{8}\right)\right| > |-\sqrt{2}| = \sqrt{2}.$$

Hence, the assertion also holds. Therefore, $M(A, B)$ has the minimum value $\sqrt{2}$, which is attained if and only if $A = B = 0$.

12. By the condition, we may assume that $a = \sin A$ and $b = \cos A$. Then

$$DE^2 = EF^2 + FD^2 = (1+\sin A)^2 \sin^2\frac{B}{2} + (1+\cos A)^2\sin^2\frac{A}{2}$$
$$= \frac{1}{2}(1+\sin A)^2(1-\sin A)$$

$$
\begin{aligned}
&+\frac{1}{2}(1+\cos A)^2(1-\cos A)\\
&=\frac{1}{2}\cos^2 A(1+\sin A)+\frac{1}{2}\sin^2 A(1+\cos A)\\
&=\frac{1}{2}+\frac{1}{2}\sin A\cos A(\sin A+\cos A)\\
&=\frac{1}{2}+\frac{\sqrt{2}}{4}\sin 2A\cos\left(\frac{\pi}{4}-A\right).
\end{aligned}
$$

Hence, $\frac{1}{2}<DE^2\le\frac{1}{2}+\frac{\sqrt{2}}{4}$. When $A\to 0$ we have $DE^2\to\frac{1}{2}$, and $DE^2=\frac{1}{2}+\frac{\sqrt{2}}{4}$ when $A=\frac{\pi}{4}$. Therefore, the value range of DE is $\left(\frac{\sqrt{2}}{2},\sqrt{\frac{1}{2}+\frac{\sqrt{2}}{4}}\right]$.

13. Make the trigonometric substitutions $a=\tan A, b=\tan B$ and $c=\tan C$, where A, B, and C are acute angles. Then from the given condition, $\tan A+\tan B+\tan C=\tan A\tan B\tan C$. Hence,

$$
\tan A=-\frac{\tan B+\tan C}{1-\tan B\tan C}=-\tan(B+C),
$$

which means that $A+B+C=\pi$. Therefore, the problem reduces to finding the maximum value of $\cos A+\cos B+\cos C$ among all acute triangles ABC. Using (2) in exercise 5 of Chapter 14, we see that the maximum value of this expression is $\frac{3}{2}$.

14. When $\theta=\frac{\pi}{4}$, we have $\left(\frac{1}{\sin^n\theta}-1\right)\left(\frac{1}{\cos^n\theta}-1\right)=2^n-2^{\frac{n}{2}+1}+1$. Next, we show that for any $\theta\in\left(0,\frac{\pi}{2}\right)$,

$$
\left(\frac{1}{\sin^n\theta}-1\right)\left(\frac{1}{\cos^n\theta}-1\right)\ge 2^n-2^{\frac{n}{2}+1}+1. \text{①}
$$

We prove it by induction. If $n=2$, then the two sides of are equal, so the inequality holds when $n=2$. Suppose ① holds for some n. Then

$$
\begin{aligned}
&\left(\frac{1}{\sin^{n+1}\theta}-1\right)\left(\frac{1}{\cos^{n+1}\theta}-1\right)\\
&=\frac{1}{\sin^{n+1}\theta\cos^{n+1}\theta}(1-\sin^{n+1}\theta)(1-\cos^{n+1}\theta)\\
&=\frac{1}{\sin^{n+1}\theta\cos^{n+1}\theta}(1-\sin^{n+1}\theta-\cos^{n+1}\theta)+1\\
&=\frac{1}{\sin\theta\cos\theta}\left(\frac{1}{\sin^n\theta\cos^n\theta}-\frac{\cos\theta}{\sin^n\theta}-\frac{\sin\theta}{\cos^n\theta}\right)+1
\end{aligned}
$$

$$= \frac{1}{\sin\theta\cos\theta}\left[\left(\frac{1}{\sin^n\theta}-1\right)\left(\frac{1}{\cos^n\theta}-1\right)\right.$$
$$\left. + \frac{1-\cos\theta}{\sin^n\theta} + \frac{1-\sin\theta}{\cos^n\theta} - 1\right] + 1$$
$$\geq \frac{1}{\sin\theta\cos\theta}\left[(2^n - 2^{\frac{n}{2}+1}) + 2\sqrt{\frac{(1-\cos\theta)(1-\sin\theta)}{\sin^n\theta\cos^n\theta}}\right] + 1 \quad ②$$

Here the inequality in ② was obtained by the inductive hypothesis and the AM-GM inequality. Note that $\sin\theta\cos\theta = \frac{1}{2}\sin 2\theta \leq \frac{1}{2}$ and

$$\frac{(1-\cos\theta)(1-\sin\theta)}{\sin^n\theta\cos^n\theta} = \left(\frac{1}{\sin\theta\cos\theta}\right)^{n-2} \cdot \frac{1}{(1+\sin\theta)(1+\cos\theta)},$$

where $(1+\sin\theta)(1+\cos\theta) = 1+\sin\theta+\cos\theta+\sin\theta\cos\theta = 1+t+\frac{t^2-1}{2} = \frac{1}{2}(t+1)^2 \leq \frac{1}{2}(\sqrt{2}+1)^2$. (Here we used $t = \sin\theta+\cos\theta = \sqrt{2}\sin(\theta+\frac{\pi}{4}) \in (1, \sqrt{2}]$.) Hence,

$$\sqrt{\frac{(1-\cos\theta)(1-\sin\theta)}{\sin^n\theta\cos^n\theta}} \geq \frac{2^{\frac{n-1}{2}}}{\sqrt{2}+1} = 2^{\frac{n}{2}} - 2^{\frac{n-1}{2}}.$$

By ② we also have

$$\left(\frac{1}{\sin^{n+1}\theta}-1\right)\left(\frac{1}{\cos^{n+1}\theta}-1\right)$$
$$\geq 2[(2^n - 2^{\frac{n}{2}+1}) + 2(2^{\frac{n}{2}} - 2^{\frac{n-1}{2}})] + 1$$
$$= 2(2^n - 2^{\frac{n+1}{2}}) + 1 = 2^{n+1} - 2^{\frac{n+1}{2}+1} + 1,$$

therefore ① holds for $n+1$. This completes the induction, so the minimum value of the given expression is $2^n - 2^{\frac{n}{2}+1} + 1$.

15. Let $t = \sqrt{\tan\frac{\alpha}{2}\tan\frac{\beta}{2}}$. Then since α and β are acute angles, $0 < t < 1$. Note that

$$\cot\alpha + \cot\beta = \frac{1-\tan^2\frac{\alpha}{2}}{2\tan\frac{\alpha}{2}} + \frac{1-\tan^2\frac{\beta}{2}}{2\tan\frac{\beta}{2}}$$
$$= \frac{\left(\tan\frac{\alpha}{2}+\tan\frac{\beta}{2}\right)\left(1-\tan\frac{\alpha}{2}\tan\frac{\beta}{2}\right)}{2\tan\frac{\alpha}{2}\tan\frac{\beta}{2}}.$$

Hence,

$$\begin{aligned}
A &= \frac{\left(1-\sqrt{\tan\frac{\alpha}{2}\tan\frac{\beta}{2}}\right)^2}{\cot\alpha+\cot\beta} \\
&= \frac{2\tan\frac{\alpha}{2}\tan\frac{\beta}{2}\left(1-\sqrt{\tan\frac{\alpha}{2}\tan\frac{\beta}{2}}\right)^2}{\left(\tan\frac{\alpha}{2}+\tan\frac{\beta}{2}\right)\left(1-\tan\frac{\alpha}{2}\tan\frac{\beta}{2}\right)} \\
&= \frac{2\tan\frac{\alpha}{2}\tan\frac{\beta}{2}\left(1-\sqrt{\tan\frac{\alpha}{2}\tan\frac{\beta}{2}}\right)}{\left(\tan\frac{\alpha}{2}+\tan\frac{\beta}{2}\right)\left(1+\sqrt{\tan\frac{\alpha}{2}\tan\frac{\beta}{2}}\right)} \\
&\le \frac{2\tan\frac{\alpha}{2}\tan\frac{\beta}{2}\left(1-\sqrt{\tan\frac{\alpha}{2}\tan\frac{\beta}{2}}\right)}{2\sqrt{\tan\frac{\alpha}{2}\tan\frac{\beta}{2}}\left(1+\sqrt{\tan\frac{\alpha}{2}\tan\frac{\beta}{2}}\right)} \\
&= \frac{t(1-t)}{1+t}.
\end{aligned}$$

Next, we compute the maximum value of $f(t)=\frac{t(1-t)}{1+t}$ for $0<t<1$. We have

$$\begin{aligned}
f(t) &= \frac{t(1-t)}{1+t} = \left(1-\frac{1}{1+t}\right)(1-t) = 1-t+\frac{t-1}{t+1} \\
&= 2-t-\frac{2}{1+t} = 3-\left[(1+t)+\frac{2}{1+t}\right] \\
&\le 3-2\sqrt{(1+t)\cdot\frac{2}{1+t}} = 3-2\sqrt{2},
\end{aligned}$$

and the equality holds when $1+t=\frac{2}{1+t}$, or equivalently $t=\sqrt{2}-1\in(0,1)$.

Now, going back to the original problem, we have $A\le 3-2\sqrt{2}$, and the equality holds when $\alpha=\beta$ and $\tan\frac{\alpha}{2}=\sqrt{2}-1$. Therefore, the maximum value of the expression A is $3-2\sqrt{2}$.

Solution 16

Inverse Trigonometric Functions and Trigonometric Equations

1. Since $-\frac{\pi}{3} \le 2\arcsin(x-2) \le \pi$, so $-\frac{\pi}{6} \le \arcsin(x-2) \le \frac{\pi}{2}$. By the monotonicity of the inverse sine function, $-\frac{1}{2} \le x-2 \le 1$. Hence, the domain of $f(x)$ is $\left[\frac{3}{2}, 3\right]$.
2. The original equation is equivalent to

$$\begin{cases} 1+\cos 2x > 0 \text{ and } 1+\cos 2x \neq 1, \text{ ①} \\ \sin x + \sin 3x > 0, \text{ ②} \\ \sin x + \sin 3x = 1 + \cos 2x. \text{ ③} \end{cases}$$

The equation ③ is equivalent to $2\sin 2x \cos x = 2\cos^2 x$, namely $\cos x(\sin 2x - \cos x) = 0$. Hence,

$$\cos^2 x(2\sin x - 1) = 0.$$

Since $\cos x \neq 0$ by ① clearly $2\sin x = 1$, that is $\sin x = \frac{1}{2}$. Therefore, the solutions are $x = 2k\pi + \frac{\pi}{6}$ and $x = 2k\pi + \frac{5}{6}\pi$ with $k \in \mathbf{Z}$.
3. Since the domains of the inverse sine function and the inverse cosine function are both $[-1, 1]$,

$$\begin{cases} -1 \le 1 - x \le 1, \\ -1 \le 2x \le 1, \end{cases}$$

so $x \in \left[0, \frac{1}{2}\right]$. Combining the above with the monotonicity of the functions, we see that $y = \arcsin(1-x)$ is decreasing on $\left[0, \frac{1}{2}\right]$, and

$y = \arccos 2x$ is also decreasing on $\left[0, \frac{1}{2}\right]$. Hence, $f(0) \geq f(x) \geq f\left(\frac{1}{2}\right)$, and the range of the function is $\left[\frac{\pi}{6}, \pi\right]$.

4. The given condition gives $\sin x = \sin \frac{\pi x}{180}$, so $x = k\pi + (-1)^k \cdot \frac{\pi x}{180}$. This implies that $x = \frac{360m\pi}{180-\pi}$ or $\frac{180(2m+1)\pi}{180+\pi}$ with $m \in \mathbf{Z}$. Hence, the desired minimum value of x is $\frac{180\pi}{180+\pi}$.
5. We have $\arcsin(\sin\alpha + \sin\beta) = \frac{\pi}{2} - \arcsin(\sin\alpha - \sin\beta)$. Taking the sine of both sides, we get

$$\sin\alpha + \sin\beta = \cos(\arcsin(\sin\alpha - \sin\beta)) = \sqrt{1 - (\sin\alpha - \sin\beta)^2}.$$

Note that we have used $\sin\alpha \pm \sin\beta \in [-1, 1] \subset \left[-\frac{\pi}{2}, \frac{\pi}{2}\right]$ here. Squaring both sides, we obtain $(\sin\alpha + \sin\beta)^2 + (\sin\alpha - \sin\beta)^2 = 1$, so $\sin^2\alpha + \sin^2\beta = 1$.
6. It follows by the condition that $f(x) = a(x-1)^2 + d$, where d is a constant. Note that $\sin 1 > \sin\frac{\pi}{4} = \frac{\sqrt{2}}{2} > \frac{2}{3} = \sin\left(\arcsin\frac{2}{3}\right)$, so $1 > \arcsin\frac{2}{3}$. Also

$$\sin\left(\arccos\frac{3}{4}\right) = \sqrt{1 - \left(\frac{3}{4}\right)^2} = \frac{\sqrt{7}}{4} < \frac{2}{3},$$

thus $\arccos\frac{3}{4} < \arcsin\frac{2}{3}$. Combining it with $f\left(\arcsin\frac{2}{3}\right) > f\left(\arccos\frac{3}{4}\right)$, we see that $f(x)$ is increasing when $x < 1$. Hence, $a < 0$ and $b = -2a > 0$. In other words, a is negative and b is positive.
7. Let $t = \arctan x$. Then $5f(t) + 3f(-t) = t - \frac{\pi}{2}$. Substituting $-x$ for x in the equation, and since $\arctan x$ is an odd function, $5f(-t) + 3f(t) = -t - \frac{\pi}{2}$. Now, we cancel $f(-t)$ in the two equations above, and get $(25-9)f(t) = 5\left(t - \frac{\pi}{2}\right) - 3\left(-t - \frac{\pi}{2}\right)$, so $f(t) = \frac{1}{2}t - \frac{\pi}{16}$. Therefore, $f(x) = \frac{1}{2}x - \frac{\pi}{16}$ for $x \in \left(-\frac{\pi}{2}, \frac{\pi}{2}\right)$ is the desired function.
8. The given condition implies that $\cos\alpha = 2\cos\alpha\cos 3\alpha$, so $\cos\alpha = 0$ or $\cos 3\alpha = \frac{1}{2}$. Also

$$\sin\alpha + \sin 2\alpha - \sin 4\alpha = \sin\alpha - 2\cos 3\alpha \sin\alpha = \sin\alpha(1 - 2\cos 3\alpha).$$

If $\cos\alpha = 0$, then $\cos 3\alpha = 0$ and $\sin\alpha = \pm 1$. In this case, the value of the given expression equals ± 1. If $\cos 3\alpha = \frac{1}{2}$, then the value of the given expression is 0. Therefore, $\sin\alpha + \sin 2\alpha - \sin 4\alpha \in \{0, \pm 1\}$ (from the argument above, these values are all obtainable).
9. Note that $1 \geq \cos 12x = 5\sin 3x + 9\tan^2 x + \cot^2 x \geq 5\sin 3x + 6|\tan x \cot x| = 6 + 5\sin 3x \geq 6 - 5 = 1$, so every inequality here has to hold as an equality. This requires that $\begin{cases} \cos 12x = 1 \\ \sin 3x = -1 \\ 3|\tan x| = |\cot x| \end{cases}$. Hence,

there exist $k, l, m \in \mathbf{Z}$ such that

$$x = \frac{k\pi}{6} = \frac{2l\pi}{3} + \frac{\pi}{2} = m\pi \pm \frac{\pi}{6}.$$

Combining the above with $x \in (0, 2\pi)$, we find that $x = \frac{7\pi}{6}$ or $\frac{11\pi}{6}$.

10. The original equation is equivalent to $|\sin x - \frac{\sin x}{x}| + \sin x + \frac{\sin x}{x} = 2$. If $\sin x \geq \frac{\sin x}{x}$, then the equation becomes $2\sin x = 2$, so $x = \frac{\pi}{2}$. If $\sin x < \frac{\sin x}{x}$, then $\frac{2\sin x}{x} = 2$, that is $\sin x = x$, which has no solution for $x \in (0, \pi)$ since $\sin x < x$ in this interval. Therefore, $x = \frac{\pi}{2}$.

11. First, we show that for all positive integers k,

$$\arctan\frac{k}{k+1} - \arctan\frac{k-1}{k} = \arctan\frac{1}{2k^2}. \quad ①$$

In fact, for each positive integer k,

$$0 < \arctan\frac{k}{k+1} - \arctan\frac{k-1}{k} < \frac{\pi}{2}, \quad 0 < \arctan\frac{1}{2k^2} < \frac{\pi}{2},$$

and since the tangent function is monotonic on $\left(0, \frac{\pi}{2}\right)$, it suffices to show that the two sides of ① have the same tangent value. Now,

$$\begin{aligned}
&\tan\left(\arctan\frac{k}{k+1} - \arctan\frac{k-1}{k}\right) \\
&= \frac{\frac{k}{k+1} - \frac{k-1}{k}}{1 + \frac{k}{k+1} \cdot \frac{k-1}{k}} = \frac{k^2 - (k-1)(k+1)}{k(k+1) + k(k-1)} \\
&= \frac{1}{2k^2} = \tan\left(\arctan\frac{1}{2k^2}\right).
\end{aligned}$$

Hence, ① holds. Next, we go back to the original problem. It follows that

$$\begin{aligned}
&\arctan\frac{1}{2} + \arctan\frac{1}{8} + \cdots + \arctan\frac{1}{2n^2} \\
&= \left(\arctan\frac{1}{2} - \arctan\frac{0}{1}\right) + \left(\arctan\frac{2}{3} - \arctan\frac{1}{2}\right) \\
&\quad + \cdots + \arctan\frac{n}{n+1} - \arctan\frac{n-1}{n} \\
&= \arctan\frac{n}{n+1}.
\end{aligned}$$

12. The equation is equivalent to

$$(\cos\pi(a-x) - 1)^2 + 1 + \cos\frac{3\pi x}{2a}\cos\left(\frac{\pi x}{2a} + \frac{\pi}{3}\right) = 0.$$

The first term of the left side is nonnegative and the sum of the next two terms is also nonnegative, so necessarily $\cos\pi(a-x) = 1$ and $\cos\frac{3\pi x}{2a}\cos\left(\frac{\pi x}{2a}+\frac{\pi}{3}\right) = -1$. Solving both equations, we see that there exist $k, l, m \in \mathbf{Z}$ such that

$$x = 2k - a = \frac{4la}{3} = 2a(2m+1) - \frac{2a}{3}$$

or

$$x = 2k - a = \frac{2(2l+1)a}{3} = 4akm - \frac{2a}{3}.$$

The existence of k, l, and m requires that $3|a$. Also by discussing whether each term is odd or even we see that a must be even, thus $6|a$. Therefore, the minimum value of a is 6.

13. If $t > 1$, then $a_2 < 0$, so $a_n < 0$ for each $n \geq 2$. If $t < 0$, then by a similar method, $a_n < 0$ for every positive integer n. Hence, $0 \leq t \leq 1$, and we may assume that $t = \sin^2\theta$ with $0 \leq \theta \leq \frac{\pi}{2}$, thus $a_2 = 4\sin^2\theta\cos^2\theta = \sin^2 2\theta$. By induction we can prove that $a_n = \sin^2 2^{n-1}\theta$. If $a_n = 0$, then $\sin^2 2^{n-1}\theta = 0$, from which $2^{n-1}\theta = k\pi$ for $k \in \mathbf{Z}$. Since $0 \leq \theta \leq \frac{\pi}{2}$, we have $0 \leq \frac{k}{2^{n-1}} \leq \frac{1}{2}$, so $0 \leq k \leq 2^{n-2}$. Therefore, the number of different values of t is $2^{n-2}+1$.

14. If we let $t = \cos x$, then the equation will become a polynomial equation of degree 7, which is hard to solve for all solutions. We should deal with this problem from the viewpoint of trigonometry.

    Considering the third term on the left side of the equation, we may multiply both sides by $4\sin x$. Thus,

$$4\sin x\cos^2 x + 4\sin x\cos^2 2x - \sin 8x = 3\sin x.$$

Then subtract $4\sin^3 x$ from both sides, and using the triple-angle formula, we have

$$4\sin x(\cos^2 x - \sin^2 x + \cos^2 2x) = \sin 3x + \sin 8x,$$

$$4\sin x(\cos 2x + \cos^2 2x) = \sin 3x + \sin 8x,$$

$$4\sin x\cos 2x(1+\cos 2x) = \sin 3x + \sin 8x,$$

$$8\sin x\cos^2 x\cos 2x = \sin 3x + \sin 8x,$$

$$2\sin 4x\cos x = \sin 3x + \sin 8x,$$

$$\sin 5x + \sin 3x = \sin 3x + \sin 8x.$$

Therefore, $\sin 5x = \sin 8x$, and the solutions to this equation are

$$8x = 2k\pi + 5x \quad \text{or} \quad (2k+1)\pi - 5x, \quad k \in \mathbf{Z}.$$

In other words, $x = \frac{2k\pi}{3}$ or $\frac{(2k+1)\pi}{13}$ with $k \in \mathbf{Z}$.

Note that we need to drop the solutions where $\sin x = 0$. If $\sin x = 0$, then each term of the left side in the original equation is an integer, so the left side cannot equal $\frac{3}{4}$. In summary, the solutions to the original equation are $x = 2k\pi + \alpha$, where

$$\alpha \in \left\{\pm\frac{2\pi}{3}, \pm\frac{\pi}{13}, \pm\frac{3\pi}{13}, \pm\frac{5\pi}{13}, \pm\frac{7\pi}{13}, \pm\frac{9\pi}{13}, \pm\frac{11\pi}{13}\right\}, \qquad k \in \mathbf{Z}.$$

Solution 17

The Law of Sines and the Law of Cosines

1. Suppose the observer stands x meters away from the wall and has the angle of observation y. Then $y = \arctan\frac{b}{x} - \arctan\frac{a}{x}$. The problem becomes finding the maximum value of y for $x > 0$ and the given $b > a > 0$. Note that

$$\tan y = \frac{\frac{b}{x} - \frac{a}{x}}{1 + \frac{b}{x}\cdot\frac{a}{x}} = \frac{(b-a)x}{x^2 + ab} \le \frac{(b-a)x}{2\sqrt{ab}x} = \frac{b-a}{2\sqrt{ab}}.$$

Hence, $y \le \arctan\frac{b-a}{2\sqrt{ab}}$, and the equality holds when $x = \sqrt{ab}$. Therefore, the angle of observation is maximal when the observer stands $\sqrt{ab}$ meters from the wall.

2. Suppose $ABCD$ is a quadrilateral that satisfies the condition. Then $S_{ABCD} = 32$. Also assume that $AD + AC + CB = 16$. Let $\angle DAC = \alpha$ and $\angle ACB = \beta$. Then

$$S_{ABCD} = \frac{1}{2}AD\cdot AC\cdot\sin\alpha + \frac{1}{2}BC\cdot CA\cdot\sin\beta,$$

so $32 = \frac{1}{2}AC(AD\sin\alpha + BC\sin\beta) \le \frac{1}{2}AC(AD+BC) = \frac{1}{2}AC(16 - AC) \le \frac{1}{2}\left(\frac{AC+16-AC}{2}\right)^2 = 32$. Hence, each inequality holds as an equality, thus $AC = 8$ and $\sin\alpha = \sin\beta = 1$, which means that $\alpha = \beta = 90°$. Therefore, the other diagonal $BD = 8\sqrt{2}$.

3. Let AD and BE be altitudes of ΔABC. Then they intersect at the orthocenter H. Since C, D, H and E are concyclic, $AH\cdot AD = AE\cdot AC$. Since $AD = \frac{bc\sin A}{a}$ (here a, b, and c are the side lengths of the triangle), $AE = c\cos A$ (since ABC is an acute triangle and $\cos A$ is positive), thus

$$AH = \frac{abc\cos A}{bc\sin A} = a\cot A.$$

By the law of sines $\frac{a}{2\sin A} = 2R$, so $AO = \frac{a}{2\sin A}$. Using $AO = AH$, we have $\cot A = \frac{1}{2\sin A}$, or equivalently $\cos A = \frac{1}{2}$. Therefore, $\angle A = 60°$.

4. Let $\angle FDA = \alpha, \angle EDA = \beta, \angle EDC = \gamma$ and $\angle BDF = \delta$. Then $\alpha + \beta = 90°$. By the law of sines $\frac{AE}{AD} = \frac{\sin\beta}{\sin\angle AED}$ and $\frac{EC}{CD} = \frac{\sin\gamma}{\sin\angle CED}$, and since $\angle AED + \angle CED = 180°$, we have $\frac{AE}{EC} = \frac{AD\sin\beta}{CD\sin\gamma}$. By the property of the angle bisector,

$$\frac{AB}{BC} = \frac{AE}{EC} = \frac{AD\sin\beta}{CD\sin\gamma}.$$

Similarly, $\frac{AC}{BC} = \frac{AD\sin\alpha}{BD\sin\delta}$. Dividing the two equations, we get $\frac{AB}{AC} = \frac{BC\sin\beta\sin\delta}{CD\sin\gamma\sin\alpha}$. Again by the property of the angle bisector, $\frac{AB}{AC} = \frac{BD}{CD}$, so $\sin\beta\sin\delta = \sin\gamma\sin\alpha$. Since $\alpha+\beta = 90°$, we also have $\gamma+\delta = 90°$. Consequently $\sin\beta\cos\gamma = \sin\gamma\cos\beta$, namely $\sin(\beta-\gamma) = 0$. Hence, $\beta = \gamma$, and DE bisects $\angle ADC$. Thus,

$$\frac{\sin C}{\sin\frac{A}{2}} = \frac{AD}{DC} = \frac{AE}{EC} = \frac{AB}{BC} = \frac{\sin C}{\sin A}.$$

That is, $\sin A = \sin\frac{A}{2}$, from which $A = 180° - \frac{A}{2}$ namely $A = 120°$.

5. It follows that $B = 60°$ and $\frac{1}{2}(c-a)b = \frac{1}{2}ab\sin C$. By the law of sines, $\sin C - \sin A = \sin A\sin C$. Hence,

$$2\cos\frac{C+A}{2}\sin\frac{C-A}{2} = \frac{1}{2}(\cos(C-A) - \cos(C+A)).$$

This implies that $4\sin\frac{B}{2}\sin\frac{C-A}{2} = 1 - 2\sin^2\frac{C-A}{2} + \cos B$, so

$$2\sin^2\frac{C-A}{2} + 2\sin\frac{C-A}{2} - \frac{3}{2} = 0.$$

Solving the equation, we have $\sin\frac{C-A}{2} = \frac{1}{2}$ (and the other solution is dropped).

6. Similar to the previous problem, $\sin C - \sin A = \sin A \sin C$, so

$$\begin{aligned}2\cos\frac{C+A}{2}\sin\frac{C-A}{2} &= \frac{1}{2}(\cos(C-A) - \cos(C+A))\\ &= \frac{1}{2}\left[1 - 2\sin^2\frac{C-A}{2} - \left(2\cos^2\frac{C+A}{2} - 1\right)\right]\\ &= 1 - \sin^2\frac{C-A}{2} - \cos^2\frac{C+A}{2}.\end{aligned}$$

Consequently, $\left(\sin\frac{C-A}{2} + \cos\frac{C+A}{2}\right)^2 = 1$, hence $\sin\frac{C-A}{2} + \cos\frac{C+A}{2} = 1$ (the negative value is dropped).

7. Let O be the circumcenter of ΔABC. Then O, K and X are collinear. Similarly, O, L and Y are collinear, and O, M and Z are collinear. Note that $\angle AOB = 2\angle C$, so $OM = OA\cos C = R\cos C$. Similarly, $OK = R\cos A$ and $OL = R\cos B$ (as shown in the figure). Hence, $r + KX + LY + MZ = r + 3R - (OM + OK + OL) = r + 3R - R(\cos A + \cos B + \cos C)$. Since

$$\cos A + \cos B + \cos C = 1 + 4\sin\frac{A}{2}\sin\frac{B}{2}\sin\frac{C}{2} = 1 + \frac{r}{R},$$

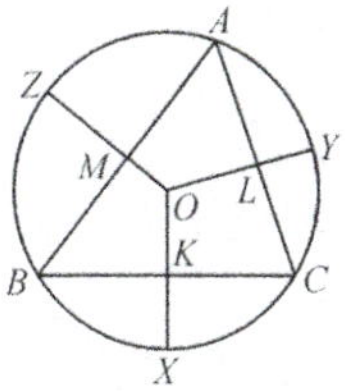

Exercise 7

we have

$$r + KX + LY + MZ = r + 3R - R\left(1 + \frac{r}{R}\right) = 2R.$$

8. As shown in the figure, assume that the side length of the square is 1, and let $BK = x$ and $DN = y$. Form the line segment AM. Then $\cot\angle BKC = x$ and $\cot\angle DNC = y$. Since $AK \cdot AN = 2BK \cdot DN$, we have $(1-x)(1-y) = 2xy$ and $x + y = -xy + 1$. Hence,

$$\tan(\angle BKC + \angle DNC) = \frac{\frac{1}{x} + \frac{1}{y}}{1 - \frac{1}{xy}} = \frac{x+y}{xy-1} = -1,$$

so $\angle BKC + \angle DNC = 135^\circ$. Thus,

$$\angle BLK = 180^\circ - 45^\circ - \angle BKL = 135^\circ - \angle BKL = \angle DNC.$$

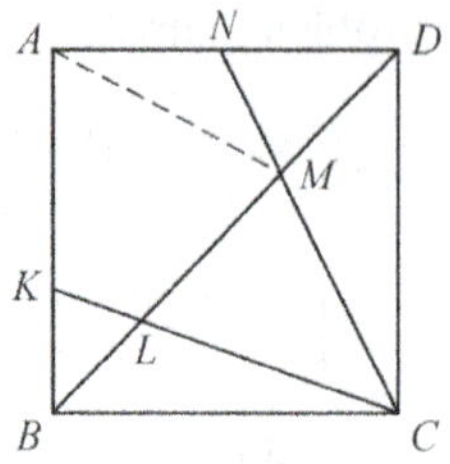

Exercise 8

Since $BC \parallel AD$, we have $\angle DNC = \angle BCM$, and by $\Delta BCM \cong \Delta BAM$ we see that $\angle BCM = \angle KAM$. Hence, K, L, M and A are concyclic. Similarly, A, N, M, and L are concyclic. Therefore, A, K, L, M, and N all lie on the same circle.

9. As shown in the figure, let $ABCD$ be a quadrilateral that satisfies the condition, and let $\angle ABC = \alpha$ and $\angle ADC = \beta$. It suffices to show that $\alpha + \beta = 180°$. By the given condition,

$$\frac{1}{2}(ab\sin\alpha + cd\sin\beta) = \sqrt{abcd} \text{ ①}$$

and $a + c = b + d$ (this is because $ABCD$ has an inscribed circle). Further, by the law of cosines

$$\begin{aligned} AC^2 &= a^2 + b^2 - 2ab\cos\alpha \\ &= c^2 + d^2 - 2cd\cos\beta. \text{ ②} \end{aligned}$$

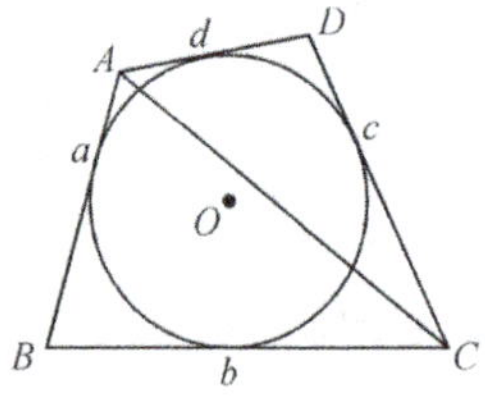

Exercise 9

Combining ② with $a + c = b + d$, we get

$$(a+b)^2 - 2ab(-1+\cos\alpha) = (c+d)^2 - 2cd(-1+\cos\beta) \quad \text{and}$$

$$ab(-1+\cos\alpha) = cd(-1+\cos\beta). \text{ ③}$$

From ①, $(ab\sin\alpha + cd\sin\beta)^2 = 4abcd$, thus

$$(ab)^2(1-\cos^2\alpha) + (cd)^2(1-\cos^2\beta) = 4abcd\left(1 - \frac{\sin\alpha\sin\beta}{2}\right).$$

Now, plug ③ into the equation above, and we have

$$ab(1-\cos\alpha)cd(1+\cos\beta)+cd(1-\cos\beta)ab(1+\cos\alpha)$$
$$=4abcd\left(1-\frac{\sin\alpha\sin\beta}{2}\right).$$

Consequently,

$$2-2\cos\alpha\cos\beta=4-2\sin\alpha\sin\beta,$$

which implies that $\cos(\alpha+\beta)=-1$. Therefore, $\alpha+\beta=180^\circ$, and A, B, C, and D are concyclic.

10. As shown in the figure, let $\angle ABC=\alpha, \angle ADC=\beta, \angle BAC=\gamma$, and $\angle CAD=\delta$. Then by the law of sines, $\frac{CD}{AC}=\frac{\sin\delta}{\sin\beta}$, $\frac{CE}{AC}=\frac{\sin\gamma}{\sin E}$, $\frac{AB}{AC}=\frac{\sin(\alpha+\gamma)}{\sin\alpha}$, and $\frac{AE}{AC}=\frac{\sin(\beta+\delta)}{\sin E}$. Note that $E=180^\circ-\angle EAD-\angle EDA=180^\circ-(180^\circ-\gamma-\delta)-(180^\circ-\beta)=\beta+\gamma+\delta-180^\circ$, so

$$\frac{CE}{AC}=-\frac{\sin\gamma}{\sin(\beta+\gamma+\delta)},\ \frac{AE}{AC}=-\frac{\sin(\beta+\delta)}{\sin(\beta+\gamma+\delta)}.$$

Since $AC^2=CD\cdot CE-AB\cdot AE$ is equivalent to $\frac{CD}{AC}\cdot\frac{CE}{AC}-\frac{AB}{AC}\cdot\frac{AE}{AC}-1=0$, it is also equivalent to $-\sin\gamma\sin\delta\sin\alpha+\sin(\alpha+\gamma)\sin(\beta+\delta)\sin\beta-\sin\alpha\sin\beta\sin(\beta+\gamma+\delta)=0$, in other words,

$$(\cos(\gamma-\delta)-\cos(\gamma+\delta))\sin\alpha-(\cos(\beta+\delta-\alpha-\gamma)$$
$$-\cos(\alpha+\beta+\delta+\gamma))\sin\beta+(\cos(\alpha-\beta)$$
$$-\cos(\alpha+\beta))\sin(\beta+\gamma+\delta)=0.$$

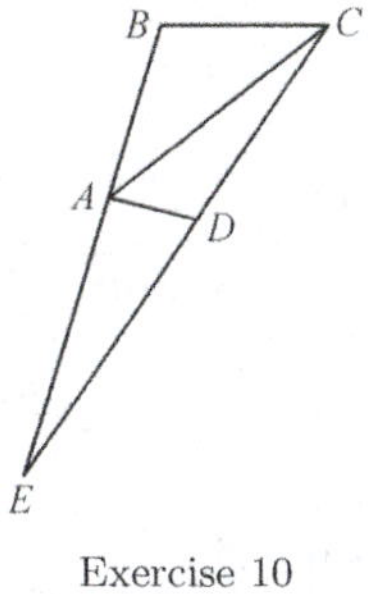

Exercise 10

We apply the product-to-sum formula to every term above to get

$$\sin(\alpha+\gamma-\delta)-\sin(\gamma-\alpha-\delta)-\sin(\alpha+\gamma+\delta)$$
$$+\sin(\gamma+\delta-\alpha)-\sin(2\beta+\delta-\alpha-\gamma)+\sin(\delta-\alpha-\gamma)$$
$$+\sin(2\beta+\alpha+\gamma+\delta)-\sin(\alpha+\gamma+\delta)+\sin(\alpha+\gamma+\delta)$$
$$-\sin(\alpha-\gamma-\delta-2\beta)-\sin(2\beta+\alpha+\gamma+\delta)+\sin(\alpha-\gamma-\delta)=0.$$

Equivalently, $\sin(\alpha+\gamma+\delta)+\sin(\gamma-\alpha-\delta)-\sin(2\beta+\delta-\alpha-\gamma)-\sin(\alpha-\gamma-\delta-2\beta)=0$. Then we apply the sum-to-product formula to the left side of this equality, and get

$$\sin\gamma\cos(\alpha+\delta)-\sin(-\gamma)\cos(2\beta+\delta-\alpha)=0,$$

so $\sin\gamma(\cos(\alpha+\delta)-\cos(\alpha-2\beta-\delta))=0$, that is

$$\sin\gamma\sin(\alpha-\beta)\sin(\beta+\delta)=0.$$

Note that $\sin\gamma\neq 0$ and $\sin(\beta+\delta)\neq 0$, so the equality is equivalent to $\sin(\alpha-\beta)=0$, or in other words $\alpha=\beta$. Therefore, the desired proposition holds.

11. By the law of cosines,

$$\begin{aligned}a^2-(b^2+c^2)(1-\cos A)&=b^2+c^2-2bc\cos A-(b^2+c^2)(1-\cos A)\\&=(b^2+c^2-2bc)\cos A=(b-c)^2\cos A\geq 0,\end{aligned}$$

so $a^2\geq(b^2+c^2)(1-\cos A)=2(b^2+c^2)\sin^2\frac{A}{2}$. Similarly,

$$b^2\geq 2(c^2+a^2)\sin^2\frac{B}{2},\quad c^2\geq 2(a^2+b^2)\sin^2\frac{C}{2}.$$

Hence,

$$a^2b^2c^2\geq 8(a^2+b^2)(b^2+c^2)(c^2+a^2)\sin^2\frac{A}{2}\sin^2\frac{B}{2}\sin^2\frac{C}{2},$$

and further,

$$2\sin\frac{A}{2}\sin\frac{B}{2}\sin\frac{C}{2}\leq\frac{abc}{\sqrt{2(a^2+b^2)(b^2+c^2)(c^2+a^2)}}.$$

On the other hand, since $S_{\triangle ABC}=\frac{1}{2}ab\sin C=\frac{1}{2}r(a+b+c)$, by the law of sines

$$\begin{aligned}\frac{r}{2R}&=\frac{\sin A\sin B\sin C}{\sin A+\sin B+\sin C}=\frac{\sin A\sin B\sin C}{4\cos\frac{A}{2}\cos\frac{B}{2}\cos\frac{C}{2}}\\&=2\sin\frac{A}{2}\sin\frac{B}{2}\sin\frac{C}{2}.\end{aligned}$$

Hence,

$$\frac{r}{2R}\leq\frac{abc}{\sqrt{2(a^2+b^2)(b^2+c^2)(c^2+a^2)}}.$$

12. As shown in the figure, let a, b and c be the side lengths of ΔABC, let r be the radius of its incircle, and let p be half its perimeter. Then $BE = BD + DE = (p-b) + DE$ and $CD = p - c$. Now, we compute the length of DE. In order to do this, we form the line segment AO and denote $\angle EFD = \alpha$, so $\frac{2r}{DE} = \cot\alpha$. In ΔAOF, we see that $AO = \frac{r}{\sin\frac{A}{2}}$ and

$$\begin{aligned}\angle AOF &= 180^\circ - \angle GOD - \angle GOA = \angle B - \angle GOA \\ &= \angle B - \left(90^\circ - \frac{\angle A}{2}\right) = B + \frac{A}{2} - 90^\circ = \frac{B-C}{2}.\end{aligned}$$

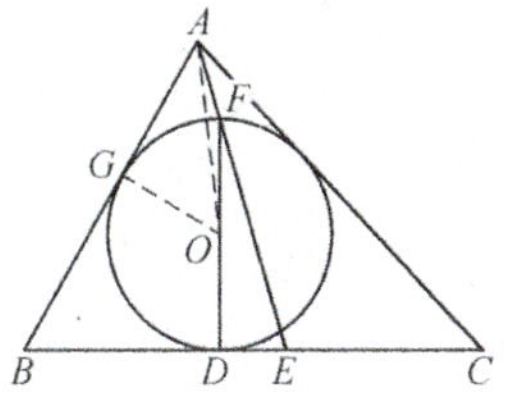

Exercise 12

By the law of sines, $\frac{OF}{\sin\angle OAF} = \frac{AO}{\sin\angle AFO}$, so

$$\frac{r}{\sin\left(\alpha - \frac{B-C}{2}\right)} = \frac{r}{\sin\alpha\sin\frac{A}{2}} = \frac{r}{\sin\alpha\cos\frac{B+C}{2}}.$$

Hence,

$$\begin{aligned}\sin\alpha\cos\frac{B+C}{2} &= \sin\left(\alpha - \frac{B-C}{2}\right) \\ &= \sin\alpha\cos\frac{B-C}{2} - \cos\alpha\sin\frac{B-C}{2},\end{aligned}$$

thus

$$\cot\alpha = \frac{\cos\frac{B-C}{2} - \cos\frac{B+C}{2}}{\sin\frac{B-C}{2}} = \frac{2\sin\frac{B}{2}\sin\frac{C}{2}}{\sin\frac{B-C}{2}}.$$

So far, we have

$$\frac{2r}{DE} = \frac{2\sin\frac{B}{2}\sin\frac{C}{2}}{\sin\frac{B-C}{2}} = \frac{2\sin\frac{A}{2}\sin\frac{B}{2}\sin\frac{C}{2}}{\sin\frac{A}{2}\sin\frac{B-C}{2}}.$$

Note that r and R satisfy the relation $\frac{r}{R} = 4\sin\frac{A}{2}\sin\frac{B}{2}\sin\frac{C}{2}$, and so $DE = 4R\sin\frac{A}{2}\sin\frac{B-C}{2}$, in other words

$$DE = 4R\cos\frac{B+C}{2}\sin\frac{B-C}{2} = 2R(\sin B - \sin C) = b - c.$$

The last step used the law of sines. Therefore, $BE = CD$.

13. As shown in the figure, let $\angle BAP = \alpha$, $\angle CBP = \beta$ and $\angle ACP = \gamma$. Then by the law of sines $\frac{AP}{BP} = \frac{\sin(B-\beta)}{\sin\alpha}$, $\frac{BP}{PC} = \frac{\sin(C-\gamma)}{\sin\beta}$, and $\frac{PC}{AP} = \frac{\sin(A-\alpha)}{\sin\gamma}$. Multiplying the three equalities, we have

$$\sin\alpha\sin\beta\sin\gamma = \sin(A-\alpha)\sin(B-\beta)\sin(C-\gamma). \text{ ①}$$

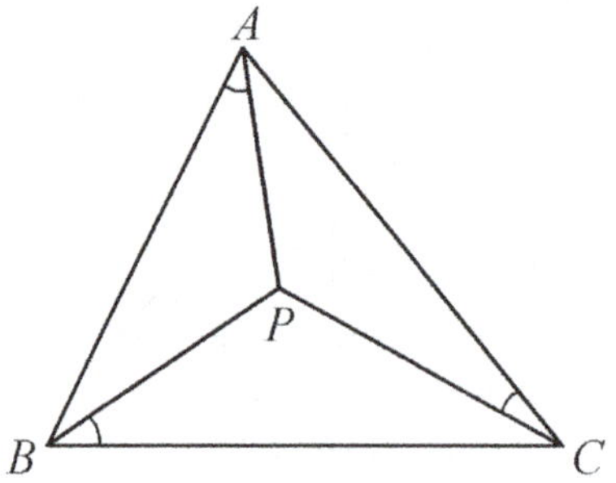

Exercise 13

Note that if one of α, β and γ is at least $150°$ (we assume it to be α), then $\beta+\gamma \leq 30°$, and the conclusion follows trivially. Therefore, in the following discussion we may assume that α, β, and γ are all less than $150°$. Note that

$$\sin\alpha\sin(A-\alpha) = \frac{1}{2}(\cos(2\alpha - A) - \cos A) \leq \frac{1}{2}(1 - \cos A) = \sin^2\frac{A}{2}.$$

Similarly, $\sin\beta\sin(B-\beta) \leq \sin^2\frac{B}{2}$ and $\sin\gamma\sin(C-\gamma) \leq \sin^2\frac{C}{2}$. Combining them with ①, we have

$$(\sin\alpha\sin\beta\sin\gamma)^2 \leq \left(\sin\frac{A}{2}\sin\frac{B}{2}\sin\frac{C}{2}\right)^2,$$

From which

$$\sin\alpha\sin\beta\sin\gamma \leq \sin\frac{A}{2}\sin\frac{B}{2}\sin\frac{C}{2} \leq \frac{1}{8}.$$

This implies that one of $\sin\alpha\sin\beta$ and $\sin\gamma$ is less than or equal to $\frac{1}{2}$. Assume that $\sin\alpha \leq \frac{1}{2}$. Then since $\alpha < 150°$, we have $\alpha \leq 30°$, and the desired proposition follows.

Remark. We can also use the Erdös-Mordell inequality to prove this result.

14. As shown in the figure, let R be the radius of the circumcircle of ΔABC. Then since $\angle B'A'C' = \angle B'A'A + \angle C'A'A = \frac{B+C}{2}$, and similarly $\angle A'B'C' = \frac{C+A}{2}$ and $\angle B'C'A' = \frac{A+B}{2}$, by the law of sines and the area formula we have

$$S_{\Delta ABC} = \frac{1}{2}ab\sin C = 2R^2 \sin A \sin B \sin C.$$

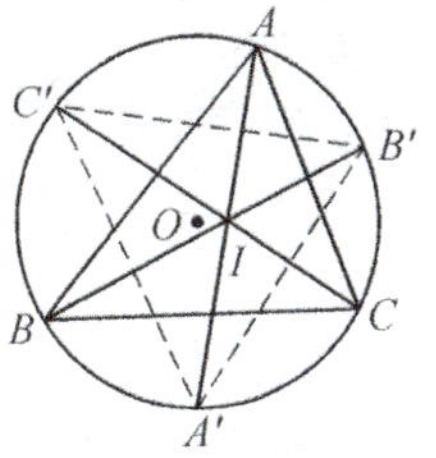

Exercise 14

The radius of the circumcircle of $\Delta A'B'C'$ is also R, thus

$$S_{\Delta A'B'C'} = 2R^2 \sin\frac{A+B}{2}\sin\frac{B+C}{2}\sin\frac{C+A}{2}.$$

By $2\sin\frac{B+C}{2} \geq 2\sin\frac{B+C}{2}\cos\frac{B-C}{2} = \sin B + \sin C$ and similar inequalities,

$$\begin{aligned} S_{\Delta A'B'C'} &\geq \frac{R^2}{4}(\sin B + \sin C)(\sin C + \sin A)(\sin A + \sin B) \\ &\geq \frac{R^2}{4}(2\sqrt{\sin B \sin C})(2\sqrt{\sin C \sin A})(2\sqrt{\sin A \sin B}) \\ &= 2R^2 \sin A \sin B \sin C = S_{\Delta ABC}. \end{aligned}$$

Therefore, the proposition is true.

15. Let G be the barycenter of ΔABC and AA' intersect BC at A''. Also let a, b and c be the side lengths of ΔABC, and m_a, m_b and m_c be the lengths of its medians. Since $S_{\Delta ABG} = S_{\Delta BCG} = S_{\Delta ACG} = \frac{1}{3}S_{\Delta ABC}$, in order to prove $S_{\Delta A'B'C'} \geq S_{\Delta ABC}$, it suffices to show that $\frac{S_{\Delta A'C'G}}{S_{\Delta ACG}} + \frac{S_{\Delta B'C'G'}}{S_{\Delta BCG}} + \frac{S_{\Delta A'B'G}}{S_{\Delta ABG}} \geq 3$. Equivalently, we only need to prove $\frac{A'G \cdot C'G}{AG \cdot CG} + \frac{B'G \cdot C'G}{BG \cdot CG} + \frac{A'G \cdot B'G}{AG \cdot BG} \geq 3$. Now, we compute $\frac{A'G}{AG}$. The equality $AA'' \cdot A''A' = BA'' \cdot A''C$ implies that $A''A' = \frac{a^2}{4m_a}$. Hence, $\frac{A'G}{AG} = \frac{A'A''+A''G}{AG} = \frac{\frac{a^2}{4m_a}+\frac{1}{3}m_a}{\frac{2}{3}m_a} = \frac{3a^2+4m_a^2}{8m_a^2}$. Then apply the median

length formula $4m_a^2 = 2b^2 + 2c^2 - a^2$, and we get

$$\frac{A'G}{AG} = \frac{3a^2 + (2b^2 + 2c^2 - a^2)}{2(2b^2 + 2c^2 - a^2)} = \frac{a^2 + b^2 + c^2}{2b^2 + 2c^2 - a^2}.$$

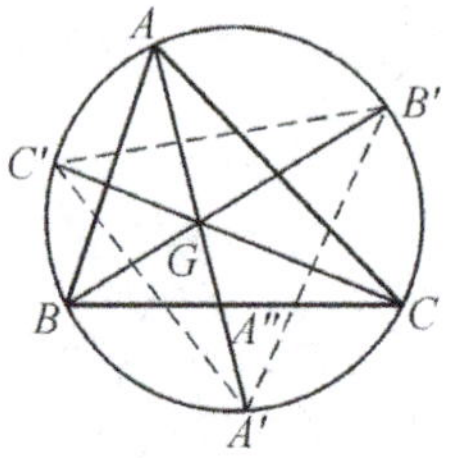

Exercise 15

Similarly, $\frac{C'G}{CG} = \frac{a^2+b^2+c^2}{2a^2+2b^2-c^2}$ and $\frac{B'G}{BG} = \frac{a^2+b^2+c^2}{2c^2+2a^2-b^2}$. Now, it suffices to prove that

$$\sum \frac{(a^2 + b^2 + c^2)^2}{(2b^2 + 2c^2 - a^2)(2a^2 + 2b^2 - c^2)} \geq 3.$$

or equivalently

$$3(a^2 + b^2 + c^2)^3 \geq 3\prod(2a^2 + 2b^2 - c^2).$$

This can be proven by using the AM-GM inequality. In fact,

$$\prod(2a^2 + 2b^2 - c^2) \leq \left(\frac{1}{3}\sum(2a^2 + 2b^2 - c^2)\right)^3 = (a^2 + b^2 + c^2)^3.$$

Therefore, the proposition follows.

Solution 18

Concepts and Operations of Vectors

1. Let $\vec{a} = 2\vec{m} + \vec{n}$ and $\vec{b} = 2\vec{n} - 3\vec{m}$. Since $|\vec{m}| = |\vec{n}| = 1$ and $\vec{m} \cdot \vec{n} = |\vec{m}| \cdot |\vec{n}| \cdot \cos 60° = \frac{1}{2}$,
$$|\vec{a}|^2 = (2\vec{m} + \vec{n})^2 = 4\vec{m}^2 + 4\vec{m} \cdot \vec{n} + \vec{n}^2 = 7,$$
$$|\vec{b}|^2 = (2\vec{n} - 3\vec{m})^2 = 4\vec{n}^2 - 12\vec{m} \cdot \vec{n} + 9\vec{m}^2 = 7.$$
$$\vec{a} \cdot \vec{b} = -6\vec{m}^2 + \vec{m} \cdot \vec{n} + 2\vec{n}^2 = -\frac{7}{2},$$
Hence, the angle θ between $\vec{a}$ and $\vec{b}$ satisfies
$$\cos\theta = \frac{\vec{a} \cdot \vec{b}}{|\vec{a}| \cdot |\vec{b}|} = -\frac{\frac{7}{2}}{7} = -\frac{1}{2},$$
so $\theta = 120°$.

2. By the condition,
$$\begin{cases} \left(\vec{a} + 3\vec{b}\right) \cdot (7\vec{a} - 5\vec{b}) = 0, \\ \left(\vec{a} - 4\vec{b}\right) \cdot (7\vec{a} - 2\vec{b}) = 0, \end{cases}$$
from which
$$\begin{cases} 7\left|\vec{a}\right|^2 + 16\vec{a} \cdot \vec{b} - 15|\vec{b}|^2 = 0, \\ 7|\vec{a}|^2 - 30\vec{a} \cdot \vec{b} + 8|\vec{b}|^2 = 0. \end{cases}$$
Hence, $\vec{a} \cdot \vec{b} = \frac{1}{2}|\vec{b}|^2$ and $|\vec{a}|^2 = |\vec{b}|^2$, so $|\vec{a}| \cdot |\vec{b}| \cdot \cos\theta = \frac{1}{2}|\vec{b}|^2$. This implies that $\cos\theta = \frac{1}{2}$, thus $\theta = 60°$. Therefore, the angle between $\vec{a}$ and $\vec{b}$ is $60°$.

3. By the law of cosines, the angle θ between $\overrightarrow{BA}$ and $\overrightarrow{BC}$ satisfies $\cos\theta = \frac{7^2+5^2-6^2}{2\times7\times5}$, so $\overrightarrow{BA}\cdot\overrightarrow{BC} = \frac{1}{2}(7^2+5^2-6^2) = 19$.
4. Note that $3^2+11^2 = 130 = 7^2+9^2$. Since $\overrightarrow{AB}+\overrightarrow{BC}+\overrightarrow{CD}+\overrightarrow{DA} = \vec{0}$,

$$\begin{aligned} DA^2 = \overrightarrow{DA}^2 &= (\overrightarrow{AB}+\overrightarrow{BC}+\overrightarrow{CD})^2 \\ &= AB^2+BC^2+CD^2+2(\overrightarrow{AB}\cdot\overrightarrow{BC}+\overrightarrow{BC}\cdot\overrightarrow{CD}+\overrightarrow{CD}\cdot\overrightarrow{AB}) \\ &= AB^2-BC^2+CD^2+2(\overrightarrow{BC}^2+\overrightarrow{AB}\cdot\overrightarrow{BC}+\overrightarrow{BC}\cdot\overrightarrow{CD} \\ &\quad +\overrightarrow{CD}\cdot\overrightarrow{AB}) \\ &= AB^2-BC^2+CD^2+2(\overrightarrow{AB}+\overrightarrow{BC})(\overrightarrow{BC}+\overrightarrow{CD}). \end{aligned}$$

Hence, $2\overrightarrow{AC}\cdot\overrightarrow{BD} = AD^2+BC^2-AB^2-CD^2 = 0$, so $\overrightarrow{AC}\cdot\overrightarrow{BD} = 0$.
5. Let D be the intersection point of BC and AO. Then

$$\frac{BD}{DC} = \frac{S_{\Delta BOD}}{S_{\Delta COD}} = \frac{S_{\Delta ABD}}{S_{\Delta ACD}} = \frac{S_{\Delta BOD}-S_{\Delta ABD}}{S_{\Delta COD}-S_{\Delta ACD}} = \frac{S_{\Delta AOB}}{S_{\Delta AOC}} = \frac{\gamma}{\beta}.$$

Hence, by the formula of definite proportion,

$$\overrightarrow{OD} = \frac{\overrightarrow{OB}+\frac{\gamma}{\beta}\overrightarrow{OC}}{1+\frac{\gamma}{\beta}} = \frac{\beta\overrightarrow{OB}+\gamma\overrightarrow{OC}}{\beta+\gamma}.$$

Also $\frac{AO}{OD} = \frac{S_{\Delta AOB}}{S_{\Delta BOD}} = \frac{S_{\Delta AOC}}{S_{\Delta COD}} = \frac{S_{\Delta AOB}+S_{\Delta AOC}}{S_{\Delta BOC}} = \frac{\beta+\gamma}{\alpha}$, hence

$$\overrightarrow{OD} = -\frac{\alpha}{\beta+\gamma}\cdot\overrightarrow{OA}.$$

Combining the two equalities above, we get the desired conclusion.
6. Since $|\vec{a}| = |\vec{b}| = |\vec{c}| = |\vec{d}|$, the quadrilateral $ABCD$ has a circumcircle. Also since $\vec{a}+\vec{b} = -(\vec{c}+\vec{d})$, we have $|\vec{a}+\vec{b}| = |\vec{c}+\vec{d}|$, so

$$(\vec{a}+\vec{b})\cdot(\vec{a}+\vec{b}) = (\vec{c}+\vec{d})\cdot(\vec{c}+\vec{d}).$$

Expanding the equality, we obtain $\vec{a}\cdot\vec{b} = \vec{c}\cdot\vec{d}$, thus $\angle AOB = \angle COD$ (here O is the common starting point of the four vectors). Similarly, $\angle BOC = \angle AOD$. Hence,

$$\begin{aligned} \angle AOC &= \angle AOB+\angle BOC \\ &= \frac{1}{2}(\angle AOB+\angle BOC+\angle COD+\angle DOA) = 180^\circ. \end{aligned}$$

This means that AC is a diameter of the circumcircle of $ABCD$. Similarly, BD is also a diameter, hence $ABCD$ is a rectangle.

7. It follows by the condition that
$$\overrightarrow{OA}+2\overrightarrow{OB}+3\overrightarrow{OC}=3(\overrightarrow{OB}-\overrightarrow{OA})+2(\overrightarrow{OC}-\overrightarrow{OB})+(\overrightarrow{OA}-\overrightarrow{OC}).$$
from which $3\overrightarrow{OA}+\overrightarrow{OB}+2\overrightarrow{OC}=\vec{0}$. Comparing it with the conclusion of Exercise 5, by a reverse argument we see that
$$(\alpha,\beta,\gamma)=\left(\frac{1}{2},\frac{1}{6},\frac{1}{3}\right).$$
Hence, the desired answer is $\frac{1}{3}+1+\frac{1}{2}=\frac{11}{6}$.
8. By the definition of scalar products and the law of cosines, $\overrightarrow{AB}\cdot\overrightarrow{AC}=bc\cos A=\frac{b^2+c^2-a^2}{2}$. Similarly, $\overrightarrow{BA}\cdot\overrightarrow{BC}=\frac{a^2+c^2-b^2}{2}$ and $\overrightarrow{CA}\cdot\overrightarrow{CB}=\frac{a^2+b^2-c^2}{2}$. Hence, the condition reduces to
$$b^2+c^2-a^2+2(a^2+c^2-b^2)=3(a^2+b^2-c^2),$$
or equivalently $a^2+2b^2=3c^2$.
By the law of cosines and the basic inequalities,
$$\cos C=\frac{a^2+b^2-c^2}{2ab}=\frac{a^2+b^2-\frac{1}{3}(a^2+2b^2)}{2ab}$$
$$=\frac{a}{3b}+\frac{b}{6a}\geq 2\sqrt{\frac{a}{3b}\cdot\frac{b}{6a}}=\frac{\sqrt{2}}{3}.$$
This implies that
$$\sin C=\sqrt{1-\cos^2 C}\leq\frac{\sqrt{7}}{3},$$
where the equality is valid if and only if $a:b:c=\sqrt{3}:\sqrt{6}:\sqrt{5}$. Therefore, the maximum value of $\sin C$ is $\frac{\sqrt{7}}{3}$.
9. As shown in the figure, F is the midpoint of the side CD of the parallelogram $ABCD$. Let $\overrightarrow{AE}=a\overrightarrow{AF}$ and $\overrightarrow{BE}=b\overrightarrow{BD}$. Then
$$\overrightarrow{AB}+b\overrightarrow{BD}=a\overrightarrow{AF}.$$

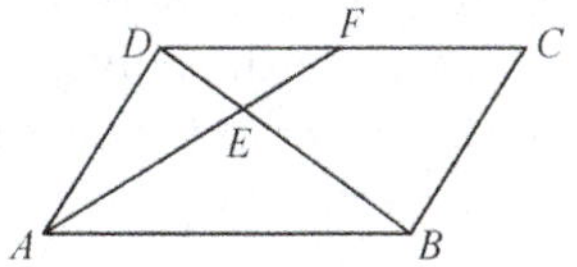

Exercise 9

Equivalently, $\overrightarrow{AB}+b(\overrightarrow{AD}-\overrightarrow{AB})=a(\overrightarrow{AD}+\frac{1}{2}\overrightarrow{DC})$, which is also equivalent to
$$\left(1-b-\frac{a}{2}\right)\overrightarrow{AB}+(b-a)\overrightarrow{AD}=\vec{0}.$$

Since $\overrightarrow{AB}$ and $\overrightarrow{AD}$ are not collinear, it follows that $1-b-\frac{a}{2}=b-a=0$, so $a=b=\frac{2}{3}$. Therefore, $\overrightarrow{BE}=b\overrightarrow{BD}=\frac{2}{3}\overrightarrow{BD}$, and E is a trisection point on BD (it is closer to D).

10. Since $\overrightarrow{AF}=\overrightarrow{AE}+\frac{1}{2}\overrightarrow{ED}$ and $\overrightarrow{BE}=\overrightarrow{BD}+\overrightarrow{DE}=\overrightarrow{DC}+\overrightarrow{DE}=2\overrightarrow{DE}-\overrightarrow{CE}$, we have $\overrightarrow{AF}\cdot\overrightarrow{BE}=2\overrightarrow{AE}\cdot\overrightarrow{DE}+\overrightarrow{ED}\cdot\overrightarrow{DE}-\overrightarrow{AE}\cdot\overrightarrow{CE}-\frac{1}{2}\overrightarrow{ED}\cdot\overrightarrow{CE}$. By $DE\perp AE$ we obtain that $\overrightarrow{AE}\cdot\overrightarrow{DE}=\overrightarrow{ED}\cdot\overrightarrow{CE}=0$. Hence,
$$\overrightarrow{AF}\cdot\overrightarrow{BE}=\overrightarrow{ED}\cdot\overrightarrow{DE}-\overrightarrow{AE}\cdot\overrightarrow{CE}=|\overrightarrow{AE}||\overrightarrow{CE}|-|\overrightarrow{DE}|^2.$$
Also by the property of isosceles triangles, $AD\perp DC$, thus
$$|\overrightarrow{DE}|^2=|\overrightarrow{AE}||\overrightarrow{CE}|.$$
Therefore, $\overrightarrow{AF}\cdot\overrightarrow{BE}=0$, from which $AF\perp BE$.

11. We choose an arbitrary point O as the origin, and use the letter X to denote the vector $\overrightarrow{OX}$. If P lies inside $A_1A_2\ldots A_n$, then we draw an arbitrary line through P. Suppose this line intersects the boundary of the polygon at P_1 and P_2, and let $P=\lambda P_1+(1-\lambda)P_2$, where $0<\lambda<1$. Then
$$\begin{aligned}f(P)&=\sum_{i=1}^{n}|\lambda P_1+(1-\lambda)P_2-A_i|\\&=\sum_{i=1}^{n}|\lambda(P_1-A_i)+(1-\lambda)(P_2-A_i)|\\&\le\lambda\sum_{i=1}^{n}|P_1-A_i|+(1-\lambda)\sum_{i=1}^{n}|P_2-A_i|\\&=\lambda f(P_1)+(1-\lambda)f(P_2)\le\max\{f(P_1),\,f(P_2)\}.\end{aligned}$$
Hence, the maximum value of $f(P)$ is attained on the boundary. Similarly, if P lies on the side A_kA_{k+1}, then by a similar argument, $f(P)\le\max\{f(A_k),\,f(A_{k+1})\}$. Therefore, the maximum value of $f(P)$ is attained when P is a vertex of the polygon.

12. We use $(i,\,j)$ to denote the center of the square in the ith row and the jth column, and let
$$S=\{(i,\,j)\,|\,1\le i,\,j\le n\}.$$
For each $X\in S$, let $\omega(X)$ denote the number in the square X. Let $O(0,\,0)$ be the origin. It suffices to show that
$$\overrightarrow{OP}=\sum_{\substack{X,Y\in S\\ \omega(X)<\omega(Y)}}(\overrightarrow{OY}-\overrightarrow{OX})=\vec{0}.$$

Note that for each $Y \in S$, there are exactly $\omega(Y) - 1$ elements $X \in S$ such that $\omega(X) < \omega(Y)$, and there are exactly $n^2 - \omega(Y)$ elements $X \in S$ such that $\omega(X) > \omega(Y)$. Hence,

$$\overrightarrow{OP} = \sum_{Y \in S} (2\omega(Y) - (n^2+1))\overrightarrow{OY}. \text{ ①}$$

Suppose the sum of the numbers in each column or row equals σ. Then by, ① the x-coordinate of $\overrightarrow{OP}$ satisfies

$$\begin{aligned} x_P &= \sum_{k=1}^{n} \sum_{Y \in T_k} k(2\omega(Y) - (n^2+1)) \\ &= \sum_{k=1}^{n} k(2\sigma - n(n^2+1)) \\ &= n(n+1)\sigma - \frac{n^2(n+1)}{2}(n^2+1) \\ &= n(n+1) \cdot \frac{1+2+\cdots+n^2}{n} - \frac{n^2(n+1)}{2}(n^2+1) \\ &= 0. \end{aligned}$$

Here T_k denotes the squares in the kth column.
Similarly, we can prove that the y-coordinate of $\overrightarrow{OP}$ is also 0, and the desired conclusion follows.

13. We use $\vec{r_1}, \vec{r_2}, \ldots, \vec{r_a}$ to denote the row vectors corresponding to the $a \times b$ subtable that are all 1. (Here we are referring to the row vectors in the original table.)
By the assumption, $\vec{r_i} \cdot \vec{r_j} = \begin{cases} n, & i = j \\ 0, & i \neq, j \end{cases}$, so

$$\sum_{1 \le i,\, j \le a} \vec{r_i} \cdot \vec{r_j} = na.$$

Equivalently, $(\vec{r_1} + \vec{r_2} + \cdots + \vec{r_a})^2 = na$.
Note that the vector $\vec{r_1} + \vec{r_2} + \cdots + \vec{r_a}$ has b components that are equal to a (since the column sums in the $a \times b$ subtable are equal to a). Hence, $(\vec{r_1} + \vec{r_2} + \cdots + \vec{r_a}) \geq ba^2$. Therefore, $ba^2 \leq na$, thus $ab \leq n$.

Solution 19

"Angles" and "Distances" in Spaces

1. $45°$. Hint: Form the line segment CF, let G be the midpoint of AC, and form the line segments EG and FG. Since $EG \parallel SA$ and $EG = \frac{1}{2}SA$, the angle $\angle FEG$ is the angle between SA and EF, so we may solve the problem in ΔEFG.
2. $\frac{13}{5}$. Hint: Let E be the projection of A onto BD, and form the line segment PE. By the theorem of three perpendiculars, PE is exactly the distance that we want to find.
3. 3. Hint: By the definition of the angle between skew lines, if we draw parallel lines $a' \parallel a$ and $b' \parallel b$ such that a' and b' both pass through P, then the two angles formed by a' and b' are $80°$ and $100°$, respectively. If a line forms the same angle with a' and b', then its projection onto the plane spanned by a' and b' must be an angle bisector (unless it is perpendicular to the plane, which is not the case here). If the projection bisects the $80°$ angle, then there are two such lines, and if the projection bisects the $100°$ angle, then there is one such line (it lies inside the plane).
4. 7. Hint: As shown in the figure, let D, E, F, G, and H be the midpoints of their respective edges. Then there are four planes of the type DEF, and three planes of the type $DFHG$. These are all the planes that are equidistant to the four vertices.

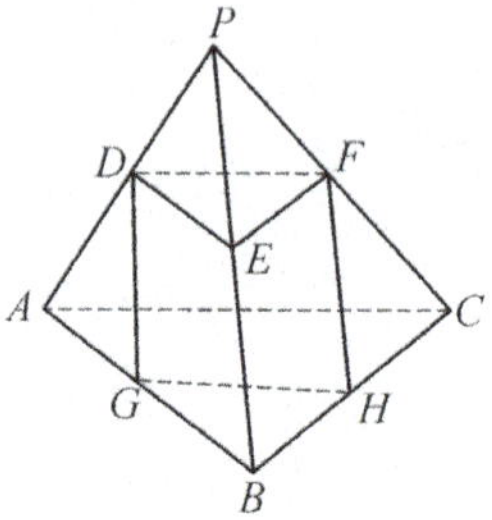

Exercise 4

5. $30°$. Hint: Let O be the midpoint of CD and form the line segments EO and PO. Since $EP \perp PD$ and $EP \perp PC$, we have $EP \perp$ plane PCD. Hence, $\angle POE$ is the plane angle of the dihedral angle $P - CD - E$.
6. $120°$. Hint: The cube has edge length 1. Let E be the projection of B onto A_1C and form the line segment DE. Then by the property of cubes, $\Delta A_1DC \cong \Delta A_1BC$. Hence, $DE \perp A_1C_1$, so $\angle BED$ is the plane angle of the dihedral angle $B - A_1C - D$. Now $BE = DE = \frac{\sqrt{2}}{\sqrt{3}}$ and $BD = \sqrt{2}$. Therefore,

$$\cos \angle BED = \frac{\frac{2}{3} + \frac{2}{3} - 2}{2 \times \frac{\sqrt{2}}{\sqrt{3}} \times \frac{\sqrt{2}}{\sqrt{3}}} = -\frac{1}{2},$$

and the dihedral angle $B - A_1C - D$ equals $120°$.

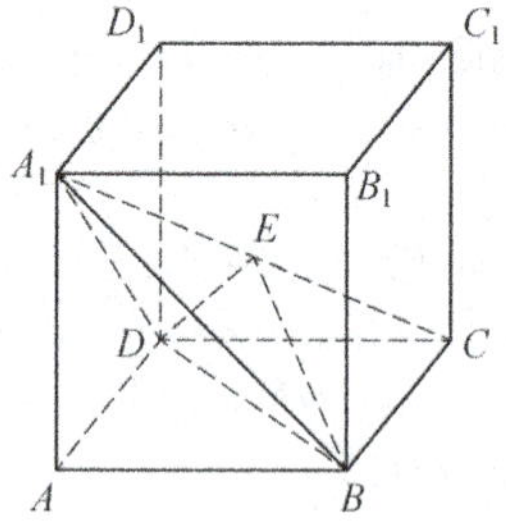

Exercise 6

7. As shown in the figure, since $MN \parallel BC_1$, we have $MN \parallel$ plane ABC_1, and the problem becomes finding the distance d between the line MN

and the plane ABC_1D_1. Form the line segment B_1C, and we have $B_1C \perp BC_1$. Also $D_1C_1 \perp$ plane BB_1C_1C, so $D_1C_1 \perp B_1C$, and hence $B_1C \perp$ plane ABC_1D_1. Therefore, the distance from B_1 to the plane ABC_1D_1 is $\frac{1}{2}CB_1 = \frac{\sqrt{2}}{2}$, thus $d = \frac{\sqrt{2}}{4}$.

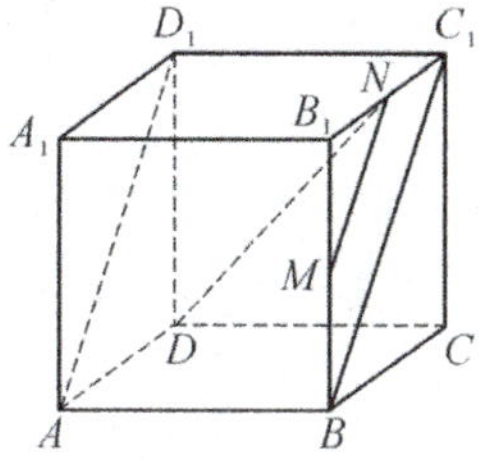

Exercise 7

8. As shown in the figure, since $AC \parallel A_1C_1$, the angle $\angle A_1C_1P$ is exactly what we want. Since the lengths of the three sides of ΔABC are all equal to 1, we have $\angle A_1AB = \angle ABC = \angle BAC = 60°$. Hence, ΔAA_1B is an equilateral triangle with $A_1B = 1$ and ΔA_1AC is an equilateral triangle with $A_1C = 1$. Let M be the midpoint of BC and N be the midpoint of B_1C_1. By a similar argument, we see that ΔA_1BC is an equilateral triangle, so $A_1M \perp BC$. Since $\Delta A_1B_1C_1$ is also equilateral, $A_1N \perp BC$. Combining the above with $BC \parallel B_1C_1$, we have $B_1C_1 \perp$ plane A_1MN, thus $B_1C_1 \perp MN$. Also $MN \parallel CC_1$, hence $B_1C_1 \perp CC_1$ and $\angle BCC_1 = 90°$. This implies that $BC_1 = \sqrt{2}$, and combining it with $A_1B = A_1C_1 = 1$, we get $\angle C_1A_1B = 90°$. Therefore, $\tan \angle A_1C_1P = \frac{\sqrt{3}}{3}$, from which $\angle A_1C_1P = 30°$.

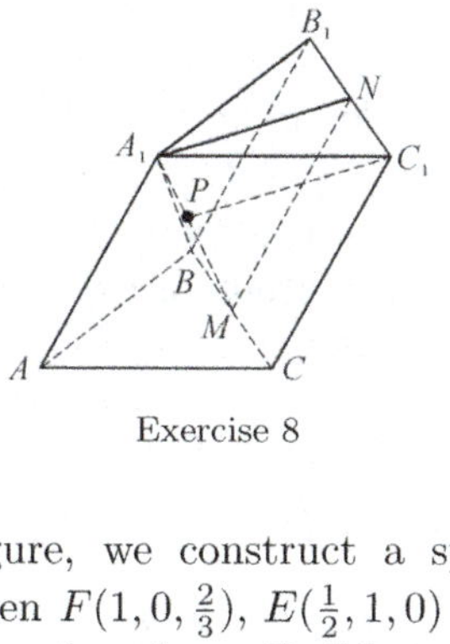

Exercise 8

9. As shown in the figure, we construct a space coordinate system. Denote $A(1,0,0)$. Then $F(1,0,\frac{2}{3})$, $E(\frac{1}{2},1,0)$ and $B_1(1,1,1)$. Suppose the normal vector to the plane B_1EF is $\vec{n} = (x_0, y_0, z_0)$. Since

$\vec{EF} = (\frac{1}{2}, -1, \frac{2}{3})$ and $\vec{EB_1} = (\frac{1}{2}, 0, 1)$,

$$\begin{cases} \vec{n} \cdot \vec{EF} = 0, \\ \vec{n} \cdot \vec{EB_1} = 0. \end{cases} \quad \text{or equivalently} \quad \begin{cases} \frac{1}{2}x_0 - y_0 + \frac{2}{3}z_0 = 0, \\ \frac{1}{2}x_0 + z_0 = 0. \end{cases}$$

Hence, $\vec{n} = (-2, -\frac{1}{3}, 1)$.

Note that a normal vector to the plane $A_1B_1C_1D_1$ is $\vec{m} = (0, 0, 1)$, so

$$\cos\langle \vec{m}, \vec{n} \rangle = \frac{\vec{m} \cdot \vec{n}}{|\vec{m}| \cdot |\vec{n}|} = \frac{1}{1 \times \sqrt{5 + \frac{1}{9}}} = \frac{3}{\sqrt{46}}.$$

Therefore, the dihedral angle between the plane face B_1EF and the face $A_1B_1C_1D_1$ equals $\arccos \frac{3\sqrt{46}}{46}$.

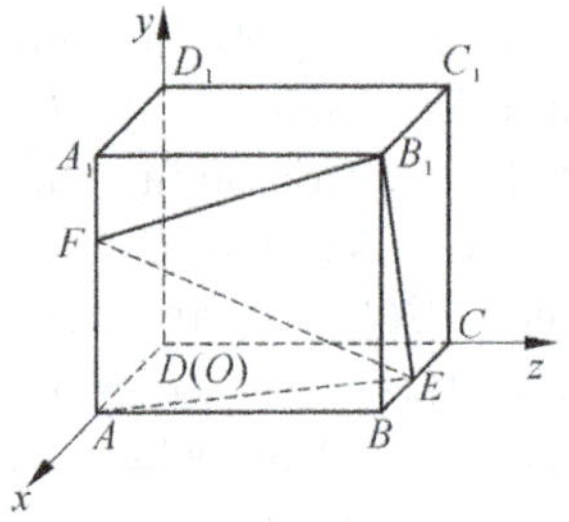

Exercise 9

10. As shown in the figure, let F be the midpoint of BD, and form the line segment EF. Then $EF \parallel CD$. Thus, the distance between CD and SE is equal to the distance from CD to the plane SEF, and it is equivalent to finding the altitude of the tetrahedron $C - SEF$ above the face SEF. On one hand,

$$\begin{aligned} V_{S-CEF} &= \frac{1}{3}SC \cdot S_{\Delta CEF} = \frac{1}{3}SC \cdot S_{\Delta BEF} \\ &= \frac{1}{3} \times 2 \times \frac{1}{2} \times 2\sqrt{2} \times \sqrt{2} \times \sin 60^\circ = \frac{2\sqrt{3}}{3}. \end{aligned}$$

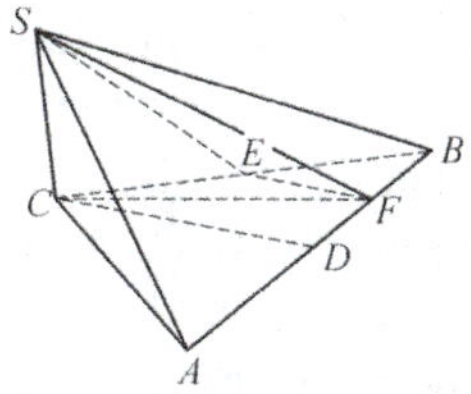

Exercise 10

On the other hand, let h be the distance from C to the face SEF. Then

$$V_{C-SEF} = \frac{1}{3}h \cdot S_{\Delta SEF}.$$

In the triangle SEF,

$$SE = \sqrt{4 + (2\sqrt{2})^2} = 2\sqrt{3},\ EF = \frac{1}{2}CD = \frac{1}{2} \times 4\sqrt{2} \times \frac{\sqrt{3}}{2} = \sqrt{6},$$

$$\begin{aligned} SF &= \sqrt{SD^2 + DF^2} = \sqrt{SA^2 - AD^2 + DF^2} \\ &= \sqrt{2^2 + (4\sqrt{2})^2 - (2\sqrt{2})^2 + (\sqrt{2})^2} = \sqrt{30}. \end{aligned}$$

Hence,

$$\cos\angle SEF = \frac{12 + 6 - 30}{2 \times 2\sqrt{3} \times \sqrt{6}} = -\frac{\sqrt{2}}{2},$$

so $\frac{2\sqrt{3}}{3} = \frac{h}{3} \cdot \frac{1}{2} \times 2\sqrt{3} \times \sqrt{6} \times \frac{\sqrt{2}}{2}$, which gives that $h = \frac{2\sqrt{3}}{3}$.

11. Let E be the midpoint of CC_1. Then $DE \parallel AC_1$, so the problem becomes finding the distance from AC_1 to the plane BDE. We apply the volume method to the tetrahedron $A-BDE$. Let d be the distance from A to the plane BDE. Then

$$\frac{1}{3}d \cdot S_{\Delta BDE} = \frac{1}{3} \times 4 \times S_{\Delta ABD} = \frac{1}{3} \times 18\sqrt{3}.$$

Note that in the triangle BDE, we have $DE = 5$, $BD = 3\sqrt{3}$, and $BE = 2\sqrt{13}$, thus

$$\cos\angle BDE = \frac{25 + 27 - 52}{30\sqrt{3}} = 0.$$

Hence, $S_{\Delta BDE} = \frac{15}{2}\sqrt{3}$. Plugging into the equation above, we get $d = \frac{15}{2}$, which is the distance that we wanted to find.

12. (1) Without loss of generality, assume that the edge lengths of the tetrahedron are all equal to 1. (In fact, it is easy to show that this is a regular tetrahedron.) Form the line segment DF, let G be the midpoint of DF, and form the line segment EG. Then $EG \parallel AF$. Also form the line segment CG (as shown in the figure). Thus, $AF = CE = \frac{\sqrt{3}}{2}$ and $EG = \frac{1}{2}AF = \frac{\sqrt{3}}{4}$. Further,

$$CG^2 = CF^2 + FG^2 = \frac{1}{4} + \frac{3}{16} = \frac{7}{16}.$$

so $\cos \angle CEG = \frac{\frac{3}{4}+\frac{3}{16}-\frac{7}{16}}{2\times\frac{\sqrt{3}}{2}\times\frac{\sqrt{3}}{4}} = \frac{2}{3}$. Therefore, the angle between AF and CE is $\arccos\frac{2}{3}$.

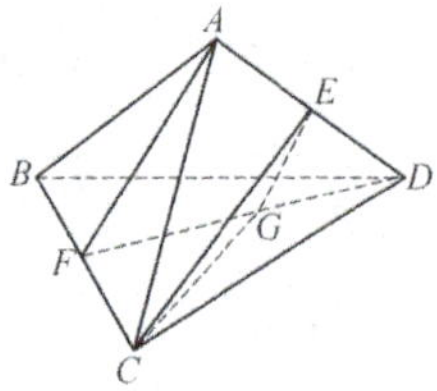

Exercise 12

(2) In order to find the angle between CE and the face BCD, let H be the projection of E onto the plane BCD. Then H lies on DF. We apply the area method to ΔEFD, so

$$\frac{1}{2}\cdot DE\cdot\sqrt{AF^2 - AE^2} = \frac{1}{2}EH\cdot DF\frac{1}{2}\cdot\frac{1}{2}\cdot\sqrt{\frac{3}{4}-\frac{1}{4}}$$
$$= \frac{1}{2}EH\cdot\frac{\sqrt{3}}{2}.$$

Hence, $EH = \sqrt{\frac{1}{6}}$, from which

$$\sin\angle ECH = \frac{EH}{CE} = \sqrt{\frac{1}{6}}\cdot\frac{2}{\sqrt{3}} = \frac{\sqrt{2}}{3}.$$

Therefore, the angle between CE and the face BCD is $\arcsin\frac{\sqrt{2}}{3}$.

13. Let LM be the common perpendicular line segment between l and m, with $L \in l$ and $M \in m$. Let P be a plane passing through M and parallel to l. Let G, H, and K be the projections of A, B, and C onto the plane P, respectively. Then G, H, and K all lie on a line l' that is parallel to l. Form the line segments CD, HE, and KF.

    Since $CK \perp$ plane P, $BH \perp$ plane P, $AG \perp$ plane P, and $AB = BC$, we have $HK = GH$. Also since CF, BE, and AD are perpendicular to m, the lines FK, EH, and DG are also perpendicular to m. Hence, $KF \parallel EH \parallel GD$ and $2EH = FK + DG$.

    Let the length of the common perpendicular line segment be x. Then $LM = CK = BH = AG = x$, $EH = \sqrt{\frac{49}{4} - x^2}$, $FK = \sqrt{10 - x^2}$, and $DG = \sqrt{15 - x^2}$. Thus,

$$2\sqrt{\frac{49}{4} - x^2} = \sqrt{15 - x^2} + \sqrt{10 - x^2}.$$

    Solving the equation gives $x = \sqrt{6}$ (here the negative solution is dropped). Therefore, the distance between l and m is $\sqrt{6}$.

    (If A, B, and C do not lie on the same side of L, then $2\sqrt{\frac{49}{4} - x^2} = \sqrt{15 - x^2} - \sqrt{10 - x^2}$, which has no solution. Hence, A, B, and C have to lie on the same side of L.)

14. Since PA, PB, and PC are pairwise perpendicular, we may construct a cuboid with edges PA, PB, and PC. This cuboid has the same circumscribed sphere as the triangular pyramid $P - ABC$, and the body diagonal of the cuboid (also a diameter of its circumscribed sphere) equals 2. Now, let $PA = x$, $PB = y$; and $PC = z$. Then $x^2 + y^2 + z^2 = 4$.

    (1) We have

$$\begin{aligned} V_{P-ABC} &= \frac{1}{6}xyz = \frac{1}{6}(x^2y^2z^2)^{\frac{1}{2}} \\ &\le \frac{1}{6}\left(\left(\frac{x^2 + y^2 + z^2}{3}\right)^3\right)^{\frac{1}{2}} = \frac{4}{27}\sqrt{3}, \end{aligned}$$

    where the equality holds when $x = y = z = \frac{2\sqrt{3}}{3}$. Hence, the maximal volume of the tetrahedron $P - ABC$ is $\frac{4}{27}\sqrt{3}$.

(2) Note that

$$h_{P-ABC} = \frac{3V_{P-ABC}}{S_{\Delta ABC}} = \frac{\frac{1}{2}xyz}{\frac{1}{2}\sqrt{x^2y^2+y^2z^2+z^2x^2}}$$

$$= \frac{1}{\sqrt{\frac{1}{x^2}+\frac{1}{y^2}+\frac{1}{z^2}}} = \frac{2}{\sqrt{(x^2+y^2+z^2)\left(\frac{1}{x^2}+\frac{1}{y^2}+\frac{1}{z^2}\right)}}$$

$$\leq \frac{2}{\sqrt{9}} = \frac{2}{3}.$$

where the equality holds when $x = y = z = \frac{2\sqrt{3}}{3}$. Therefore, the maximum value of the distance from P to the plane ABC is $\frac{2}{3}$.

15. If A and B lie on the different sides of α, then P is the intersection point of AB and α. In this case, $\angle APB = 180°$.
Suppose A and B lie on the same side of α, and the line AB intersects α at O. We consider the following cases.

(1) If $AB \perp \alpha$, then the trajectory of P is the circle centered at O with radius $\sqrt{OA \cdot OB}$.
(2) If AB is not perpendicular to α, then there is a unique plane β passing through AB and perpendicular to α. Let l be the intersection line of α and β. Then the circle centered at O with radius $\sqrt{OA \cdot OB}$ has two intersection points with l, which we denote as M and N. If $\angle AOM$ is acute, then M is the point P, otherwise N is the point P.
If $AB \parallel \alpha$, then P is the midpoint of the projection $A'B'$ of AB onto α.

To summarize, the angle $\angle APB$ is maximized when A, P, and B are collinear with P lying between A and B, or when the circle passing through A, P, and B is tangent to the line l.

Now, we prove the above claim.

If there is a point P' on l such that $\angle AP'B > \angle APB$ and P' does not lie on the circle determined by A, P, and B, then $P'A$ and $P'B$ have at least one intersection point K with the circle. Since $\angle AKB = \angle APB$ and $\angle AKB > \angle AP'B$, so $\angle APB > \angle AP'B$, which is a contradiction.

If there is a point P'' inside α and outside l such that $\angle AP''B > \angle APB$, then we form the line segments AP'' and BP'', and rotate the plane determined by these three points with the axis AB until it coincides with β, with P'' moving to Q. It follows that $\angle AQB = \angle AP''B$. Since Q does not lie on the circle, $\angle APB > \angle AQB = \angle AP''B$, a contradiction.

Therefore, $\angle APB$ is the maximal angle.

Solution 20

Cross Section, Folding, and Unfolding

1. $\frac{\sqrt{6}}{2}$. Hint: Suppose the cross section intersects AA_1 and CC_1 at E and F, respectively. Then $S_{\text{cross}} = S_{\Delta BD_1E} + S_{\Delta BD_1F} = 2S_{\Delta BD_1E}$. The minimal distance from a point E on AA_1 to BD_1 is the distance between the skew lines AA_1 and BD_1 (it is the length of their common perpendicular line segment), and the distance equals $\frac{\sqrt{2}}{2}$. Hence, $S_{\text{cross}} = 2 \times \frac{1}{2} \times \sqrt{3} \times \frac{\sqrt{2}}{2} = \frac{\sqrt{6}}{2}$.
2. $\sqrt{10}$. Hint: Let O be the projection of H onto EG, and form the line segment OF. We apply the law of cosines to the triangle EFO, and get $FO = \sqrt{7}$. Then by the area formula, $OH = \sqrt{3}$, so $FH = \sqrt{10}$.
3. $\frac{\sqrt{3}}{3}$. Hint: Apparently ΔACD is an equilateral triangle with side length 2. Let F be the midpoint of CD. Then $AF = \sqrt{3}$, $EF = 2$, and $AE = 1$, and CD is perpendicular to the plane AEF. By Pythagorean theorem, ΔAEF is a right triangle, and its area is $S = \frac{1}{2}AE \cdot AF = \frac{\sqrt{3}}{2}$. The cross section AEF cuts the tetrahedron into two parts with an equal volume, so

$$V = 2 \cdot \frac{1}{3} S \cdot CF = 2 \cdot \frac{1}{3} \cdot \frac{\sqrt{3}}{2} \cdot 1 = \frac{\sqrt{3}}{3}.$$

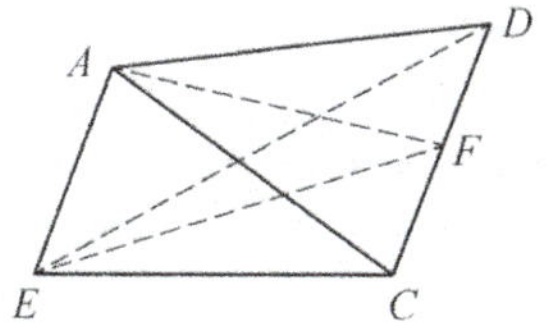

Exercise 3

4. $(0, \sqrt{2}a^2]$. Hint: The minimal area of the quadrilateral tends to 0 and the maximum value is attained when it passes through a pair of opposite edges.

5. $S_{\max} = \begin{cases} \frac{1}{2}l^2, & \alpha \in \left[\frac{\pi}{2}, \pi\right) \\ \frac{1}{2}l^2 \sin\alpha, & \alpha \in \left(0, \frac{\pi}{2}\right) \end{cases}$.

6. $\frac{\sqrt{3}}{4}a$. Hint: Let O be the midpoint of BD and form the line segments OA and OC. Then $BD \perp$ plane AOC. Let E be the midpoint of AC. Then OE is the distance between AC and BD. Since $\angle AOC$ is the plane angle of the dihedral angle $A - BD - C$, we have $OE = \frac{\sqrt{3}}{4}a$.

7. As shown in the figure, $BA \perp AD$, so $S_{\Delta ABD} = \frac{1}{2}AB \cdot AD = a \cdot AD$. Since AD is minimal when $AD \perp PC$, the minimal area of ΔABD is

$$a \cdot \frac{PA \cdot AC}{PC} = \frac{3}{10}\sqrt{10}a^2.$$

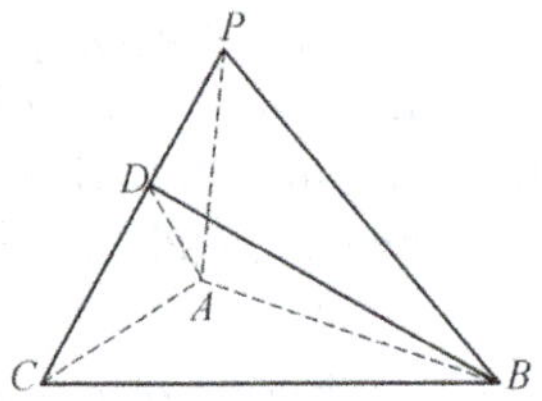

Exercise 7

8. Let V denote the volume of the triangular frustum. Then it follows that $V = V_{F-ABC} + V_{B-EFA} + V_{A-EFD}$. Note that $BC \parallel EF$, so the distance from BC to the base plane AEF is exactly the altitude of the triangular frustum, which we denote as h. By the given condition,

$$\frac{1}{3}h(a^2 + b^2 + ab) = \frac{1}{3}h \cdot b^2 + \frac{1}{3}h \cdot S_{\Delta EFA} + \frac{1}{3}h \cdot a^2.$$

Hence, $S_{\Delta EFA} = ab$.

9. As shown in the figure, suppose the side length of the square is 4 and AC is its diagonal. Then

$$AE = EF = \sqrt{2},\ AF = \sqrt{2 + 2^2} = \sqrt{6}.$$

Hence, $\cos\angle AEF = \frac{2+2-6}{2\cdot\sqrt{2}\cdot\sqrt{2}} = -\frac{1}{2}$, so $\angle AEF = 120°$.

Also it is easy to see that AE and FG are perpendicular. Therefore, the angle between any two of these line segments is $120°$ or $90°$.

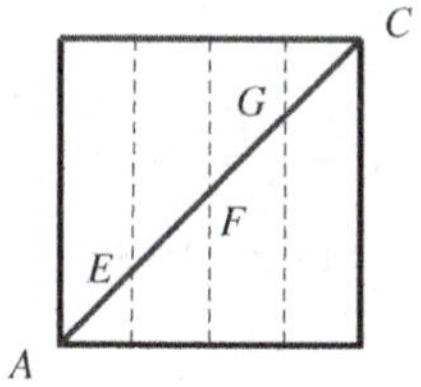

Exercise 9(1)

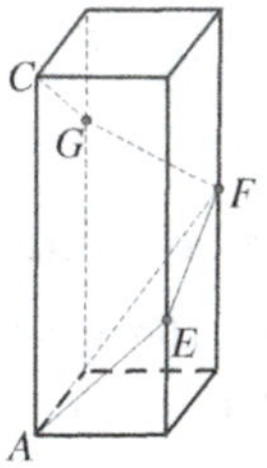

Exercise 9(2)

10. As shown in the figure, suppose the bottom face of the cone (a circle) has radius R and the generatrix is l. Let α denote the angle between the generatrix and the bottom face and let r be the radius of the inscribed sphere. Then the condition implies that $\pi R(R+l) = n \cdot 4\pi r^2$. In ΔO_1OB we have $r = R\tan\frac{\alpha}{2}$ and in ΔBOC we have $l = \frac{R}{\cos\alpha}$. Plugging them into the equation above and cancelling πR^2, we get

$$1 + \frac{1}{\cos\alpha} = 4n\tan^2\frac{\alpha}{2}.$$

Since $\cos\alpha = \frac{1-\tan^2\frac{\alpha}{2}}{1+\tan^2\frac{\alpha}{2}}$,

$$\frac{2}{1-\tan^2\frac{\alpha}{2}} = 4n\tan^2\frac{\alpha}{2}.$$

Let $\tan\frac{\alpha}{2} = z$. Then

$$z^4 - z^2 + \frac{1}{2n} = 0.$$

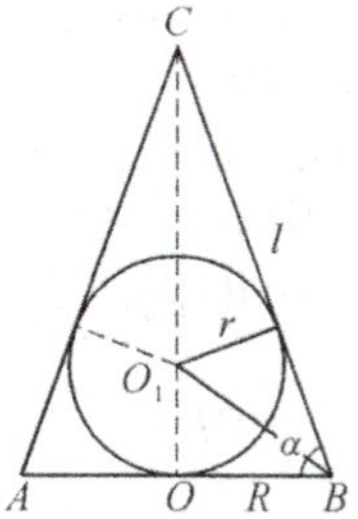

Exercise 10

Hence, when $\Delta = 1^2 - 4 \times \frac{1}{2n} = 1 - \frac{2}{n} \geq 0$, or equivalently $n \geq 2$,

$$z^2 = \frac{1}{2} \pm \sqrt{\frac{1}{4} - \frac{1}{2n}} > 0.$$

Since z needs to be positive, $z_{1,2} = \sqrt{\frac{1}{2} \pm \sqrt{\frac{1}{4} - \frac{1}{2n}}}$ (here $z_1 = z_2$ if $n = 2$). Note that the solution also needs to satisfy $0 < \alpha < 90°$, hence $0 < z = \tan\frac{\alpha}{2} < 1$. Since

$$0 < \frac{1}{2} + \sqrt{\frac{1}{4} - \frac{1}{2n}} < \frac{1}{2} + \sqrt{\frac{1}{4}} = 1,\ 0 < \frac{1}{2} - \sqrt{\frac{1}{4} - \frac{1}{2n}} < \frac{1}{2},$$

both solutions satisfy this condition. Therefore the value of n can be any integer greater than or equal to 2.

11. Let π be the cross section of S cut by the plane spanned by AB and AC. Then π is a circle. Suppose the circle intersects AB at X_1 and Y_1, and intersects AC at X_2 and Y_2. Then $AX_1 \cdot AY_1 = AX_2 \cdot AY_2$, so $\frac{2}{9}AB^2 = \frac{2}{9}AC^2$ and $AB = AC$. Thus, we can prove that every edge of the polyhedron has an equal length. Further, let O be the center of S. Then $OX_1 = OY_1 = R$ (here R is the radius of the sphere). Hence, the midpoint Z_1 of AB is also the midpoint of AB, so $OZ_1 \perp AB$, and

$$OZ_1 = R^2 - \left(\frac{1}{2}X_1Y_1\right)^2 = R^2 - \frac{1}{36}AB^2.$$

Combining the above two results, we see that the line segment between O and the midpoint of each edge has an equal length, and is perpendicular to the edge. Therefore, the sphere centered at O with the radius $r = \sqrt{R^2 - \frac{1}{36}AB^2}$ is tangent to every edge.

12. Let D be the midpoint of BC, and AM intersects the plane PBC at K. Then K lies on PD. Suppose the line passing through K and parallel to BC intersects PB and PC at E and F, respectively. Then Menelaus' theorem implies that $\frac{PK}{KD} \cdot \frac{DA}{AO} \cdot \frac{OM}{MP} = 1$, so $\frac{PK}{KD} = \frac{2}{3}$. Hence,

$$\frac{PE}{PB} = \frac{PF}{PC} = \frac{PK}{PD} = \frac{2}{5},\ \frac{V_{P-AEF}}{V_{P-ABC}} = \frac{4}{25}.$$

Therefore, the volume ratio between the upper and lower parts is 4:21.

13. As shown in the figure, suppose A_1BC_1D is the tetrahedron, and we extend it to a cuboid $ABCD-A_1B_1C_1D_1$. Then the three edges on one face of the tetrahedron are exactly the diagonals of three nonparallel faces of the cuboid. Let $AB = \alpha$, $AD = \beta$, $AA_1 = \gamma$, $A_1D = a$, $A_1B = b$, and $DB = c$. Then $\alpha^2 + \beta^2 = c^2$, $\beta^2 + \gamma^2 = a^2$, and $\gamma^2 + \alpha^2 = b^2$. Hence,

$$\alpha^2 = \frac{1}{2}(b^2 + c^2 - a^2),\ \beta^2 = \frac{1}{2}(a^2 + c^2 - b^2),$$
$$\gamma^2 = \frac{1}{2}(a^2 + b^2 - c^2),$$

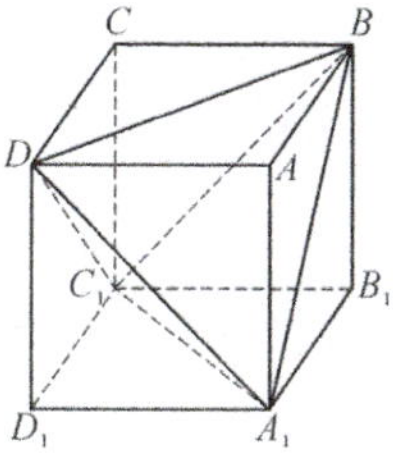

Exercise 13

from which

$$\begin{aligned} V_{A_1BC_1D} &= V_{ABCD-A_1B_1C_1D_1} - V_{AA_1BD} - V_{BCC_1D} - V_{A_1BB_1C_1} \\ &\quad - V_{A_1C_1DD_1} \\ &= \alpha\beta\gamma - 4 \times \frac{1}{3} \times \frac{1}{2}\alpha\beta\gamma = \frac{1}{3}\sqrt{\alpha^2\beta^2\gamma^2} \\ &= \frac{1}{12}\sqrt{2(a^2 + b^2 - c^2)(a^2 + c^2 - b^2)(b^2 + c^2 - a^2)}. \end{aligned}$$

14. Let M be the midpoint of BC, and AM intersects EF at N. Since ΔABC is equilateral and $EF \parallel BC$, we have $AM \perp BC$ and $AN \perp EF$, and N is the midpoint of EF. Note that after folding we still have $A'N \perp EF$.

Since $A' - EF - M$ is a right dihedral angle, $A'N \perp$ plane $BCFE$, as well as $A'N \perp MN$. Also since $BC \perp MN$, we see that MN is the common perpendicular line segment between $A'N$ and BC.
Let $MN = x$. Then $A'N = AN = \frac{\sqrt{3}}{2}a - x$, and by the distance formula of two points on skew lines,

$$A'B^2 = MN^2 + A'N^2 + BM^2 - 2A'N \cdot BM \cos 90^\circ$$

$$= x^2 + \left(\frac{\sqrt{3}}{2}a - x\right)^2 + \left(\frac{a}{2}\right)^2 = 2x^2 - \sqrt{3}ax + a^2$$

$$= 2\left(x - \frac{\sqrt{3}}{4}a\right)^2 + \frac{5}{8}a^2.$$

Therefore, when $x = \frac{\sqrt{3}}{4}a$, the length of $A'B$ attains the minimum value $\frac{\sqrt{10}}{4}a$, in which case E and F are the midpoints of AB and AC, respectively.

Solution 21

Projections and the Area Projection Theorem

1. $\angle CEB > \angle DEB$. Hint: Since $\cos\angle CED \cdot \cos\angle DEB = \cos\angle CEB$, we have $\cos\angle DEB > \cos\angle CEB$, so $\angle DEB < \angle CEB$.
2. Orthocenter. Hint: By the theorem of three perpendiculars, the projection of each side edge onto the bottom face is an altitude of the bottom triangle.
3. $\frac{\pi}{3}$. Hint: Let O be the projection of A onto BC and form the line segment OD. By the theorem of three perpendiculars, $\angle AOD$ is the plane angle of the dihedral angle $A - BC - D$. Let $AD = x$. Then $AB = 2x$ and $AC = \sqrt{2}x$, so $BC = \sqrt{6}x$ and $AO = \frac{2}{\sqrt{3}}x$. Thus, $\sin\angle AOD = \frac{AD}{AO} = \frac{\sqrt{3}}{2}$, from which $\angle AOD = \frac{\pi}{3}$.
4. 1. Hint: As shown in the figure, let A_1 be the projection of A onto l and let B_1 be the projection of B onto l. We find a point B_2 such that AB_1BB_2 is a parallelogram, and form the line segments A_1B_2 and AB_2. Then $\cos^2 x + \cos^2 y - \cos^2 z = \frac{AB_1^2}{AB^2} + \frac{A_1B^2}{AB^2} - \frac{A_1B_1^2}{AB^2} = 1$.

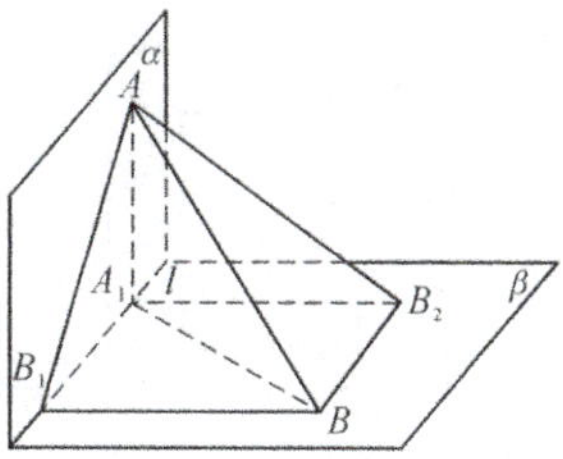

Exercise 4

5. $\sqrt{3}$.
6. $2\sqrt{33}$. Hint: By the condition, $EF \parallel BC$. Since the four faces of the regular tetrahedron $ABCD$ are congruent equilateral triangles, $AE = AF = EF = 4$ and $AD = AB = AE + BE = 7$. By the law of cosines,

$$DE = \sqrt{AD^2 + AE^2 - 2AD \cdot AE \cdot \cos 60°}$$
$$= \sqrt{49 + 16 - 28} = \sqrt{37}.$$

Similarly, $DF = \sqrt{37}$. Now, let DH be an altitude of ΔDEF. Then $EH = \frac{1}{2}EF = 2$, from which

$$DH = \sqrt{DE^2 - EH^2} = \sqrt{33},$$

and $S_{\Delta DEF} = \frac{1}{2}EF \cdot DH = 2\sqrt{33}$.
7. Note that the projection of P onto the bottom face has an equal distance to the three sides of ΔABC (it is the altitude times $\cot\theta$), and the result follows.
8. Let S be the area of the cross section. Since the projection of the cross section onto the bottom face is the pentagon $CBMND$, we have $S_{CBMND} = \frac{7}{8}a^2$.

Suppose the cross section forms an angle α with the bottom face. Then $\tan\alpha = \frac{\frac{a}{2}}{\frac{3}{4}\sqrt{2}a} = \frac{2}{3\sqrt{2}}$, so $\cos\alpha = \frac{3}{\sqrt{11}}$. By the area projection theorem, $\cos\alpha = \frac{S_{CBMND}}{S}$, hence

$$S = \frac{\frac{7}{8}a^2}{\frac{3}{\sqrt{11}}} = \frac{7\sqrt{11}}{24}a^2.$$

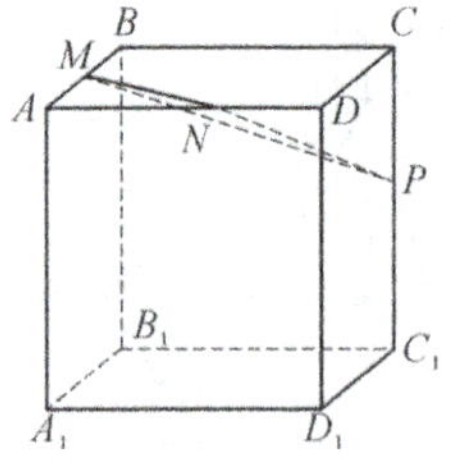

Exercise 8

9. $S_{\text{top}} = \frac{\sqrt{3}}{4} \times 36 = 9\sqrt{3}$ and $S_{\text{bottom}} = \frac{\sqrt{3}}{4} \times 100 = 25\sqrt{3}$. Suppose the side face and the bottom face form an angle θ. Then by the area projection theorem, $\cos\theta = \frac{S_{\text{bottom}} - S_{\text{top}}}{S_{\text{side}}}$, so $\frac{\sqrt{2}}{2} = \frac{16\sqrt{3}}{S_{\text{side}}}$, from which $S_{\text{side}} = 16\sqrt{6}$. In summary, the total surface area is $16\sqrt{6} + 34\sqrt{3}$.

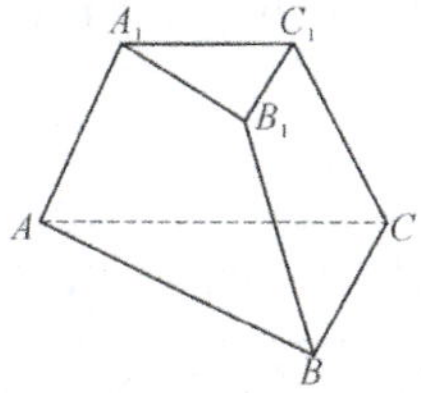

Exercise 9

10. Let $AP = x$ and let P_1 be the projection of P onto the face $A_1B_1C_1D_1$. Then P_1 lies on A_1C_1 and $P_1A_1 = x$. In $\Delta P_1A_1B_1$, the distance from P_1 to A_1B_1 is $\frac{1}{\sqrt{2}}x$, so $\tan\alpha = \frac{\sqrt{2}}{x}$. Similarly, $\tan\beta = \frac{\sqrt{2}}{\sqrt{2}-x}$, hence $\tan(\alpha+\beta) = \frac{2}{x(\sqrt{2}-x)-2}$. Note that $x(\sqrt{2}-x) \leq (\frac{\sqrt{2}}{2})^2 = \frac{1}{2}$ for $0 < x < \sqrt{2}$, where the equality holds when $x = \frac{\sqrt{2}}{2}$. Thus,

$$\tan(\alpha+\beta) \geq -\frac{4}{3}.$$

Therefore, the minimum value of $\tan(\alpha+\beta)$ is $-\frac{4}{3}$, attained when P is the midpoint of AC.

11. Since the projection of D onto the face ABC is the orthocenter, the projection of BD is then an altitude of ΔABC, so by the theorem of three perpendiculars, $BD \perp AC$. Since $BD \perp CD$, we have $BD \perp \text{plane } ACD$. Hence, $BD \perp AD$, and similarly $CD \perp AD$. Let $AD = x$, $BD = y$ and $CD = z$. Then by Pythagorean theorem,

$$AB = \sqrt{x^2+y^2},\ AC = \sqrt{x^2+z^2},\ BC = \sqrt{y^2+z^2},$$

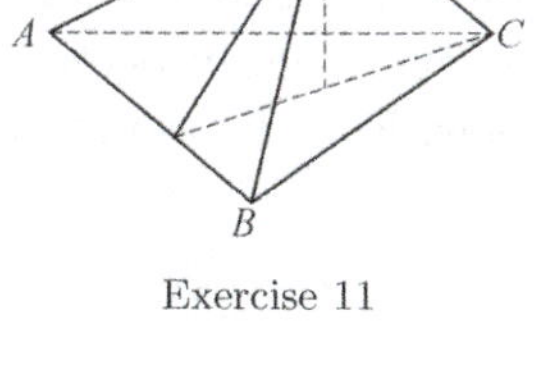

Exercise 11

and it suffices to show that

$$(\sqrt{x^2+y^2}+\sqrt{x^2+z^2}+\sqrt{y^2+z^2})^2 \le 6(x^2+y^2+z^2).$$

This follows by Cauchy's inequality, since

$$\begin{aligned}
&6(x^2+y^2+z^2)\\
&\quad = (1+1+1)((\sqrt{x^2+y^2})^2+(\sqrt{x^2+z^2})^2+(\sqrt{y^2+z^2})^2)\\
&\quad \ge (\sqrt{x^2+y^2}+\sqrt{x^2+z^2}+\sqrt{y^2+z^2})^2.
\end{aligned}$$

Therefore, the proposition is true.

12. Let Q be the projection of P onto the bottom face. Then Q is the incenter of ΔABC. Let r be the radius of the incircle of ΔABC. Then since the dihedral angle between any side face and the bottom face is $45°$, we have $PQ = r$. Further, suppose ΔABC has the side lengths $a < b < c$. Then $a = b-2$ and $c = b+2$, and b is an even number. In particular, $b-2 > 0$ and $b-2+b > b+2$, from which $b \ge 6$. Since ΔABC is an obtuse triangle, $b^2+(b-2)^2 < (b+2)^2$, which implies that $b < 8$, thus $b = 6$. By Heron's formula,

$$S_{\Delta ABC} = \sqrt{p(p-a)(p-b)(p-c)} = 3\sqrt{15},$$

where $p = \frac{1}{2}(a+b+c)$. Since $rp = S_{\Delta ABC}$, we get $PQ = r = \frac{1}{3}\sqrt{15}$. Therefore,

$$V_{P-ABC} = \frac{1}{3}PQ \cdot S_{\Delta ABC} = 5.$$

13. Without loss of generality, assume that π passes through AC, and let S', B', and D' be the projections of S, B, and D onto π. Since plane $SBD \perp AC$, the points S', B', and D' are collinear, and $S'D' \perp AC$. Also since $ABCD$ is a square, its projection onto π is a rhombus. As shown in the figure below, let O be the center of $ABCD$. Then

$$OB' = OB \cdot \cos\alpha = \sqrt{2}\cos\alpha$$

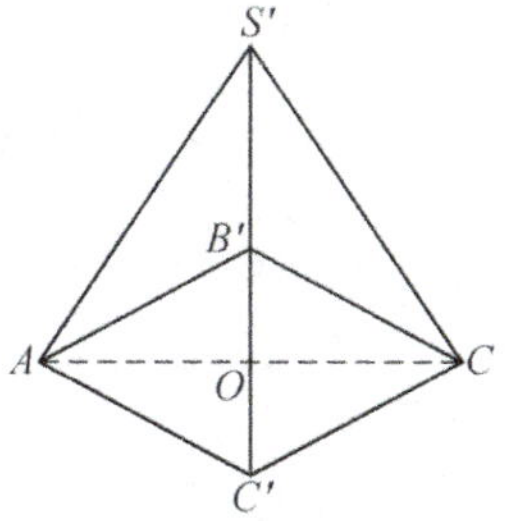

Exercise 13

and $S'O = SO \cdot \cos\left(\frac{\pi}{2} - \alpha\right) = h \sin\alpha$. If $S'O \le B'O$, then the projection is $AB'CD'$, whose area is $4\cos\alpha \le 4$ and the equality holds when $\alpha = 0$. If $S'O > B'O$, then the area of the projection is

$$\begin{aligned} S_{\Delta S'AC} + S_{\Delta AD'C} &= 2\cos\alpha + \sqrt{2}h\sin\alpha \\ &= \sqrt{2h^2+4}\sin(\alpha+\varphi) \le \sqrt{2h^2+4}. \end{aligned}$$

Therefore, the maximal area of the projection is $\max\{4, \sqrt{2h^2+4}\}$. In other words, if $h < \sqrt{6}$, then the maximal area is 4, attained when $\alpha = 0$; if $h > \sqrt{6}$, then the maximal area is $\sqrt{2h^2+4}$, attained when $\alpha = \frac{\pi}{2} - \arctan\frac{\sqrt{2}}{h}$; if $h = \sqrt{6}$, then the maximal area is 4, attained when $\alpha = 0$ or $\frac{\pi}{6}$.

14. Suppose the bottom face is $A_1A_2\cdots A_n$ and O is its center. Assume that $OA_1 = 1$ and the altitude $SO = x$ (here S is the apex). Then

$$\tan\beta = x, \ \tan\alpha = \frac{2}{\cos\frac{\pi}{n}}$$

Hence,

$$\begin{aligned} \sin^2\alpha - \sin^2\beta = \cos^2\beta - \cos^2\alpha &= \frac{1}{\sec^2\beta} - \frac{1}{\sec^2\alpha} \\ &= \frac{1}{1+x^2} - \frac{1}{1+\left(\dfrac{x}{\cos\dfrac{\pi}{n}}\right)^2}. \end{aligned}$$

Let $u = \sin^2\alpha - \sin^2\beta$, $y = x^2$, and $t = \cos^2\frac{\pi}{n}$. Then $u = \frac{1}{1+y} - \frac{t}{t+y}$, so $uy^2 + [(1+t)u + t - 1]y + tu = 0$. This is a quadratic equation for

y, which has a positive real root. Hence,

$$\begin{cases} (tu+u+t-1)^2 - 4tu^2 \geq 0 \\ tu+u+t-1 < 0 \end{cases},$$

thus $u < \frac{1-t}{1+t}$ and $u^2 - \frac{2(1+t)}{1-t}u + 1 \geq 0$. These imply that

$$u \leq \frac{1+t-2\sqrt{t}}{1-t},$$

hence $u \leq \frac{1-\sqrt{t}}{1+\sqrt{t}}$. Therefore,

$$\sin^2\alpha - \sin^2\beta \leq \frac{1-\cos\frac{\pi}{n}}{1+\cos\frac{\pi}{n}} = \tan^2\frac{\pi}{2n}.$$

Solution 22

Partitions of Sets

1. (1) Suppose the sum of the numbers in each T_i is k. Clearly the sum of the numbers in T is pk. Then $1+2+\cdots+n = pk$, so $2pk = n(n+1)$. This means that $p|n(n+1)$. Since p is prime, $p|n$ or $p|n+1$.
   (2) Let $n = 2pk$. Then we may pair up the numbers i and $n-i+1$ ($i = 1, 2, \ldots, pk$), so the sum in each pair is equal. Since there are pk pairs, we may choose k pairs to constitute each T_j (so each T_j contains $2k$ numbers) such that its sum of numbers is $(n+1)k$. This satisfies the requirement.
2. Let $A_k = \{2^\alpha \cdot (2k-1) \,|\, \alpha = 0, 1, \ldots\}$. We claim that $A_1, A_2, \ldots$ is a valid partition. In fact, for x, y, z, and w in the same A_i, we assume that

$$x = 2^\alpha \cdot (2i-1), \quad y = 2^\beta \cdot (2i-1),$$
$$z = 2^\gamma \cdot (2i-1), \quad w = 2^\delta \cdot (2i-1).$$

   Here we may also assume that $\alpha > \beta$ and $\gamma > \delta$. Then $x - y = 2^\beta \cdot (2i-1)(2^{\alpha-\beta}-1)$ and $z - w = 2^\delta(2i-1)(2^{\gamma-\delta}-1)$, so $x-y$ and $z-w$ are in the same A_k if and only if

$$(2i-1)(2^{\alpha-\beta}-1) = (2i-1)(2^{\gamma-\delta}-1).$$

   Equivalently, $\alpha - \beta = \gamma - \delta$, which is also equivalent to $\frac{x}{y} = \frac{z}{w}$.
3. The smallest positive integer is $h(n) = 2n$. On one hand, consider the n-partition $\{1, n\}, \{2, n+2\}, \ldots, \{n-1, 2n-1\}, \{n\}$, of the set $\{1, 2, \ldots, 2n-1\}$. Note that if $a + x \le a + y < a + x + y$ are in the same subset, then $a + x = a + y$ and there exists $i \in \{1, 2, \ldots, n-1\}$ such that

$$a + x = a + y = i, \; a + x + y = i + n.$$

However, this implies that $x = y = n$, which is a contradiction. Hence, $h(n) \geq 2n$. On the other hand, for every n-partition of $\{1, 2, \ldots, 2n\}$, two of the numbers $n, n+1, \ldots, 2n$ should be in the same subset. Let them be $n \leq u < v(\leq 2u)$. Then we let $y = x = v - u(>0)$ and $a = u - x = 2u - v \geq 0$. Then $a + x = a + y = u$ and $a + x + y = a + 2x = v$ are in the same subset.

4. The smallest positive integer n is 96. First, we show that $n = 96$ has the described property. In fact, let $\{1, 2, \ldots, 96\} = A_1 \cup A_2$, where $A_1 \cap A_2 = \emptyset$, and the product of any two numbers in the same subset is not in that subset. Then we assume that $2 \in A_1$, and the locations of 3 and 4 have four cases.

   Case 1: $3 \in A_1$ and $4 \in A_1$. Then $6, 8, 12 \in A_2$, so $48(=6 \times 8) \in A_1$. Since $96 = 2 \times 48 = 8 \times 12$, we have $96 \notin A_1$ and $96 \notin A_2$, a contradiction.
   Case 2: $3 \in A_1$ and $4 \in A_2$. Then $6 \in A_2$, $24 \in A_1$, $48 \in A_2$, and $8 \in A_2$ (since $3 \times 8 = 24$), but in this case $6 \times 8 = 48$, where the three numbers all belong to A_2, a contradiction.
   Case 3: $3 \in A_2$ and $4 \in A_1$. Then $8 \in A_2$, $24 \in A_1$, $96 \in A_2$, $32 \in A_1$ (since $3 \times 32 = 96$), $16 \in A_2$, and $48 \in A_1$, so $2 \times 24 = 48$ in A_1, a contradiction.
   Case 4: $3 \in A_2$ and $4 \in A_2$. Then $12 \in A_1$, $6 \in A_2$ (since $2 \times 6 = 12$), and $24 \in A_2$ (since $2 \times 12 = 24$), so $4 \times 6 = 24$ in A_2, a contradiction. Therefore, when $n = 96$, there are always three distinct numbers in the same A_i such that one is the product of the other two.

   Next, for $n = 95$ we choose the following partition:

$$A_1 = \{1, 2, 3, 4, 5, 7, 9, 11, 13, 17, 19, 23, 48, 60, 72, 80, 84, 90\}$$

$$A_2 = \{1, 2, \ldots, 95\} \backslash A_1.$$

   Then A_1 and A_2 compose a two-partition of $\{1, 2, \ldots, 95\}$, and no three distinct elements x, y, and z in the same A_i satisfy $xy = z$. In summary, the minimum value of n is 96.

5. The answer is affirmative. For example, let $A = \{2! + 1; 3! + 1, 4! + 2; 5! + 1, 6! + 2, 7! + 3; 8! + 1, 9! + 2, 10! + 3, 11! + 4, \ldots\}$. In other words, we construct the set in the following way: if the number before the nth semicolon is $m! + n$, then the next group of numbers is: $(m+1)! + 1, (m+1)! + 2, \ldots, (m+1)! + (n+1)$, followed by the $(n+1)$th semicolon.

Since every element in A (except the first one) is greater than twice the element before it, no three elements of A form an arithmetic sequence. Then let B be the complement of A. We show that B does not contain a nonconstant infinite arithmetic sequence.

In fact, if $\{a + rd\}$ with $r = 0, 1, \ldots$, is such a sequence in B, where $a, d \in \mathbf{N}^*$, then by the construction of A, there are infinitely many $m \in \mathbf{N}^*$ such that $m! + a \in A$. Hence, there exists $m \geq d$ such that $m! + a \in A$. Now, choose $r = \frac{m!}{d}$. Then $a + rd \in A$, which is a contradiction.

Therefore, the partition $\{A, B\}$ satisfies the conditions.

6. We can choose two sets at most. First, since the difference between any two perfect squares is not 2, we see that $S_{0,0} \cap S_{0,2} = \emptyset$. Next, by a translation we may assume that $a \in \{0, 1\}$ (for example, if $a = 2k + 1$, then $n^2 + an + b = (n + k)^2 + (n + k) + (b - k^2 - k)$, so replacing n by $n + k$ gives $a = 1$).

   Note that for $x, y \in \mathbf{Z}$, by $(x - y)^2 + x = (x - y)^2 + (x - y) + y$, we see that $S_{0,x} \cap S_{1,y} \neq, \emptyset$. This means that all the sets that we chose have the same type, either $S_{0,x}$ or $S_{1,y}$.

   If $a = 0$, then since $(m+1)^2 - m^2 = 2m+1$ and $(m+1)^2 - (m-1)^2 = 4m$, we have $S_{0,b} \cap S_{0,c} \neq \emptyset$ when $b - c \not\equiv 2 \pmod 4$. Hence, we can find at most two sets of the type $S_{0,c}$, since we cannot find three numbers x, y, and z such that $x - y \equiv y - z \equiv z - x \equiv 2 \pmod 4$.

   If $a = 1$ and $b - c$ is even, then let $x = \frac{c-b}{2}$ and $y = \frac{b-c}{2}$, and we have

$$c - b = (x - y)(x + y + 1),$$

   so $x^2 + x + b = y^2 + y + c$, which means that $S_{1,b} \cap S_{1,c} \neq \emptyset$. Therefore, in this case we can also choose at most two sets.

7. It suffices to show that every positive integer has a unique location.

   By $1 + 2 = 2^0 + 2$, we have $2 \in B$. Suppose the locations of all positive integers less than $n(\geq 3)$ have been determined. Then consider n. Suppose $2^k \leq n < 2^{k+1}(k \in \mathbf{N}^*)$. If $n = 2^k (k \geq 2)$, then since $2 \in B$ we have $n \in A$. Now, for any $m \in A$ with $m < n$ we have $2^k < n + m < 2^{k+1}$ and $n + m \neq 2^k + 2$ (because $m \neq 2$), so $n \in A$ is compatible with the previous elements.

   If $n = 2^k + 1$, then since $1 + (2^k + 1) = 2^k + 2$, necessarily $n \in B$. In this case, for each $m \in B$ with $m < n$ we have $2^k + 1 < n + m < 2^{k+1} + 2$, which means that $n \in B$ is compatible with the previous elements.

   If $2^k + 1 < n < 2^{k+1}$, then $2^{k+1} + 2 - n < n$, thus n does not belong to the set containing $2^{k+1} + 2 - n$ (so that the location of n is uniquely determined). In this case, for every m in the same set with n and less than n, we have $2^k + 1 < n + m < 2^{k+2} + 2$ and $n + m \neq 2^{k+1} + 2$, hence n is compatible with the previous numbers.

Therefore, the proposition follows.

8. We have $\frac{n}{r} > 1$, so $[\frac{in}{r}]$ is strictly increasing with respect to i and the greatest element in A is $[\frac{(r-1)n}{r}] = n + [-\frac{n}{r}] \leq n - 2$. This means that $A \subseteq \{1, 2, \ldots, n-2\}$ and $|A| = r - 1$. Similarly, $B \subseteq \{1, 2, \ldots, n-2\}$ and $|B| = s - 1$. Since $r + s = n$, we see that $\{A, B\}$ is a two-partition of $\{1, 2, \ldots, n-2\}$ if and only if $A \cap B = \emptyset$. Hence, it suffices to show that $A \cap B = \emptyset$ if and only if $(r, n) = 1$.

   In fact, if $(r, n) = 1$, then by $r + s = n$ we also have $(s, n) = 1$. In this case, for $1 \leq i \leq r - 1$ and $1 \leq j \leq s - 1$, none of the numbers $\frac{in}{r}$ and $\frac{jn}{s}$ are integers. If $[\frac{in}{r}] = [\frac{jn}{s}] (= m)$, then $m < \frac{in}{r} < m + 1$ and $m < \frac{jn}{s} < m + 1$. Thus, $mr < in < (m+1)r$ and $ms < jn < (m+1)s$. Adding the two inequalities, we get

$$mn = mr + ms < (i + j)n < (m + 1)r + (m + 1)s = (m + 1)n,$$

   so $m < i + j < m + 1$, which is impossible. Hence, $A \cap B = \emptyset$.

   On the other hand, if $(r, n) = k > 1$, then let $i = \frac{r}{k}$ and $j = \frac{s}{k}$, and we have $[\frac{in}{r}] = [\frac{jn}{s}] = \frac{n}{k}$, thus $A \cap B \neq \emptyset$. Therefore, the proposition is true.

9. Without loss of generality, assume that $1 \in A$, and if B and C are both nonempty, assume that the smallest element of B is smaller than the smallest element of C.

   First, consider the location of 2. Since 2 belongs to either A or B, there are two options. Now, suppose all the numbers less than $a(\geq 3)$ have two possible locations (given the configuration of the previous numbers), and have been positioned, then we consider the location of a. There are following cases:

   (1) $B = C = \emptyset$. In this case, a belongs to A or B, so there are two options.

   (2) $B \neq \emptyset$ and $C = \emptyset$. In this case, a may belong to the set containing $a - 1$ (we assume it to be A). If the other option for $a - 1$ is C, then $a - 1$ has a different parity with the smallest element of B (i.e., one is odd and the other is even). Thus, a has the same parity with the smallest number of B, so by the second condition, a cannot belong to C. Since $a - 1$ cannot belong to B, we see that the greatest number of B has the same parity with $a - 1$, thus a can be placed in B. Therefore, there are also two options for a.

   If the second option for $a - 1$ is B, then the argument is similar.

(3) $B \neq \emptyset$ and $C \neq \emptyset$. The analysis is similar to (2), and we see that one option for a is the set containing $a-1$, and the other option is the set that $a-1$ cannot belong to (when determining the location of $a-1$). Hence, a has two possible locations.

To summarize, there are 2^{n-1} such (unordered) partitions.

10. If $p+q \equiv 0 \pmod 3$, then since p and q are coprime and $p \neq q$, we may assume without loss of generality that $p \equiv 1 \pmod 3$, and $q \equiv 2 \pmod 3$. Then let $A = \{a \mid a \equiv 0 (\bmod 3)\}$, $B = \{b \mid b \equiv 1 (\bmod 3)\}$ and $C = \{c \mid c \equiv 2 (\bmod 3)\}$, and for $z \in \mathbf{N}^*$, since z, $z+p$, and $z+q$ compose a complete system of residues modulo 3, exactly one of them belongs to each of A, B, and C.

On the other hand, if there exists a three-paritition of $\mathbf{N}^*$ into A, B, and C with the desired property, then for each positive integer z, if $z \in A$, $z+p \in B$, and $z+q \in C$, then $(z+p)+q \notin B$ and $(z+q)+p \notin C$, so $z+p+q \in A$. Hence, if $z_1, z_2 \in \mathbf{N}^*$ and $z_1 \equiv z_2 (\bmod p+q)$, then z_1 and z_2 belong to the same subset.

Now, let $I = \{1, 2, \ldots, p+q\}$, and we show that $|A \cap I| = |B \cap I| = |C \cap I|$, so that $3 | p+q$. In fact, for each $z \in A \cap I$, let

$$z + p \equiv p(z) (\bmod p+q),$$

$$z + q \equiv q(z) (\bmod p+q),$$

where $p(z)$, $q(z) \in I$. Then by the argument above, $p(z) \notin A$ and $q(z) \notin A$, and also $p(z) \neq q(z)$. For z_1, $z_2 \in A \cap I$, if $z_1 \neq z_2$, then $p(z_1) \neq p(z_2)$ and $q(z_1) \neq q(z_2)$. If $p(z_1) = q(z_2)$, then

$$z_1 + 2p \equiv p(z_1) - q = q(z_2) - q \equiv z_2 (\bmod p+q).$$

This requires that $z_1 + 2p \in A$, but if we consider the numbers $z_1 + p$, $(z_1+p)+p$, and $(z_1+p)+q$, then since they should belong to different subsets, $(z_1+p)+q \in A$, thus $(z_1+p)+p \notin A$, a contradiction. Hence, $p(z_1) \neq q(z_2)$.

These results imply that the number of integers in $I \cap (B \cup C)$ is at least twice that of $I \cap A$, from which

$$p + q = |I| = |I \cap A| + |I \cap (B \cup C)| \geq 3|I \cap A|.$$

Similarly, $|I| \geq 3|I \cap B|$ and $|I| \geq 3|I \cap C|$, so

$$|I \cap A| = |I \cap B| = |I \cap C| = \frac{p+q}{3},$$

and the conclusion follows.

11. Let A, B, C, and D be the vertices of the rectangle R. We construct a plane coordinate system, with A as the origin, AB as the x-axis, and AD as the y-axis. It suffices to show that at least one of B, C, and D is a grid point.

Since each small rectangle R_i has a side with integer length, and its sides are parallel to the coordinate axes, the number of the grid points in the vertices of R_i is 0, 2, or 4. This means that among the vertices of all rectangles R_i, the number of grid points is even (here the number of the vertices is counted with some multiplicity).

Note that in the counting method above, A is a grid point, which is counted once. If a point lies inside or on the boundary of $ABCD$ (but different from A, B, C, and D), then it is the vertex of 0, 2, or 4 small rectangles, so it is counted with an even multiplicity. Therefore, since the total number of the grid vertices is even, at least one of B, C, and D is a grid point, and the proposition follows.

Solution 23

Synthetical Problems of Quadratic Functions

1. We have $PC = CA$ and $BC = CA + AB = PC + 4$, so $\frac{PC}{PC+4} = \frac{3}{7}$, and

$$PC = 3, \quad OA = CA - CO = 2.$$

Hence, the coordinates of A and B are $(2, 0)$ and $(6, 0)$, respectively. The equation of the parabola is $y = -\frac{1}{7}(x-2)(x-6) = -\frac{1}{7}x^2 + \frac{8}{7}x - \frac{12}{7}$, and $b + c = -\frac{4}{7}$.

2. The graph of $y = x^2 - |x| - 12$ intersects the x-axis at $A(4, 0)$ and $B(-4, 0)$, so we can assume that

$$y = ax^2 + bx + c = a(x+4)(x-4).$$

Since ΔPAB is an isosceles right triangle, $|16a| = 4$, so $a = \pm\frac{1}{4}$. Therefore, $a = -\frac{1}{4}$, $b = 0$, and $c = 4$ or $a = \frac{1}{4}$, $b = 0$, and $c = -4$.

3. $y = \left(x + \frac{1}{2}\right)^2 + p\left(x + \frac{1}{2}\right)$. When $x = -\frac{1}{2}$, we have $y = 0$, so the parabola passes through a fixed point $\left(-\frac{1}{2}, 0\right)$. Also the vertex of the parabola has coordinates $x = -\frac{p+1}{2}$ and $y = -\frac{1}{4}p^2$. Cancelling p, we have

$$y = -\left(x + \frac{1}{2}\right)^2.$$

4. Let $f(x) = \frac{a}{2}x^2 + bx + c$. Then

$$\begin{aligned} f(x_1)f(x_2) &= \left(\frac{a}{2}x_1^2 + bx_1 + c\right)\left(\frac{a}{2}x_2^2 + bx_2 + c\right) \\ &= \left(\frac{a}{2}x_1^2 - ax_1^2\right)\left(\frac{a}{2}x_2^2 + ax_2^2\right) = -\frac{3}{4}a^2x_1^2x_2^2 < 0. \end{aligned}$$

Hence, $f(x)$ has a root between x_1 and x_2.

5. Let $u = \cos x$. Then $f(u) = 4u^2 + (a-5)u + 1$ has two distinct real roots in $(0, 1)$, so

$$\begin{cases} \Delta = (a-5)^2 - 16 > 0, \\ 0 < -\dfrac{a-5}{8} < 1, \\ f(0) = 1 > 0, \\ f(1) = 4 + (a-5) + 1 > 0. \end{cases}$$

Solving the inequalities, we get $0 < a < 1$.

6. Subtracting the two equations and completing the square, we have $b + c = \pm(a-1)$. Since $bc = a^2 - 8a + 7$, we see that b and c are two real roots of the equation $x^2 \pm (a-1)x + a^2 - 8a + 7 = 0$. This implies that

$$\Delta = (a-1)^2 - 4(a^2 - 8a + 7) \geq 0.$$

Hence, $1 \leq a \leq 9$.

7. By the condition, the system of equations $\begin{cases} y = x^2 + 2bx + 1 \\ y = 2a(x+b) \end{cases}$ has no real solutions. Hence, the quadratic equation

$$x^2 + 2(b-a)x + (1-2ab) = 0$$

has the discriminant $\Delta = 4(b-a)^2 - 4(1-2ab) < 0$. This is equivalent to $a^2 + b^2 < 1$, so $S = \{(a, b) \mid a^2 + b^2 < 1\}$ is the unit circle centered at the origin, and its area is π.

8. (1) Let $f(x) = x^2 + (2m-1)x + (m-6)$. Since $f(-1) \leq 0$ and $f(1) \leq 0$, we get $-4 \leq m \leq 2$.
   (2) Let x_1 and x_2 be the roots of the equation. Then

$$x_1^2 + x_2^2 = 4m^2 - 6m + 13 = 4\left(m - \frac{3}{4}\right)^2 + 10\frac{3}{4}.$$

   Hence, the minimum value of $x_1^2 + x_2^2$ is $10\frac{3}{4}$ and the maximum value is 101.

9. $f(-2) = 4a - 2b + c = 3f(-1) + f(1) - 3f(0)$ and $f(2) = f(-1) + 3f(1) - 3f(0)$. Hence, $|f(-2)| \leq 3|f(-1)| + |f(1)| + 3|f(0)| \leq 7$. Similarly, $|f(2)| \leq 7$. If $|-\frac{b}{2a}| > 2$, then $|f(x)| \leq \max\{|f(-2)|, |f(2)|\} \leq 7$ for $x \in [-2, 2]$. If $\left|-\frac{b}{2a}\right| \leq 2$, then since

$$|b| = \left|\frac{1}{2}f(1) - \frac{1}{2}f(-1)\right| \leq \frac{1}{2} + \frac{1}{2} = 1,$$

it follows that $\left|f\left(-\frac{b}{2a}\right)\right| = \left|\frac{4ac-b^2}{4a}\right| \le |c| + \left|\frac{b}{2a}\right| \cdot \frac{|b|}{2} \le 1 + 2 \cdot \frac{1}{2} = 2$. Thus, we also have $|f(x)| \le 7$.

10. It is easy to see that the parabola intersects the x-axis at $A(3, 0)$ and $B\left(\frac{4}{3m}, 0\right)$, and intersects the y-axis at $C(0, 4)$. If $AC = BC$, then $m = -\frac{4}{9}$ and $u = -\frac{4}{9}x^2 + 4$. If $AC = AB$, then $m_1 = \frac{1}{6}$ and $m_2 = -\frac{2}{3}$, from which $y = \frac{1}{6}x^2 - \frac{11}{6}x + 4$ or $y = -\frac{2}{3}x^2 + \frac{2}{3} + 4$. If $AB = BC$, then $m = -\frac{8}{7}$, and $y = -\frac{8}{7}x^2 + \frac{44}{21}x + 4$.
11. By $\frac{x^2}{10} - \frac{x}{10} + \frac{9}{5} \le |x|$, we can solve that $-6 \le x \le -3$ or $2 \le x \le 9$. Hence, the grid points that we want are $(-6, 6)$, $(-3, 3)$, $(2, 2)$, $(4, 3)$, $(7, 6)$, and $(9, 9)$.
12. (1) The function $f(x)$ on I_k is obtained by translating $y = x^2$ to the right by $2k$ units, so the formula is $f(x) = (x - 2k)^2$ for $x \in I_k$.

    (2) The graph of $f(x)$ looks like the first picture below. For a positive integer k, we only consider the part as in the second picture. For each k, if the line $y = ax$ has two intersection points with the graph of $f(x)$ over I_k, then from the graph we can see that $0 < a \le \frac{1}{2k+1}$. Therefore, $M_k = \left\{a \,|\, 0 < a \le \frac{1}{2k+1}\right\}$.

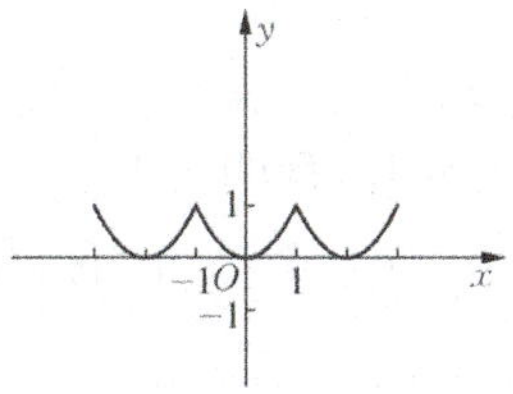

Exercise 12(1)

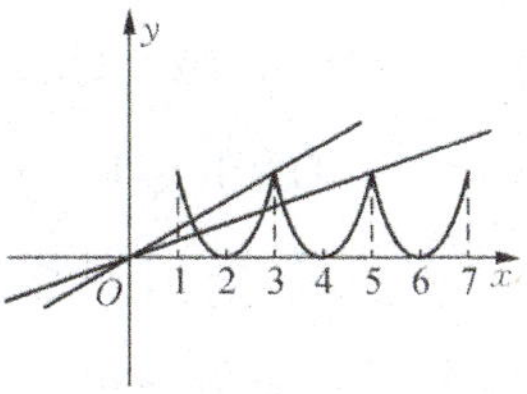

Exercise 12(2)

13. Denote $A(x_1, 0)$ and $B(x_2, 0)$. Then since the parabola opens upwards, and $|x_1| < 1$ and $|x_2| < 1$, we have $b^2 - 4ac > 0$ and $f(-1) = a - b + c > 0$. Hence, $b^2 > 4ac$ and $a + c \ge b + 1$. Since $b^2 > 4$, we have $b \ge 3$. If $b = 3$, then $4ac < 9$, so $a + c \le 3$ and $a + c \ge b + 1 = 4$, a contradiction. If $b = 4$, then $4ac < 16$ and $a + c \le 4$ but $a + c \ge 5$, also a contradiction. Hence, $b \ge 5$ and $a + c \ge 6$, from which $a + b + c \ge 11$. On the other hand, the function $y = 5x^2 + 5x + 1$ satisfies the condition. Therefore, the minimum value of $a + b + c$ is 11.

    Alternative Solution. By the condition, the function $f(x) = ax^2 + bx + c$ has two different zeros x_1 and x_2 such that $-1 < x_1 < x_2 < 0$ (here we assumed $x_1 < x_2$ without loss of generality). Thus,

$$f(x) = a(x - x_1)(x - x_2).$$

Then $f(0) = c > 0$ and $f(-1) = a(-1 - x_1)(-1 - x_2) > 0$, so

$$\begin{aligned} 1 &\le f(0)f(-1) = a^2(-x_1)(1 + x_1)(-x_2)(1 + x_2) \\ &\le a^2\left(\frac{-x_1 + 1 + x_1}{2}\right)^2\left(\frac{-x_2 + 1 + x_2}{2}\right)^2 = \frac{a^2}{16}, \end{aligned}$$

where the equality holds if and only if $x_1 = x_2 = -\frac{1}{2}$, which is impossible. This means that $\frac{a^2}{16} > 1$, namely $a \ge 5$. Further, since $b^2 > 4ac$ we have $b \ge 5$, thus $a + b + c \ge 5 + 5 + 1 = 11$, and since $y = 5x^2 + 5x + 1$ satisfies the condition, the minimum value of $a + b + c$ is 11.

14. Since $f(-1) = a - b + c$, $f(0) = c$, and $f(1) = a + b + c$, we can solve that $a = \frac{1}{2}f(1) + \frac{1}{2}f(-1) - f(0)$ and $b = \frac{1}{2}f(1) - \frac{1}{2}f(-1)$. Thus,

$$\begin{aligned} |2ax + b| &= \left|\left(x + \frac{1}{2}\right)f(1) + \left(x - \frac{1}{2}\right)f(-1) - 2f(0)x\right| \\ &\le \left|x + \frac{1}{2}\right| \cdot |f(1)| + \left|x - \frac{1}{2}\right| \cdot |f(-1)| + 2|x| \cdot |f(0)| \\ &\le \left|x + \frac{1}{2}\right| + \left|x - \frac{1}{2}\right| + 2 \le 4. \end{aligned}$$

15. Let $f(x) = ax^2 + bx + c$. Then

$$|f(0)| \le 1, \quad |f(-1)| \le 1, \ |f(1)| \le 1.$$

Hence, $|c| \le 1$, $|a - b + c| \le 1$, and $|a + b + c| \le 1$. For $x \in [-1, 1]$,

$$\begin{aligned} |cx^2 \pm bx + a| &= |cx^2 - c + c \pm bc + a| \\ &\le |c| \cdot |x^2 - 1| + |c \pm bx + a| \\ &\le 1 + |a \pm bx + c|. \end{aligned}$$

Note that $g(x) = a \pm bx + c$ is a linear function of x, so its maximum and minimum values are both attained at the endpoints of the domain. Therefore,

$$|a \pm bx + c| \le \max\{|a - b + c|, |a + b + c|\} \le 1.$$

Solution 24

Maximum and Minimum Values of Discrete Quantities

1. Since $4a + b = abc$, we have $c = \frac{1}{a} + \frac{4}{b}$ and

$$a + b + c = a + b + \frac{1}{a} + \frac{4}{b} \geq 6.$$

Hence, $a + b + c$ has its minimum value 6, attained if and only if $a = 1$ and $b = 2$.

2. The minimum value of k is 51. Note that the 50 numbers $2, 4, 6, \ldots, 100$ satisfy that no two of them are coprime, and any collection of 51 numbers among $1, 2, \ldots, 100$ contains two consecutive integers, which are necessarily coprime.

3. It is equivalent to the following problem: For $n \in \mathbf{N}^*$ (not fixed) and positive integers $a_1 < a_2 < \cdots < a_n$ such that $a_1 + a_2 + \cdots + a_n = 200$, find the maximum value of $a_1 a_2 \cdots a_n$.

When $a_1 a_2 \ldots a_n$ reaches the maximum value, $a_1 \geq 2$, since otherwise we may replace the original sequence by $a_2, a_3, \ldots, a_{n-1}, a_n + 1$, which still satisfies the condition, and the product is greater. It also follows that $a_{i+1} - a_i \leq 2$ for $1 \leq i \leq n - 1$, since otherwise we replace a_i, a_{i+1} with $a_i + 1, a_{i+1} - 1$, and the product will increase. Finally, if there are two indices $1 \leq i < j \leq n - 1$ such that $a_{i+1} - a_i = a_{j+1} - a_j = 2$, then we replace the original sequence with $a_1, \ldots, a_{i-1}, a_i + 1, a_{i+1}, \ldots, a_j, a_{j+1} - 1, a_{j+2}, \ldots, a_n$, and the product will increase.

Therefore, the maximum value is $(2 \times 3 \times \cdots \times 8) \times (10 \times 11 \times \cdots \times 20) = \frac{20!}{9}$.

4. Note that $n(X) = 2^{|X|}$, so $2^{100} + 2^{100} + 2^{|C|} = 2^{|A \cup B \cup C|}$. The only possibility is $|C| = 101$ and $|A \cup B \cup C| = 102$. By $A \cup B \subseteq A \cup B \cup C$ and other similar relations, $|A \cap B| \geq 98$, $|A \cap C| \geq 99$, and $|B \cap C| \geq 99$. On the other hand, choose

$$A = \{a_1, a_2, \ldots, a_{100}\}, \quad B = \{a_1, a_2, \ldots, a_{98}, a_{101}, a_{102}\},$$
$$C = \{a_1, a_2, \ldots, a_{97}, a_{99}, a_{100}, a_{101}, a_{102}\}.$$

Then these three sets satisfy the conditions, and

$$|A \cap B| = 98, \quad |A \cap C| = 99, \quad |B \cap C| = 99.$$

Hence, $|A \cap B \cap C| = |A \cup B \cup C| - |A| - |B| - |C| + |A \cap B| + |B \cap C| + |C \cap A|$ has the minimum value 97.

5. Since $n^2 + (n+1)^2 + (n+2)^2 = 2(a+b+c)$ is even, we see that n is odd. Apparently $n > 1$, and assume that $a < b < c$. If $n = 3$, then $a+b = 9$, $a+c = 16$, and $b+c = 25$, which gives $a = 0$, a contradiction.

   Hence, $n \geq 5$. If $n = 5$, we have $a+b = 25$, $a+c = 36$, and $b+c = 49$, so $a = 6$, $b = 19$, and $c = 30$, thus $a^2+b^2+c^2 = 1297$, which is minimal.

6. We discuss the following cases, based on the number of columns containing 1.

   (1) If all 1's lie in the same column, then $M = 9$.

   (2) If the 1's are distributed in two columns, then the sum of the numbers in these two columns is at least $2M$, while the sum is at most $9 \times 1 + 9 \times 4 = 45$ by the condition. Hence, $M \leq 22$.

   (3) If the 1's are distributed in three columns, then by a similar argument we have $3M \leq 9 \times 1 + 9 \times 4 + 9 \times 3$, so $M \leq 24$.

   (4) If the 1's are distributed in four columns, then $4M \leq 9 \times 1 + 9 \times 4 + 9 \times 3 + 9 \times 2$, so $M \leq 22$.

   (5) If the 1's are distributed in more than four columns, then since there are only 27 numbers 2, 3, and 4, some column containing 1 will have a number greater than 4, which is contradictory to the assumption, so this case cannot happen.

   Consequently, the minimum values of all column-sums do not exceed 24. Next, we give an example to show that 24 can be reached:

$$\begin{matrix} 1 & 1 & 1 & 2 & 2 & 6 & 7 & 8 & \cdots & 2000 \\ 1 & 1 & 1 & 2 & 2 & 6 & 7 & 8 & \cdots & 2000 \\ 1 & 1 & 1 & 2 & 2 & 6 & 7 & 8 & \cdots & 2000 \end{matrix}$$

$$\begin{array}{cccccccccc} 3 & 3 & 3 & 2 & 2 & 6 & 7 & 8 & \cdots & 2000 \\ 3 & 3 & 3 & 2 & 5 & 6 & 7 & 8 & \cdots & 2000 \\ 3 & 3 & 3 & 5 & 5 & 6 & 7 & 8 & \cdots & 2000 \\ 4 & 4 & 4 & 5 & 5 & 6 & 7 & 8 & \cdots & 2000 \\ 4 & 4 & 4 & 5 & 5 & 6 & 7 & 8 & \cdots & 2000 \\ 4 & 4 & 4 & 5 & 5 & 6 & 7 & 8 & \cdots & 2000 \end{array}$$

Therefore, the maximum value of the minimum values of all column-sums is 24.

7. Since A has k elements, the number a_i has k different values, $a_i + a_j$ has at most $\frac{k(k+1)}{2}$ different values, and $|a_i - a_j|$ has at most $\frac{k(k-1)}{2}$ different values, so there are at most $k(k+1)$ different values for numbers of these forms. Thus, $k(k+1) \geq 19$ with $k \geq 4$.

   If $k = 4$, then $k(k+1) = 20$. Assume that $a_1 < a_2 < a_3 < a_4$. Then $a_4 \geq 10$.
   If $a_4 = 10$, then necessarily $a_3 = 9$. Now, $a_2 = 8$ or 7. If $a_2 = 8$, then $10 + 10 = 20$ and $10 - 9 = 9 - 8 = 1$, so we don't have enough combinations for 1 through 19. If $a_2 = 7$, then $a_1 = 6$ or $a_1 = 5$. Since $10 + 10 = 20$, $10 - 9 = 1$, and $7 - 6 = 1$ or $9 - 7 = 2$ and $7 - 5 = 2$, we still don't have enough combinations.
   If $a_4 = 11$, then $a_3 = 8$, so $a_2 = 7$ and $a_1 = 6$.
   If $a_4 = 12$, then $a_3 = 7$, and $a_2 = 6$ and $a_1 = 5$.
   If $a_4 = 13$, then $a_3 = 6$, $a_2 = 5$, and $a_1 = 4$.
   If $a_4 = 14$, then $a_3 = 5$, $a_2 = 4$, and $a_3 = 3$.
   If $a_4 = 15$, then $a_3 = 4$, $a_2 = 3$, and $a_1 = 2$.
   If $a_4 = 16$, then $a_3 = 3$, $a_2 = 2$, and $a_1 = 1$.

   The case $a_4 \geq 17$ is also impossible.

   None of these cases satisfy the condition. Hence, $k \geq 5$. On the other hand, let $A = \{1, 3, 5, 9, 16\}$. We can see that A satisfies the condition. Therefore, the minimum value of k is 5.

8. Suppose the sequence has n terms at most, and let $a_1, a_2, \ldots, a_n$ be such a sequence. If $n \geq 17$, then consider the following matrix:

$$\begin{array}{c} a_1\, a_2\, a_3 \cdots a_{11} \\ \\ a_2\, a_3\, a_4 \cdots a_{12} \\ \\ \cdots \qquad \cdots \\ \\ a_7\, a_8\, a_9 \cdots a_{17}. \end{array}$$

By the assumption, the sum of every column is negative, while the sum of every row is positive. This is a contradiction since the total sum of the numbers in the matrix is the same.

Hence, $n \leq 16$. On the other hand, consider the sequence $5, 5, -13, 5, 5, 5, -13, 5, 5, -13, 5, 5, 5, -13, 5, 5$. It is easy to verify that this sequence satisfies the condition. Therefore, the sequence has 16 terms at most.

9. We first show that twelve rows are enough. We arrange all the students in a line such that the students from the same school stand consecutively. Then we choose the students with numbers $200, 400, 600, \ldots, 1800$ as well as all other students in their respective schools. For the remaining students, the line is cut into ten pieces such that every piece has no more than 200 students, so they can be seated in ten rows. The chosen students come from nine schools and each school has at most 40 students, so they can be seated in at most two rows. Therefore, twelve rows are enough.

   On the other hand, the following case shows that eleven rows are not enough. Let $n = 59$ such that each of the first 58 schools has 34 students and the last school has 28 students. Since $34 \times 6 > 200$, we see that except for at most one row (it may contain six schools), all other rows contain at most five schools. Thus, if there are only eleven rows, at most $5 \times 10 + 6 = 56$ schools can be seated, which means that eleven rows are not enough.

   Therefore, the stadium has to prepare at least twelve rows.

10. The answer is 11. We generalize the problem, and let $f(n)$ be the minimal amount of the money needed in order to find a number in $\{1, 2, \ldots, n\}$. Then $f(n)$ is non-decreasing, and if we choose an n-number set at the first step, then $f(n) \leq \max\{f(m) + 2, f(n - m) + 1\}$.

    Consider the Fibonacci sequence $\{F_n\}$, where $F_1 = 1$, $F_2 = 2$, and $F_{n+2} = F_{n+1} + F_n$ for $n = 1, 2, \ldots$. We claim that $f(F_n) = n$. First, we show that $f(F_n) \leq n$ for every positive integer n. In fact, when $n = 1$ and 2, apparently $f(F_n) \leq n$. Suppose we already have $f(F_{n-2}) \leq n-2$ and $f(F_{n-1}) \leq n-1$. Then consider F_n. We first choose a subset with F_{n-2} numbers, so $f(F_n) \leq \max\{f(F_{n-2}) + 2, f(F_n - F_{n-2}) + 1\} = \max\{f(F_{n-2} + 2), f(F_{n-1}) + 1\} \leq n$. Hence, by induction we have $f(F_n) \leq n$.

    On the other hand, we show that for any integer x with $F_{n-1} < x \leq F_n$, we have $f(x) \geq n$ (here $n \geq 2$ is an integer). Note that $x = F_2$ when

$n = 2$, and apparently $f(x) \geq 2$. Suppose the result holds for integers not exceeding n. Consider $n+1$, in which case $F_n < x \leq F_{n+1}$. If we choose a subset with at most F_{n-2} numbers in the first step, then the answer may be "No," in which case $f(x) \geq f(x - F_{n-2}) + 1 \geq f(F_{n-1}+1)+1 \geq n+1$. Here, we have applied the monotonicity result and the inductivie hypothesis. If we choose a subset with more than F_{n-2} numbers in the first step, then the answer may be "Yes," in which case $f(x) \geq f(F_{n-2}+1) + 2 \geq n - 1 + 2 = n + 1$. Therefore, for an integer x such that $F_n < x \leq F_{n+1}$, we have $f(x) = n + 1$.

Since $144 = F_{11}$, the minimum amount of the money that Bob needs to prepare is 11.

11. We claim that if we choose the smallest and second smallest numbers for the operation at each step, then the last remaining number is maximal. We use the notation $a * b$ to denote the operation of erasing a and b and adding the number $ab+1$ to the blackboard. Suppose in some operation we do not choose the smallest and second smallest numbers x and y. Then in the following operations, before x and y meet each other they must have met some other a_i and b_j, respectively, where these a_i and b_j are all greater than or equal to y and x, respectively. Hence, in some operation the two numbers are $(((x * a_1) * a_2) * \cdots) * a_k$ and $(((y * b_1) * b_2) * \cdots) * b_n$. Here these two numbers can be written explicitly as

$$\begin{aligned} X &= a_1a_2\cdots a_k\left(x + \frac{1}{a_1} + \frac{1}{a_1a_2} + \cdots + \frac{1}{a_1a_2\cdots a_k}\right) \\ &= a_2\cdots a_k(a_1x + 1 + u), \\ Y &= b_1b_2\cdots b_n\left(y + \frac{1}{b_1} + \frac{1}{b_1b_2} + \cdots + \frac{1}{b_1b_2\cdots b_n}\right) \\ &= b_1b_2\cdots b_n(y + v), \end{aligned}$$

where u and v are nonnegative numbers. We may assume that $b_1 \leq b_2 \leq \cdots \leq b_n$ (in fact, if $b_j > b_{j+1}$, then interchanging b_j and b_{j+1} in Y will make Y increase). Now, interchange a_1 and y, and get $X' = a_2\cdots a_k(xy + 1 + u)$ and $Y' = b_1b_2\cdots b_n(a_1 + v)$. Thus,

$$\begin{aligned} &X'Y' + 1 - (XY + 1) \\ &\quad = a_2a_3\cdots a_kb_1b_2\cdots b_n[(xy + 1 + u)(a_1 + v) - (a_1x + 1 + u)(y + v)] \\ &\quad = a_2a_3\cdots a_kb_1b_2\cdots b_n(a_1 - y)(u + 1 - ux). \end{aligned}$$

Note that $a_1 > y$ and $v = \frac{1}{b_1} + \frac{1}{b_1 b_2} + \cdots + \frac{1}{b_1 b_2 \cdots b_n} \leq \frac{1}{b_1} + \left(\frac{1}{b_1}\right)^2 + \cdots + \left(\frac{1}{b_1}\right)^n < \frac{1}{b_1 - 1} \leq \frac{1}{x}$ (because $b_1 > x$). This shows that $X' * Y' > X * Y$. Therefore, the number A is obtained by consecutively operating the smallest and second smallest numbers. In this process, consider the numbers modulo 10, and we will experience the following steps: 128 copies of "1"; 64 copies of "2"; 32 copies of "5"; 16 copies of "6"; 8 copies of "7"; 4 copies of "0"; 2 copies of "1"; 1 number "2". Therefore, the last digit of A is 2.

12. We claim that the number of bad cells in A does not exceed 25. Suppose not. Then there is at most 1 cell that is not bad. By symmetry we can assume that all the cells in the first row are bad.

    Let a_i, b_i, and c_i for $i = 1, 2, \ldots, 9$ be the numbers in the cells of the ith column (from top to bottom). Denote

$$S_k = \sum_{i=1}^{k} a_i, \quad T_k = \sum_{i=1}^{k} (b_i + c_i), \quad k = 0, 1, \ldots, 9.$$

    Here, we assume that $S_0 = T_0 = 0$. We show that each of the three sequences $S_0, S_1, \ldots, S_9$; $T_0, T_1, \ldots, T_9$; $S_0 + T_0$, $S_1 + T_1, \ldots, S_9 + T_9$ composes a complete system of residues modulo 10. In fact, if there exist m and n with $0 \leq m < n \leq 9$ such that $S_m \equiv S_n \pmod{10}$, then

$$\sum_{i=m+1}^{n} a_i = S_n - S_m \equiv 0 \pmod{10}.$$

    so the $(m+1)$-th through nth cells in the first row compose a good rectangle, from which at least two cells are not bad, a contradiction. Similarly, if there exist m and n with $0 \leq m < n \leq 9$ such that $T_m \equiv T_n \pmod{10}$, then the $(m+1)$-th through nth column by the second and third rows compose a good rectangle. Also if there exist m and n with $0 \leq m < n \leq 9$ such that $S_m + T_m \equiv S_n + T_n \pmod{10}$, then the $(m+1)$-th through n-th column by the first through third rows compose a good rectangle.

    Hence, the claim holds, from which

$$\sum_{k=0}^{9} S_k \equiv \sum_{k=0}^{9} T_k \equiv \sum_{k=0}^{9} (S_k + T_k) \equiv 0 + 1 + \cdots + 9 \pmod{10}.$$

    But this is impossible, since $0 + 1 + \cdots + 9 \equiv 5 \pmod{10}$ and $5 + 5 \equiv 0 \pmod{10}$. Therefore, there cannot be more than 25 bad cells.

On the other hand, we construct an example to show that 25 is reachable:

$$\begin{array}{c} 1\,1\,1\,2\ \ 1\,1\,1\,1\,10 \\ 1\,1\,1\,1\ \ 1\,1\,1\,1\,1 \\ 1\,1\,1\,10\,1\,1\,1\,1\,2. \end{array}$$

Therefore, the maximal number of bad cells is 25.

13. Without loss of generality, assume that $a_1 < a_2 < \cdots < a_n$. Then any two different subsets of M have different sums of integers (if two subsets intersect, then we discard their common integers to get two disjoint subsets, and regard the sum for the empty set as 0) Hence, for $1 \le k \le n$, if we consider all subsets of M that are contained in $\{a_1, a_2, \ldots, a_k\}$ (there are 2^k such subsets), then we have the following inequalities:

$$\begin{aligned} &\frac{1}{a_1} + \frac{1}{a_2} + \cdots + \frac{1}{a_n} - \left(1 + \frac{1}{2} + \cdots + \frac{1}{2^{n-1}}\right) \\ &= \left(\frac{1}{a_1} - 1\right) + \left(\frac{1}{a_2} - \frac{1}{2}\right) + \cdots + \left(\frac{1}{a_n} - \frac{1}{2^{n-1}}\right) \\ &= \frac{1-a_1}{a_1} + \frac{2-a_2}{2a_2} + \cdots + \frac{2^{n-1}-a_n}{2^{n-1}} \\ &\le \frac{1-a_1}{2a_2} + \frac{2-a_2}{2a_2} + \cdots + \frac{2^{n-1}-a_n}{2^{n-1}a_n} \\ &\le \frac{(1+2)-(a_1+a_2)}{2^2 a_3} + \frac{2^2-a_3}{2^2 a_3} + \cdots + \frac{2^{n-1}-a_n}{2^{n-1}} \\ &\le \cdots \\ &\le \frac{1}{2^{n-1}a_n}[(1+2+\cdots+2^{n-1}) - (a_1+a_2+\cdots+a_n)] \le 0. \end{aligned}$$

Hence, $\frac{1}{a_1} + \frac{1}{a_2} + \cdots + \frac{1}{a_n} \le 1 + \frac{1}{2} + \cdots + \frac{1}{2^{n-1}} = 1 - \frac{1}{2^{n-1}}$. Apparently, when $a_i = 2^{i-1}$ the equality holds, and M has the desired property. Therefore, the maximum value of $\frac{1}{a_1} + \frac{1}{a_2} + \cdots + \frac{1}{a_n}$ is $1 - \frac{1}{2^{n-1}}$.

14. Let $f(n)$ be the desired maximum value. Without loss of generality, assume that $a \le c$. If $a \ge n+1$, then $\frac{b}{a} + \frac{d}{c} \le \frac{b+d}{a} \le \frac{n}{n+1}$, so $f(n) \le \frac{n}{n+1}$. If $a \le n$, then we fix a, and show that $\frac{b}{a} + \frac{d}{c} \le \frac{a-1}{a} + \frac{n-(a-1)}{a(n-a+1)+1} = 1 - \frac{1}{a[a(n-a+1)+1]}$.

In fact, let $x = a(n-a+1)+1$. Then if $c \le x$, we have $\frac{b}{a}+\frac{d}{c} = \frac{bc+ad}{ac} \le \frac{ac-1}{ac} = 1-\frac{1}{ac} \le 1-\frac{1}{ax}$. If $c \ge x$, then

$$\left(\frac{b}{a}+\frac{d}{c}\right)-\left(\frac{a-1}{a}+\frac{b+d-a+1}{c}\right) = (b-a+1)\left(\frac{1}{a}-\frac{1}{c}\right) \le 0.$$

Here, we have used $\frac{b}{a}<1$, so $a \ge b+1$ and $a \le c$. Hence,

$$\begin{aligned}\frac{b}{a}+\frac{d}{c} &\le \frac{a-1}{a}+\frac{b+d-a+1}{c} \le \frac{a-1}{a}+\frac{n-a+1}{c}\\ &\le \frac{a-1}{a}+\frac{n-a+1}{x} = 1-\frac{1}{ax}.\end{aligned}$$

We can see that all the equalities here can be reached simultaneously, so the problem reduces to finding the maximum value of $g(a) = 1-\frac{1}{a[a(n-a+1)+1]}$ for $a \in [2, n] \cap \mathbf{N}^*$. Now, denote $h(x) = x[x(n-x+1)+1]$, and we consider the difference between consecutive integer values. Since $h(x+1)-h(x) = -3x^2+(2n-1)x+n+1$, the equation $h(x+1)-h(x) = 0$ has a unique positive real root $x_0 = \frac{2n-1+\sqrt{(2n-1)^2+12(n+1)}}{6}$. This implies that $h(x)\,(x \in [2, n]\cap)\mathbf{N}^*)$ attains the maximum value at the greatest integer not exceeding x_0, in which case $g(x)$ is also maximal. Note that $\frac{(2n-1)+2n+2}{6} < x_0 < \frac{(2n-1)+2n+3}{6}$, so $\frac{4n+1}{6} < x_0 < \frac{4n+2}{6}$. Therefore, $g(a)$ is maximal when $a = \left[\frac{2n}{3}\right]$. In summary,

$$f(n) = 1-\frac{1}{a[a(n-a+1)+1]},$$

where $a = \left[\frac{2n}{3}\right]$.

15. We denote $r_1 = 2$, $r_2 = 3$, and $r_{n+1} = r_1 r_2 \cdots r_n + 1$ for $n \ge 2$. First, $\frac{1}{r_1}+\frac{1}{r_2}+\cdots+\frac{1}{r_n} = 1-\frac{1}{r_1 r_2 \cdots r_n}$. In fact, this equality holds obviously when $n = 1$ and 2, and if it holds for k, then for $k+1$,

$$\frac{1}{r_1}+\frac{1}{r_2}+\cdots+\frac{1}{r_{k+1}} = 1-\frac{1}{r_1 r_2 \cdots r_k}+\frac{1}{r_{k+1}} = 1-\frac{1}{r_1 r_2 \cdots r_{k+1}}.$$

Hence, the equality holds for all n. Next, we prove that for all $a_1, a_2, \ldots, a_n$ that satisfy the condition,

$$\frac{1}{a_1}+\frac{1}{a_2}+\cdots+\frac{1}{a_n} \le \frac{1}{r_1}+\frac{1}{r_2}+\cdots+\frac{1}{r_n}.$$

Apparently, the inequality is valid when $n = 1$. Suppose the inequality already holds for integers $n \le k$, and consider the case $n = k+1$. Let $a_1, a_2, \ldots, a_{k+1}$ be $k+1$ numbers with the given property. Without loss of generality, assume that $a_1 \le a_2 \le \cdots \le a_{k+1}$. Then since

$$\frac{1}{a_1} < \frac{1}{a_1} + \frac{1}{a_2} < \cdots < \frac{1}{a_1} + \cdots + \frac{1}{a_k} < \frac{1}{a_1} + \cdots + \frac{1}{a_{k+1}} < 1,$$

by the inductive hypothesis we have

$$\begin{cases} \frac{1}{a_1} \le \frac{1}{r_1}, & (1) \\ \frac{1}{a_1} + \frac{1}{a_2} \le \frac{1}{r_1} + \frac{1}{r_2}, & (2) \\ \cdots & \\ \frac{1}{a_1} + \cdots + \frac{1}{a_k} \le \frac{1}{r_1} + \cdots + \frac{1}{r_k}. & (k) \end{cases}$$

Suppose

$$\frac{1}{a_1} + \cdots + \frac{1}{a_{k+1}} > \frac{1}{r_1} + \cdots + \frac{1}{r_{k+1}}. \quad (k+1)$$

Then we consider the sum $(1) \times (a_1 - a_2) + (2) \times (a_2 - a_3) + \cdots + (k) \times (a_k - a_{k+1}) + (k+1) \times a_{k+1}$. Combining it with $a_1 \le a_2 \le \cdots \le a_{k+1}$, we get

$$k + 1 > \frac{a_1}{r_1} + \frac{a_2}{r_2} + \cdots + \frac{a_{k+1}}{r_{k+1}}.$$

By the AM-GM inequality,

$$k + 1 > (k+1)\sqrt[k+1]{\frac{a_1 a_2 \cdots a_{k+1}}{r_1 r_2 \cdots r_{k+1}}},$$

From which $a_1 a_2 \cdots a_{k+1} < r_1 r_2 \cdots r_{k+1}$. Thus,

$$\frac{1}{a_1} + \frac{1}{a_2} + \cdots + \frac{1}{a_{k+1}} = \frac{M}{a_1 a_2 \cdots a_{k+1}} \le 1 - \frac{1}{a_1 a_2 \cdots a_{k+1}}$$
$$< 1 - \frac{1}{r_1 r_2 \cdots r_{k+1}} = \frac{1}{r_1} + \frac{1}{r_2} + \cdots + \frac{1}{r_{k+1}},$$

a contradiction. Therefore, the proposition is true, and the desired maximum value is $1 - \frac{1}{r_1 r_2 \cdots r_n}$, where $r_1 = 2$ and $r_{k+1} = r_1 r_2 \cdots r_k + 1$ $(k = 1, 2, \ldots, n-1)$.

Solution 25

Simple Function Iteration and Functional Equations

1. Let β be one of the common real solutions to the equations $f(x) = 0$ and $f(f(x)) = 0$. Then $f(\beta) = 0$ and $f(0) = 0$. Since $f(0) = 0$, we have $b = 0$, so $f(x) = x^2 + ax$, and the solutions to $f(x) = 0$ are 0 and $-a$. Thus, $f(f(x)) = (x^2 + ax)^2 + a(x^2 + ax) = (x^2 + ax)(x^2 + ax + a)$. If the solution set to $f(f(x)) = 0$ is still $\{0, -a\}$ and $a \neq 0$, then we should have $\Delta = a^2 - 4a < 0$. Hence, $0 \leq a < 4$ (in the case $a = 0$, the solution sets to $f(x) = 0$ and $f(f(x)) = 0$ are both $\{0\}$). Therefore, the set of (a, b) that satisfy the condition is $\{(a, 0) \mid 0 \leq a < 4\}$.
2. Apply the result in Example 7, and we have $f^{(100)} = 3^{100}(x + 1) - 1 = 3^{100}x + 3^{100} - 1$. Hence, it suffices to show that there exists $m \in \mathbf{N}^*$ such that $2000 | (3^{100}m + 3^{100} - 1)$.

   Note that $3^{100} = 81^{25} \equiv 1^{25} \equiv 1 (\bmod 16)$ and $3^{100} = 243^{20} \equiv (-7)^{20} = 2401^5 \equiv 26^5 = (25 + 1)^5 = \sum_{i=1}^{5} \binom{5}{i} 5^{2i} + 1 \equiv 1 (\bmod 125)$, and since $\gcd(16, 125) = 1$, we have $2000 | 3^{100} - 1$. Therefore, if $m \equiv 1 (\bmod 2000)$, then $2000 | f^{(100)}(m)$.

   Remark. In fact we do not need to prove that $2000 | 3^{100} - 1$, since by Bezout's identity, the fact $\gcd(3^{100}, 2000) = 1$ already implies the result.
3. We apply the fixed point method. Consider the equation $\sqrt{20x^2 + 10} = x$. We have $x^2 = -\frac{10}{19}$. Then we can rewrite the function as

$$f(x) = \sqrt{20\left(x^2 + \frac{10}{19}\right) - \frac{10}{19}},$$

from which

$$f^{(2)}(x) = \sqrt{20\left(f(x)^2 + \frac{10}{19}\right) - \frac{10}{19}}$$
$$= \sqrt{20\left(20\left(x^2 + \frac{10}{19}\right)\right) - \frac{10}{19}} = \sqrt{20^2\left(x^2 + \frac{10}{19}\right) - \frac{10}{19}}.$$

By this method we can show that $f^{(n)}(x) = \sqrt{20^n\left(x^2 + \frac{10}{19}\right) - \frac{10}{19}}$.

4. Let $f(x) = x^2 + 2x$, $\varphi(x) = x + 1$, and $h(x) = x^2$. Then $\varphi^{-1}(x) = x - 1$ and $\varphi^{-1}(h(\varphi(x))) = \varphi^{-1}((x+1)^2) = x^2 + 2x = f(x)$. By the bridge function method, if we let $u(x) = x^{\sqrt[8]{2}}$, then $h(x) = x^2 = u^{(8)}(x)$. Hence, if $g(x) = \varphi^{-1}(u(\varphi(x))) = (x+1)^{\sqrt[8]{2}} - 1$, then $g^{(8)}(x) = f(x) = x^2 + 2x$.
5. The polynomials $P_1(x, y) = 1$ and $P_2(x, y) = xy + x + y - 1$ are both symmetric polynomials in x and y. Suppose $P_{n-1}(x, y)$ and $P_n(x, y)$ $(n \geq 2)$ are both symmetric polynomials in x and y. Consider $P_{n+1}(x, y)$. We have

$$\begin{aligned}
P_{n+1}(x, y) &= (x + y - 1)(y + 1)P_n(x, y + 2) + (y - y^2)P_n(x, y) \\
&= (x + y - 1)(y + 1)P_n(y + 2, x) + (y - y^2)P_n(y, x) \\
&= (x + y - 1)(y + 1)[(x + y + 1)(x + 1)P_{n-1}(y + 2, x + 2) \\
&\quad + (x - x^2)P_{n-1}(y + 2, x)] + (y - y^2)[(y + x - 1)(x + 1) \\
&\quad \times P_{n-1}(y, x + 2) + (x - x^2)P_{n-1}(y, x)] \\
&= (x + y - 1)(y + 1)(x + y + 1)(x + 1)P_{n-1}(y + 2, x + 2) \\
&\quad + (y - y^2)(x - x^2)P_{n-1}(y, x) + [(x + y - 1)(y + 1) \\
&\quad \times (x - x^2)P_{n-1}(y + 2, x) + (y - y^2)(x + y - 1) \\
&\quad \times (x + 1)P_{n-1}(x + 2, y)].
\end{aligned}$$

Note that in the expression above, the first two terms are both symmetric in x and y, and the third term (inside the bracket) is also symmetric (since replacing (x, y) with (y, x) does not change its value). Therefore, $P_{n+1}(x, y)$ is also a symmetric polynomial, and by induction every $P_n(x, y)$ is symmetric, i.e., $P_n(x, y) = P_n(y, x)$.

6. We substitute x for $\frac{x-1}{x}$ and $\frac{1}{1-x}$ in the equation, and get

$$\begin{cases} f(x) + f\left(\dfrac{x-1}{x}\right) = 1 + x, \\ f\left(\dfrac{x-1}{x}\right) + f\left(\dfrac{1}{1-x}\right) = \dfrac{2x-1}{x}, \\ f\left(\dfrac{1}{1-x}\right) + f(x) = \dfrac{2-x}{1-x}. \end{cases}$$

Cancelling $f\left(\frac{1}{1-x}\right)$ and $f\left(\frac{x-1}{x}\right)$ in the equations above, we have $f(x) = \frac{1+x^2-x^3}{2x(1-x)}$.

7. Let $f(1) = a$. Put $x = y = 1$ in the equation, and we get $f(1+a) = 1+a$. Also letting $(x, y) = (1, 1+a)$ gives $f(2+a) = 2a+1$. Since

$$f(2+a) = f(2+f(1)) = f(2)+1.$$

it follows that $f(2) = 2a$.

Now, $f(1+2a) = f(1+f(2)) = f(1)+2 = a+2$, and also

$$f(1+2a) = f(a+f(1+a)) = f(a)+1+a.$$

Hence, $f(a) = 1$. Further, $f(2a) = f(a+f(1)) = f(a)+1 = 2$, so

$$2 = f(1+2a-1) = f(2a-1+f(a)) = a+f(2a-1).$$

Combining the above with the fact that a, $f(2a-1) \in \mathbf{N}^*$, we get $a = 1$, so $f(1) = 1$.

By this result and the original equation, for every $x \in \mathbf{N}^*$,

$$f(x+1) = f(x+f(1)) = f(x)+1.$$

and since $f(1) = 1$, we conclude by induction that $f(x) = x$. Apparently, this function satisfies the equation, so it is the unique solution to the functional equation.

8. Let $x = y$ in the equation, and we have $f(x)^2 = f(0)$. Then let $x = 0$, and we get $f(0)^2 = f(0)$, so $f(0) = 0$ or 1. If $f(0) = 0$, then $f(x) = 0$ for all x since $f(x)^2 = f(0)$, which is contradictory to the assumption (i.e., it is nonzero). Hence, $f(0) = 1$ and $f(x) \in \{-1, 1\}$ for every $x \in \mathbf{R}$. In particular, this shows that $f(x)$ has no zero. Thus, let $y = \frac{x}{2}$ in

$f(x)f(y) = f(x-y)$, and we have $f(x)f(\frac{x}{2}) = f(\frac{x}{2})$, which implies that $f(x) = 1$ since $f(\frac{x}{2}) \neq 0$. Therefore, the only solution to the functional equation is $f(x) = 1$ for $x \in \mathbf{R}$, which apparently satisfies the equation.

9. We let $y = 1$ in the condition (2), and get

$$f(x) + f(1) = 1 + f(x+1)\,(x \neq -1, x \neq 0). \text{ ①}$$

Also if we substitute x for $-x$ and substitute y for $x+1$ in ②, then

$$f(-x) + f(x+1) = 1 + f(1)\,(x \neq -1, x \neq 0). \text{②}$$

By ①+ ②we have

$$f(x) + f(-x) = 2\,(x \neq -1, x \neq 0). \text{ ③}$$

Note that ③ holds when $x = 1$, so it also holds when $x = -1$, thus $f(x) + f(-x) = 2$ for all $x \neq 0$. By the condition (1) and the equation ③, we see that $f(x) = xf\left(\frac{1}{x}\right) = x\left(2 - f\left(-\frac{1}{x}\right)\right) = 2x + (-x)f\left(\frac{1}{-x}\right) = 2x + f(-x)$ when $x \neq 0$, or equivalently $f(x) - f(-x) = 2x$. Combining it with ③, we obtain $f(x) = x + 1$.

10. First, for $x, y \in S$, if $f(x) = f(y)$, then $f^{(50)}(x) = f^{(50)}(y)$, so $x = y$. Hence, f is an injection on S. For each $n \in \mathbf{N}^*$, we have $f^{(n)}(S) = f(f^{(n-1)}(S)) \subseteq f(S)$, so the fact that $f^{(50)}(x)$ is a surjection implies that $f(x)$ is a surjection. Hence, $f(x)$ is a bijection from S to itself.

Next, for $x \in S$, let r be the smallest positive integer such that $f^{(r)}(x) = x$ (it exists since $f^{(50)}(x) = x$), and we call r the cycle length of x. If r is the cycle length for some $x \in S$, then we have the following properties: $r \leq 5$ and $r|50$.

In fact, since $f(x)$ is a bijection and r exists, we see that $x, f(x), \ldots, f^{(n-1)}(x)$ are distinct numbers, so $r \leq 5$. On the other hand, if $r \nmid 50$, then $50 = rq + m$ with $0 < m < r$, from which

$$f^{(50)}(x) = f^{(m)}(f^{(qr)}(x)) = f^{(m)}(f^{((q-1)r)}(x)) = \cdots = f^{(m)}(x) \neq x,$$

a contradiction. Therefore, for each $x \in S$, the cycle length of x can only be 1, 2, or 5. Now, we consider the following cases:

(1) The cycle length of every $x \in S$ is 5. In this case, $f(x)$ can be viewed as a circle permutation of 1, 2, 3, 4, 5, and there are $4! = 24$ such functions.

(2) The cycle length of every $x \in S$ is 1. Then there is only one such function $f(x) = x$.

(3) Three elements of S have cycle length 1 and the other two have cycle length 2. Then there are $\mathrm{C}_5^3 = 10$ such functions.

(4) One element has cycle length 1 and the others have cycle length 2. In this case, there are $\mathrm{C}_5^1 \cdot \frac{\mathrm{C}_4^2}{2} = 15$ such functions.

In summary, there are 50 such functions.

11. First, we see that $f(x) = x$ is a function that satisfies the equation. Then we let $g(x) = f(x) - x$, and by the given condition,

$$g(x) = 2g(g(x) + x).$$

If there exists $x \in \mathbf{Z}$ such that $g(x) > 0$, then among all such integers there is one such that $g(x)$ is minimal (here we have applied the least number principle to the set $\{g(x) \mid x \in \mathbf{Z},\ g(x) > 0\}$). We choose x_0 such that $g(x_0)$ is minimal and positive. Then $g(g(x_0)+x_0) = \frac{1}{2}g(x_0) < g(x_0)$ is a smaller positive value of $g(x)$, which gives a contradiction. Hence, $g(x) \leq 0$ for all $x \in \mathbf{Z}$. Similarly, we also have $g(x) \leq 0$ for all $x \in \mathbf{Z}$. Therefore, $g(x) = 0$, that is $f(x) = x$.

Alternative Solution. For each $x \in \mathbf{Z}$, let $a_0 = x$, $a_1 = f(x)$, and $a_n = f^{(n)}(x)$ for $n = 2, 3, \ldots$. Then $\{a_n\}$ is a sequence of integers, and by the condition that $3f(x) - 2f(f(x)) = x$, we have $3f(a_n) - 2f(f(a_n)) = a_n$, so $3a_{n+1} - 2a_{n+1} = a_n$ for $n = 0, 1, 2, \ldots$. This implies that $a_{n+2} - a_{n+1} = \frac{1}{2}(a_{n+1} - a_n)$, thus $a_{n+2} - a_{n+1} = \frac{1}{2^{n+1}}(a_1 - a_0)$. Note that the left side is always an integer, hence $2^{n+1} | (a_1 - a_0)$ for all positive integers n, which can only happen when $a_1 = a_0$. Therefore, $f(x) = x$ is the only function that satisfies the equation.

12. Let A be the range of $f(x)$ and let $f(0) = c$. Putting $x = y = 0$ in the equation gives

$$f(x) = \frac{c+1}{2} - \frac{x^2}{2}. \text{①}$$

Also let $y = 0$ in the equation, and we get $f(x-c) - f(x) = cx + f(c) - 1$. Since $x \neq 0$, for each $y \in \mathbf{R}$, there exist $y_1, y_2 \in A$ such that $y = y_1 - y_2$ (since the function $y = cx + f(c) - 1$ is a surjection on $\mathbf{R}$ when $c \neq 0$). Hence, by ① and the given equation we have

$$\begin{aligned} f(y) &= f(y_1 - y_2) = f(y_2) + y_1y_2 + f(y_1) - 1 \\ &= \frac{c+1}{2} - \frac{y_2^2}{2} + y_1y_2 + \frac{c+1}{2} - \frac{y_1^2}{2} - 1 \\ &= c - \frac{(y_1 - y_2)^2}{2} = c - \frac{y^2}{2}. \end{aligned}$$

Comparing this result with ①, we obtain $\frac{c+1}{2} = c$, namely $c = 1$. Therefore, $f(x) = 1 - \frac{x^2}{2}$ for all real x, and apparently this function satisfies the equation.

13. It is easy to see that the function $f(x) = x + 1$ satisfies the equation. Next, we show that it is the only solution.

Let $h(x) = \frac{f(x)-x-1}{x}$. Then since $g(x) = \frac{f(x)}{x}$ is a bounded function, $h(x)$ is also bounded, and by $f(x) \geq 1$ we have $h(x) \geq -1$. From the condition (1),

$$(x+1)h(x+1) + x + 2 = \frac{(xh(x) + x + 2)(xh(x) + x)}{x},$$

$$h(x+1) = \frac{x}{x+1}h(x)^2 + 2h(x). \text{ ①}$$

If there exists x such that $h(x) > 0$, then $h(x+1) > 2h(x) > 0$ by ①, so $h(x+n) > 2^n h(x)$ for all positive integers n. When $n \to +\infty$, we see that $2^n h(x) \to \infty$, which is contradictory to the assumption that h is bounded.

If there exists x such that $h(x) < 0$, then by ①,

$$\frac{h(x+1)}{h(x)} = \frac{x}{x+1}h(x) + 2.$$

Combining it with $h(x) \geq -1$, we get $\frac{h(x+1)}{h(x)} \geq 2 - \frac{x}{x+1} = \frac{x+2}{x+1}$, which means that $h(x+1) < 0$. By induction, we can show that for all positive integers n,

$$\frac{h(x+n)}{h(x)} > \frac{x+n+1}{x+1}.$$

Note that $\frac{x+n+1}{x+1} \to +\infty$ when $n \to +\infty$, so for some positive integer n we have $h(x+n) < -1$, which is a contradiction. Therefore, $h(x) = 0$ for all x, and $f(x) = x + 1$.

14. Note that the functions $f(x) = 2$ and $f(x) = x$ both satisfy the equation. Next, we show that these are the only two solutions.

In fact, suppose f is a function that satisfies the condition. Then for all $x, y > 0$,

$$f(x+y) + f(x)f(y) = f(xy) + f(x) + f(y) \text{ ①}$$

Let $x = y = 2$ in ①. Then $f(4) + f(2)^2 = f(4) + 2f(2)$, so $f(2) = 2$. Also let $x = y = 1$, and we get $f(2) + f(1)^2 = 3f(1)$, so $f(1)^2 - 3f(1) + 2 = 0$, which means that $f(1) = 1$ or $f(1) = 2$.

(1) If $f(1) = 2$, then let $y = 1$ in ①, and we have $f(x+1) = 2f(x) + f(1) - f(x)f(1) = 2$. This implies that $f(x) = 2$ for all $x \in (1, +\infty)$. For $x \in (0, 1)$, choose $n \in \mathbf{N}^*$ such that $2^n x > 1$, and let $y = 2$ in ①, so $f(x+2) + 2f(x) = f(2x) + f(x) + 2$. Since $f(x+2) = 2$, we have $f(2x) = f(x)$, thus $f(x) = f(2x) = \cdots = f(2^n x) = 2$. Therefore, $f(x) = 2$ in this case.

(2) If $f(1) = 1$, then let $y = 1$ in ①, and we get

$$f(x+1) + f(x) = f(x) + f(x) + 1.$$

so $f(x+1) = f(x) + 1$. Combining it with $f(1) = 1$, by induction we have $f(n) = n$ for all positive integers n.
Now, for $n \in \mathbf{N}^*$, let $y = n$ in ①. Then

$$f(x+n) = f(nx) + f(x) + f(n) - f(n)f(x).$$

By $f(x+1) = f(x) + 1$, we see that $f(x+n) = f(x) + n$, and since $f(n) = n$, it follows that $f(nx) = f(x+n) - f(x) + nf(x) - f(n) = f(x) + n - f(x) + nf(x) - n = nf(x)$. Hence, for $x \in \mathbf{R}_+$ and $y \in \mathbf{N}^*$,

$$f(xy) = yf(x). \text{ ②}$$

Consider $x \in \mathbf{Q}_+$, and let $x = \frac{p}{q}$ with $p, q \in \mathbf{N}^*$ and $\gcd(p, q) = 1$. By ② we have $f(qx) = qf(x)$, and since $f(p) = p$, we get $f(x) = \frac{p}{q} = x$, so $f(x) = x$ for all positive rational numbers x.

Further, for $x \in \mathbf{R}_+$ and $y \in \mathbf{Q}_+$, let $y = \frac{p}{q}$ with $p, q \in$ and $\gcd(p, q) = 1$. By the argument above,

$$f(x+y) + yf(x) = f(xy) + f(x) + y.$$

Since $pf(x) = f(px) = f(qxy) = qf(xy)$, we get $f(xy) = \frac{p}{q}f(x) = yf(x)$, and also

$$f(x+y) = f(x) + y. \text{ ③}$$

Then we concern the value of $f(x)$ when $x \in \mathbf{R}_+$. If there exists $x_0 \in \mathbf{R}_+$ such that $f(x_0) < x_0$, then assume that $f(x_0) = x_0 - t$ with $t \in \mathbf{R}_+$. Choose $c \in \mathbf{Q}_+$ such that $x_0 - t < c < x_0$. Then by ③we have $f(x_0) = f(x_0 - c) + c$, so $f(x_0 - c) = f(x_0) - c < 0$, a contradiction. Hence, $f(x) \geq x$ for all positive x.

On the other hand, we show that $f(x) \leq x$ for all positive x, so that $f(x) = x$ for all x.

For $x \in (0, 1)$, we claim that $f(x) \leq 1$. In fact, if there exists $x_0 \in (0, 1)$ with $f(x_0) > 1$, then let $y = \frac{1}{x_0}$ in ①, and we see that

$$f\left(x_0 + \frac{1}{x_0}\right) + f(x_0)f\left(\frac{1}{x_0}\right) = f(1) + f(x_0) + f\left(\frac{1}{x_0}\right).$$

Equivalently, $(f(x_0) - 1)\left(f\left(\frac{1}{x_0}\right) - 1\right) = 2 - f(x_0 + \frac{1}{x_0}) \leq 2 - \left(x_0 + \frac{1}{x_0}\right) < 2 - 2\sqrt{x_0 \cdot \frac{1}{x_0}} = 0$. (Here, we have used the fact that the equality in the involved AM-GM inequality cannot hold for $x \in (0, 1)$.) Since $f(x_0) > 1$, we have $f\left(\frac{1}{x_0}\right) < 1$. However, $\frac{1}{x_0} > 1$, which implies that $f\left(\frac{1}{x_0}\right) \geq \frac{1}{x_0} > 1$, a contradiction. Hence, $f(x) \leq 1$ for $x \in (0, 1)$.

Then we show that $f(x) \leq x$ for all $x \in \mathbf{R}_+$.

Suppose there exists $x_0 \in \mathbf{R}_+$ such that $f(x_0) > x_0$. Let $f(x_0) = tx_0$ with $t > 1$. Then choose $c \in \mathbf{Q}_+$ such that $x_0 < c < tx_0$, and we obtain that

$$f\left(\frac{x_0}{c}\right) = \frac{1}{c}f(x_0) = \frac{tx_0}{c} > 1$$

while $\frac{x_0}{c} \in (0, 1)$, which is contradictory to our previous result.

To summarize, there are exactly two functions that satisfy the equation, which are $f(x) = 2$ and $f(x) = x$.

15. If there exists a function $f(x)$ that satisfies the condition, then by the given inequality (we denote it as ①), we see that $f(x)$ is monotonically decreasing on $\mathbf{R}_+$.

Next, we show that for all $x > 0$,

$$f(x) - f(x+1) \geq \frac{1}{2}. \quad ②$$

In fact, for $x > 0$, we choose a positive integer n such that $nf(x+1) \geq 1$. Then for $k = 0, 1, \ldots, n-1$

$$f\left(x + \frac{k}{n}\right) - f\left(x + \frac{k+1}{n}\right) \geq \frac{f\left(x + \frac{k}{n}\right) \cdot \frac{1}{n}}{f\left(x + \frac{k}{n}\right) + \frac{1}{n}}.$$

Note that $f\left(x + \frac{k}{n}\right) \geq f(x+1) \geq \frac{1}{n}$, so the inequality above implies that

$$f\left(x + \frac{k}{n}\right) - f\left(x + \frac{k+1}{n}\right) \geq \frac{\frac{1}{n}}{1 + \frac{1}{nf\left(x + \frac{k}{n}\right)}} \geq \frac{1}{2n}.$$

Adding the n inequalities, we get that $f(x) - f(x+1) \geq \frac{1}{2}$.

For $x > 0$, choose $m \in \mathbf{N}^*$ such that $f(x) \leq \frac{m}{2}$, and then

$$\begin{aligned} f(x) - f(x+m) &= \sum_{i=0}^{m-1} (f(x+i) - f(x+i+1)) \\ &\geq \sum_{i=0}^{m-1} \frac{1}{2} = \frac{m}{2} \geq f(x). \end{aligned}$$

Hence, $f(x+m) \leq 0$, which is a contradiction.

Therefore, such a function does not exist.

Solution 26

Constructing Functions to Solve Problems

1. Let x_1 and x_2 be two roots of the equation $ax^2+bx+c=0$, and

$$f(x)=ax^2+bx+c+k\left(x+\frac{b}{2a}\right).$$

Then

$$\begin{aligned}f(x_1)f(x_2)&=k^2\left(x_1+\frac{b}{2a}\right)\left(x_2+\frac{b}{2a}\right)\\&=k^2\left[x_1x_2+\frac{b}{2a}(x_1+x_2)+\frac{b^2}{4a^2}\right]\\&=\frac{k^2(4ac-b^2)}{4a^2}<0.\end{aligned}$$

Hence, the equation $f(x)=0$ has a real root between x_1 and x_2.

2. Let $f(x)=\frac{x+(a+b)}{a+ab}$. Since $|a|<1$ and $|b|<1$, we have $1+ab>0$, so $f(x)$ is increasing. Let $x_1=-(1+a)(1+b)<0$ and $x_2=(1-a)(1-b)>0$. Since $f(x_1)=-1$ and $f(x_2)=1$, we have $-1<f(0)<1$. Hence, $-1<\frac{a+b}{1+ab}<1$.

3. Let $f(x)=x^2-ax+b$. Then

$$\begin{cases}f(-1)=1+a+b\geq 0, & ①\\ f(1)=1-a+b\leq 0, & ②\\ f(2)=4-2a+b\geq 0. & ③\end{cases}$$

By ①+③ we get $5 - a + 2b \geq 0$, so $a - 2b \leq 5$. By ②we have

$$a - b - 1 \geq 0.$$

Combining the above with ③we have $a - 2b \geq -1$. Therefore, the value range of $a - 2b$ is $[-1, 5]$.

4. Let $ab + bc + ca = m$. Then $\frac{1}{2} = \frac{1}{a} + \frac{1}{b} + \frac{1}{c} = \frac{ab+bc+ca}{abc} = \frac{m}{abc}$. Hence, $abc = 2m$. Now,

$$\begin{aligned} f(x) &= (x-a)(x-b)(x-c) \\ &= x^3 - 2x^2 + mx - 2m = (x-2)(x^2+m). \end{aligned}$$

   Therefore, one of a, b and c must be 2.

5. The original equation is equivalent to $(x-b)(x-c)+(x-c)(c-a)+(x-a)(x-b) = 0$. Let $f(x) = (x-b)(x-c)+(x-c)(x-a)+(x-a)(x-b)$. Then $f(\frac{2a+b}{3}) = \frac{1}{3}(a-b)(b-c) > 0$ and $f(\frac{a+2b}{3}) = \frac{1}{3}(b-a)(a-c) < 0$. Hence, $f(x) = 0$ has a real root between $\frac{2a+b}{3}$ and $\frac{a+2b}{3}$. Since $a < \frac{2a+b}{3} < \frac{a+2b}{3} < b < c$, this root is a solution to the original equation (since it does not make the denominator zero). Similarly, we also have a root between $\frac{2b+c}{3}$ and $\frac{2c+b}{3}$.

6. The original equation is equivalent to $(3x+y)^5 + (3x+y) = -(x^5+x)$. We construct the function $f(t) = t^5+t$. Then $f(t)$ is an odd increasing function. Since $f(3x+y) = -f(x) = f(-x)$, it follows that $3x+y = -x$, hence $4x + y = 0$.

7. The original equation is equivalent to $2a^2x^2 + 2ax + 1 - a^2 = 0$. Let

$$f(x) = 2a^2x^2 + 2ax + 1 - a^2.$$

   Then the graph of $f(x)$ is a parabola that opens upwards. Since $f(0) = 1 - a^2 < 0$, the equation $f(x) = 0$ has a positive root and a negative root. Since

$$\begin{aligned} f(-1) &= 2a^2 - 2a + 1 - a^2 = (a-1)^2 > 0 \\ f(1) &= 2a^2 + 2a + 1 - a^2 = (a+1)^2 > 0, \end{aligned}$$

   the real roots lie in the intervals $(-1, 0)$ and $(0, -1)$, respectively. This is the desired result.

8. Let $v = w = 1$ and $u = z + \epsilon\,(\epsilon > 0)$. Then $u^2 > 4vw$. Thus, we have

$$((2+\epsilon)^2 - 4)^2 > k\epsilon^2.$$

so $(4\epsilon + \epsilon^2)^2 > k\epsilon^2$, namely $16 + 8\epsilon + \epsilon^2 > k$. Let ϵ tends to 0, and we have $k \leq 16$.

Next, we show that $k = 16$ satisfies the required property.

Suppose positive real numbers u, v and w satisfy the condition. By symmetry, we may assume that $u \leq w$. Consider the function

$$f(x) = 2(2v^2 - uw)x^2 + (u^2 - 4vw)x + 2(2w^2 - uv).$$

We have

$$\begin{aligned} f(1) &= 4v^2 - 2uw + u^2 - 4vw + 4w^2 - 2uv \\ &= u^2 - 2u(v+w) + 4v^2 - 4vw + 4w^2 \\ &= (u - (v+w))^2 + 3v^2 - 6vw + 3w^2 \\ &= (u - (v+w))^2 + 3(v-w)^2 \geq 0. \end{aligned}$$

where the equality holds if and only if $v = w$ and $u = v + w = 2v$. However, this is contradictory to $u^2 > 4vw$. Hence, $f(1) > 0$.

Also since $uw > \sqrt{4vw} \cdot w \geq \sqrt{4v^2}w = 2vw \geq 2v^2$, the parabola graph of $f(x)$ opens downwards. Combining the fact with $f(1) < 0$, we have

$$\Delta = (u^2 - 4vw)^2 - 16(wv^2 - uw)(2w^2 - uv) > 0.$$

9. If $x_1^2 + x_2^2 + x_3^2 = 1$, then the inequality holds obviously.
 If $x_1^2 + x_2^2 + x_3^2 < 1$, then we consider the quadratic function

$$\begin{aligned} f(t) &= (x_1^2 + x_2^2 + x_3^2 - 1)t^2 - 2(x_1y_1 + x_2y_2 + x_3y_3 - 1)t \\ &\quad + (y_1^2 + y_2^2 + y_3^2 - 1) \\ &= (x_1t - y_1)^2 + (x_2t - y_2)^2 + (x_3t - y_3)^2 - (t-1)^2. \end{aligned}$$

The graph of the function is a parabola that opens downwards. Since

$$f(1) = (x_1 - y_1)^2 + (x_2 - y_2)^2 + (x_3 - y_3)^2 \geq 0,$$

the parabola intersects the x-axis. Hence,

$$\begin{aligned} \Delta = 4(x_1y_1 + x_2y_2 + x_3y_3 - 1)^2 - 4(x_1^2 + x_2^2 + x_3^2 - 1) \\ \times (y_1^2 + y_2^2 + y_3^2 - 1) \geq 0 \end{aligned}$$

Equivalently, $(x_1y_1 + x_2y_2 + x_3y_3 - 1)^2 \geq (x_1^2 + x_2^2 + x_3^2 - 1)(y_1^2 + y_2^2 + y_3^2 - 1)$.

10. Since $a+b+c=2$, we have $a^2+b^2+c^2=4-2(ab+bc+ca)$. The original inequality is equivalent to $ab+bc+ca-abc>1$. Let

$$f(x)=(x-a)(x-b)(x-c)=x^3-2x^2+(ab+bc+ca)x-abc.$$

Then $f(1)=1-2+(ab+bc+ca)-abc=ab+bc+ca-abc-1$. On the other hand, since $0<a,\ b,\ c<1$,

$$f(1)=(1-a)(1-b)(1-c)>0.$$

Therefore, $ab+bc+ca-abc-1>0$.

11. Such a function exists. We construct the following function:

$$f(x)=\begin{cases}2, & x=1\\ 3, & x=2\\ 1, & x=3\\ x, & x\neq 1,2,3\end{cases}.$$

12. Let $f(x)=(x-a)(x-b)(x-c)=x^3-(a+b+c)x^2+(ab+bc+ca)x-abc$. If $x\le 0$, then $f(x)<0$, which means that $f(x)$ has no negative real roots. Therefore, a, b and c are all positive.

13. Construct the function

$$f(x)=\left(\sum_{i=1}^{n}\frac{a_i}{\lambda_i}\right)x^2-\left(\frac{\lambda_1+\lambda_n}{\sqrt{\lambda_1\lambda_n}}\right)x+\sum_{i=1}^{n}a_i\lambda_i.$$

We have $f\left(\sqrt{\lambda_1\lambda_n}\right)=\sum_{i=1}^{n-1}a_i\frac{(\lambda_1-\lambda_i)(\lambda_n-\lambda_i)}{\lambda_i}\le 0$ and $\sum_{i=1}^{n}\frac{a_i}{\lambda_i}>0$. Hence, the graph of $f(x)$ opens upwards, and it intersects the $x-$axis. Therefore,

$$\Delta=\left(\frac{\lambda_1+\lambda_n}{\sqrt{\lambda_1\lambda_n}}\right)^2-4\left(\sum_{i=1}^{n}\frac{a_i}{\lambda_i}\right)\left(\sum_{i=1}^{n}a_i\lambda_i\right)\ge 0.$$

14. Let $a=x_2+x_3+x_4$ and $b=x_2x_3x_4$. Then the original inequality is equivalent to $(x_1+a)^2\le 4bx_1$, that is $x_1^2+2(a-2b)x_1+a^2\le 0$. We construct the quadratic function $f(x)=x^2+2(a-2b)x+a^2$. Since $\frac{a}{b}=\frac{x_2+x_3+x_4}{x_2x_3x_4}=\frac{1}{x_2x_3}+\frac{1}{x_3x_4}+\frac{1}{x_4x_2}\le\frac{1}{4}+\frac{1}{4}+\frac{1}{4}<1$, we have $b>a$. Thus, $\Delta=4(a-2b)^2-4a^2=16b(b-a)>0$, and the graph of $f(x)$ has two different intersection points with the $x-$axis, which are $(2b-a-2\sqrt{b(b-a)},0)$ and $(2b-a+2\sqrt{b(b-a)},0)$. Next, we show that $2b-a-2\sqrt{b(b-a)}\le x_1\le 2b-a+2\sqrt{b(b-a)}$.

Since $a \le 3x_1 \le 3a$, we see that $x \in [\frac{a}{3}, a]$. Also

$$2b - a + 2\sqrt{b(b-a)} \ge 2b - a \ge a$$

and

$$\begin{aligned}
&2b - a - 2\sqrt{b(b-a)} \\
&\quad = (\sqrt{b} - \sqrt{b-a})^2 = \left(\frac{a}{\sqrt{b} + \sqrt{b-a}}\right)^2 \\
&\quad = \frac{a}{\left(\sqrt{\frac{b}{a}} + \sqrt{\frac{b}{a} - 1}\right)^2} \le \frac{a}{\left(\sqrt{\frac{4}{3}} + \sqrt{\frac{1}{3}}\right)^2} = \frac{a}{3}.
\end{aligned}$$

Therefore, $2b - a - 2\sqrt{b(b-a)} \le \frac{a}{3} \le x_1 \le a \le 2b - a + 2\sqrt{b(b-a)}$, from which $f(x_1) \le 0$.

15. We construct the quadratic function

$$\begin{aligned}
f(x) &= x^2 - \sqrt{4-k^2}(ac - bd)x + 1 \\
&= (a^2 + b^2 - kab)x^2 - \sqrt{4-k^2}(ac - bd)x + (c^2 + d^2 - kcd) \\
&= \left[\frac{\sqrt{2-k}}{2}(a+b)x - \frac{\sqrt{2+k}}{2}(c-d)\right]^2 \\
&\quad + \left[\frac{\sqrt{2+k}}{2}(a-b)x - \frac{\sqrt{2-k}}{2}(c+d)\right]^2.
\end{aligned}$$

Since $f(x) \ge 0$ for all x, we see that $\Delta = (4-k^2)(ac-bd)^2 - 4 \le 0$. Equivalently, $|ac - bd| \le \frac{2}{\sqrt{4-k^2}}$.

Solution 27

Vectors and Geometry

1. Let G be the origin. Since G is the barycenter of ΔABC, we have $\overrightarrow{GA}+\overrightarrow{GB}+\overrightarrow{GC}=\vec{0}$, $\overrightarrow{AB}=\overrightarrow{GB}-\overrightarrow{GA}$, $\overrightarrow{BC}=\overrightarrow{GC}-\overrightarrow{GB}$ and $\overrightarrow{CA}=\overrightarrow{GA}-\overrightarrow{GC}$. Hence,

$$\overrightarrow{AB}^2+\overrightarrow{BC}^2+\overrightarrow{CA}^2=2\sum\overrightarrow{GA}^2-2\sum\overrightarrow{GA}\cdot\overrightarrow{GB}.$$

Since $(\sum\overrightarrow{GA})^2=0$, we have $\sum\overrightarrow{GA}^2=-2\sum\overrightarrow{GA}\cdot\overrightarrow{GB}$, and combining it with the above relation gives

$$\sum AB^2=3\sum GA^2.$$

2. Suppose the center of the circumcircle is O, and the radius is R. Then

$$A_iG\cdot GB_i=(R+GO)(R-GO)=R^2-GO^2. \text{①}$$

Since G is the barycenter of the polygon, $\sum_{i=1}^n\overrightarrow{A_iG}=\vec{0}$, so $(\sum_{i=1}^n\overrightarrow{A_iG})\cdot\overrightarrow{GO}=0$. Hence,

$$\begin{aligned}nR^2&=\sum_{i=1}^n\overrightarrow{A_iO}^2=\sum_{i=1}^n(\overrightarrow{A_iG}+\overrightarrow{GO})^2\\&=\sum_{i=1}^n\overrightarrow{A_iG}^2+2\left(\sum_{i=1}^n\overrightarrow{A_iG}\right)\cdot\overrightarrow{GO}+nGO^2.\end{aligned}$$

Therefore, $\sum_{i=1}^n\overrightarrow{A_iG}^2=n(R^2-GO^2)$, and combining it with ① we get $\sum_{i=1}^n\frac{A_iG}{B_iG}=n$.

3. Choose an arbitrary point O in the plane as the origin, and using position vectors, we may assume that

$$\overrightarrow{OA_1} = \alpha\overrightarrow{OB} + (1-\alpha)\overrightarrow{OC},$$
$$\overrightarrow{OB_1} = \beta\overrightarrow{OC} + (1-\beta)\overrightarrow{OA},$$
$$\overrightarrow{OC_1} = \gamma\overrightarrow{OA} + (1-\gamma)\overrightarrow{OB},$$

where $\alpha, \beta, \gamma \in (0, 1)$. Thus,

$$\overrightarrow{OG} = \frac{1}{3}\sum\overrightarrow{OA},\ \overrightarrow{OG_1} = \frac{1}{3}\sum\overrightarrow{OA_1} = \frac{1}{3}\sum(\gamma-\beta+1)\overrightarrow{OA}$$
$$\overrightarrow{OG_a} = \frac{1}{3}((2-\beta+\gamma)\overrightarrow{OA} + (1-\gamma)\overrightarrow{OB} + \beta\overrightarrow{OC})$$
$$\overrightarrow{OG_b} = \frac{1}{3}(\gamma\overrightarrow{OA} + (2-\gamma+\alpha)\overrightarrow{OB} + (1-\alpha)\overrightarrow{OC})$$
$$\overrightarrow{OG_c} = \frac{1}{3}((1-\beta)\overrightarrow{OA} + \alpha\overrightarrow{OB} + (2-\alpha+\beta)\overrightarrow{OC}).$$

Further,

$$\overrightarrow{OG_2} = \frac{1}{3}\sum\overrightarrow{OG_a} = \frac{1}{9}\sum(3-2\beta+2\gamma)\overrightarrow{OA}.$$

By the results above, we obtain that

$$\overrightarrow{GG_1} = \overrightarrow{OG_1} - \overrightarrow{OG} = \frac{1}{3}\sum(\gamma-\beta)\overrightarrow{OA},$$
$$\overrightarrow{GG_2} = \overrightarrow{OG_2} - \overrightarrow{OG} = \frac{2}{3}\sum(\gamma-\beta)\overrightarrow{OA}.$$

Hence $\overrightarrow{GG_1} = \frac{3}{2}\overrightarrow{GG_2}$, and G, G_1, and G_2 are collinear.

4. Let $|\overrightarrow{OA_1}| = |\overrightarrow{OA_2}| = |\overrightarrow{OA_3}| = |\overrightarrow{OA_4}| = R$, and C is a point in the plane such that $\overrightarrow{OA_1} + \overrightarrow{OA_2} + \overrightarrow{OA_3} + \overrightarrow{OA_4} = \overrightarrow{OC}$. Since H_1 is the orthocenter of $\Delta A_2A_3A_4$ and $\overrightarrow{OH_1} = \overrightarrow{OA_2} + \overrightarrow{OA_3} + \overrightarrow{OA_4} = \overrightarrow{OC} - \overrightarrow{OA_1}$, so $|\overrightarrow{H_1C}| = |\overrightarrow{OC} - \overrightarrow{OH_1}| = |\overrightarrow{OA_1}| = R$. Similarly, $|\overrightarrow{H_2C}| = |\overrightarrow{H_3C}| = |\overrightarrow{H_4C}| = R$, thus they all lie on the circle centered at C with radius R.

5. Let the circumcenter O of ΔABC be the origin, and using the result of example 2 in Chapter 18, we may assume that $\overrightarrow{OP} = \alpha\overrightarrow{OA} + \beta\overrightarrow{OB} +$

$\gamma\overrightarrow{OC}$, where α, β, $\gamma \in [0, 1]$ and $\alpha + \beta + \gamma = 1$. Then

$$|\overrightarrow{PA}| = |\overrightarrow{OP} - \overrightarrow{OA}| = |(\alpha - 1)\overrightarrow{OA} + \beta\overrightarrow{OB} + \gamma\overrightarrow{OC}|$$
$$\leq (1 - \alpha)|\overrightarrow{OA}| + \beta|\overrightarrow{OB}| + \gamma|\overrightarrow{OC}|.$$

Hence, $\sum|\overrightarrow{PA}| \leq \sum((1 - \alpha)|\overrightarrow{OA}| + \beta|\overrightarrow{OB}| + \gamma|\overrightarrow{OC}|) = \sum(1 - \alpha + \beta + \gamma)R = 4R$.

6. Since $\overrightarrow{OG} = \frac{1}{3}(\overrightarrow{OA} + \overrightarrow{OB} + \overrightarrow{OC})$,

$$OG^2 = \overrightarrow{OG}^2 = \frac{1}{9}\left(\sum\overrightarrow{OA}\right)^2 = \frac{1}{3}R^2 + \frac{2}{9}\sum\overrightarrow{OA}\cdot\overrightarrow{OB}$$
$$= \frac{1}{3}R^2 + \frac{2}{9}R^2\sum\cos 2C.$$

Hence,

$$R^2 - OG^2 = \frac{2}{3}R^2\left(1 - \frac{1}{3}\sum\cos 2C\right)$$
$$= \frac{2}{3}R^2\left(1 - \frac{1}{3}\sum(1 - 2\sin^2 C)\right)$$
$$= \frac{4}{9}R^2\sum\sin^2 C$$
$$= \frac{1}{9}(a^2 + b^2 + c^2).$$

On the other hand, since $S_{\triangle ABC} = \frac{1}{2}ab\sin C = \frac{abc}{4R} = \frac{1}{2}r(a+b+c)$, we have $2Rr = \frac{abc}{a+b+c}$. Also $(a^2+b^2+c^2)(a+b+c) \geq (3\sqrt[3]{a^2b^2c^2})(3\sqrt[3]{abc}) = 9abc$, which means that $2Rr \leq R^2 - OG^2$, hence $OG \leq \sqrt{R(R - 2r)}$.

Remark. Euler's formula gives that $OI = \sqrt{R(R - 2r)}$, so this result is equivalent to $OG \leq OI$.

7. We need the following statement: Suppose in a convex quadrilateral $ABCD$, the four side lengths are a, b, c and d, the diagonals have lengths e and f, and the distance between the midpoints of the two diagonals is m. Then

$$a^2 + b^2 + c^2 + d^2 = e^2 + f^2 + 4m^2. \quad ①$$

In fact,

$$\sum a^2 = \sum(\overrightarrow{OA} - \overrightarrow{OB})^2 = 2\sum\overrightarrow{OA}^2 - 2\sum\overrightarrow{OA}\cdot\overrightarrow{OB} \quad ②$$

and

$$\begin{aligned}
&e^2+f^2+4m^2\\
&=(\overrightarrow{OA}-\overrightarrow{OC})^2+(\overrightarrow{OB}-\overrightarrow{OD})^2+4\left(\frac{\overrightarrow{OA}+\overrightarrow{OC}}{2}-\frac{\overrightarrow{OB}+\overrightarrow{OD}}{2}\right)^2\\
&=(\overrightarrow{OA}-\overrightarrow{OC})^2+(\overrightarrow{OB}-\overrightarrow{OD})^2+(\overrightarrow{OA}+\overrightarrow{OC}-\overrightarrow{OB}-\overrightarrow{OD})^2\\
&=2\sum\overrightarrow{OA}^2-2\overrightarrow{OA}\cdot\overrightarrow{OC}-2\overrightarrow{OB}\cdot\overrightarrow{OD}+2\overrightarrow{OA}\cdot\overrightarrow{OC}+2\overrightarrow{OB}\cdot\overrightarrow{OD}\\
&\quad-2\overrightarrow{OA}\cdot\overrightarrow{OB}-2\overrightarrow{OA}\cdot\overrightarrow{OD}-2\overrightarrow{OC}\cdot\overrightarrow{OB}-2\overrightarrow{OC}\cdot\overrightarrow{OD}\\
&=2\sum\overrightarrow{OA}^2-2\sum\overrightarrow{OA}\cdot\overrightarrow{OB}.\ ③
\end{aligned}$$

Comparing ② and ③ we get ①. Now, go back to the original problem. In the figure below we mark the sides and diagonals of the pentagon, and let m_{xy} be the distance between the midpoints of the diagonals x and y. By the result above,

$$\begin{cases}
a^2+b^2+h^2+e^2=f^2+j^2=4m_{fj}^2,\\
a^2+b^2+c^2+i^2=f^2+g^2+4m_{fg}^2,\\
a^2+d^2+e^2+g^2=i^2+j^2+4m_{ij}^2,\\
c^2+d^2+e^2+f^2=h^2+i^2+4m_{hi}^2,\\
b^2+c^2+d^2+j^2=h^2+g^2+4m_{hg}^2.
\end{cases}$$

We sum up the five equalitiess, and since m_{fj} and m_{fg} are not both zero, we conclude that $3(a^2+b^2+c^2+d^2+e^2)>f^2+g^2+h^2+i^2+j^2$.

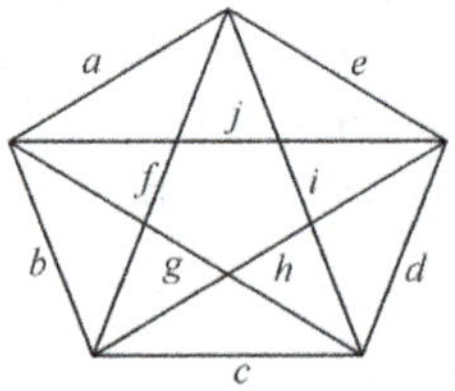

Exercise 7

8. As shown in the figure, let O be the center of the equilateral triangle ABC, and P is a point on its incircle such that $\overrightarrow{PA}$, $\overrightarrow{PB}$, $\overrightarrow{PC}$, and $\overrightarrow{PO}$

are pairwise noncollinear. Then

$$\begin{aligned}
&(\overrightarrow{PA}+\overrightarrow{PB})\cdot(\overrightarrow{PC}+\overrightarrow{PO})\\
&=(\overrightarrow{PO}+\overrightarrow{OA}+\overrightarrow{PO}+\overrightarrow{OB})\cdot(\overrightarrow{PO}+\overrightarrow{OC}+\overrightarrow{PO})\\
&=(2\overrightarrow{PO}+\overrightarrow{OA}+\overrightarrow{OB})\cdot(2\overrightarrow{PO}+\overrightarrow{OC})\\
&=(2\overrightarrow{PO}-\overrightarrow{OC})\cdot(2\overrightarrow{PO}+\overrightarrow{OC})\\
&=4\overrightarrow{PO}^2-\overrightarrow{OC}^2=4PO^2-OC^2=0.
\end{aligned}$$

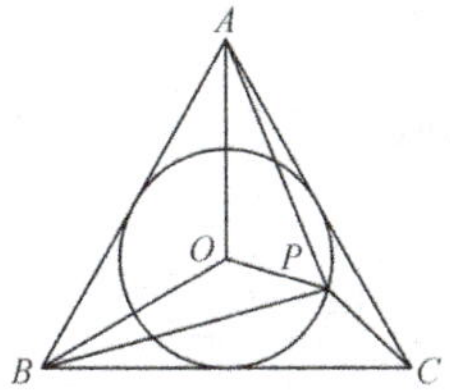

Exercise 8

Hence, $(\overrightarrow{PA}+\overrightarrow{PB})$ is perpendicular to $(\overrightarrow{PC}+\overrightarrow{PO})$. Other relations are similar. Therefore, such four vectors exist.

9. Let T be the common point of AP, BQ, CR and DW. We use the outer product of vectors to solve this problem. (If $\vec{a}\times\vec{b}=\vec{c}$, then $|\vec{c}|=|\vec{a}|\cdot|\vec{b}|\cdot\sin\theta$, where θ is the angle between $\vec{a}$ and $\vec{b}$.)
By the given condition, $\overrightarrow{TA}\times\overrightarrow{TP}=\vec{0}$, so $\overrightarrow{TA}\times(\overrightarrow{TC}+\overrightarrow{TD})=\vec{0}$. Hence,

$$\overrightarrow{TA}\times\overrightarrow{TC}=\overrightarrow{TD}\times\overrightarrow{TA}.$$

Similarly, we get $\overrightarrow{TB}\times\overrightarrow{TD}=\overrightarrow{TE}\times\overrightarrow{TB}$, $\overrightarrow{TC}\times\overrightarrow{TE}=\overrightarrow{TA}\times\overrightarrow{TC}$, and $\overrightarrow{TD}\times\overrightarrow{TA}=\overrightarrow{TB}\times\overrightarrow{TD}$. We sum up the four equalities, obtaining that $\overrightarrow{TE}\times\overrightarrow{TB}=\overrightarrow{TC}\times\overrightarrow{TE}$. Equivalently, $\overrightarrow{TE}\times(\overrightarrow{TB}+\overrightarrow{TC})=\vec{0}$, hence the line EN passes through T.

10. We draw a folded line $B_0B_1\ldots B_n$ such that $\overrightarrow{B_{i-1}B_i}=\alpha_i\overrightarrow{OA_i}(1\le i\le n)$. Construct a plane coordinate system such that B_0 is the origin, B_1 lies on the positive half of the $x-$axis, and B_2 lies above the $x-$axis. For $1\le i\le n$, let $(x_i,\, y_i)$ be the coordinates of B_i. Then $0=y_1<y_2<\cdots<y_{\left[\frac{(n+1)}{2}\right]}$, $y_n<y_{n-1}<\cdots<y_{\left[\frac{n}{2}\right]+1}$. For $1\le j\le\left[\frac{n-1}{2}\right]$, the slope of the line B_jB_{j+1} has the same absolute value as the slope of $B_{n-j}B_{n-j+1}$, and the line segment B_jB_{j+1} has a greater length than $B_{n-j}B_{n-j+1}$, so

$$(y_{j+1}-y_j)+(y_{n-j+1}-y_{n-j})>0.$$

Hence, $y_n = y_n - y_1 = (y_n - y_{n-1}) + (y_{n-1} - y_{n-2}) + \cdots + (y_2 - y_1) > 0$, thus $B_0 \neq B_n$ and

$$\alpha_1\overrightarrow{OA_1} + \alpha_2\overrightarrow{OA_2} + \cdots + \alpha_n\overrightarrow{OA_n} \neq \vec{0}.$$

11. Assume that the radius of Γ is 1. The following proof can be compared with the result and method in Example 6.

(1) By the conclusion (2) in Example 6,

$$\sum_{i=1}^{n} PA_i^2 = n(1 + \overrightarrow{OP}^2), \quad ①$$

where O is the center of Γ. Since $PA_i \cdot PB_i = 1 - OP^2$ with $i = 1, 2, \ldots, n$,

$$\begin{aligned} PB_i^2 &= \frac{(1-OP^2)^2}{PA_i^2} = \frac{(1-OP^2)^2}{(\overrightarrow{OP} - \overrightarrow{OA_i})^2} \\ &= \frac{(1-OP^2)^2}{1 + OP^2 - 2\overrightarrow{OP} \cdot \overrightarrow{OA_i}}. \quad ② \end{aligned}$$

In order to prove the proposition, we need the following inequality:

$$(1-OP^2)^2 \leq (1 + OP^2 - 2\overrightarrow{OP} \cdot \overrightarrow{OA_i})(1 + OP^2 + 2\overrightarrow{OP} \cdot \overrightarrow{OA_i}). \quad ③$$

In fact, ③ is equivalent to

$$\begin{aligned} &(1-OP^2)^2 \leq (1+OP^2)^2 - 4(\overrightarrow{OP} \cdot \overrightarrow{OA_i})^2 \\ &\Leftrightarrow 4(\overrightarrow{OP} \cdot \overrightarrow{OA_i})^2 \leq (1+OP^2)^2 - (1-OP^2)^2 = 4OP^2 \\ &\Leftrightarrow \left|\overrightarrow{OP} \cdot \overrightarrow{OA_i}\right| \leq OP \end{aligned}$$

and $|\overrightarrow{OP} \cdot \overrightarrow{OA_i}| \leq |\overrightarrow{OP}| \cdot |\overrightarrow{OA_i}| = OP^2$, so ③ holds.
By ③ and ②, we get

$$\begin{aligned} \sum_{i=1}^{n} PB_i^2 &\leq \sum_{i=1}^{n} (1 + OP^2 + 2\overrightarrow{OP} \cdot \overrightarrow{OA_i}) \\ &= n(1+OP^2) + 2\overrightarrow{OP} \cdot \sum_{i=1}^{n} \overrightarrow{OA_i} \\ &= n(1+OP^2). \end{aligned}$$

Comparing the above with ①, we have proved (1).

(2) By the conclusion (1) in Example 6,

$$\sum_{i=1}^{n} PA_i \geq n,$$

so

$$\begin{aligned}\sum_{i=1}^{n} PB_i &= \sum_{i=1}^{n}(A_iB_i - PA_i)\\ &\leq \sum_{i=1}^{n}(2 - PA_i) = 2n - \sum_{i=1}^{n} PA_i\\ &\leq n \leq \sum_{i=1}^{n} PA_i.\end{aligned}$$

12. We use $(i,\, j)$ to denote the center of the square in the ith row and the jth column, and let

$$S = \{(i,\, j)\,|\,1 \leq i,\, j \leq n\}.$$

For each $X \in S$, let $\omega(X)$ denote the number in the square X. Let $O(0,\, 0)$ be the origin. It suffices to show that

$$\overrightarrow{OP} = \sum_{\substack{X,Y\in S\\ \omega(X)<\omega(Y)}} (\overrightarrow{OY} - \overrightarrow{OX}) = \vec{0}.$$

Note that for each $Y \in S$, there are exactly $\omega(Y) - 1$ points $X \in S$ such that $\omega(X) < \omega(Y)$, and there are exactly $n^2 - \omega(Y)$ points $X \in S$ such that $\omega(X) > \omega(Y)$. Hence,

$$\overrightarrow{OP} = \sum_{Y\in S} (2\omega(Y) - (n^2+1))\overrightarrow{OY}. \text{①}$$

Suppose the sum of the numbers in each column or row equals σ. Then by ①, the x-coordinate of $\overrightarrow{OP}$ satisfies

$$\begin{aligned}x_P &= \sum_{k=1}^{n}\sum_{Y\in T_k} k(2\omega(Y) - (n^2+1))\\ &= \sum_{k=1}^{n} k(2\sigma - n(n^2+1))\end{aligned}$$

$$= n(n+1)\sigma - \frac{n^2(n+1)}{2}(n^2+1)$$

$$= n(n+1)\cdot\frac{1+2+\cdots+n^2}{n} - \frac{n^2(n+1)}{2}(n^2+1)$$

$$= 0.$$

Here, T_k denotes the squares in the kth column.
Similarly, we can prove that the y-coordinate of $\overrightarrow{OP}$ is also 0, and the desired conclusion follows.

Solution 28

Tetrahedrons

1. $\sqrt{3}$. Hint: As shown in the figure, let $AB = a$. Then the surface area of the cube is $S_1 = 6a^2$, and the surface area of the tetrahedron A_1BDC_1 is

$$S_2 = 4S_{\Delta A_1BD} = 2\sqrt{3}a^2.$$

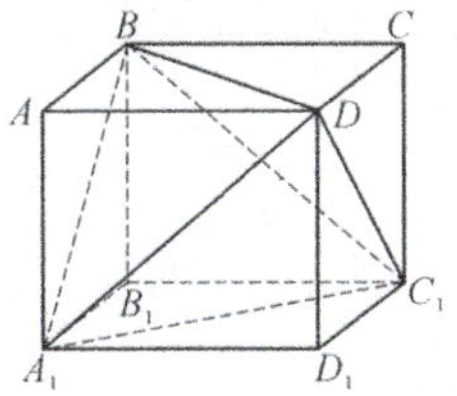

Exercise 1

2. $\frac{h}{3}\sqrt{S_tS_b}$. Hint: The volume of the triangular frustum is

$$V_1 = \frac{h}{3}(S_t + \sqrt{S_tS_b} + S_b).$$

Hence, the desired volume is $V = \frac{h}{3}(S_t + \sqrt{S_tS_b} + S_b) - \frac{h}{3}S_t - \frac{h}{3}S_b = \frac{h}{3}\sqrt{S_tS_b}$.

3. $\pi - \arccos\frac{1}{3}$. Hint: Assume that the regular tetrahedron has edge length 1, and $AO = R$. Let E be the intersection point of the line AO and the face BCD. Then E is the center of ΔBCD, so $BE = \frac{2}{3} \times \frac{\sqrt{3}}{2}BC = \frac{\sqrt{3}}{3}$ and $AE = \sqrt{1 - \left(\frac{\sqrt{3}}{3}\right)^2} = \frac{\sqrt{6}}{3}$. Let $\angle AOB = \alpha$. Then

$$\cos(\pi - \alpha) = \frac{OE}{OB} = \frac{\sqrt{6}}{3R} - 1.$$

In the right triangle BEO, we have $R^2 = \left(\frac{\sqrt{3}}{3}\right)^2 + \left(\frac{\sqrt{6}}{3} - R\right)^2$. Hence, $R = \frac{\sqrt{6}}{4}$ and $\cos(\pi - \alpha) = \frac{1}{3}$, from which $\alpha = \pi - \arccos\frac{1}{3}$.

4. $90°$. Hint: Suppose the line passing through E and parallel to BD intersects AD at G. Then $\angle FEG = \alpha$ and $\frac{AE}{EB} = \frac{AG}{GD} = \frac{CF}{FD}$, so $FG \parallel AC$ and $\angle EFG = \beta$. Hence,

$$\alpha + \beta = 180° - \angle EGF = 90°.$$

5. Since I is the incenter of the tetrahedron, the distances from I to the faces ACD and BCD are equal. Since I lies on EF, the distances from E to the faces ACD and BCD are also equal (note that E lies on the bisector of the dihedral angle $A-CD-B$). Let h be the distance from E to these two faces, let h_1 be the distance from A to the face BCD, and let h_2 be the distance from B to the face ACD. Then since E is the midpoint of AB, we have $h_1 = 2h$ and $h_2 = 2h$, so $h_1 = h_2$. By the volume method, $S_{\triangle ACD} = S_{\triangle BCD}$. This implies that A and B are equidistant to CD. Similarly, C and D are equidistant to AB. Let C' and D' be the projections of C and D onto the line AB, respectively.

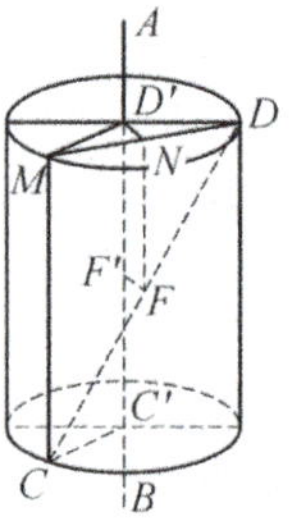

Exercise 5

From the proven results, we see that C and D lie on the same cylinder with the axis AB (as shown in the figure). Let π be the plane passing through DD' and perpendicular to AB, and let M and N be the projections of C and F onto π, respectively. Then N is the midpoint of DM. Since $DD' = D'M$, we have $D'N \perp MD$, and since $CM \perp D'N$, we also have $D'N \perp$ plane CMD. Further, if we draw a plane through the midpoint of $C'D'$ (denoted as F') and perpendicular to AB, then this plane bisects CD. Thus, the quadrilateral $D'NFF'$ is a rectangle, and $FF' \parallel D'N$ and $FF' \perp$ plane CMD, so FF' is the common perpendicular line segment between AB and CD.

Let π^* be the perpendicular bisecting plane of BC. We claim that π^* intersects EF. Suppose not. Then $\pi^* \parallel EF$ and $BC \perp EF$, so $EF \perp$ plane ABC and $EF \perp$ plane BCD, which implies that plane $ABC \parallel$ plane BCD, a contradiction. Now, let O be the intersection point of π^* and EF. Then $OB = OC$. Since $OE \perp AB$ and E is the midpoint of AB, we also have $OA = OB$. Similarly, $OC = OD$, and hence $OA = OB = OC = OD$. Therefore, O is the circumcenter of the tetrahedron $ABCD$, which lies on the line EF.

6. As shown in the figure, $V_{P-QRS} = V_{P-QOR} + V_{P-QOS} + V_{P-ROS}$. Note that $V_{P-QOR} = \frac{PQ \cdot PR}{PA \cdot PB} V_{P-AOB}$, $V_{P-QOS} = \frac{PS \cdot PQ}{PC \cdot PA} V_{P-AOC}$, $V_{P-ROS} = \frac{PR \cdot PS}{PB \cdot PC} V_{P-BOC}$, $V_{P-QRS} = \frac{PQ}{AP} \cdot \frac{PR \cdot PS}{PB \cdot PC} V_{A-PBC}$, $V_{P-AOB} = V_{P-AOC} = V_{P-BOC} = \frac{1}{3} V_{A-PBC}$, and $PA = PB = PC$. Hence,

$$\frac{PQ \cdot PR}{PA^2} + \frac{PS \cdot PQ}{PA^2} + \frac{PR \cdot PS}{PA^2} = \frac{3PQ \cdot PR \cdot PS}{PA^3}.$$

Therefore, $\frac{1}{PQ} + \frac{1}{PR} + \frac{1}{PS} = \frac{3}{PA}$ is a constant.

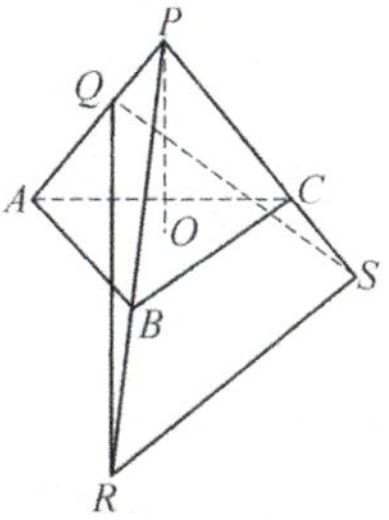

Exercise 6

7. By the condition that $PD \parallel SC$, the points S, C, D, and P lie in the same plane. Since M lies on CD, the lines SM and PD lie in the same plane. The line SM does not intersect SC, so it must intersect PD, and we denote their intersection point as O. Since $SA \perp SC$ and $SB \perp SC$, we have $SC \perp$ plane SAB, so $SC \perp AB$ and $SC \perp SD$. Since $OD \parallel SC$, it follows that $OA = OB = \sqrt{AD^2 + OD^2} = \sqrt{SD^2 + OD^2} = OS$. Further, $OD \parallel SC$ and M is the barycenter of ΔABC, so $\frac{OD}{SC} = \frac{DM}{MC} = \frac{1}{2}$. Thus,

$$OC = \sqrt{SD^2 + (SC - OD)^2} = \sqrt{SD^2 + OD^2} = OS.$$

Therefore, O is the circumcenter of this tetrahedron.

8. Let α be the dihedral angle between two faces of a regular tetrahedron and β be the angle between an edge and its adjacent altitude. Suppose the distances from E to BC, CA and AB are x, y, and z, respectively. Then $\angle FEG = \pi - \alpha$, $EF = z \sin \alpha$, and $EG = y \sin \alpha$. It follows that $S_{\Delta EGF} = \frac{1}{2} EF \cdot EG \cdot \sin \alpha = \frac{1}{2} yz \sin^3 \alpha$. Since plane $EFG \perp AD$, the distance from O to the plane EGF equals the length of the projection of OE onto AD. Let P be the projection of E onto BC. Then we easily see that $AD \perp$ plane OBC, so the length of the projection of OE onto AD equals the length of the projection of EP onto AD. Let $EP = x$.

Then the angle between AD and EP is β, thus $V_{OEFG} = \frac{1}{3}x\cos\beta \cdot \frac{1}{2}yz\sin^3\alpha = \frac{1}{6}xyz\sin^3\alpha\cos\beta$. Similarly, $V_{OEGH} = \frac{1}{6}xyz\sin^3\alpha\cos\beta$ and $V_{OEFH} = \frac{1}{6}xyz\sin^3\alpha\cos\beta$, so $V_{OEFG} = V_{OEGH} = V_{OEFH}$. By $V_{OEFG} = V_{OEGH}$ we get $V_{F-OEG} = V_{H-OEG}$, hence F and H have an equal distance to the plane OEG. Suppose the intersection line of the plane OEG and the plane GFH is GQM. Then M is the midpoint of FH, therefore GQ passes through the midpoint of FH. Similarly, FQ passes through the midpoint of FG, and thus Q is the barycenter of ΔFGH.

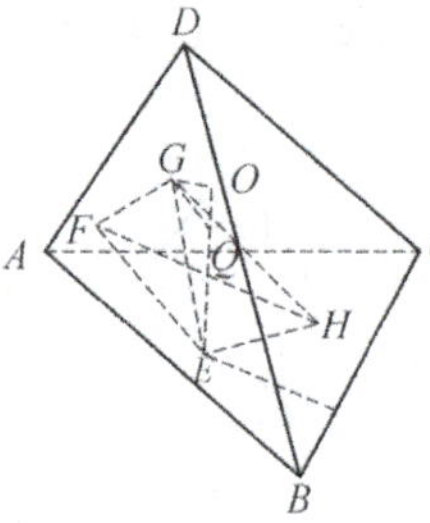

Exercise 8

9. (1) As shown in the figure, let O be the projection of P onto the plane ABC and let D be the intersection point of the lines CO and AB. By the condition, O is the orthocenter of ΔABC, so $PD \perp AB$ and $CP \perp PD$. Let $S_{\Delta PAB} = S_1$, $S_{\Delta BPC} = S_2$, $S_{\Delta CPA} = S_3$, $S_{\Delta ABC} = S$, $S_{\Delta OAB} = S_1{}'$, $S_{\Delta OBC} = S_2{}'$, and $S_{\Delta OCA} = S_3{}'$. Then $\frac{1}{3}S_1c = \frac{1}{3}S_2a = \frac{1}{3}S_3b = \frac{1}{3}Sh$. Hence, $S_1 = \frac{h}{c}S$, $S_2 = \frac{h}{a}S$, and $S_3 = \frac{h}{b}S$. Since $PC \perp PD$, we have $S_1 = \frac{S_1{}'}{\cos\angle PDC} = \frac{S_1{}'}{\frac{h}{c}}$. Similarly, $S_2 = S_2{}' \cdot \frac{a}{h}$ and $S_3 = S_3{}' \cdot \frac{b}{h}$, so $S_1{}' = \frac{h^2}{c^2}S$, $S_2{}' = \frac{h^2}{a^2}S$, and $S_3{}' = \frac{h^2}{b^2}S$. Since $S_1{}' + S_2{}' + S_3{}' = S$, we obtain that $h^{-2} = a^{-2} + b^{-2} + c^{-2}$.

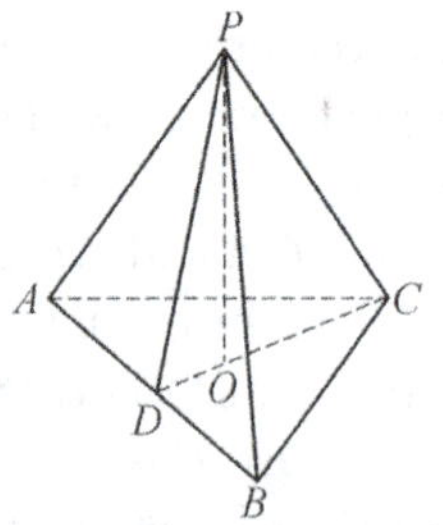

Exercise 9

(2) We have $V_{M-PAC} + V_{M-PAB} + V_{M-PBC} = V_{C-PAB}$, so $\frac{x}{a} + \frac{y}{b} + \frac{z}{c} = 1$. Hence,

$$\frac{xyz}{abc} \le \left(\frac{\frac{x}{a} + \frac{y}{b} + \frac{z}{c}}{3}\right)^3 = \frac{1}{27},$$

and the equality holds when $\frac{x}{a} = \frac{y}{b} = \frac{z}{c}$.

10. Let S be the barycenter of the tetrahedron $C - ABD$, and let EFG be a plane passing through S and parallel to the plane ABD, which intersects CA, CB and CD at G, E, and F, respectively (as shown in the figure). Then $SC : SO = 3 : 1$.

If P lies inside the tetrahedron $SABD$, then $PP_1 \le \frac{1}{4}h_C$ and $PC \ge \frac{3}{4}h_C$ (here h_C is the distance from C to the plane ABD). Hence, $PP_1 \le \frac{1}{3}PC$, so $H \ge 3h$.

Similarly, if P lies in $SABC$, $SBCD$, and $SACD$, then we also have $H \ge 3h$, and thus the proposition follows.

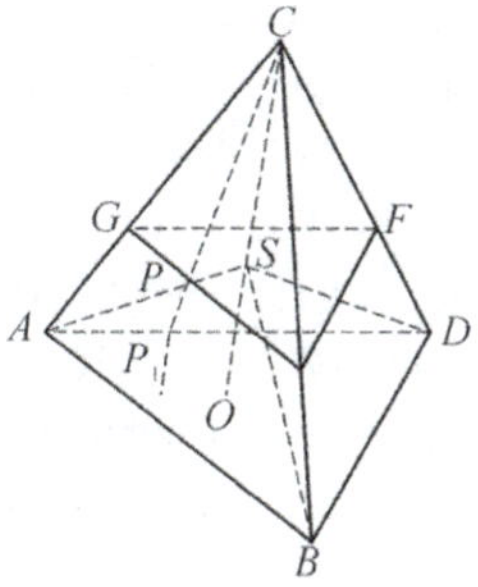

Exercise 10

Remark. In a tetrahedron, the line segment between a vertex and the barycenter of its opposite face is called a median of the tetrahedron. We invite the readers to prove the following fact: The four medians of a tetrahedron intersect at one point (called the barycenter of the tetrahedron), and the barycenter divides each median into two parts with ratio $3 : 1$.

11. Suppose the plane π passing through the points A, P, and Q intersects the tetrahedron at the triangle AEF, and assume that E lies on BC and F lies on BD. Then $\angle PAQ < \angle EAF$. We shall prove that $\angle EAF \le 60°$. Note that at least one of AE and AF is not an edge of the tetrahedron. Assume that AE is not an edge. Then by symmetry $CF = AF$, and since $\angle CEF > \angle CBD = 60° = \angle DCE \ge \angle FCE$,

we have $CF > EF$, so $AF > EF$. Therefore, we always have $\angle FAE \leq 60°$, which implies that $\cos\angle PAQ > \frac{1}{2}$.

12. We use analogy to think. Consider an equilateral triangle with side length 1, and for each side we draw a half-circle whose diameter is that side. Let S' be the intersection of the three regions enclosed by the half-circles, and we examine the properties of S'.

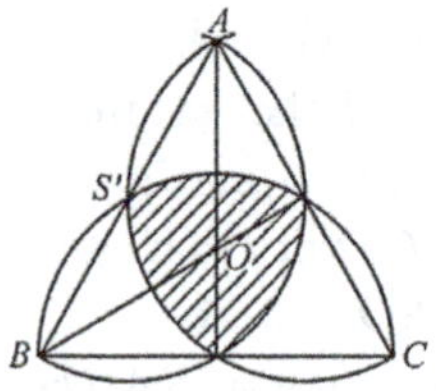

Exercise 12(1)

As shown in the first figure, it is easy to see that S' is contained in a circle with radius $\frac{\sqrt{6}}{3}$, whose center is the center of the equilateral triangle. Hence, the distance between any two points in S' is not greater than $\frac{\sqrt{3}}{3}$.

We deal with the problem via similar methods. As shown in the second figure, let M and N be the midpoints of BC and AD, respectively. Let G be the center of the face BCD, and $MN \cap AG = O$. Then apparently O is the center of the regular tetrahedron.

By the volume method, $OG = \frac{1}{4}AG = \frac{1}{2\sqrt{6}}$. Thus, the distance between any two points in the ball with center O and radius OG is not greater than $\frac{1}{\sqrt{6}}$. Next, we show that this ball contains S.

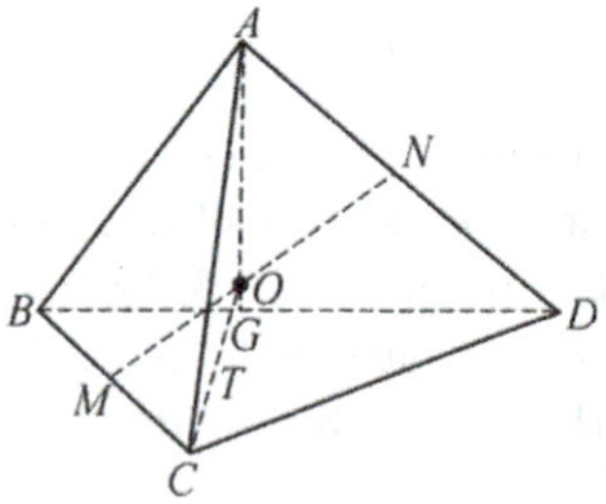

Exercise 12(2)

By symmetry, we may assume that P lies in the tetrahedron $OMCG$, and we claim that if $PO > \frac{1}{\sqrt{6}}$, then $P \notin S$. Suppose the ball (centered at O with radius OG) intersects OC at T. Then in the triangle TON,

$$ON = \frac{\sqrt{2}}{4}, \quad OT = \frac{1}{2\sqrt{6}}, \quad \cos\angle TON = \cos(\pi - \angle TOM) = -\frac{OM}{OC}.$$

By the law of cosines, $TN^2 = ON^2 + OT^2 + 2ON \cdot OT \cdot \frac{OM}{OC} = \frac{1}{4}$, so $TN = \frac{1}{2}$. Also N is the midpoint of AD in the right triangle AGD so $GN = \frac{1}{2}$. Since $GN = TN = \frac{1}{2}$, $DG = OT$ and $ON = ON$, we see that $\Delta GON \cong \Delta TON$. Now, P lies in the tetrahedron $OMCG$, from which $\angle PON \geq \angle TON > \frac{\pi}{2}$ and $PO > TO$. Thus, by the law of cosines,

$$\begin{aligned} PN^2 &= PO^2 + ON^2 - 2PO \cdot ON \cdot \cos\angle PON \\ &> TO^2 + ON^2 - 2TO \cdot ON \cdot \cos\angle TON = TN^2 = \frac{1}{4}. \end{aligned}$$

Hence, P lies outside the ball with diameter AD, which means that $P \notin S$. Therefore, the ball centered at O with radius OG contains S, and our proof is completed.

Solution 29

The Five Centers of a Triangle

1. As shown in the figure, let K be the midpoint of BC. Then $\angle KAC = \angle NKC = 30°$, so

$$\angle ANK = \angle NKC + \angle ACB = 60°.$$

Hence, A, K, N, and M all lie on Ω, and

$$\angle KMN = \angle KAN = 30°,$$

$$\angle AMK = \angle ANK = 60°.$$

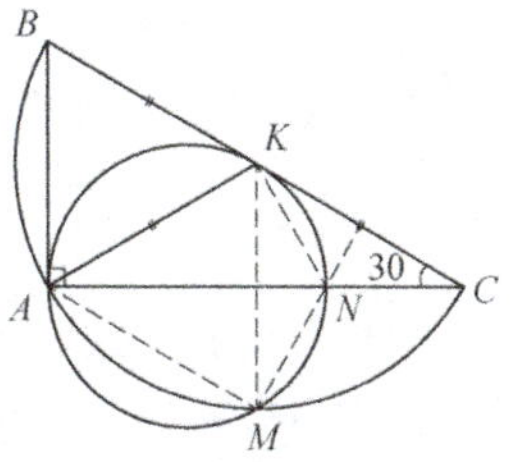

Exercise 1

Apparently, K is the circumcenter of ΔABC, so $KM = KC = AK$. Also $\angle AMK = 60°$, so ΔAKM is an equilateral triangle, which means that $\angle AKM = 60°$. Since $\angle AKB = 60°$, it follows that $\angle MKC = 60°$ and $\angle KMN = 30°$, hence $MN \perp BC$.

2. As shown in the figure, let M be the midpoint of CD. Then AM is the perpendicular bisector of CD. Suppose the line AM intersects the lines DE and BE at P and Q, respectively. Then $PC = PD$. By the given condition,

$$\begin{aligned}\angle EBA + \angle CBA &= \angle A + \angle B + \angle A \\ &= 180° - \angle C + \angle A = 90°.\end{aligned}$$

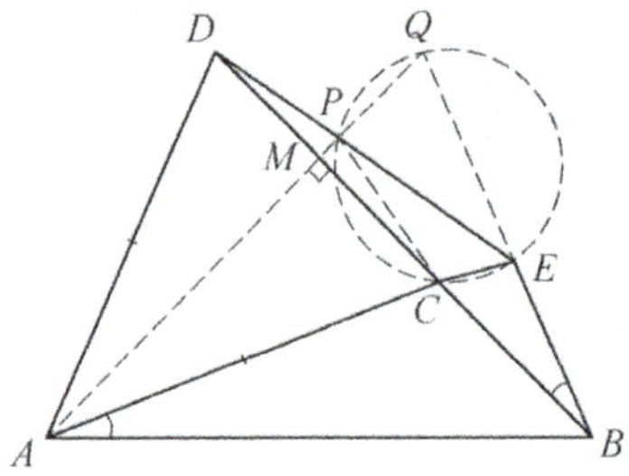

Exercise 2

Hence, $AC \perp BE$. Note that BC and AC are altitudes of the triangle ABQ, so C is the orthocenter of ΔABQ, and thus

$$\angle CQE = \angle CQB = \angle A = \frac{1}{2}\angle A + \frac{1}{2}\angle A = \angle PDC + \angle PCD = \angle CPE.$$

Therefore, C, P, Q and E are concyclic, hence

$$\angle CED = \angle CEP = \angle CQP = \angle CQA = \angle CBA = \angle B.$$

3. As shown in the figure, let I be the incenter of the triangle ABC. Then

$$\angle BFK = 90^\circ - \angle B, \angle BFD = 90^\circ - \frac{1}{2}\angle B.$$

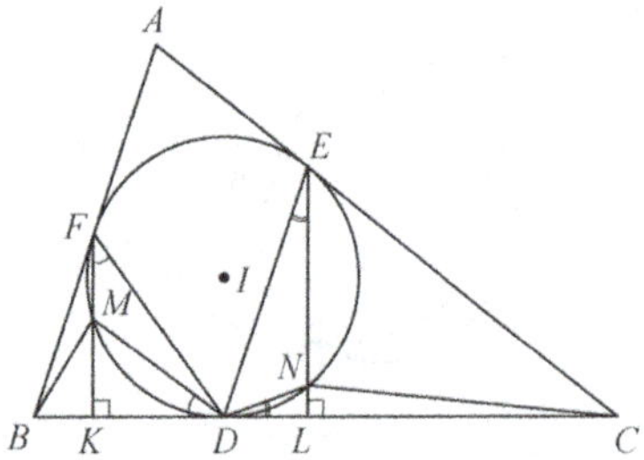

Exercise 3

Thus, $\angle DFM = \angle BFD - \angle BFK = \frac{1}{2}\angle B$. Also since BD is tangent to the incircle, $\angle DFM = \angle MDK$, so $\angle MDK = \frac{1}{2}\angle B$. Hence, $\Delta MDK \sim \Delta BID$, from which $\frac{MK}{DK} = \frac{r}{BD}$ (here r is the radius of the incircle). Equivalently, $r = \frac{MK \cdot BD}{DK}$. Similarly, $r = \frac{NL \cdot CD}{DL}$, so $\frac{MK \cdot BD}{DK} = \frac{NL \cdot CD}{DL}$, and consequently

$$\frac{S_{\Delta BMD}}{S_{\Delta CND}} = \frac{MK \cdot BD}{NL \cdot CD} = \frac{DK}{DL}.$$

4. If $AB = BC$, then the conclusion holds obviously. Now, we assume that $AB < BC$, and let I_1 and I_2 denote the incenters of ΔAKB

and ΔCLB, respectively (as shown in the figure). Suppose the lines BI_1 and BI_2 intersect the circumcircle of ΔABC again at P and Q, respectively, and let R be the midpoint of the arc ABC.

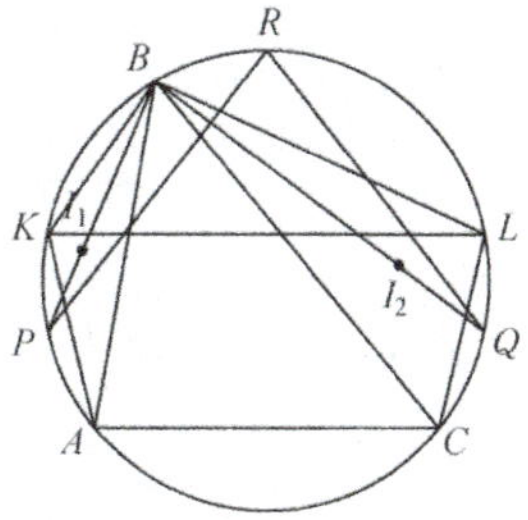

Exercise 4

Since $KL \parallel AC$, we have $\widehat{AK} = \widehat{CL}$, and since P and Q are the midpoints of these two arcs, $RP = RQ$ and $PA = QC$. Hence, by the property three of the incenter, $PI_1 = PA = QC = QI_2$. Also $\angle I_1PR = \angle I_2QR$, so $\Delta I_1PR \cong \Delta I_2QR$ and $RI_1 = RI_2$.

5. Let M and N be the midpoints of AB and AC, respectively. Then B, M, N and C are concyclic. Since $\angle BPC = 120° > 90°$, the point P lies inside the circle passing through B, M, N, and C. Hence, $\angle MPN > \angle MBN = 30°$. Also since K, M, O, and P are concyclic, and L, N, O, and P are concyclic, $\angle MKO = \angle MPO$ and $\angle NLO = \angle NPO$, thus

$$\angle AKO + \angle ALO = \angle MPN > 30°.$$

Therefore, $\angle KOL = \angle A + \angle AKO + \angle ALO > 90°$.

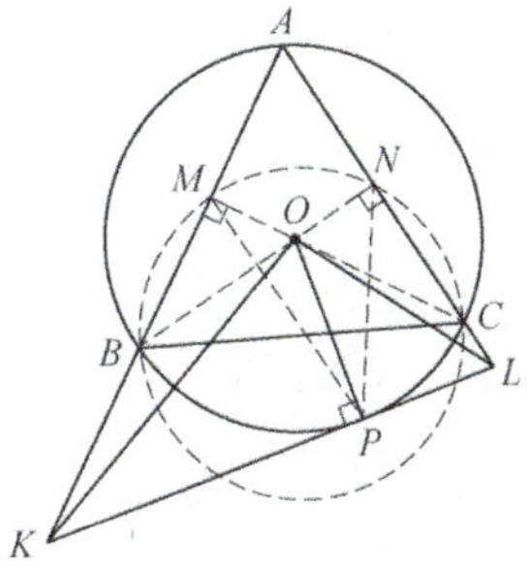

Exercise 5

6. As shown in the figure, we form the line segments DE and DC, and suppose the angle bisector of $\angle BAC$ intersects DE and DC at I and G, respectively. Since $AD = AC$, we get that $AG \perp DC$ and $ID = IC$.

Since D, C and E lie on $\odot A$, we have $\angle IAC = \frac{1}{2}\angle DAC = \angle IEC$, so A, I, E, and C are concyclic, and $\angle CIE = \angle CAE = \angle ABC$. Since $\angle CIE = 2\angle ICD$, it follows that $\angle ICA = \frac{1}{2}\angle ABC$, and hence

$$\angle AIC = \angle IGC + \angle ICG = 90^\circ + \frac{1}{2}\angle ABC,$$

$$\angle ACI = \frac{1}{2}\angle ACB.$$

These imply that I is the incenter of ΔABC.

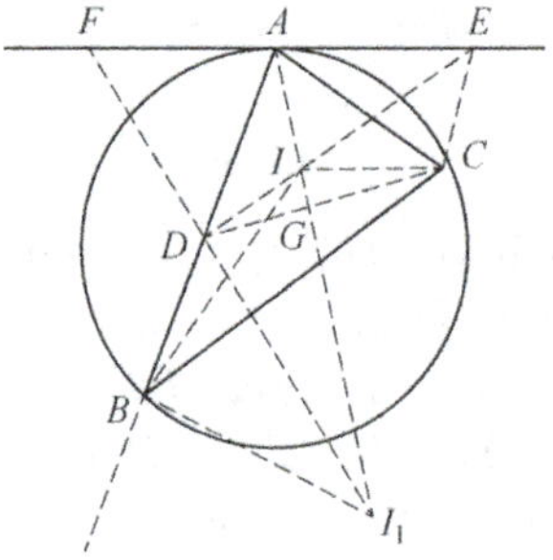

Exercise 6

7. As shown in the figure, let T be the intersection point of AO and l, and form the line segment $B'T$. Obviously AO is the axis of symmetry of ΔABC, and B' and C' are symmetric with respect to AT, so

$$\angle B'TO = \angle C'TO. \text{ ①}$$

Since $l \parallel AC$, we also see that

$$\angle C'TO = \angle OAC = \angle OCA. \text{ ②}$$

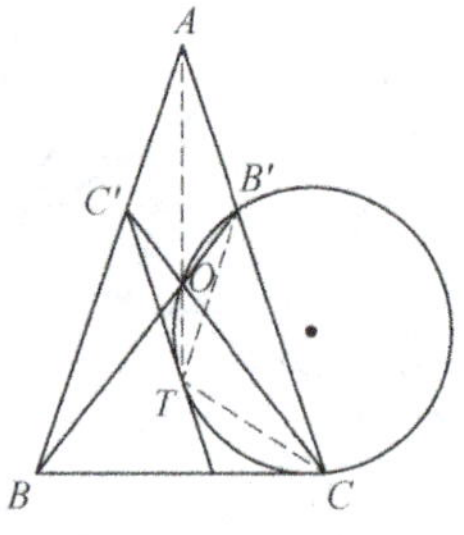

Exercise 7

From ① and ② we get $\angle B'TO = \angle OCA$, so T lies on the circumcircle of $\Delta B'OC$. By symmetry, $\angle OB'T = \angle OC'T = \angle OCA = \angle OTC'$, hence $C'T$ is tangent to the circumcircle of $\Delta B'OC$.

8. Let $\angle A = \alpha$, $\angle B = \beta$, and $\angle C = \gamma$. Since $\angle PBA + \angle PCA + \angle PBC + \angle PCB = \alpha + \beta$, by the condition,

$$\angle PBC + \angle PCB = \frac{\alpha + \beta}{2}.$$

Hence, $\angle BPC = \pi - \frac{\alpha+\beta}{2} = \angle BIC$. Since P and I lie on the same side of BC, the points B, C, I and P are concyclic, in other words, P lies on the circumcircle of ΔBCI (denoted as ω). Let Ω denote the circumcircle of ΔABC. Then the center of ω is the midpoint of the arc BC (on Ω), which is the intersection point of the angle bisector of $\angle BAC$ with Ω.

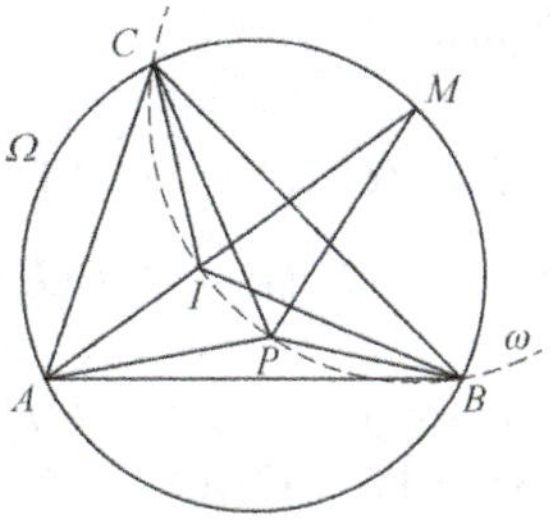

Exercise 8

In ΔAPM, we have

$$AP + PM \geq AM = AI + IM = AI + PM.$$

so $AP \geq AI$. The equality holds if and only if P lies on the line segment AI, in other words $P = I$.

9. As shown in the figure, let I be the incenter of ΔABD and form the line segment BI. Let l be a line passing through I and tangent to $\odot I_1$ (but different from AC) at E. Suppose l intersects AB at M.

By well-known results, $CI = CB$ and $CI + MB = CB + MI$ (this is a property of a quadrilateral having an incircle). Hence, $MB = MI$ and $\angle MBI = \angle MIB$. Note that I is the incenter of ΔABD, so $\angle MBI = \angle DBI$, which implies that $\angle MIB = \angle DBI$, and $IE \parallel BD$.

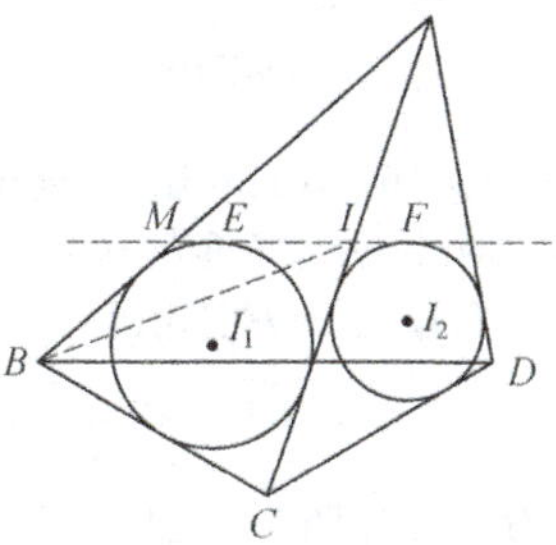

Exercise 9

Similarly, if we draw the second tangent line from I to $\odot I_2$, and let F be the tangent point, then we also have $IF \parallel BD$, so E, I, and F are collinear, and this line is an external common tangent line of $\odot I_1$ and $\odot I_2$.

10. As shown in the figure, since $AF = AG$ and AO bisects $\angle FAG$, the points F and G are symmetric with respect to AO. In order to prove that $X \in AO$, it suffices to show that $\angle AFK = \angle AGL$.

First, we have $\angle AFK = \angle DFG + \angle GFA - \angle DFK$, so D, F, G, and E are concyclic, and $\angle DFG = \angle CEG$. Since A, F, B, and G are concyclic, $\angle GFA = \angle GBA$. Since D, B, F, and K are concyclic, $\angle DFK = \angle DBK$. Hence,

$$\begin{aligned}\angle AFK &= \angle CEG + \angle GBA - \angle DBK \\ &= \angle CEG - \angle CBG.\end{aligned}$$

Since C, E, L, and G are concyclic, $\angle CEG = \angle CLG$, and since C, B, A, and G are concyclic, $\angle CBG = \angle CAG$. Thus,

$$\angle AFK = \angle CLG - \angle CAG = \angle AGL,$$

and the conclusion follows.

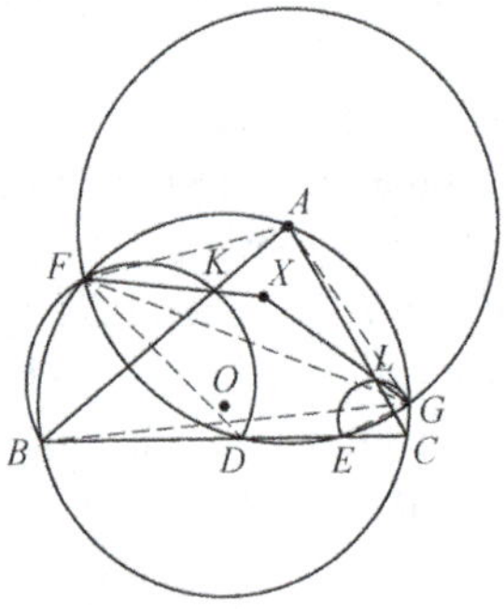

Exercise 10

11. As shown in the figure, we form the line segments I_1B, I_1C, I_2B, and I_2C. Since $PE = PF$, we have $\angle PEF = \angle PFE$. Note that $\angle PEF = \angle I_2I_1B - \angle EBI_1 = \angle I_2I_1B - \angle I_1BC$ and $\angle PFE = \angle I_1I_2C - \angle FCI_2 = \angle I_1I_2C - \angle I_2CB$, so

$$\angle I_2I_1B - \angle I_1BC = \angle I_1I_2C - \angle I_2CB.$$

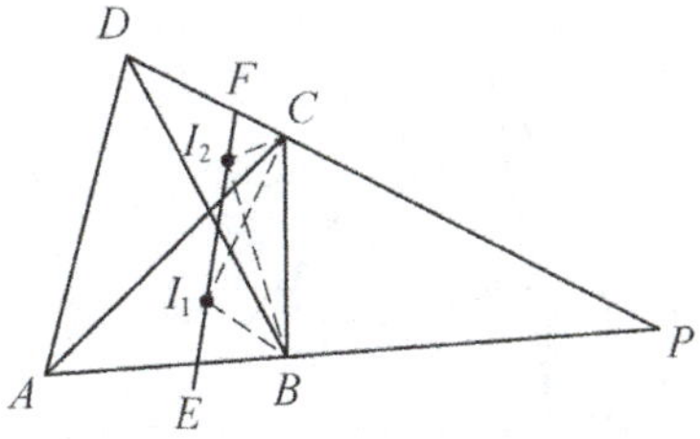

Exercise 11

Hence,

$$\angle I_2I_1B + \angle I_2CB = \angle I_1I_2C + \angle I_1BC.$$

Since

$$\angle I_2I_1B + \angle I_2CB + \angle I_1I_2C + \angle I_1BC = 2\pi,$$

we get

$$\angle I_2I_1B + \angle I_2CB = \pi,$$

and I_1, I_2, C, and B are concyclic. Thus, $\angle BI_1C = \angle BI_2C$, and

$$\angle I_1BC + \angle I_1CB = \angle I_2BC + \angle I_2CB$$

$$\angle ABC + \angle ACB = \angle DBC + \angle DCB,$$

from which $\angle BAC = \angle BDC$. Therefore, A, B, C, and D are concyclic.

12. Let O be the circumcenter of ΔAPQ, and M and N be the midpoints of AP and AQ, respectively. It suffices to show that $\angle OA'A = 90°$.

We claim that $\Delta A'NY \sim \Delta A'XM$. To prove this assertion, note that AA' is a diameter of $\odot AXY$, which implies that $\angle A'XM = \angle A'YN$, so we only need to show that

$$\frac{A'X}{MX} = \frac{A'Y}{NY} \text{ ①}$$

Let H, K, M', and N' be the projections of A, A', M, and N onto BC, respectively. By the condition, M and N are the midpoints of AP and AQ, respectively, so $BM' = HM'$ and $CN' = HN'$. Hence,

$$XM' = HM' - XH = \frac{1}{2}BH - XH,$$

$$YN' = HY - HN' = HY - \frac{1}{2}CH = (XY - XH) - \frac{BC - BH}{2}, \text{ ②}$$

$$YN' = \left(XY - \frac{1}{2}BC\right) + \left(\frac{1}{2}BH - XH\right) = \frac{1}{2}BH - XH. \text{ ③}$$

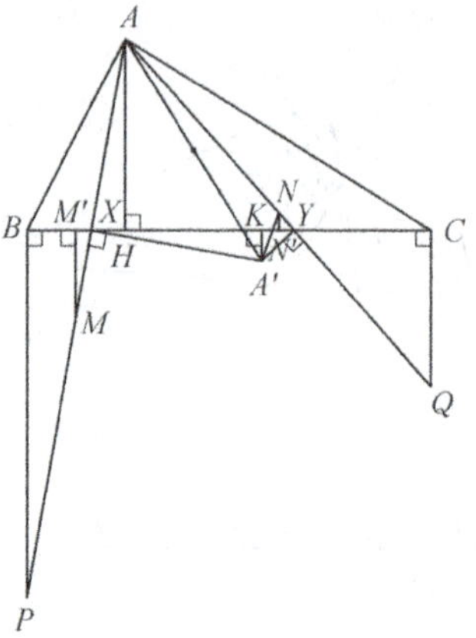

Exercise 12

By ② and ③ we obtain $XM' = YN'$. Since $\angle A'YA = \angle A'XA = 90°$, we have $\angle KA'Y = \angle NYN'$ and $\angle KA'X = \angle MXM'$, so $\Delta A'KX \sim \Delta XM'M$ and $\Delta A'KY \sim \Delta YN'N$. Thus, $\frac{A'X}{MX} = \frac{A'K}{XM'}$, $\frac{A'K}{N'Y} = \frac{A'Y}{NY}$, and dividing these two equalities we get $\frac{A'X}{MX} = \frac{A'Y}{NY}$, and consequently ① holds.

Now, $\Delta A'NY \sim \Delta A'XM$, so $\angle A'MX = A'NY$, and A, M, A', and N are concyclic. Also $\angle AMO = \angle ANO = 90°$, so A, M, O, and N are also concyclic, and the five points A, O, M, A', and N all lie on the same circle. Therefore, $\angle OA'A = 90°$.

13. Let S be the second intersection point of BI and the circumcircle of ΔABC. Then $SA = SC = SI$. Suppose the lines SB' and CA intersect at T. Then by symmetry $IB = IB'$, so $\angle IB'B = \angle IB\,B'$, which we denote as φ.

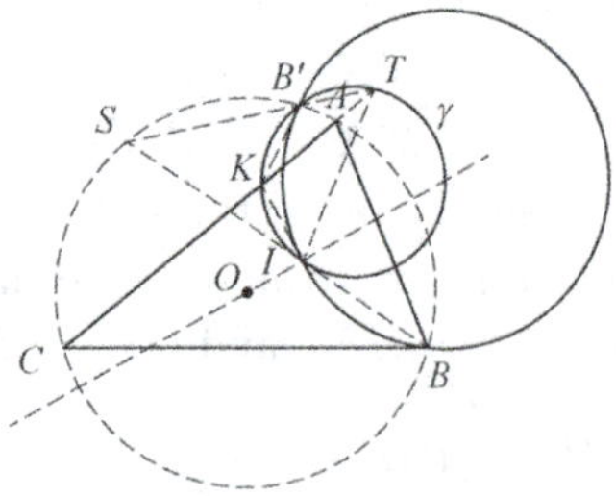

Exercise 13

Since $OB = OB'$, the points A, B', S, and B all lie on the circumcircle of ΔABC, so $\angle SAB' = \angle SBB' = \varphi$. Hence,

$$\angle ATS = \angle SAC - \angle ASB' = \angle SBC - \angle ABB'$$
$$= \angle SBA - \angle ABB' = \angle SBB' = \varphi,$$

or equivalently $\angle ATS = \angle B'AS$. Therefore, $\Delta SAB' \sim \Delta STA$, from which $SB' \cdot ST = SA^2 = SI^2$.

Let γ denote the circumcircle of $\Delta TB'I$. Then SI is tangent to γ, thus $\angle ITB' = \angle B'IS = 2\varphi$ and

$$\angle ITA = \angle ITB' - \varphi = 2\varphi - \varphi = \varphi.$$

Suppose AC intersects γ again at K. Then since

$$\angle KB'I = \angle KTI = \varphi = \angle IB'B,$$

$$\angle KIB' = \angle KTB' = \varphi = \angle IB'B,$$

we conclude that KI and KB' are both tangent to the circumcircle of $\Delta BB'I$, and the proposition follows.

14. By the condition, $\angle ABI_B = \frac{1}{2}\angle ABC = \frac{1}{2}\angle ABD + \frac{1}{2}\angle DBC = \angle I_ABD + \angle DBI_C = \angle I_AAI_D$. Similarly, $\angle BAI_A = \angle I_BAI_D$, so $\angle BAI_B = \angle I_AAI_D$.

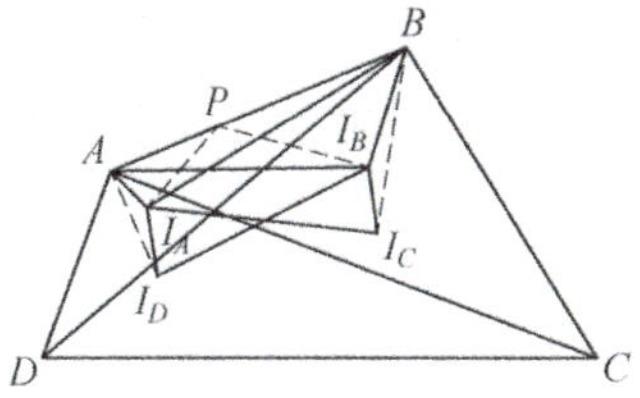

Exercise 14

Now, let P be a point on the ray AB such that $\angle API_B = \angle AI_AI_D$. This means that $\Delta API_B \sim \Delta AI_AI_D$, which also implies that $\Delta API_A \sim \Delta AI_BI_D$. By the condition, $\angle BI_AI_C + \angle AI_AI_D = 180°$, so

$$\begin{aligned}\angle BI_AI_C &= 180° - \angle AI_AI_D \\ &= 180° - \angle API_B = \angle BPI_B.\end{aligned}$$

The last step follows since P lies on the line segment AB (note that $180° > \angle I_ABI_C + \angle BI_AI_C = \angle ABI_B + 180° - \angle API_B$, so $\angle API_B > \angle ABI_B$). Combining the above with $\angle I_ABI_C = \angle ABI_B$, we see that $\Delta BPI_A \sim \Delta BI_BI_C$, which also implies that $\Delta BPI_A \sim \Delta BI_BI_C$. Therefore, $\angle BI_BI_C + \angle AI_BI_D = \angle BPI_A + \angle API_A = 180°$, or equivalently $\angle BI_BA + \angle I_CI_BI_D = 180°$.

Solution 30

Some Famous Theorems in Plane Geometry

1. By Menelaus' theorem, $\frac{AD}{DB} \cdot \frac{PB}{PC} \cdot \frac{MC}{AM} = 1$, and since D is the midpoint of AB, we have $AD = DB$. Since PA is tangent to $\odot O$, we have $PA^2 = PB \cdot PC$, or equivalently $PB = \frac{PA^2}{PC}$. Combining these results, we obtain $\frac{PA^2}{PC^2} = \frac{AM}{MC}$.

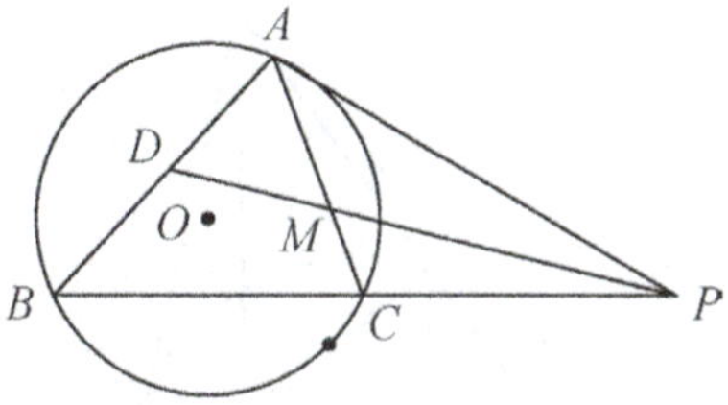

Exercise 1

2. Suppose $\frac{AG}{GB} = \lambda_1$ and $\frac{AH}{HB} = \lambda_2$. We shall show that λ_2 is uniquely determined by λ_1 (in other words H is uniquely determined by G). In the triangle ABC, since AE, BF and CG are concurrent, by Ceva's theorem $\frac{AG}{GB} \cdot \frac{BE}{EC} \cdot \frac{CF}{FA} = 1$, so

$$\lambda_1 \cdot \frac{BE}{EC} \cdot \frac{CF}{FA} = 1. \text{①}$$

Also we apply Menelaus' theorem to the triangle ABC and the line HEF, getting

$$\lambda_2 \cdot \frac{BE}{EC} \cdot \frac{CF}{FA} = 1. \text{②}$$

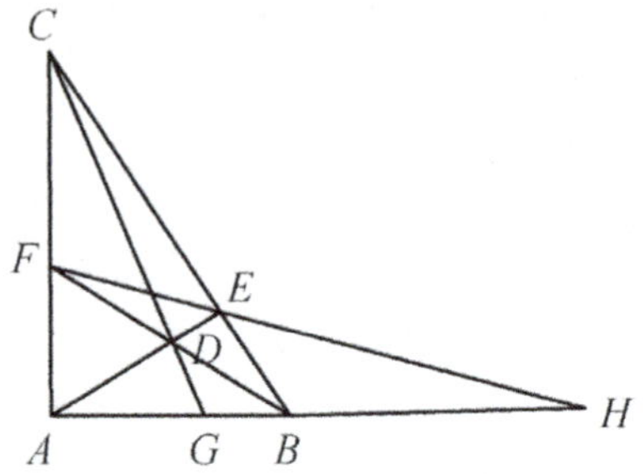

Exercise 2

By ① and ② we conclude that $\lambda_1 = \lambda_2$, hence the position of H is independent of the choice of D.

3. As shown in the figure, let H be the intersection point of the lines AM and BC, and form the line segments BE and CD. Then $\angle BEC = \angle BDC = 90°$. Apply Menelaus' theorem to the triangle AHC with the line FME, and the triangle ABH with the line GMD, and we obtain that

$$\frac{AM}{MH} \cdot \frac{HF}{FC} \cdot \frac{CE}{EA} = 1, \quad \frac{AM}{MH} \cdot \frac{HG}{GB} \cdot \frac{BD}{DA} = 1.$$

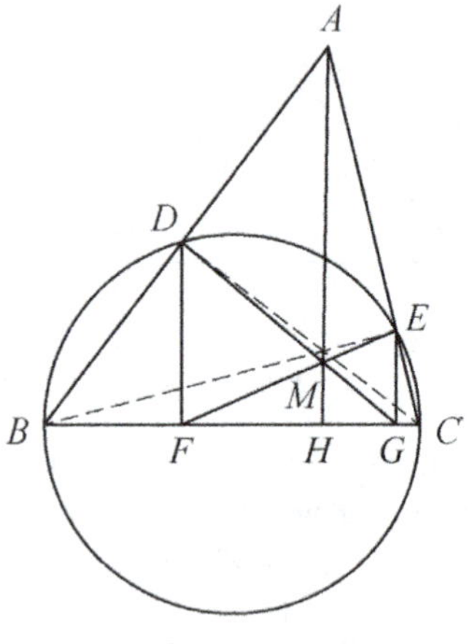

Exercise 3

Hence,

$$\frac{FH}{HG} = \frac{CF \cdot AE \cdot BD}{CE \cdot BG \cdot AD}. \text{①}$$

In the right triangles DBC and EBC, we see that $CD^2 = BC \cdot FC$ and $BE^2 = BC \cdot BG$, thus

$$\frac{CF}{BG} = \frac{CD^2}{BE^2}. \text{②}$$

Plug ② into ①, resulting in

$$\frac{FH}{HG} = \frac{CD^2 \cdot AE \cdot BD}{BE^2 \cdot CE \cdot AD}. \quad ③$$

Since $\Delta ABE \sim \Delta ACD$, we have $\frac{CD}{BE} = \frac{AD}{AE}$. Plugging into ③ gives that

$$\frac{FH}{HG} = \frac{CD \cdot BD}{BE \cdot CE} = \frac{S_{\Delta DBC}}{S_{\Delta EBC}} = \frac{DF}{EG} = \frac{DM}{MG}.$$

Hence, $MH \parallel DF$. Since $DF \perp BC$, so $MH \perp BC$, which means that $AM \perp BC$.

4. As in the figure, suppose the lines BD and AC intersect at H, and we apply Ceva's theorem to the triangle BCD and the point F, so

$$\frac{CG}{GB} \cdot \frac{BH}{HD} \cdot \frac{DE}{EC} = 1.$$

By the property of angle bisectors,

$$\frac{BH}{HD} = \frac{AB}{AD}.$$

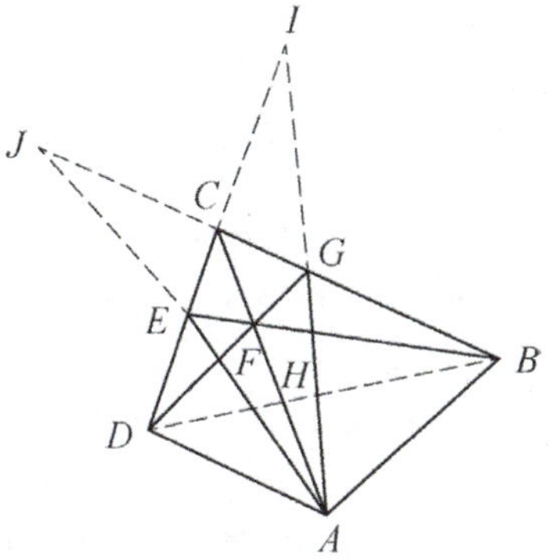

Exercise 4

Hence,

$$\frac{CG}{GB} \cdot \frac{AB}{AD} \cdot \frac{DE}{EC} = 1.$$

Let I be a point on the line AG such that $CI \parallel AB$ and let J be a point on the line AE such that $CJ \parallel AD$. Then

$$\frac{CG}{GB} = \frac{CI}{AB}, \frac{DE}{EC} = \frac{AD}{CJ},$$

which implies that

$$\frac{CI}{AB} \cdot \frac{AB}{AD} \cdot \frac{AD}{CJ} = 1.$$

Hence, $CI = CJ$. Since $CI \parallel AB$ and $CJ \parallel AD$, we have $\angle ACI = 180^\circ - \angle BAC = 180^\circ - \angle DAC = \angle ACJ$, hence $\Delta ACI \cong \Delta ACJ$, thus $\angle IAC = \angle JAC$, which completes the proof.

5. Let D be the second intersection point of the line AI and the circumcircle of ΔABC. Since AE and AD are the outer and inner angle bisectors of $\angle BAC$, respectively, $\angle EAD = 90^\circ$, so DE is a diameter of the circumcircle and O is the midpoint of DE. Note that by well-known results, $BD = DI = DC$.

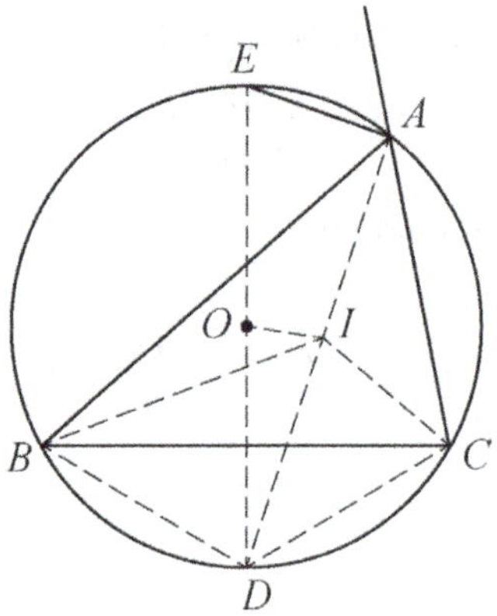

Exercise 5

On the other hand, we apply Ptolemy's theorem to the quadrilateral $ABCD$, thus

$$CD \cdot AB + BD \cdot AC = BC \cdot AD,$$

or equivalently

$$CD \cdot (AB + AC) = BC \cdot AD.$$

This implies that $AD = 2CD = 2DI$. Hence, O and I are the midpoints of DE and DA, respectively, from which $OI = \frac{1}{2}AE$.

6. By Ptolemy's theorem, $AB \cdot CD + AD \cdot BC = AC \cdot BD$. Since $AB \cdot CD = AD \cdot BC$ and $AE = EC$, we have $AB \cdot CD = AE \cdot BD = EC \cdot BD$.

In the triangles CED and ABD, we have $\angle ABD = \angle ECD$ and $AB \cdot CD = EC \cdot BD$, so $\Delta CED \sim \Delta ABD$, and hence $\angle CED = \angle BAD$. Similarly, $\angle AEB = \angle DCB$, and

$$\angle AEB = \angle DCB = 180^\circ - \angle BAD = 180^\circ - \angle CED = \angle AED.$$

Therefore, EP bisects $\angle BED$, so $\frac{BE}{ED} = \frac{BP}{PD}$ by the property of angle bisectors.

7. Let BD and AM intersect at E, and CD and AM intersect at F. Apply Menelaus' theorem to the triangle DBC, resulting in $\frac{BE}{DE} \cdot \frac{DF}{CF} \cdot \frac{CM}{BM} = 1$. Since $BE = AE$, $CF = AF$, and $BM = CM$, it follows that $\frac{DE}{DF} = \frac{AE}{AF}$, so AD is the outer bisector of $\angle EDF$.

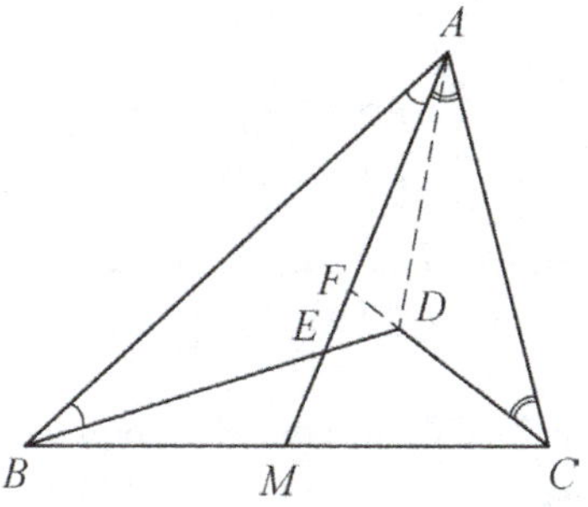

Exercise 7

Hence, $\angle DAM = \frac{1}{2}\angle DFE - \frac{1}{2}\angle DEF = \angle CAM - \angle BAM$ and

$$\angle DAB = \angle DAM + \angle BAM = \angle CAM.$$

This implies that $\angle DAB = \angle ACD$ and $\angle DAC = \angle ABD$, thus $\Delta ABD \sim \Delta CAD$. Therefore,

$$\frac{AB^2}{AC^2} = \frac{BD}{AD} \cdot \frac{AD}{CD} = \frac{BD}{CD}.$$

8. Let $BC = a$, $CA = b$, and $AB = c$, and suppose BE intersects AC at F. Form the line segments AI and CE.

   Applying Menelaus' theorem to the triangle BCK with the line MID, we get $\frac{BM}{MC} \cdot \frac{CD}{DF} \cdot \frac{FI}{IB} = 1$. But $BM = MC$, so $\frac{CD}{DF} = \frac{IB}{FI}$.

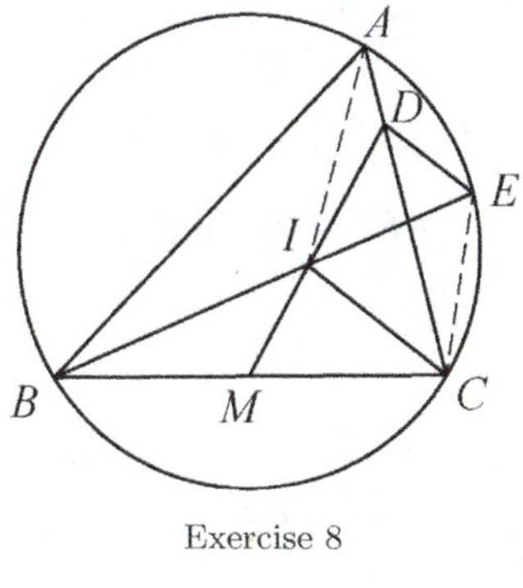

Exercise 8

Since AI bisects $\angle BAC$, we see that $\frac{IB}{FI} = \frac{AB}{AF}$. Since $\Delta ABF \sim \Delta ECF$ we obtain $\frac{AB}{AF} = \frac{EC}{EF}$. Also note that

$$\angle EBC = \angle EBA = \angle ECA. \text{①}$$

Hence, $\angle EIC = \angle EBC + \angle BCI = \angle ECA + \angle ICA = \angle ECI$, hence $EC = EI$.

By the results above,

$$\frac{CD}{DF} = \frac{IB}{FI} = \frac{AB}{AF} = \frac{EC}{EF} = \frac{EI}{EF},$$

which means that $ED \parallel IC$. Thus, $\angle BCI = \angle ICD = \angle CDE$. Combining the angle equalities with ①, we have $\Delta BCI \sim \Delta CDE$, from which $\frac{IC}{IB} = \frac{ED}{EC} = \frac{ED}{EI}$.

9. First, we show that $YX \parallel BC$, which is equivalent to $\frac{AX}{XC} = \frac{AY}{YB}$.

Form the line segments BD and CD. Since

$$\frac{S_{\Delta ACQ}}{S_{\Delta ABC}} \cdot \frac{S_{\Delta ABC}}{S_{\Delta ABP}} = \frac{S_{\Delta ACQ}}{S_{\Delta ABP}}.$$

$$\frac{\frac{1}{2}AC \cdot CQ \cdot \sin\angle ACQ}{\frac{1}{2}AB \cdot BC \cdot \sin\angle ABC} \cdot \frac{\frac{1}{2}AC \cdot BC \cdot \sin\angle ACB}{\frac{1}{2}AB \cdot BP \cdot \sin\angle ABP}$$

$$= \frac{\frac{1}{2}AC \cdot AQ \cdot \sin\angle CAQ}{\frac{1}{2}AB \cdot AP \cdot \sin\angle BAP}.$$

By the given condition, BP and CQ are tangent to the circle ω, so $\angle ACQ = \angle ABC$ and $\angle ACB = \angle ABP$. Also since $\angle CAQ = \angle DBC = \angle DCB = \angle BAP$ (note that D is the midpoint of the arc BC), combining the angle equalities with the previous equality we obtain

$$\frac{AB \cdot AQ}{AC \cdot AP} = \frac{CQ}{BP}. \text{①}$$

Since $\angle CAQ = \angle BAP$, we have $\angle BAQ = \angle CAP$, so

$$\frac{S_{\Delta ABQ}}{S_{\Delta ACP}} = \frac{\frac{1}{2}AB \cdot AQ \cdot \sin\angle BAQ}{\frac{1}{2}AC \cdot AP \cdot \sin\angle CAP} = \frac{AB \cdot AQ}{AC \cdot AP}. \text{②}$$

Since we also have

$$\frac{S_{\Delta BCQ}}{S_{\Delta BCP}} = \frac{\frac{1}{2}BC \cdot CQ \cdot \sin\angle BCQ}{\frac{1}{2}BC \cdot BP \cdot \sin\angle CBP} = \frac{CQ}{BP}, \text{③}$$

by ①, ② and ③, it follows that

$$\frac{S_{\Delta ABQ}}{S_{\Delta ACP}} = \frac{S_{\Delta CBQ}}{S_{\Delta BCP}},$$

or equivalently $\frac{S_{\Delta ABQ}}{S_{\Delta CBQ}} = \frac{S_{\Delta ACP}}{S_{\Delta BCP}}$. Since

$$\frac{S_{\Delta ABQ}}{S_{\Delta CBQ}} = \frac{AX}{XC}, \frac{S_{\Delta ACP}}{S_{\Delta BCP}} = \frac{AY}{YB},$$

we obtain $\frac{AX}{XC} = \frac{AY}{YB}$. Let M be the midpoint of BC. Then

$$\frac{AX}{XC} \cdot \frac{CM}{MB} \cdot \frac{BY}{YA} = 1,$$

thus by (the inverse) Ceva's theorem, the lines AM, BX and CY are concurrent. Let T be their intersection point. From $YX \parallel BC$ we conclude that AT bisects the line segment XY.

10. We need two lemmas.

Lemma 1. $AD^2 = AB \cdot AC - DB \cdot DC$.

Proof of Lemma 1. From the property of angle bisectors, $AB \cdot CD = AC \cdot BD$. Hence, by Stewart's theorem,

$$\begin{aligned} AD^2 &= \frac{AB^2 \cdot CD + AC^2 \cdot DB}{BC} - DB \cdot DC \\ &= \frac{AB \cdot AC \cdot DB + AC \cdot AB \cdot DC}{BC} - DB \cdot DC \\ &= AB \cdot AC \cdot \frac{DB + DC}{BC} - DB \cdot DC \\ &= AB \cdot AC - DB \cdot DC. \end{aligned}$$

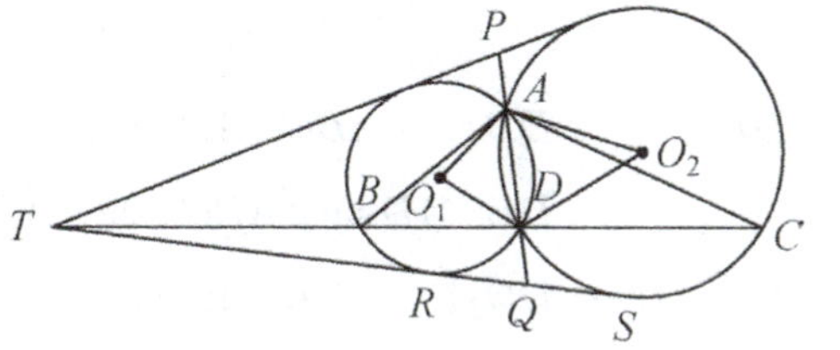

Exercise 10

Lemma 2. $DB \cdot DC = RS^2$.

Proof of Lemma 2. Let $\odot O_1$ and $\odot O_2$ denote the circumcircles of ΔABD and ΔACD with radii R_1 and R_2, respectively. From the law of sines,

$$DB \cdot DC = 2R_1 \sin \frac{A}{2} \cdot 2R_2 \sin \frac{A}{2} = 2R_1R_2(1 - \cos A).$$

On the other hand,

$$\begin{aligned}\angle O_1AO_2 &= 180^\circ - \frac{1}{2}\angle AO_1D - \frac{1}{2}\angle AO_2D \\ &= 180^\circ - \angle ABD - \angle ACD = \angle BAC.\end{aligned}$$

Hence, by the law of cosines and the formula for the common tangent length,

$$\begin{aligned}RS^2 &= O_1O_2^2 - (R_1 - R_2)^2 = R_1^2 + R_2^2 - 2R_1R_2\cos A - (R_1 - R_2)^2 \\ &= 2R_1R_2(1 - \cos A)\end{aligned}$$

which means that $DB \cdot DC = RS^2$.

Now, go back to the original problem. Since $QR^2 = QD \cdot QA = QS^2$ from the circle-power theorem, and by symmetry, $QD = PA$.It follows that

$$\begin{aligned}RS^2 &= 4QR^2 = 2QD \cdot 2QA \\ &= (PQ - AD)(PQ + AD) = PQ^2 - AD^2. \text{ ①}\end{aligned}$$

From ① and the two lemmas,

$$PQ^2 = AD^2 + RS^2 = AB \cdot AC - DB \cdot DC + DB \cdot DC = AB \cdot AC.$$

11. Let $\odot K$ be the circle with diameter AD and let G be a point on $\odot K$ such that $AG \parallel BC$. Then $\angle GAE = 180^\circ - \angle ABC$ and $\angle GAF = \angle ACB$. Hence,

$$\begin{aligned}DE \cdot GF &= DB \sin \angle ABC \cdot AD \sin \angle GAF \\ &= DB \cdot AD \sin \angle ABC \cdot \sin \angle ACB, \\ DF \cdot GE &= DC \sin \angle ACB \cdot AD \sin \angle GAE \\ &= DC \cdot AD \sin \angle ACB \cdot \sin \angle ABC.\end{aligned}$$

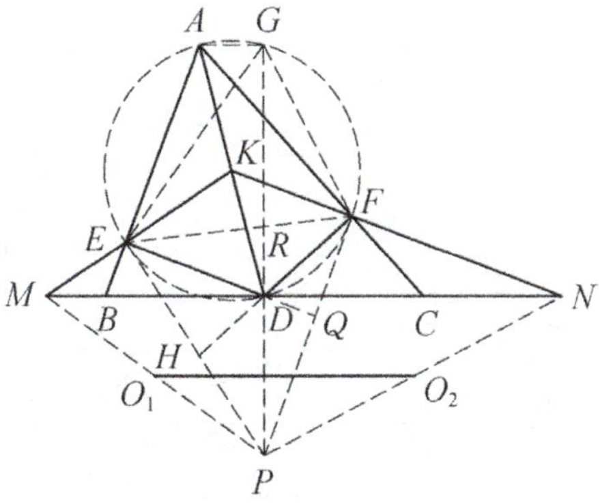

Exercise 11

Since $DB = DC$, we have $DE \cdot GF = DF \cdot GE$.

Let P be the intersection point of the tangent lines of $\odot K$ at E and F, and let EF and DG intersect at R. Suppose the lines ED and FD intersect the lines PF and PE at Q and H, respectively. Then in the triangle DEF

$$\frac{ER}{EF} \cdot \frac{FH}{HD} \cdot \frac{DQ}{QE} = \frac{ED \sin \angle EDG}{FD \sin \angle FDG} \cdot \frac{EF \sin \angle FEH}{ED \sin \angle DEH} \cdot \frac{FD \sin \angle QFD}{EF \sin \angle QFE}$$

$$= \frac{\sin \angle EDG \cdot \sin \angle FED}{\sin \angle FDG \cdot \sin \angle DFE} = \frac{GE \cdot DF}{GF \cdot DE} = 1$$

(Note that $\angle FEH = \angle QFE$, $\angle QFD = \angle FED$, and $\angle DEH = \angle DFE$.)

Therefore, by (the inverse) Ceva's theorem, the lines EH, FQ, and RD are concurrent. In other words, G, D, and P are collinear. Since $DG \perp AG$ and $BC \parallel AG$, we have $PD \perp BC$. Also, $ME \perp PE$, so the points P, E, D and M all lie on the circle with diameter PM, and the circumcenter of ΔDEM (denoted as O_1) is the midpoint of PM. Similarly, the circumcenter O_2 of ΔDFN is the midpoint of PN. Therefore, $O_1O_2 \parallel MN$, or equivalently $O_1O_2 \parallel BC$.

12. Let AL be an altitude of ΔABC, and Z is the second intersection point of the circles ω_1 and ω_2 (different from W). We show that X, Y, Z, and H are collinear.

Since $\angle BNC = \angle BMC = 90°$, the points B, C, M, and N are concyclic, and we denote this circle as ω_3. Since WZ is the radical axis of the circles ω_1 and ω_2, BN is the radical axis of ω_1 and ω_3, and CM is the radical axis of ω_2 and ω_3, so by Monge's theorem they are concurrent. Since BN and CM intersect at A, it follows that A lies on the line WZ.

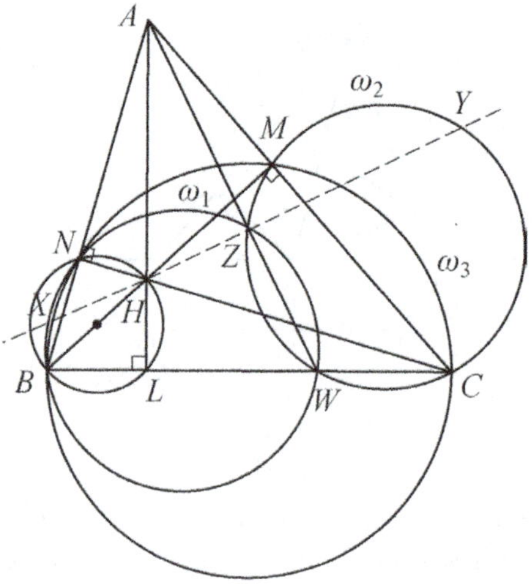

Exercise 12

Since $\angle BNH = \angle BLH = 90°$, the points B, L, H, and N are concyclic, from which

$$AL \cdot AH = AB \cdot AN = AW \cdot AZ. \text{①}$$

If H lies on the line AW, then $H = W$. Otherwise by ①

$$\frac{AZ}{AH} = \frac{AL}{AW},$$

so $\Delta AHZ \sim \Delta AWL$ and $\angle HZA = \angle WLA = 90°$. Therefore, H lies on the line XYZ.

13. We first show that A, B, D, and P are concyclic. In fact, let P' be the midpoint of the arc ADB. Then P' lies on the perpendicular bisector of the line segment AB. Let X be a point on the extension of BD. Then since $P'A = P'B$ and the points A, B, D, and P' lie on the same circle, $\angle P'DA = \angle P'BA = \angle P'AB = \angle P'DX$. Hence, P' lies on the outer angle bisector of $\angle ADB$, so $P = P'$. Therefore, A, B, D, and P are concyclic.

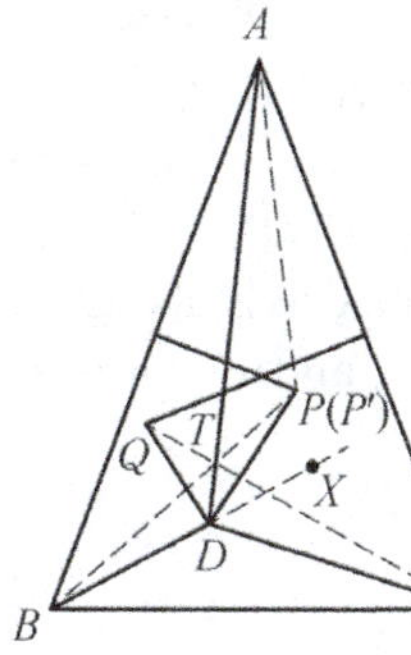

Exercise 13

By Ptolemy's theorem, $AB \cdot DP + BD \cdot AP = AD \cdot BP$. Combining it with $PA = PB$ and $AD = BD + CD$, we obtain

$$\begin{aligned} AB \cdot DP &= AD \cdot BP - BD \cdot AP \\ &= AP \cdot (AD - BD) = AP \cdot CD, \end{aligned}$$

thus $\frac{AP}{DP} = \frac{AB}{CD}$. Let T be the intersection point of BP and AD. Then since $\angle BAP + \angle BDP = 180°$,

$$\begin{aligned} \frac{AT}{TD} &= \frac{S_{\Delta ABP}}{S_{\Delta DBP}} = \frac{\frac{1}{2}AB \cdot AP \cdot \sin \angle BAP}{\frac{1}{2}DB \cdot DP \cdot \sin \angle BDP} \\ &= \frac{AB}{DB} \cdot \frac{AP}{DP} = \frac{AB}{DB} \cdot \frac{AB}{CD} = \frac{AB^2}{BD \cdot CD}. \end{aligned}$$

Similarly, A, C, D, and Q are concyclic, and if T' denotes the intersection point of CQ and AD, then $\frac{AT'}{T'D} = \frac{AC^2}{BD \cdot CD}$. Since $AB = AC$, we have $\frac{AT'}{T'D} = \frac{AT}{TD}$, hence $T' = T$.

Finally, by the circle-power theorem, $TB \cdot TP = TA \cdot TD = TC \cdot TQ$, and the points B, C, P, and Q are concyclic.

14. Since $\angle ECD = \angle EBD = \angle EBP = 180^\circ - \angle PQE$, the points C, E, Q and H are concyclic. Similarly, H, Q, F and D are concyclic.

Since C, E, F and D are concyclic, HQ is the radical axis of the circles $\odot(CEQH)$ and $\odot(HQFD)$, CE is the radical axis of $\odot(CEQH)$ and $\odot O$, and FD is the radical axis of $\odot(HQFD)$ and $\odot O$. By Monge's theorem, the lines PQ, CE, and DF are either concurrent or pairwise parallel.

Solution 31

The Extreme Principle

1. Suppose the distance between A and B is the greatest among all distances between two of the 100 points. Let Γ be the circle with diameter AB. Then since $AB \leq 1$, the radius of Γ is not greater than $\frac{1}{2}$. For any point C in the remaining 98 points, since $AB \geq BC$ and $AB \geq AC$, and by the condition that ΔABC is an obtuse triangle, it follows that $\angle C$ is obtuse, so C lies inside Γ. Therefore, all 100 points are covered by the circle Γ.
2. (1) Consider two boys A and B with the minimal distance between them (among all distances between any two boys). By the condition, these two pass their balls to each other. If someone else passes his ball to one of the two chosen boys, then by the pigeonhole principle, at least one boy has no ball in his hands. Otherwise, we can ignore A and B, and consider the remaining 13 boys. It follows that everyone of them passes his ball to another boy among these 13. Proceed the discussion, and since 15 is odd, we will eventually obtain one boy, who receives no balls.

   (2) If at least six boys pass their balls to the same person O, let them be A, B, C, D, E, and F, respectively. Then since OA, OB, OC, OD, OE, and OF are six rays starting at the same point, by the pigeonhole principle the angle between some two of them (assumed to be $\angle AOB$) is not greater than $60°$. Then AB is not the greatest side of ΔOAB, so one of OA and OB is greater than AB, which means that one of A and B should not pass his ball to O, giving a contradiction.

3. Suppose the equation has positive integer solutions. Then we assume that (x_1, y_1, z_1) is a solution such that x is minimal (among all positive integer solutions to the equation). Since $x_1^3 + 2y_1^3 = 4z_1^3$, we see that x_1^3 is even, so x_1 is even. Let $x_1 = 2x_2$. Then $8x_2^3 + 2y_1^3 = 4z_1^3$ and $4x_2^3 + y_1^3 = 2z_1^3$. This implies that y_1 is also even, and let $y_1 = 2y_2$. Thus, $4x_2^3 + 8y_2^3 = 2z_1^3$, which again shows that z_1 is even. Hence, if we let $z_1 = 2z_2$, then we have $x_2^3 + 2y_2^3 = 4z_2^3$. This means that (x_2, y_2, z_2) is also a positive integer solution to the equation and $x_2 < x_1$, which is contradictory to our assumption. Therefore, the original equation has no positive integer solutions.
4. Apparently, $a_1 = 1$, and $k > 2$ since $n > 2$ (please consider why). We claim that a_2 is prime. Note that a_2 is the smallest integer greater than 1 and relatively prime to n. Suppose a_2 is not prime. Then $a_2 = xy$ since $a_2 > 1$, where x and y are positive integers greater than 1. Since a_2 is relatively prime to n, we know that x and y are also relatively prime to n and $1 < x < a_2$, which is contradictory to our assumption. Therefore, a_2 must be prime.
5. Consider the person with the greatest number of friends. Suppose he has friends $A_1, A_2, \ldots, A_n$. Then each of $A_1, A_2, \ldots, A_n$ has at most n friends in S and no two of them have the same number of friends. This means that their numbers of friends are exactly a permutation of $1, 2, \ldots, n$, so someone has exactly one friend.
6. We call a list of 23 nonnegative integers $\{x_1, x_2, \ldots, x_{23}\}$ "balanced" if these numbers satisfy the required condition. Let S be the sum of these numbers. If x_1 serves as the referee, then $S - x_1$ must be even, since the remaining numbers can be divided into two parts with the same sum. Similarly, $S - a_2, S - a_3, \ldots, S - a_{23}$ are all even, so we see that the numbers in a "balanced" list all have the same parity. Next, we show that the numbers in a "balanced" list are all equal. Let a be the smallest number in the list and define $b_i = x_i - a$. Then the new list $\{b_1, b_2, \ldots, b_{23}\}$ is also balanced (note that every number is nonnegative) and at least one of them is 0. Since the numbers in the list have the same parity, they are necessarily all even. Hence, if we let $c_i = \frac{b_i}{2}$, then the list $\{c_1, c_2, \ldots, c_{23}\}$ is also a "balanced" one and at least one of the numbers is 0. Such operations can be done infinitely many times, and if there is a nonzero number in the list, then after some steps it will no longer be an integer, which is a contradiction. Therefore, the list $\{b_1, b_2, \ldots, b_{23}\}$ is the list of 23 zeros, which means that $a_1 = a_2 = \cdots = a_{23}$.

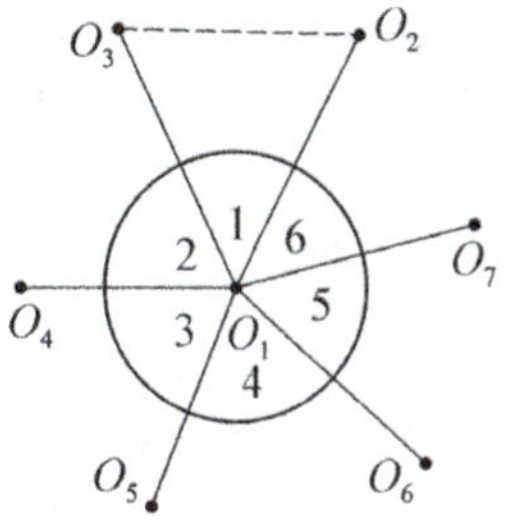

Exercise 7

7. Choose the circle with the smallest radius (and if there are multiple circles with this radius, choose an arbitrary one). Let this circle be $\odot O_1$. If $\odot O_1$ intersects six or more other circles, say $\odot O_1$ intersects $\odot O_2$, $\odot O_3$, $\odot O_4$, $\odot O_5$, $\odot O_6$, and $\odot O_7$, then at least one of $\angle 1$, $\angle 2$, $\angle 3$, $\angle 4$, $\angle 5$, and $\angle 6$ is not greater than $60°$. Assume that $\angle 1 = \angle O_2O_1O_3 \le 60°$. Then form the line segment O_2O_3, as shown in the figure. Suppose the radii of $\odot O_1$, $\odot O_2$ and $\odot O_3$ are R_1, R_2 and R_3, respectively, with $R_1 \le R_2$ and $R_1 \le R_3$. Since $\odot O_1$ intersects $\odot O_2$, we have $O_1O_2 \le R_1 + R_2 \le R_2 + R_3$. Similarly, $O_1O_3 \le R_1 + R_3 \le R_2 + R_3$. Also since $\angle 1 \le 60°$ in $\Delta O_1O_2O_3$, at least one of $\angle O_2$ and $\angle O_3$ is not less than $60°$. Assume that $\angle O_2 \ge 60° \ge \angle 1$. Then $O_1O_3 \ge O_2O_3$, so $R_2 + R_3 \ge O_2O_3$. Thus, $\odot O_2$ intersects $\odot O_3$, and the circles $\odot O_1$ $\odot O_2$ and $\odot O_3$ are pairwise intersecting, which is contradictory to the condition. Therefore, there is one circle that intersects at most five other circles.
8. Consider the person who has the greatest number of acquaintances in the party (and choose an arbitrary one if multiple people have this number of acquaintances). Let this person be A, and A knows n other people, denoted as $B_1B_2, \ldots, B_n$.

   Since any two persons B_i and B_j have a common acquaintance A, it follows that they have different numbers of acquaintances. Since the number of acquaintances of B_i does not exceed n, their numbers of acquaintances form exactly a permutation of $1, 2, \ldots, n$.

   Now, from the given condition, $n \ge 2008$. By the argument above, we see that one of $B_1, B_2, \ldots, B_n$ has exactly 2008 acquaintances.
9. First, it is easy to see that the absolute value of any entry in the matrix is unchanged after the operations. Hence, the sum of all mn numbers in the matrix (denoted as S) has at most 2^{mn} possible values, so S has the maximum value, among all the sums that can be obtained through a sequence of operations.

Choose a sequence of operations such that S is maximal. If the sum of numbers in a row (or column) is negative, then we can change the sign of every entry in this row (or column), so that S increases, which is a contradiction since S is maximal.

Therefore, the proposition follows.

10. For a fixed k, we choose a pair (a, b) such that b is minimal among all (a, b) satisfying $\frac{b+1}{a} + \frac{a+1}{b} = k$. Then $x = a$ is a root of the equation $x^2 + (1 - kb)x + b^2 + b = 0$. Let the other root be $x = a'$. Then by $a + a' = kb - 1$ we see that a' is an integer, and since $aa' = b(b+1)$, we have $a' > 0$.

    Now, $\frac{b+1}{a'} + \frac{a'+1}{b} = k$, so by the assumption, $a \geq b$ and $a' \geq b$. If $a \geq b+1$ and $a' \geq b+1$, then $aa' \geq b^2 + 2b + 1 > b^2 + b$, so it follows that at least one of a and a' is equal to b. Assume that $a = b$. Then $k = 2 + \frac{2}{b}$, so $b = 1$ or 2. Hence, $k = 3$ or 4. When $a = b = 1$ we have $k = 4$, and when $a = 1$ and $b = 2$ we have $k = 3$. Therefore, all possible values of k are 3 and 4.

11. We call a player H "permissible" if we can discard H and the remaining players have different sets of opponents (among the remaining players). The problem is to prove the existence of "permissible" players.

    Suppose A is a player who has played against the greatest number of opponents (chosen arbitrarily if there are more than one such player). If there is no "permissible" player, then A is not "permissible," so there exist players B and C such that the set of the opponents whom they have played against are the same if A is discarded. In particular this means that B and C have not played against each other and exactly one of them has played against A.

    Since C is not "permissible," there exist players D and E with the similar property. We assume that D has played against C and E has not played against C. Apparently, $D \neq A$ and $D \neq B$, so D has played against C implies that D has played against B. Since B has played against D, we deduce that B must have played against E. Since C has not played against E, the only possibility is $E = A$. Therefore, the set of D's opponents is the set of A's opponents plus C, which is contradictory to the maximality of A.

    Thus, the proposition is proven.

12. Let $g(E) = f(E) - 1990$. Then by the given properties (1) and (2), we have the following facts:

    (I) There exists an even subset D with $g(D) > 0$.
    (II) For any two disjoint even subsets A and B, $g(A \cup B) = g(A) + g(B)$.

    Since X is a finite set, the number of even subsets of X is even, so there is an even subset $P_1 \subseteq S$ such that $g(P_1)$ is maximal. By (I), we see that $g(P_1) > 0$. If there exists a nonempty even subset A_1 of P_1 with $g(A_1) = 0$, then we may replace P_1 with $P_1 - A_1$, since $g(P_1 - A_1) = g(P_1) - g(A_1) = g(P_1)$. Such a process will terminate, so we have some even subset P such that $g(P)$ is maximal and any nonempty even subset of P has a nonzero value under g.

    Let $Q = X - P$. Then we verify that the sets P and Q satisfy the required condition. In fact, if S is a nonempty even subset of P and $g(S) < 0$ (here we have already proven that $g(S) \neq 0$), then $g(P - S) = g(P) - g(S) > g(P)$, which is contradictory to the maximality of $g(P)$. If T is an even subset of Q with $g(T) > 0$, then $g(P \cup T) = g(P) + g(T) > g(P)$, also contradictory to the assumption. Therefore, such a pair of subsets (P, Q) exists.